ENCYCLOPEDIA OF PHYSICAL SCIENCE

VOLUME II

ENCYCLOPEDIA OF
PHYSICAL SCIENCE

VOLUME II

JOE ROSEN, Ph.D., and

LISA QUINN GOTHARD

Katherine Cullen, Ph.D., Managing Editor

✔ Facts On File
An imprint of Infobase Publishing

ENCYCLOPEDIA OF PHYSICAL SCIENCE

Facts On File, Inc.
An Imprint of Infobase Publishing
132 West 31st Street
New York NY 10001

Library of Congress Cataloging-in-Publication Data

Rosen, Joe, 1937–
Encyclopedia of physical science / Joe Rosen and Lisa Quinn Gothard.
p. cm.
Includes bibliographical references and index.
ISBN-13: 978-0-8160-7011-4
ISBN-10: 0-8160-7011-3
1. Physical sciences—Encyclopedias. I. Gothard, Lisa Quinn. II. Title.
Q121.R772009
500.203—dc22   2008036444

Text design by Annie O'Donnell
Illustrations by Richard Garratt and Melissa Ericksen
Photo research by Suzanne M. Tibor

Printed in China

CP Hermitage 10 9 8 7 6 5 4 3 2 1

This book is printed on acid-free paper.

CONTENTS

ACKNOWLEDGMENTS

We would like to express appreciation to Frank K. Darmstadt, executive editor, and to Katherine E. Cullen, managing editor, for their critical review of this manuscript, wise advice, patience, and professionalism. Thank you to the graphics department for creating the line illustrations that accompany the entries in this work, to Alana Braithwaite for expertly assembling all of the manuscript materials, and to Suzie Tibor for performing the expert photo research. The guest essayist, Dr. Amy J. Heston, deserves recognition for generously donating time to share her expert knowledge about forensic chemistry. Appreciation is also extended to the production and copyediting departments and the many others who helped in the production of this project. Thank you all.

INTRODUCTION

Encyclopedia of Physical Science is a two-volume reference intended to complement the material typically taught in high school physics and chemistry and in introductory college physics and chemistry courses. The substance reflects the fundamental concepts and principles that underlie the content standards for science identified by the National Committee on Science Education Standards and Assessment of the National Research Council for grades 9–12. Within the category of physical science, these include structure of atoms, structure and properties of matter, chemical reactions, motions and forces, conservation of energy and increase in disorder, and interactions of energy and matter. The National Science Education Standards (NSES) also place importance on student awareness of the nature of science and the process by which modern scientists gather information. To assist educators in achieving this goal, other subject matter discusses concepts that unify the physical sciences with life science and Earth and space science: science as inquiry, technology and other applications of scientific advances, science in personal and social perspectives including topics such as natural hazards and global challenges, and the history and nature of science. A listing of entry topics organized by the relevant NSES Content Standards and an extensive index will assist educators, students, and other readers in locating information or examples of topics that fulfill a particular aspect of their curriculum.

Encyclopedia of Physical Science provides historical perspectives, portrays science as a human endeavor, and gives insight into the process of scientific inquiry by incorporating biographical profiles of people who have contributed significantly to the development of the sciences. Instruments and methodology-related entries focus on the tools and procedures used by scientists to gather information, conduct experiments, and perform analyses. Other entries summarize the major branches and subdisciplines of physical science or describe selected applications of the information and technology gleaned from physical science research. Pertinent topics in all categories collectively convey the relationship between science and individuals and science and society.

The majority of this encyclopedia comprises more than 200 entries covering NSES concepts and topics, theories, subdisciplines, biographies of people who have made significant contributions to the physical sciences, common methods, and techniques relevant to modern science. Entries average approximately 2,000 words each (some are shorter, some longer), and most include a cross-listing of related entries and a selection of recommended further readings. In addition, 12 essays are included covering a variety of subjects—contemporary topics of particular interest and specific themes common to the physical sciences. More than 150 photographs and more than 150 line art illustrations accompany the text, depicting difficult concepts, clarifying complex processes, and summarizing information for the reader. A chronology outlines important events in the history of the field, and a glossary defines relevant scientific terminology. The back matter of *Encyclopedia of Physical Science* contains a list of additional print and Web resources for readers who would like to explore the discipline further. In the appendixes readers can find a periodic table of the elements, some common conversions, and tables of physical constants, astronomical data, abbreviations and symbols for physical units, the Greek alphabet, and multipliers and dividers for use with SI units.

Lisa Gothard has 10 years experience teaching high school and college level physical science, biology, and chemistry. She obtained a bachelor of science degree in molecular biology from Vanderbilt University in Nashville, Tennessee, where she received a Howard Hughes Fellowship for undergraduate research. She holds a master of science degree in biochemistry from the University of Kentucky. She has coauthored peer-reviewed journal articles in the field of gene expression and the role of chaperones in protein folding. She is currently teaching at East Canton High School in East Canton, Ohio, where she lives with her husband and four children.

Joe Rosen has been involved in physics research and teaching for more than four decades. After obtaining his doctorate in theoretical physics from the Hebrew University of Jerusalem, he did one year of postdoctoral work in the physics department of Boston University and three at Brown University. He then joined the faculty of the School of Physics and Astronomy of Tel Aviv University, where he spent the largest portion of his academic career. He has taught most of the standard undergraduate physics courses as well as many graduate-level courses. His last full-time job was chair of the Department of Physics and Astronomy of the University of Central Arkansas. Since retiring from Tel Aviv and Central Arkansas, Joe Rosen has been living in the Washington, D.C., area, where he does visiting and adjunct teaching at colleges and universities in the area, currently at The George Washington University. He has written or edited 11 books and continues writing and carrying out research in his current fields of interest: symmetry, space, time, space-time, and the quantum. His career has allowed him to keep abreast of, and even contribute to, the exciting developments in physics that have been taking place since the late 20th century.

Both authors hope that this encyclopedia will serve you as a valuable reference, and that you will learn as much from referring to it as we have from writing it.

Entries Categorized by National Science Education Standards for Content (Grades 9–12)

When relevant, an entry may be listed under more than one category. For example, Stephen Hawking, who studies cosmology, general relativity, quantum theory, and much more, is listed under all of Physical Science (Content Standard B): Motions and Forces, Conservation of Energy and Increase in Disorder, Interactions of Energy and Matter; Earth and Space Science (Content Standard D); and History and Nature of Science (Content Standard G). Biographical entries, topical entries, and entries that summarize a subdiscipline may all appear under History and Nature of Science (Content Standard G), when a significant portion of the entry describes a historical perspective of the subject. Subdisciplines are listed separately under the category Subdisciplines, which is not a NSES category, but are also listed under the related content standard category.

SCIENCE AS INQUIRY (CONTENT STANDARD A)

analytical chemistry
buffers
calorimetry
centrifugation
chromatography
classical physics
colligative properties
concentration
conservation laws
Dalton, John
DNA fingerprinting
duality of nature
enzymes
Ertl, Gerhard
inorganic chemistry
inorganic nomenclature
invariance principles
laboratory safety
Lauterbur, Paul
Lavoisier, Antoine-Laurent
Mach's principle
Mansfield, Sir Peter
mass spectrometry
measurement

Mullis, Kary
nanotechnology
nuclear chemistry
organic chemistry
pH/pOH
physics and physicists
Ramsay, Sir William
rate laws/reaction rates
ribozymes (catalytic RNA)
scientific method
separating mixtures
simultaneity
solid phase peptide synthesis
solid state chemistry
stoichiometry
surface chemistry
symmetry
theory of everything
toxicology

PHYSICAL SCIENCE (CONTENT STANDARD B): STRUCTURE OF ATOMS

accelerators
acid rain
acids and bases

analytical chemistry
atomic structure
big bang theory
biochemistry
blackbody
black hole
Bohr, Niels
bonding theories
Bose-Einstein statistics
Broglie, Louis de
buffers
chemistry and chemists
citric acid cycle
classical physics
clock
Compton effect
conservation laws
Curie, Marie
duality of nature
Einstein, Albert
electron configurations
energy and work
EPR
Fermi, Enrico
Fermi-Dirac statistics
Feynman, Richard

theory of everything
Thomson, Sir J. J.
time
vectors and scalars
waves

PHYSICAL SCIENCE (CONTENT STANDARD B): CONSERVATION OF ENERGY AND INCREASE IN DISORDER

blackbody
black hole
Broglie, Louis de
calorimetry
center of mass
chemistry and chemists
classical physics
Compton effect
conservation laws
Davy, Sir Humphrey
degree of freedom
Einstein, Albert
elasticity
electrical engineering
electricity
electron emission
energy and work
engines
equilibrium
Faraday, Michael
Fermi, Enrico
fission
fluid mechanics
force
fusion
harmonic motion
Hawking, Stephen
heat and thermodynamics
Heisenberg, Werner
lasers
Lavoisier, Antoine-Laurent
machines
mass
matter and antimatter
Maxwell, James Clerk
mechanical engineering
Meitner, Lise
momentum and collisions
Mössbauer effect
Newton, Sir Isaac
nuclear physics
Oppenheimer, J. Robert
particle physics
photoelectric effect
physics and physicists
Planck, Max

power
quantum mechanics
radioactivity
rotational motion
Schrödinger, Erwin
special relativity
speed and velocity
states of matter
statistical mechanics
theory of everything
time
vectors and scalars

PHYSICAL SCIENCE (CONTENT STANDARD B): INTERACTIONS OF ENERGY AND MATTER

acoustics
alternative energy sources
blackbody
black hole
Bohr, Niels
Bose-Einstein statistics
Broglie, Louis de
calorimetry
carbohydrates
chemistry and chemists
classical physics
Compton effect
conservation laws
cosmic microwave background
Davy, Sir Humphrey
degree of freedom
diffusion
duality of nature
Einstein, Albert
elasticity
electrical engineering
electricity
electrochemistry
electromagnetic waves
electromagnetism
electron emission
energy and work
equilibrium
Faraday, Michael
Fermi-Dirac statistics
Feynman, Richard
fission
fluid mechanics
force
fossil fuels
fusion
Gell-Mann, Murray
general relativity
glycogen metabolism
glycolysis

gravity
harmonic motion
Hawking, Stephen
heat and thermodynamics
holography
lasers
Lavoisier, Antoine-Laurent
lipids
magnetism
mass
materials science
matter and antimatter
Maxwell, James Clerk
Maxwell-Boltzmann statistics
mechanical engineering
metabolism
Mössbauer effect
Newton, Sir Isaac
nuclear chemistry
nuclear magnetic resonance (NMR)
Oppenheimer, J. Robert
optics
particle physics
photoelectric effect
physical chemistry
physics and physicists
Planck, Max
quantum mechanics
radioactivity
Rutherford, Sir Ernest
Schrödinger, Erwin
special relativity
states of matter
statistical mechanics
superconductivity
telescopes
theory of everything
Thomson, Sir J. J.
vectors and scalars
X-ray crystallography

LIFE SCIENCE (CONTENT STANDARD C)

acid rain
acids and bases
acoustics
agrochemistry (agricultural chemistry)
air pollution (outdoor/indoor)
amino acid metabolism and the urea cycle
analytical chemistry
aspirin
biochemistry
biophysics

green chemistry
greenhouse effect
Lauterbur, Paul
Mansfield, Sir Peter
Molina, Mario
nuclear chemistry
nucleic acids
nutrient cycling
Oppenheimer, J. Robert
oxidation-reduction reactions
Pasteur, Louis
pharmaceutical drug
    development
photosynthesis
pH/pOH
polymers
radical reactions
radioactivity
Rowland, F. Sherwood
textile chemistry
toxicology

HISTORY AND NATURE OF SCIENCE (CONTENT STANDARD G)

alternative energy sources
atomic structure
big bang theory
biophysics
Bohr, Niels
bonding theories
Boyle, Robert
Broglie, Louis de
Cavendish, Henry
classical physics
clock
cosmic microwave background
cosmology
Crick, Francis
Crutzen, Paul
Curie, Marie
Dalton, John
Davy, Sir Humphry
duality of nature
Einstein, Albert
electrical engineering
Ertl, Gerhard
Faraday, Michael

Fermi, Enrico
Feynman, Richard
fission
fossil fuels
Franklin, Rosalind
fusion
Galilei, Galileo
gas laws
Gell-Mann, Murray
general relativity
graphing calculators
green chemistry
greenhouse effect
Hawking, Stephen
Heisenberg, Werner
Hodgkin, Dorothy Crowfoot
invariance principles
Kepler, Johannes
Kornberg, Roger
laboratory safety
lasers
Lauterbur, Paul
Lavoisier, Antoine-Laurent
Mansfield, Sir Peter
materials science
matter and antimatter
Maxwell, James Clerk
mechanical engineering
Meitner, Lise
Mendeleyev, Dmitry
Molina, Mario
motion
Mullis, Kary
Newton, Sir Isaac
Nobel, Alfred
nuclear chemistry
nuclear magnetic resonance
    (NMR)
Oppenheimer, J. Robert
particle physics
Pasteur, Louis
Pauling, Linus
Pauli, Wolfgang
periodic table of the elements
Perutz, Max
physics and physicists
Planck, Max

polymers
Priestley, Joseph
quantum mechanics
radioactivity
Ramsay, Sir William
representing structures/molecular
    models
ribozymes (catalytic RNA)
Rowland, F. Sherwood
Rutherford, Sir Ernest
Schrödinger, Erwin
telescopes
textile chemistry
theory of everything
Thomson, Sir J. J.
Watson, James
Wilkins, Maurice

SUBDISCIPLINES

agrochemistry (agricultural
    chemistry)
analytical chemistry
atmospheric and environmental
    chemistry
biochemistry
biophysics
chemical engineering
computational chemistry
cosmology
electrical engineering
electrochemistry
green chemistry
industrial chemistry
inorganic chemistry
inorganic nomenclature
materials science
mechanical engineering
nuclear chemistry
nuclear physics
organic chemistry
particle physics
physical chemistry
solid phase peptide synthesis
solid state chemistry
surface chemistry
textile chemistry
toxicology

ENTRIES M–Z

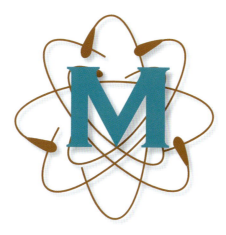

machines Machines make it easier to do physical work by using the force applied to an object and modifying its effect. Some machines are simple tools like bottle openers and screwdrivers; others are complex objects such as bicycles and hydroelectric turbines. A machine eases the load by changing either the magnitude or the direction of the force but does not change the amount of work done. The six types of simple machine are the inclined plane, lever, pulley, wheel and axle, wedge, and screw. Complex machines, such as drills and cranes, are combinations of these fundamental machines. The law of conservation of energy underlies the actions of both simple and compound machines. When a machine transmits force and applies it, the amount of energy put into the system in the form of work is the amount of energy available to expend. Overall, the total energy always remains the same. Springs may store energy, and friction converts energy into heat, but no energy is created or destroyed. Without this principle, machines would not work.

TYPES OF SIMPLE MACHINES

Raising an object to a particular height requires a certain amount of work. A machine does not reduce the total amount of work needed to move an object, but it changes the way the work is accomplished. Work has two aspects: the force applied and the distance over which the force is maintained. For a given amount of work, if the force increases, the distance decreases. If the force decreases, the distance to carry out the work increases. Climbing a hill by the steepest route requires the greatest effort (force), but the distance one must cover to reach the top is the shortest. Making the way to the top by following the gentlest slope takes less effort (requires less force),

but the distance is greater. The amount of work W is the same in both cases and equals the exerted force F multiplied by the distance d over which the force is maintained, summarized by the following equation:

$$W = Fd$$

The inclined plane is one simple machine that follows these basic rules of physics. A ramp is an example of an inclined plane that has been used since ancient times to reduce the effort required to achieve a given amount of work. Egyptians used ramps to build pyramids from huge pieces of stone. When the sloping face of the ramp is twice as long as its vertical face, the force needed to move a stone up the ramp is half of the force needed to lift the load up the vertical face. Devices that use inclined planes today include electric trimmers, zippers, and plows.

A lever is a second type of simple machine; it consists of a bar or rod that rotates about a fixed point, the fulcrum, making it easier to perform a task. Applying force at one part of the lever causes the lever to pivot at the fulcrum, producing a useful action at another point on the lever. The force might cause the lever to raise a weight or overcome a resistance, both of which are called the load. A relationship exists between the force applied to a lever and the location on the lever where one exerts the force. Less force will move the same load provided that the force is applied farther from the fulcrum.

Three classes of levers exist: first-, second-, and third-class levers. In first-class levers, fulcrums are positioned between the effort and the load. A balance is a first-class lever with the load at one end and a series of weights balancing the load on the other end. In second- and third-class levers, the fulcrum is

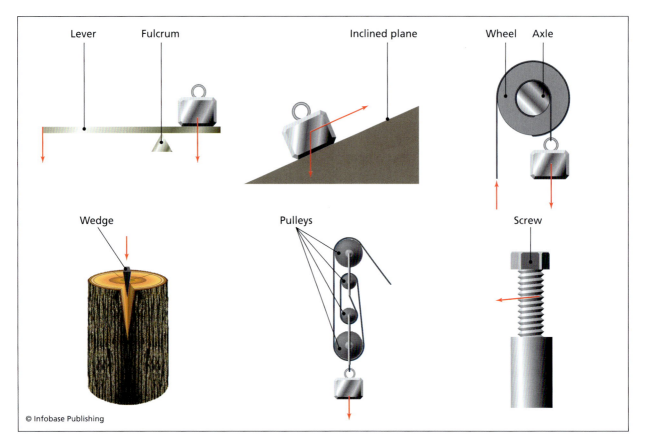

Lever　Fulcrum　Inclined plane　Wheel　Axle

Wedge　Pulleys　Screw

© Infobase Publishing

The six types of simple machines are the lever, the inclined plane, the wheel and axle, the wedge, the pulley, and the screw. Complex machines, such as escalators and tractors, are combinations of these basic machines.

located at one end. A wheel barrow is an example of a second-class lever with the wheel acting as the fulcrum, the load in the bed, and the effort exerted by lifting the handles on the other end. The load and effort reverse positions in third-class levers. A hammer acts as a third-class lever when it is used to drive a nail into a board. In this case, the wrist is the fulcrum, the effort is along the hammer handle, and the load is the resistance of the wood.

A pulley is a third type of simple machine, which uses one or more wheels with a rope or cable around them to lift up loads. In a single-pulley system, a rope attaches to a load then runs around a wheel fixed to a support above. Pulling downward on the rope, the load rises to the height of the support. In this system, no decrease in effort occurs. In reality, the amount of effort slightly increases because friction in the pulley wheel and the load must be overcome before any work occurs. An elevator operates with a single pulley system.

A compound pulley system, where more than one pulley are connected, enables one person to raise loads much greater than the person's weight. A tower crane often has a block and tackle system, an example of a compound pulley that enables substantial loads

to be raised and moved. In a block and tackle system, two sets of pulleys have a single rope tying them all together. Each individual set of pulleys rotates freely around a single axle. An upper set of pulleys links the boom (a long beam projecting from the crane that supports or guides cargo) to a lower set of pulleys that hold the load. As the crane operator pulls the rope, the lower set of pulleys raises the load. If 10 pulleys are part of this lower set, a 10-fold magnification of the original force exerted by the crane on the pulleys occurs. Although more pulleys amplify the force, friction still reduces the effect, making the overall output of work less than the input.

A fourth type of simple machine is the wheel and axle, a device that transmits force by acting as a rotating lever around a fixed point. The center of the wheel and axle forms the fulcrum of this rotating lever. The wheel is the outer part of the lever, and the axle is the inner part of the lever nearest to the fulcrum. As this machine rotates around the fulcrum, the wheel moves a greater distance than the axle but turns with less force. If effort is applied to the wheel, the axle moves with greater force. Winches use the wheel and axle to generate force to move heavy objects in this manner. Water wheels are another example of a wheel and axle machine. A large vertical wheel captures flowing water in its paddles, causing the wheel to turn. As the wheel turns, the axle that is attached to the wheel turns, converting this energy into the force necessary to operate a grindstone or to drive a pump. Today hydroelectric power stations use hydroelectric turbines, direct descendants of the water wheel, to exploit as much energy as possible from water passing around large curved blades of horizontal turbines.

Many machines make use of a fifth type of machine called the wedge, which acts as a moving inclined plane. Instead of an object moving up an inclined plane, the plane itself moves to raise the object. A door wedge works in this way. As it jams under a door, the wedge raises the door slightly and exerts a strong force on it. The door in turn forces the wedge into the floor, and friction makes the wedge grip the floor to keep the door open. Nearly all cutting machines make use of wedges as well. An axe is simply a wedge attached to a long handle. As one swings an axe toward a log, the long movement downward, which is a forward movement, converts to a powerful perpendicular, or sideways, force acting at right angles to the blade that splits open the piece of wood.

The last type of simple machine, the screw, is another disguised form of the inclined plane with a series of planes wrapping around a cylinder like a spiral staircase. Screws alter force and distance parameters just as the inclined plane does. When something like a nut moves over a bolt, the nut turns several times to travel a short distance. As in a linear inclined plane, force increases when distance decreases. A nut thus moves along the bolt with a much greater force than the effort used to turn it. A nut and bolt hold objects together by gripping them with this great force. Just as a wedge produces a strong force at right angles to its movement, a screw does so, but at right angles to its rotation as it moves into an object. Fric-

tion stops the nut from working loose, and it acts to hold the screw in a piece of material. With drills and augers, screws carry away material. As a drill cuts forward into material with its sharp point, the waste moves backward to the surface along the grooves of the screw. Hand drills, power drills, meat grinders, construction augers, and even mechanical moles use this aspect of the screw to remove waste material.

MECHANICAL ADVANTAGE OF MACHINES

The force exerted on a machine is the effort force F_e, and the force exerted by the machine is the resistance force F_r. The ratio of resistance force to effort force is the mechanical advantage, MA, of the machine.

$$MA = \frac{F_r}{F_e}$$

When the mechanical advantage is greater than 1, the machine increases the force applied.

To find the mechanical advantage of a machine, one calculates the input work as the product of the exerted effort force, F_e, and the displacement of the machine operator, d_e. In the same way, the output work is the product of the resistance force, F_r, and the displacement caused by the machine, d_r. A machine can increase force, but it cannot increase energy. An ideal machine transfers all the energy, so the output work equals the input work,

$$W_o = W_i$$

or

$$F_r d_r = F_e d_e$$

which can be rewritten

$$\frac{F_r}{F_e} = \frac{d_e}{d_r}$$

Thus, the mechanical advantage of a machine, which is defined as

$$MA = \frac{F_r}{F_e}$$

and can now be expressed in terms as the distances moved,

$$MA = \frac{d_e}{d_r}$$

For an ideal machine, one measures distances moved to calculate the ideal mechanical advantage, IMA,

A see-saw is an example of a simple machine, a first-class lever. Here two children are enjoying a see-saw. *(Sunstar/Photo Researchers, Inc.)*

but one measures the forces exerted to find the actual mechanical advantage, MA, of a machine.

In a real machine, the input work never equals the output work because more work goes into a machine than goes out. The efficiency of a machine relates output work W_o and input work W_i through a ratio,

$$\text{Efficiency} = \frac{W_o}{W_i} \times 100\%$$

An ideal machine has equal output and input work and has 100 percent efficiency.

$$\frac{W_o}{W_i} = 1, \quad \text{Efficiency} = 100\%$$

Because of friction, all real machines have efficiencies less than 100 percent. Another way to think about efficiency is to consider that a machine's design dictates its ideal mechanical advantage, while its operation in practice determines its actual mechanical advantage. With some algebraic manipulations, an alternative way to calculate efficiency becomes

$$\text{Efficiency} = \frac{\text{MA}}{\text{IMA}} \times 100\%$$

An efficient machine has an MA almost equal to its IMA. A less efficient machine has a smaller MA. Lower efficiency means that the operator of the machine must exert greater effort force to achieve the same resistance force.

All machines, no matter how complex, are combinations of the six simple machines. A compound machine consists of two or more simple machines linked so that the resistance force of one machine becomes the effort force of the second. The bicycle is an example of a compound machine. In the bicycle, the pedal and the gears act as a wheel and axle. The effort force is the force exerted on the pedal by someone's foot. The resistance is the force that the gears exert on the chain attached to the rear wheel. The chain and the rear wheel act as a second wheel and axle. The chain exerts the effort force on the wheel, and the rear wheel exerts a resistance force on the road. By Newton's third law, the ground exerts an equal force on the wheel, and this force accelerates the bicycle forward.

See also CONSERVATION LAWS; ENERGY AND WORK; FORCE; MECHANICAL ENGINEERING; POWER.

FURTHER READING

Serway, Raymond A., Jerry S. Faughn, Chris Vuille, and Charles A. Bennett. *College Physics,* 7th ed. Belmont, Calif.: Thomson Brooks/Cole, 2006.

Mach's principle   Named for the 19–20th-century Austrian physicist Ernst Mach, Mach's principle deals with the origin of mechanical inertia. Mechanical inertia is expressed by Newton's first and second laws of motion. According to the first law, in the absence of forces acting on a body (or when the forces on it cancel out), the body remains at rest or moves with constant velocity (i.e., at constant speed in a straight line). The second law states how a body resists changes in its force-free motion, or inertial motion. Such a change is an acceleration, and according to the second law, the acceleration resulting from a force is inversely proportional to a body's mass: the greater the mass, the less change a force can effect. Thus, mass serves as a measure of a body's inertia.

The question immediately arises: in what way do inertial motion and accelerated motion differ? Since in an appropriately moving reference frame any motion can appear to be inertial, the question can be restated as: What distinguishes inertial reference frames—those in which Newton's laws of motion are valid—from noninertial ones? According to Newton, the difference lies in the state of motion relative to absolute space. Newton assumed the existence of an absolute space (as well as an absolute time) that is unobservable except for its inertial effects. Inertial and accelerated motions occur with respect to absolute space. Accordingly, inertial reference frames are reference frames moving at constant velocity (or at rest) with respect to absolute space.

Mach's philosophical approach to science, however, did not allow entities that are inherently unobservable, including absolute space. How then, according to Mach, is inertial motion, or are inertial reference frames, different from all other motions or reference frames? With no absolute space, what serves as reference for this difference? Considerations, such as of the fundamental meaning of motion for a single body, two bodies, and three bodies, in an otherwise empty universe, led Mach to his principle: the origin of inertia lies in all the matter of the universe. Since the matter of the universe consists overwhelmingly of distant stars, galaxies, and so forth, Mach's principle can also be given the formulation that the origin of inertia lies in the distant stars.

When the motions of the distant stars are averaged out, they define a reference frame for rest. Stated in more detail, this reference frame—the universal reference frame—is that in which the average linear momentum of all the matter in the universe is 0. Then, according to Mach, inertial reference frames are those, and only those, that are at rest or moving at constant velocity relative to the just-defined reference frame. All other reference frames, which are in relative acceleration or rotation, are noninertial. In terms of motions of bodies, the rest, velocity, and acceleration that Newton's first two laws of motion are concerned with are relative to the universal rest frame defined by the matter in the universe.

Albert Einstein, in his development of the general theory of relativity, was greatly influenced by Mach's thinking and principle, in particular in his avoidance of unobservable entities, including absolute space, and in his aim of explaining inertia as being caused by the matter in the universe. Nevertheless, Einstein's theory does not succeed in fully implementing the principle, in showing just how distant matter engenders inertial effects. Neither has Mach's principle yet been successfully implemented by other theories in a way that is generally accepted. So, at present Mach's principle can best be viewed as a guiding principle.

Mach's principle—if it is indeed valid, and even if the question of just how the distant matter of the universe manages to influence a body and endow it with inertia so that it obeys Newton's first and second laws of motion is not answered—would indicate a certain holistic aspect of the universe. Since the distant matter forms practically all the universe, the inertial behavior of bodies would be ascribed to the universe as a whole. Rather than being a local effect, brought about by the situation in a body's immediate neighborhood, inertia would result from a body's forming part of the whole. In a similar vein, one can extend the principle to what might be called the extended Mach principle, stating that it is not only inertia that is due to the distant matter, but so are all the other laws of nature as well. Just as Mach's principle does not explain inertia, so does the extended Mach principle not explain the laws of nature. Both principles offer a holistic guiding framework for the operation of science in search of such explanations.

See also ACCELERATION; EINSTEIN, ALBERT; GENERAL RELATIVITY; MASS; MOMENTUM AND COLLISIONS; MOTION; ROTATIONAL MOTION; SPEED AND VELOCITY.

FURTHER READING

Rosen, Joe. *The Capricious Cosmos: Universe beyond Law*. New York: Macmillan, 1991.
Schutz, Bernard. *Gravity from the Ground Up: An Introductory Guide to Gravity and General Relativity*. Cambridge: Cambridge University Press, 2003.

magnetism   At its most fundamental, magnetism is the effect that electric charges in relative motion have on each other over and above their mutual electric attraction or repulsion. In other words, magnetism is that component of the mutual influence of electric charges that depends on their relative velocity. Magnetic forces are described by means of the magnetic field, which is a vector quantity that possesses a value

at every location in space and can vary over time. The magnetic field is considered to mediate the magnetic force, in the sense that any moving charge, as a *source charge,* contributes to the magnetic field, while any moving charge, as a *test charge,* is affected by the field in a manner that causes a force to act on the charge.

MAGNET

Any object that when at rest can produce and be affected by a magnetic field is a magnet. Every magnet possesses two poles, a north pole (N) and a south pole (S), and is thus a magnetic dipole, characterized by its magnetic dipole moment (discussed later).

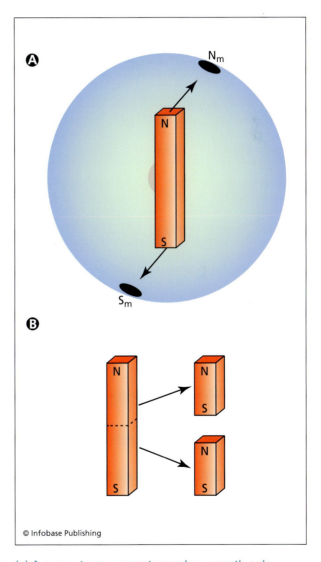

(a) A magnet possesses two poles, a north pole, N, and a south pole, S. The magnet's north pole is attracted to Earth's north geomagnetic pole N_m (near the north geographic pole) and its south pole to Earth's south geomagnetic pole S_m. (b) A magnet's poles cannot be separated. When a magnet is cut into two parts, each part is found to have a pair of north and south poles.

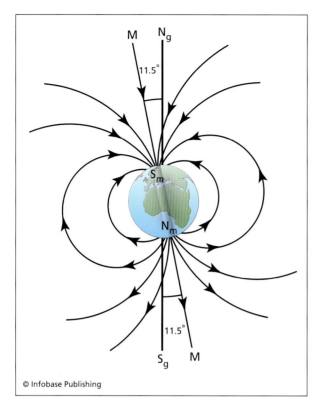

Earth behaves as if a bar magnet were embedded at its center, where the magnet's axis M-M is presently tilted about 11.5° from Earth's rotation axis N_g-S_g. Magnetic field lines outside Earth are shown. The two locations where the field lines are perpendicular to Earth's surface are Earth's geomagnetic poles. The north (south) geomagnetic pole, located near the north (south) geographic pole N_g (S_g), is actually a south (north) magnetic pole, as indicated on the imaginary bar magnet by S_m (N_m).

Same-type poles (N-N or S-S) of different magnets repel each other, while opposite-type poles (N-S) mutually attract. The planet Earth forms a magnet, a phenomenon called geomagnetism. Earth's magnetic poles wander geographically but are in the general vicinity of the geographic poles. The north pole of any magnet is defined as the pole that is attracted to Earth's magnetic pole that is near its north geographic pole. So Earth has a *south* magnetic pole at its *north* geographic pole, and vice versa. The polarity of Earth's geomagnetism undergoes a reversal at very irregular intervals of a thousand to millions of years, when the north and south geomagnetic poles flip. A magnetic pole cannot be isolated; poles always appear in north-south pairs. If a magnet is cut in two in an attempt to isolate its poles, each part is found to possess a pair of poles.

The source of a magnet's magnetism is ultimately either an electric current, the intrinsic magnetism of electrons in atoms, or a combination of both. In the first case, a current in a coil becomes a magnet, an electromagnet. The second case is represented by a permanent magnet, in which the aligned magnetic dipole moments of the material's atoms, which are due to the atoms' possessing unpaired electrons, combine to produce the magnetic field of the magnet in an effect that is an aspect of ferromagnetism (see later discussion). The two sources of magnetism come into combined play when ferromagnetic material is used for the core of an electromagnet. Then the magnetic field of the current induces the core to become a magnet.

FORCE AND FIELD

The force **F**, in newtons (N), that acts on an electric charge q, in coulombs (C), as test charge moving with velocity **v**, in meters per second (m/s), at a location where the magnetic field has the value **B**, in teslas (T), given by

$$\mathbf{F} = q\,\mathbf{v} \times \mathbf{B}$$

This formula gives the magnitude and the direction of the force. Its direction is perpendicular to both the velocity and the magnetic field. Since it is perpendicular to the velocity (i.e., to the displacement), the magnetic force performs no work, and the acceleration caused by the magnetic force changes the direction of the velocity but not the speed. The magnitude of the force, F, is

$$F = |q|vB \sin\theta$$

where v and B denote the magnitudes of the velocity and the magnetic field, respectively, and θ is the

A collection of permanent magnets of various kinds. Whatever its size, shape, or material composition, a magnet always possesses a pair of north and south poles. *(Andrew Lambert Photography/Photo Researchers, Inc.)*

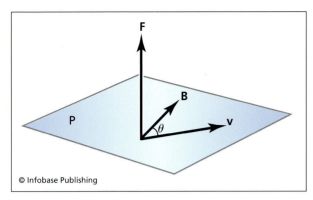

© Infobase Publishing

The force, **F**, on a moving point charge due to a magnetic field is given by $\mathbf{F} = q\mathbf{v} \times \mathbf{B}$. Here q, **v**, and **B** denote, respectively, the amount of charge (including its sign), the velocity of the charge, and the magnetic field at the location of the charge. Plane P contains **v** and **B**. For positive charge, **F** is perpendicular to P in the sense shown. If q is negative, the force is in the opposite direction. The smaller angle between **v** and **B** is denoted θ. The magnitude of the force equals $|q|vB \sin \theta$, where v and B denote the magnitudes of **v** and **B**, respectively.

smaller angle (less than 180°) between **v** and **B**. This equation indicates that the magnetic force on a moving charged particle is stronger for greater charge, higher speed, and more intense magnetic field. All other factors being equal, the force is greatest when the particle is moving perpendicular to the magnetic field, weakens as the angle between the velocity and field vectors decreases from 90°, and vanishes when the particle's motion is parallel to the direction of the field. Again, all other factors being equal, the forces on particles with charges of equal magnitude and opposite sign have equal magnitude and opposite direction.

For an electric current in a magnetic field, the infinitesimal force $d\mathbf{F}$ that acts on an infinitesimal directed element of conductor length $d\mathbf{s}$, in meters (m), that carries current i, in amperes (A), and is pointing in the sense of the current (i.e., the force that acts on an infinitesimal current element $i \, d\mathbf{s}$) is

$$d\mathbf{F} = i \, d\mathbf{s} \times \mathbf{B}$$

The magnitude of this force, dF, is

$$dF = i \, ds \, B \sin \theta$$

where ds denotes the magnitude of $d\mathbf{s}$ (i.e., the infinitesimal length of the conductor) and θ is the smaller angle (less than 180°) between the direction of the current (the direction of $d\mathbf{s}$) and **B**. The direction of the force is perpendicular to both the direction of the conductor and the magnetic field. It follows that the force **F** on a finite-length straight conductor, repre-

sented by the vector **L**, in meters (m), in a uniform magnetic field **B** is given by

$$\mathbf{F} = i\mathbf{L} \times \mathbf{B}$$

with the current flowing in the direction of **L**. The magnitude of the force is

$$F = iLB \sin \theta$$

where L denotes the length of the conductor. Note that the force is stronger with higher current, longer conductor, and more intense field. Also, for the same current, length, and field strength, the force is greatest when the conductor is aligned perpendicular to the magnetic field, weakens as the angle between the conductor and the field decreases from 90°, and vanishes when the conductor is parallel to the field. All other factors being equal, the magnetic forces on oppositely directed currents are in opposite directions.

As an example of use of this equation, assume that a current of 2.0 A is flowing through a straight wire 0.95 m long that is immersed in a uniform magnetic field of 1.5 T at an angle of 35° to the field. The magnitude of the magnetic force acting on the wire is

$$\begin{aligned} F &= iLB \sin \theta \\ &= (2.0 \text{ A})(0.95 \text{ m})(1.5 \text{ T}) \sin 35° \\ &= 1.6 \text{ N} \end{aligned}$$

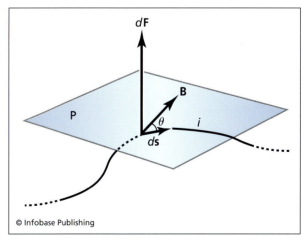

© Infobase Publishing

The infinitesimal force, $d\mathbf{F}$, on an infinitesimal length of current-carrying conductor due to a magnetic field is given by $d\mathbf{F} = i \, d\mathbf{s} \times \mathbf{B}$, where i, $d\mathbf{s}$, and **B** denote, respectively, the current, the infinitesimal directed element of conductor length (pointing in the sense of the current), and the magnetic field at the location of the conductor element. Plane P contains $d\mathbf{s}$ and **B**. $d\mathbf{F}$ is perpendicular to P in the sense shown. The smaller angle between $d\mathbf{s}$ and **B** is denoted θ. The magnitude of the force equals $i \, ds \, B \sin \theta$, where ds and B denote the magnitudes of $d\mathbf{s}$ and **B**, respectively.

FIELD PRODUCTION

As mentioned, the source of the magnetic field is moving charges. The Biot-Savart law, named for the 18–19th-century French physicists Jean-Baptiste Biot and Félix Savart, shows how all electric currents contribute to the magnetic field at every point in space. Specifically, it relates the magnitude and direction of an infinitesimal current element to the magnitude and direction of its contribution to the magnetic field at any point. Then the total magnetic field at that point is obtained by summing up, or integrating, over all current elements. The equation is

$$dB = \frac{\mu}{4\pi} \frac{i\, ds \times \mathbf{r}}{r^3}$$

where ds is a differential vector element of conductor length, in meters (m), in the direction of the current i, in amperes (A), flowing through it; $\mathbf{r}$ is the vector pointing from that element to the point where the magnetic field is being calculated, the field point, whose magnitude r is the distance between the two, in meters (m); dB is the infinitesimal vector contribution to the magnetic field at the field point, in teslas (T), due to the infinitesimal current element (the current flowing in the infinitesimal element of conductor length); and μ is the magnetic permeability of the medium in which this is taking place, in tesla-meter per ampere (T·m/A) (see later discussion). This equation gives both the magnitude and the direction of the contribution.

For the magnitude alone, the equation reduces to

$$dB = \frac{\mu}{4\pi} \frac{i\, ds\, \sin\theta}{r^2}$$

where dB is the magnitude of the infinitesimal contribution to the magnetic field at the field point, in teslas (T); ds is the magnitude of the differential element of conductor length, in meters (m); and θ is the angle between the vectors ds and $\mathbf{r}$. The direction of the contribution is determined in this manner. Consider a straight line along the vector ds (i.e., the tangent to the conductor at the point of the current element), and a circle, perpendicular to and centered on this line, that passes through the field point. Consider the tangent to the circle at the field point and endow this tangent with a direction as follows. Imagine grasping the circle at the field point with your right hand, in such a way that your fingers pass through the plane of the circle in the same direction as does the current in the current element. Then your thumb is pointing in the direction of the tangent. (Alternatively, you can place your right fist at the current element with thumb extended in the direction of the current. Then your curled fingers define a rotation sense for the circle and thus a direction of the tangent to the circle at the field point.) The direction of the tangent is the direction of dB, the infinitesimal contribution to the magnetic field at the field point.

When the field is being produced in vacuum (or, to a good approximation, in air), μ is then the magnetic permeability of the vacuum, denoted μ_0, whose value is $4\pi \times 10^{-7}$ T·m/A. The corresponding vacuum forms of the previous two equations are

$$dB = \frac{\mu_0}{4\pi} \frac{i\, ds \times \mathbf{r}}{r^3}$$

$$dB = \frac{\mu_0}{4\pi} \frac{i\, ds\, \sin}{r^2}$$

An example of use of the Biot-Savart law is to find the magnitude of the magnetic field, B, in teslas (T), at perpendicular distance r, in meters (m), from a long straight wire carrying current i, in amperes (A). Integration gives the following result:

$$B = \frac{\mu_0}{4\pi} \frac{i}{r}$$

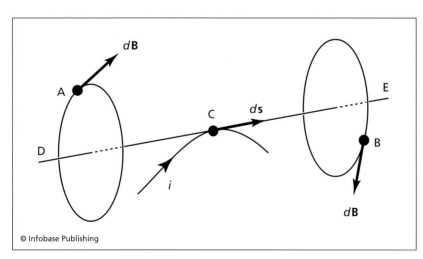

© Infobase Publishing

The direction of the contribution to the magnetic field, dB, at arbitrary field points A and B due to an infinitesimal element of electric current at point C, according to the Biot-Savart law. The symbol ds denotes the differential vector element of conductor length. It is tangent to the conductor at C and points in the sense of the current i. The straight line DE is a continuation of ds. The direction of dB is tangent to the circle that passes through the field point, is perpendicular to line DE, and has its center on DE. The sense of dB for the illustrated sense of the current is shown.

Note that the magnitude of the field is proportional to the current and decreases as 1 over the distance from the wire. As an example of use of the formula, the magnitude of the magnetic field at a distance of 0.24 m from a long straight wire carrying a current of 15 A is

$$B = \frac{4\pi \times 10^{-7}}{4\pi} \frac{15}{0.24}$$
$$= 6.3 \times 10^{-6} \text{ T}$$

A device that acts as a magnet because of an electric current flowing through it is an electromagnet. A typical electromagnet is constructed of a coil of conducting wire uniformly wound around a core of ferromagnetic material (discussed later), such that a magnetic field results when an electric current flows through the coil. In the case of a long helical coil, called a solenoid, the magnitude of the magnetic field inside the coil is approximately

$$B = \mu n i$$

and at its ends about

$$B = \frac{1}{2} \mu n i$$

where B denotes the magnitude of the magnetic field, in teslas (T); i the current, in amperes (A); n the number of turns per unit length of coil, in inverse meters (m^{-1}); and μ the permeability of the core material, in tesla-meter per ampere (T·m/A) (discussed later). In order to achieve especially high magnetic fields, the coil can be made of a superconductor, allowing the flow of very high currents.

As a numerical example, find the magnitude of the magnetic field inside a 6.0-cm-long coil of 200 turns wound around a core whose permeability is 650 times vacuum permeability, when a current of 0.15 A flows through the coil. The data are

$$i = 0.15$$

$$n = 200/(0.060 \text{ m}) = 3.33 \times 10^3 \text{ m}^{-1}$$

$$\mu = 650 \times 4\pi \times 10^{-7} \text{ T·m/A} = 8.17 \times 10^{-3} \text{ T·m/A}$$

and the result is

$$B = (8.17 \times 10^{-3} \text{ T·m/A})(3.33 \times 10^3 \text{ m}^{-1})(0.15 \text{ A})$$
$$= 4.1 \text{ T}$$

The Biot-Savart law can also be expressed in terms of the magnetic field, **B**, produced by an electric charge, q, moving with velocity **v**, giving

$$\mathbf{B} = \frac{\mu_0}{4\pi} \frac{q \, \mathbf{v} \times \mathbf{r}}{r^3}$$

By running an electric current through the large electromagnet hanging at the end of the cable, the crane operator in this junkyard easily grabs steel objects for lifting. The operator releases the objects by turning off the current. *(E. R. Degginger/Photo Researchers, Inc.)*

where **r** is the separation vector from the charge to the field point.

MAGNETIC DIPOLE

A pair of equal-magnitude and opposite magnetic poles at some distance from each other, or any equivalent situation, is known as a magnetic dipole. A *bar magnet* is a straightforward example of a magnetic dipole. The term includes all magnets, as well as electric current configurations—typically loops and coils—that behave as magnets. Indeed, an electric current flowing in a loop creates a magnetic dipole and behaves as a magnet, effectively possessing a pair of north and south magnetic poles. Such a current loop is characterized by its magnetic dipole moment, which is a measure of its strength as a magnet. Magnetic dipole moment, denoted μ_m, is a vector quantity, whose SI unit is ampere-meter^2 (A·m^2). The magnitude of the magnetic dipole moment μ_m

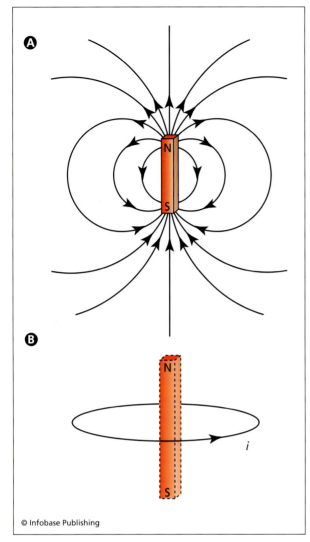

© Infobase Publishing

A magnetic dipole is a pair of separated equal-magnitude and opposite magnetic poles. (a) A bar magnet serves as a typical example of a magnetic dipole. Its north and south poles are labeled N and S, respectively. A number of magnetic field lines are shown for the dipole. (b) An electric current loop i is equivalent to a magnetic dipole, shown in dashed lines. The dipole's polarity relates to the direction of the current in the loop as shown in the figure.

loop(s) and pointing such that if the right thumb is aimed in the same direction, the curved fingers of the right hand will indicate the direction of the electric current in the loop(s).

A torque acts on a magnetic dipole in a magnetic field, tending to align the magnetic dipole moment with the field. The vector relation of torque to magnetic dipole moment and magnetic field is

$$\boldsymbol{\tau} = \boldsymbol{\mu}_m \times \mathbf{B}$$

where $\boldsymbol{\tau}$ denotes the torque vector in newton-meter (N·m) and $\mathbf{B}$ the magnetic field at the location of the dipole in teslas (T). The magnitude of the torque, τ, is

$$\tau = \mu_m B \sin \phi$$

where B denotes the magnitude of the magnetic field and ϕ is the smaller angle (less than 180°) between $\boldsymbol{\mu}_m$ and $\mathbf{B}$. The direction of $\boldsymbol{\tau}$ is perpendicular to both $\boldsymbol{\mu}_m$ and $\mathbf{B}$ and points such that if the right thumb is aimed in the same direction, the curved fingers of the right hand will indicate a rotation from the direction of $\boldsymbol{\mu}_m$ to that of $\mathbf{B}$ through the smaller angle between them. Although the sole effect of a *uniform* magnetic field on a magnetic dipole is a torque, as just described, in a *nonuniform* field, a net force acts on it in addition to the torque.

As an example of a magnetic-torque calculation, assume that a flat circular coil of 50 loops and radius 1.6 cm, carrying a current of 0.29 A, is immersed in a uniform 0.68-T magnetic field, where the angle between the magnetic dipole moment of coil current (which is perpendicular to the plane of the coil) and the magnetic field is 40°. Find the magnitude of the resulting torque acting on the coil. The data are $N = 50$, $i = 0.29$ A, $B = 0.68$ T, $\phi = 40°$, and the radius of the coil $r = 0.016$ m. To find the magnitude of

produced by an electric current i, in amperes (A), flowing in a flat loop is

$$\mu_m = iA$$

where A denotes the area, in square meters (m²), enclosed by the loop. For a coil of N loops, the magnitude of the magnetic dipole moment is

$$\mu_m = NiA$$

The direction of the magnetic dipole moment vector $\boldsymbol{\mu}_m$ is perpendicular to the plane of the current

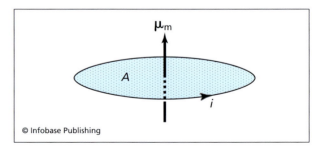

© Infobase Publishing

The magnetic dipole moment of a flat electric current loop is a vector $\boldsymbol{\mu}_m$ that is perpendicular to the plane of the current loop in the direction shown. Its magnitude is given by $\mu_m = iA$, where i denotes the current and A is the area of the loop. For a flat coil of N loops, $\mu_m = NiA$.

the torque, the magnitude of the magnetic dipole moment μ_m is needed, and for that the area of the coil A is required. Start with the area:

$$A = \pi r^2$$

Then the magnitude of the magnetic dipole moment is

$$\mu_m = NiA = \pi Nir^2$$

Now the magnitude of the torque can be found:

$$\begin{aligned}\tau &= \mu_m B \sin \phi \\ &= \pi Nir^2 B \sin \phi \\ &= \pi \times 50(0.29 \text{ A})(0.016 \text{ m})^2(0.68 \text{ T}) \sin 40° \\ &= 5.1 \times 10^{-3} \text{ N·m}\end{aligned}$$

A potential energy is associated with a dipole in a magnetic field. Its value in joules (J) is

$$\begin{aligned}\text{Potential energy} &= -\boldsymbol{\mu}_m \cdot \mathbf{B} \\ &= -\mu_m B \cos \phi\end{aligned}$$

The preceding equation means the potential energy of the system is lowest when the magnetic dipole moment and the field are parallel ($\phi = 0$, $\cos \phi = 1$), which is thus the orientation of stable equilibrium, and highest when they are antiparallel ($\phi = 180°$, $\cos$

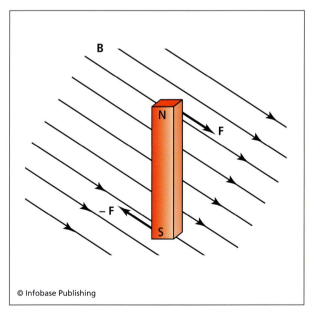

© Infobase Publishing

A uniform magnetic field exerts no net force on a magnetic dipole. Rather, the dipole's poles are affected by equal-magnitude and opposite forces **F** and **-F**, with the north pole N pulled in the direction of the field **B** and the south pole S pulled in the opposite direction. That results in a torque, whose magnitude depends on the dipole's orientation in the field. Some of the field lines are indicated.

$\phi = -1$). The potential-energy difference between the two states is $2\mu_m B$.

As an example, the configuration of the previous example possesses this amount of potential energy:

$$\begin{aligned}\text{Potential energy} &= -\mu_m B \cos \phi \\ &= -\pi Nir^2 B \cos \phi \\ &= -\pi \times 50(0.29 \text{ A})(0.016 \text{ m})^2(0.68 \text{ T}) \cos 40° \\ &= -8.5 \times 10^{-2} \text{ J}\end{aligned}$$

Atoms might also be magnetic dipoles, as a result of the net angular momentum of their electrons or their possession of unpaired electrons (or both). In the former case, electrons in the atom act as electric current loops. In the latter, an unpaired electron itself—like many other kinds of elementary particle—behaves as a magnetic dipole.

Note that there are no real magnetic poles. So, a magnet, or any other magnetic dipole, is always the result of an electric current or currents, either macroscopic or atomic, or of the intrinsic magnetic dipoles of elementary particles.

MAGNETIC FIELD LINES AND ENERGY

A magnetic field line is a directed line in space whose direction at every point on it is the direction of the magnetic field at that point. Magnetic field lines make a very useful device for describing the spatial configurations of magnetic fields. For example, the field lines for the magnetic field produced by a single positive point charge moving in a straight line are circles centered on the line of motion, whose planes are perpendicular to that line. The sense of the circles is determined by the right-hand rule, whereby if the thumb of the right hand lies along the line of motion and points in the direction of motion, the other fingers curve in the sense of the circles. For a similarly moving negative point charge, the magnetic field lines are as above, but with reversed sense. The magnetic field that is produced in this case is in the opposite direction to that produced by the moving positive charge. The field lines associated with a magnetic dipole, including a magnet, emanate from the north magnetic pole and enter the south pole. Only a single magnetic field line can pass through any point in space.

Energy is stored in the magnetic field. The energy density (i.e., the energy per unit volume), in joules per cubic meter (J/m³), at a location where the magnitude of the magnetic field is B is given by

$$\text{Energy density of magnetic field} = \frac{B^2}{2\mu_0}$$

where μ_0 is the magnetic permeability of the vacuum, whose value is $4\pi \times 10^{-7}$ T·m/A.

PERMEABILITY

More specifically called *magnetic permeability*, permeability is the constant μ that appears in the Biot-Savart law for the production of a magnetic field by an electric current or by a moving electric charge and in all other formulas for magnetic field production. When the field production takes place in a vacuum, the permeability is that of the vacuum, denoted μ_0, whose value is $4\pi \times 10^{-7}$ tesla-meter per ampere (T·m/A). When a magnetic field is produced in a material, however, the material might respond to the field by developing an induced magnetic dipole moment, or induced magnetization. The net field results from the combined effects of the applied field and the field of the induced magnetic dipole moment. The latter might enhance the applied field or might partially cancel it. In any case, the magnitude of the net magnetic field relates to that of the applied field by a dimensionless factor called the relative permeability of the material and denoted μ_r. This factor then multiplies the μ_0 in the vacuum version of the Biot-Savart law and other formulas, giving the permeability of the material

$$\mu = \mu_r \mu_0$$

Then the Biot-Savart law and all other formulas for the production of a magnetic field in a vacuum become valid for the net field in a material by the replacement of μ_0 with μ.

When the induced field partially cancels the inducing field, the effect is called diamagnetism. This is a weak effect, so μ_r has a value slightly less than 1 and the permeability of the material is not much less than the permeability of the vacuum. The effect occurs in materials that do not possess intrinsic atomic magnetic dipoles. The applied field then creates in the material an induced magnetic dipole moment that is oppositely directed to the applied field. A sample of diamagnetic material is weakly repelled from a pole of a magnet and thus behaves in a manner opposite to that of ferromagnetic and paramagnetic materials, which are attracted to magnetic poles. Actually, all materials are subject to diamagnetism, but the much stronger effect of ferromagnetism masks the effect.

The two enhancing effects are paramagnetism and ferromagnetism. The former is also a weak effect, so for it the value of μ_r is slightly greater than 1 and the permeability of the material is not much more than that of the vacuum. A sample of paramagnetic material is weakly attracted to a pole of a magnet. This effect is the result of the material's atoms' each possessing one or a few unpaired electrons, which make the atom a magnetic dipole and endow it with a magnetic dipole moment. That occurs because each electron is itself a magnetic dipole, a property related to the electron's spin. The applied magnetic field tends to align the atomic magnetic dipole moments against the disorienting effect of the atoms' random thermal motion. The result is a weak net orientation that gives the material a weak net magnetic dipole moment in the direction of the applied field. When the applied field is removed, the material immediately loses its magnetization.

Ferromagnetism is the strong response of certain materials to an applied magnetic field, by which the material becomes polarized as a magnet and may even remain a magnet when the external magnetic field is removed. In other words, ferromagnetic materials become strongly magnetized in the presence of a magnetic field and might even become permanent magnets. They are strongly attracted to the magnet producing the applied field.

The mechanism of ferromagnetism is this. Each atom of a ferromagnetic material has a relatively large number of unpaired electrons, on the average, whose magnetic dipoles endow it with a considerable magnetic dipole moment. There is strong coupling among the magnetic dipole moments of nearby atoms, which causes the atoms to form local clusters of atoms or molecules with similarly aligned magnetic dipoles spontaneously. Such clusters are called magnetic domains and are observable under a microscope. Thus a domain serves as an elementary magnet. With no application of an external magnetic field the domains are oriented randomly, so their contributions to the magnetic field cancel, and the bulk material is not magnetized.

In the presence of an applied magnetic field, two processes can occur. The one that occurs in weaker applied fields is the growth of domains whose magnetic dipole moments are aligned with the applied field at the expense of the other domains. The magnetic fields of the aligned domains reinforce each other, and the material thus becomes a magnet. When the applied field is turned off, however, the domains randomize again, and the material loses its magnetism. This process occurs when paper clips or iron nails, for instance, are placed near magnets. They become magnets themselves, but only when the inducing magnet is nearby.

A process that requires stronger applied fields is the irreversible rotation of the alignment of domains. In this process, domains that are not aligned with the applied field have their magnetic dipole moments rotated toward alignment. The stronger the applied field, the better the resulting alignment. Again, the fields of the aligned magnetic dipole moments reinforce each other. In this process, the new alignments

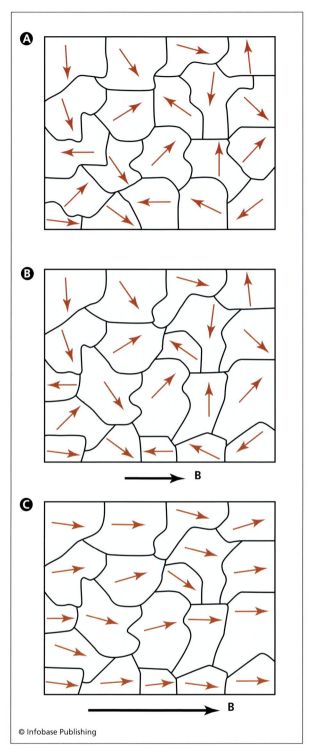

A magnetic domain is a volume in a ferromagnetic material in which all the atomic magnetic dipoles are aligned and that serves as an elementary magnet. (a) In an unmagnetized sample, the domains are randomly oriented. (b) The effect of an applied magnetic field **B** is to cause the domains whose magnetic dipole moment is oriented in the general direction of the field to grow at the expense of the other domains. (c) Sufficiently strong fields irreversibly reorient domains to align their magnetic dipole moments with the applied field.

© Infobase Publishing

are retained when the applied field is removed; the material becomes a permanent magnet. This is the process by which permanent magnets, such as compass needles, refrigerator magnets, and loud speaker magnets, are produced. It is also the process of magnetic recording on audio and video tapes and on magnetic data storage devices, such as computer diskettes and hard drives.

Since the induced magnetization is such that a ferromagnetic material strongly enhances the applied magnetic field, the magnetic permeability of a ferromagnetic material is much larger, even as much as thousands times larger, than the permeability of the vacuum. The value of the permeability of a ferromagnetic material depends on the magnitude of the applied field.

Since heat is the random motion of the atoms that constitute a material, heat acts in opposition to the coupling that aligns atoms to form magnetic domains. So, heat is an enemy of magnetism. That is why we are warned to keep video cassettes and computer diskettes away from heat sources. Depending on the material, there is a temperature, called the Curie temperature and named for the 19th-century French physicist Pierre Curie, above which ferromagnetism disappears.

MAGNETIC LEVITATION (MAGLEV)

The use of magnetic forces to overcome gravity is called magnetic levitation, or *maglev* for short. The term is commonly used in connection with transportation, such as maglev trains. But the effect can be seen in other circumstances. The use of an electromagnet to lift and move steel scrap is an example of magnetic levitation. Magnetic induction causes the steel, a ferromagnetic material, to become a magnet and be attracted to the electromagnet. A more exotic example is the floating of a small magnet above a superconductor. As the magnet falls, or is carried, toward the superconductor, electric currents are induced in the superconductor and produce magnetic fields that repel the magnet. In an ordinary conductor, because of its electric resistance, such induced currents disappear when the motion of the magnet ceases. In a superconductor, however, the induced currents continue to flow even when the magnet comes to rest. At a characteristic equilibrium height above the surface of the superconductor, the magnet can oscillate vertically or remain at rest.

In the application of magnetic levitation to vehicles, there are numerous and various proposed designs. They involve combinations from among permanent magnets, electromagnets, superconducting electromagnets (in which the magnet coils are superconducting), and conductors in various configurations. The design might involve one or more among a diverse range of guideways, lower rails, and overhead rails. Proposed levitation heights vary in the range of around one to a few centimeters. In addition to levitation, maglev systems can provide noncontact magnetic guidance and propulsion. Maglev totally eliminates the friction that is involved in conventional propulsion and guidance, as well as the rolling friction of wheeled travel. The drag forces caused by air remain, however. These can, at least in principle, be greatly reduced by operating the vehicle in an evacuated tube. Propulsion might then be achieved by introducing a small amount of air pressure behind the rear of the vehicle. But maglev inherently creates magnetic drag, which is absent in conventional modes of transportation. This is a resistance to motion caused by magnetic induction. Magnetic levitation offers the potential for achieving considerably higher vehicular speeds and significantly greater energy efficiency than conventional transportation.

MAGNETIC RESONANCE IMAGING (MRI)

Magnetic resonance imaging, commonly referred to by its initials as *MRI,* is a noninvasive imaging technique, widely used in medicine. Magnetic resonance imaging is based on nuclear magnetic resonance. This effect occurs in a magnet field for atomic nuclei that possess spin and, accordingly, a magnetic dipole moment. In that situation, the magnetic dipole moment vector tends to align itself with the magnetic field but can precess about the field's direction, much as a gyroscope precesses when a torque acts upon it. (Precession is the wobbling motion of a spinning body such that its axis of rotation sweeps out a cone.) The magnitudes of the magnetic dipole moment and the magnetic field determine the frequency of precession, which acts as a resonant frequency for the effect of radio waves on the nucleus. Such nuclei absorb energy from radio waves that impinge upon them at their resonant frequency and emit such waves.

For imaging, the object being imaged, such as part of a human body, is immersed in a very strong magnetic field. That necessitates using a very strong electromagnet, which might be a superconducting magnet (an electromagnet with superconducting coils). A magnetic pulse "kicks" the magnetic nuclei of the object into precession, and the radio waves they emit are detected. From the frequency of the waves, one can infer the identity and environment of the emitting nuclei. Careful control of the magnetic field allows for the determination of the location of the nuclei as well. In this way, a three-dimensional image of the object can be constructed. A computer performs the mathematical analysis required for that purpose.

MAGNETIC FLUX

Magnetic flux is a scalar quantity that is closely related to the magnetic field. Its definition in a simple situation is as follows. Consider a flat bounded surface whose area is A, in square meters (m^2), in a region of space where the magnetic field **B,** in teslas (T), is uniform, that is, has the same magnitude B and direction at all points on the surface. Let θ denote the smaller angle (less than 180°) between the direction of the magnetic field and that of a unit vector (a vector of magnitude 1) **n** that is normal (perpendicular) to the surface. For a given surface, **n** can point in either of two directions relative to the surface, and the choice of direction is arbitrary but must remain fixed. The flux Φ, in webers (Wb), through the surface is then given by

$$\Phi = A\mathbf{B}\cdot\mathbf{n} = AB \cos \theta$$

Note that for given magnetic field magnitude and surface area, the flux is maximal when the field is perpendicular to the surface and parallel to **n** ($\theta = 0$, $\cos \theta = 1$), is minimal when the field is antiparallel to **n** ($\theta = 180°$, $\cos \theta = -1$), and vanishes when the field is parallel to the surface ($\theta = 90°$, $\cos \theta = 0$) and no field lines penetrate the surface.

The most important application of magnetic flux is in connection with electromagnetic induction, whereby change of magnetic flux through an electric circuit induces in the circuit an emf and, if the circuit is closed, a current. This effect is governed by Ampère's law, according to which the induced emf, in volts (V), is given by

$$\text{Induced emf} = -N \frac{d\Phi}{dt}$$

where N denotes the number of times the circuit loops upon itself, such as the number of turns of a coil, and time t is in seconds (s). The derivative is the rate of change in time of the magnetic flux through the circuit. For a constant rate of change, Ampère's law can be written in the form

$$\text{Induced emf} = -N \frac{\Delta\Phi}{\Delta t} \quad \text{for constant rate of change of flux}$$

where $\Delta\Phi$ denotes the change of flux during time interval Δt. The same form can be used also to find the average induced emf during time interval Δt for general rate of change of flux:

$$\text{Average induced emf} = -N \frac{\Delta\Phi}{\Delta t}$$

The negative sign is a reminder of Lenz's law, which states that the emf induced in an electric circuit by a change of magnetic flux through the circuit is such that if the circuit is closed, the magnetic field produced by the resulting current tends to oppose the change of flux.

As an example, let a flat circular coil of diameter 3.0 cm (about 1.2 inches) and 50 turns lie in a uniform magnetic field of 0.050 T, perpendicular to the field. During 0.20 s the coil is flipped 180°. Find the average emf induced in the coil during that time interval. Let the vector **n** initially point in the direction of the magnetic field. Then the initial value of θ is $\theta_1 = 0$. When the coil flips, **n,** which must maintain its direction with respect to the coil, flips with it. So the final value of θ is $\theta_2 = 180°$. The diameter of the coil in SI units is $d = 0.030$ m. The coil's area is

$$A = \pi \left(\frac{d}{2}\right)^2$$

The initial magnetic flux through the coil is

$$\Phi_1 = AB \cos \theta_1 = \pi \left(\frac{d}{2}\right)^2 B \cos \theta_1$$

and the final flux

$$\Phi_2 = AB \cos \theta_2 = \pi \left(\frac{d}{2}\right)^2 B \cos \theta_2$$

So the change of flux is

$$\Delta\Phi = \Phi_2 - \Phi_1 = \pi \left(\frac{d}{2}\right)^2 B(\cos \theta_2 - \cos \theta_1)$$

The average induced emf during the given time interval is

Average induced emf

$$= -N \frac{\Delta\Phi}{\Delta t}$$

$$= -\frac{\pi N B}{\Delta t} \left(\frac{d}{2}\right)^2 (\cos \theta_2 - \cos \theta_1)$$

$$= -\frac{\pi \times 50 \times (0.050 \text{ T})}{(0.20 \text{ s})} \left(\frac{0.030 \text{ m}}{2}\right)^2 (\cos 180° - \cos 0)$$

$$= -0.0088 \times [(-1) - 1]$$

$$= 0.018 \text{ V}$$

See also ACCELERATION; ATOMIC STRUCTURE; ELECTRICITY; ELECTROMAGNETIC WAVES; ELECTROMAGNETISM; ENERGY AND WORK; EQUILIBRIUM; FLUX; FORCE; ROTATIONAL MOTION; SPEED AND VELOCITY; SUPERCONDUCTIVITY; VECTORS AND SCALARS.

FURTHER READING

Young, Hugh D., and Roger A. Freedman. *University Physics*, 12th ed. San Francisco: Addison Wesley, 2007.

Mansfield, Sir Peter (1933–) Scottish *Physicist* Peter Mansfield is a Scottish physicist who is best known for his development of a functional magnetic resonance imaging (MRI) machine for use in diagnostic testing of human diseases. Mansfield received the Nobel Prize in physiology or medicine in 2003 in conjunction with Paul Lauterbur for their developments of functional MRI.

EARLY YEARS AND EDUCATION

Peter Mansfield was born on October 5, 1933, in London, England, to Sidney George Mansfield and Rose Lillian Mansfield. World War II significantly impacted Mansfield's early years, particularly through the interruption the war caused to his schooling. The children of London were evacuated regularly, and Mansfield was evacuated three times. He only stayed in school until he was 15 years old, when he began working as an assistant to a printer. Mansfield worked at this position until he was 18 and then entered the army until the age of 21. In 1956 at the age of 23, Mansfield gained entrance into Queen Mary College at the University of London, and in 1959 he received his degree. While he was an undergraduate he was charged with the goal of creating a portable nuclear magnetic resonance (NMR) machine to measure the magnetic field of the Earth. He continued his work at the college and received a Ph.D. from Queen Mary College in September 1962. His graduate studies concerned the creation of a pulsed NMR spectrometer to study the structure of solids.

After completion of his degree, Mansfield took a position in the physics department at the University of Illinois in Urbana. He and his wife moved to Urbana and lived there for two years. There Mansfield continued his work on the NMR spectrum of metals. The Mansfields returned to England in 1964, and Peter took a position at the University of Nottingham. Although university politics shook up the physics department and many MRI groups left Nottingham, Mansfield remained at the university for the rest of his career, retiring in 1994.

NMR DIFFRACTION

Much of the work on NMR was carried out in the 1940s and 1950s. NMR is a technique that utilizes the fact that the atoms rotate in a magnetic

Sir Peter Mansfield is a Scottish physicist who received the Nobel Prize in physiology or medicine in 2003 for his development of magnetic resonance imaging (MRI). Mansfield shared this prize with American chemist Paul Lauterbur. Mansfield is shown in front of an MRI scanner at the Sir Peter Mansfield Magnetic Resources Centre at Nottingham University, Nottingham, England, October 6, 2003. *(AP Images)*

field. Every atom is made up of protons (positively charged particles), neutrons (neutral particles), and electrons (negatively charged particles). The protons and neutrons are arranged in a central nucleus, while the electrons rotate in an electron cloud outside the nucleus. When placed in a magnetic field, these particles rotate and can align either with or opposite to the magnetic field. When the particles are under the magnetic field, they are then subjected to radio waves. As the energized atoms lose their energy, they give off energy that is representative of the structure and location of the atoms. The frequency of energy given off depends on the atom type and its neighboring atoms as well as the strength of the applied magnetic field. Mansfield worked on the application of NMR technology for his entire career. He was not initially looking for biomedical applications of his work but was interested in discovering information about the structure of solids by using NMR techniques.

MAGNETIC RESONANCE IMAGING (MRI)

The most well-known application of NMR is in magnetic resonance imaging (MRI), a technique that applies the principles of NMR to diagnosing biological injuries, diseases, and conditions. Unlike X-rays, which can be damaging, MRI uses radio waves, which have no known negative side effects. MRI also differs from X-rays by visualizing soft tissue rather than bones. The application of NMR techniques to imaging tissue was successful as a result of the large percentage of water molecules present in living tissues. Water's hydrogen atoms are ideal subjects for NMR technology. By determining the location and quantity of water in tissues, it is possible to locate such conditions as tissue damage, tumors, and degeneration of tissues. This makes MRI irreplaceable in diagnosing brain and spinal cord diseases and injuries, cardiovascular issues, aneurysms, strokes, and multiple sclerosis, as well as sports injuries and other types of soft tissue damage. MRI has become a common medical technique with approximately 60 million procedures taking place per year.

When undergoing an MRI, the patient lies on a table that slides into a large tube encircled by the magnet. When the magnetic field is applied, the patient hears pounding noises that indicate radio waves are being pulsed through the chamber. MRI scans last approximately 20–90 minutes depending on the areas of the body being tested. Patients must remove all metal objects prior to the MRI. The intense magnetic field used during the test could be very dangerous. The following preexisting conditions are reasons for disqualification for a MRI:

- pacemaker
- metal implants
- insulin pump
- ear implant
- bullets
- tattoos
- metal fragments
- prosthesis
- hearing aids
- claustrophobia
- brain surgery
- pregnancy
- epilepsy
- history as a machinist or welder

The technician maintains communication with the patient via headphones during the procedure. The subject must remain perfectly still during the course of the test to prevent blurred images.

MANSFIELD'S ROLE IN MRI

The development of NMR technology into this common medical diagnostic procedure required some modifications and complex mathematical treatments in order for the imaging technique to be practically useful in biological science. The first problem to overcome was the mathematical synthesis of the NMR signal into a visual image that would be readable to physicians and technicians. Mansfield's work, along with that of Paul Lauterbur and others, was able to apply the Fourier transform NMR developed by Richard Ernst to produce useful images. The second problem was that of time frame. Traditional NMR technology would have taken hours in order to obtain an image of the human body. The time required to perform the NMR scan was not practical given that the patient would have to be held still within the enclosed structure throughout the entire scan. The process that Mansfield developed that truly made MRI a useful technique was echo planar imaging (EPI). EPI allows for multiple sections to be taken at the same time by using magnetic gradients. Echo planar imaging allows the images to be taken much more quickly in a reasonable time frame.

Mansfield's group also worked on the principle known as active magnetic screening, containing the original magnetic field within one screen, causing the magnetic field that emerges from the screen to be somewhat decreased (flattened) without altering the internal magnetic field that the patient would experience. The addition of a second screen surrounding the first allows for a magnetic field of 0 on the outside of the MRI machine while the patient inside experiences the full magnetic field.

LARGE-SCALE MRI MACHINES

Much of the original work on MRI was done on a very small scale. The successful imaging of the makeup of a thumb was a pivotal moment in the development of MRI technology. This was a noteworthy event shared at major conferences on NMR and MRI technology. Mansfield believed that it would soon be possible to image an entire body. The safety of this process was not fully understood. Many believed that the time it would take to produce the image would be too long to have a person under such a strong magnetic field. Others also worried that because the creation of such a strong magnetic field leads to the creation of an electric field, the procedure could be detrimental to individuals with heart problems.

Prior to a large conference on the application of MRI to human imaging, Mansfield subjected himself to the imaging technique. He successfully produced images and survived the treatment without any apparent harm. The era of the noninvasive imaging technique was just beginning. Presently there are approximately 60 million MRI scans done on patients every year. They have reduced the number of invasive and damaging procedures that need to be carried out and increased the positive outcomes of surgeries.

ACOUSTIC CONTROL IN MRI MACHINES

Peter Mansfield has always been concerned with the safety of those who are utilizing the MRI machine. Currently researchers are unaware of any harmful side effects (except a feeling of claustrophobia) of using MRI. Mansfield has focused some of his research from the late 1990s to the present day on controlling the sound in MRI machines. He finished his career at the Sir Peter Mansfield Magnetic Resonance Center (SPMMRC) at the University of Nottingham, where he retired in 1994 and presently is professor emeritus. The fundamental research carried out at the center involves EPI and slice selection (determining which spin and plane are used for the image).

Peter Mansfield is a dedicated scientist who used his knowledge and experience in NMR for the biological application of disease detection and tissue damage. The Nobel Prize he received in 2003 was one of many awards he has been given in his lifetime including knighthood in 1993. Mansfield, by his part in the creation of MRI, has impacted the daily lives and health of all humankind.

See also LAUTERBUR, PAUL; NUCLEAR MAGNETIC RESONANCE (NMR).

FURTHER READING

Conlan, Roberta, with the assistance of Richard Ernst, Erwin L. Hahn, Daniel Kleppner, Alfred G. Redfield, Charles Slichter, Robert G. Shulman, and Sir Peter Mansfield. "A Lifesaving Window on the Mind and Body: The Development of Magnetic Resonance Imaging." In *Beyond Discovery: The Path from Research to Human Benefit.* The National Academy of Sciences. Available online. URL: http://www.beyonddiscovery. org/content/view.article.asp?a=129. Last updated August 16, 2002.

Mansfield, Peter. *MRI in Medicine: The Nottingham Conference.* New York: Springer-Verlag, 1995.

The Nobel Foundation. "The Nobel Prize in Physiology or Medicine 2003." Available online. URL: http:// nobelprize.org/nobel_prizes/medicine/laureates/2003/. Accessed March 26, 2008.

mass　The notion of mass is complex, because mass has a number of aspects that seem to be quite different in concept, yet are apparently interconnected in ways that are still not fully understood. The SI unit of mass is the kilogram (kg). Various aspects of mass include amount of matter, inertial mass, gravitational mass, and energy.

AMOUNT OF MATTER

Mass serves as a measure of the amount of matter. For instance, a six-kilogram (6-kg) block of iron contains twice as many iron atoms as does a three-kilogram (3-kg) block.

INERTIAL MASS

This is mass as a measure of the inertia of a body, as a measure of a body's resistance to changes in its velocity (or linear momentum), that is, its resistance to acceleration. Newton's second law of motion deals with this aspect of mass.

GRAVITATIONAL MASS

Mass measures participation in the gravitational force. In this regard, it is further categorized as *active* or *passive*. Active gravitational mass is a measure of the capability of whatever possesses it to serve as a source of the gravitational field. Passive gravitational mass measures the degree to which the gravitational field affects its possessor. Albert Einstein's general theory of relativity is a theory of gravitation and thus deals with gravitational mass.

ENERGY

According to Einstein's special theory of relativity, mass and energy are related and can even be considered to reveal two faces of the same essence. The well-known formula relating mass and energy is

$$E = mc^2$$

where m denotes the mass, in kilograms (kg); E the energy, in joules (J); and c the speed of light in a vacuum, whose value is 2.99792458×10^8 meters per second (m/s), rounded to 3.00×10^8 m/s. What this means is that every system possesses an intrinsic internal energy due to its mass, its mass energy.

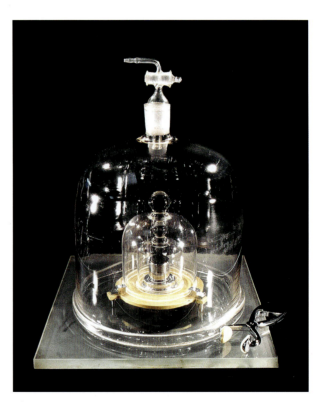

One of the kilogram mass prototypes kept and maintained by the International Bureau of Weights and Measures (BIPM, for Bureau International des Poids et Mesures) in Sèvres, France. It is stored in an atmosphere of inert gas inside three glass bell jars, in order to minimize chemical reactions that might affect its mass. (Bureau International des Poids et Mesures. This document is reproduced after permission had been obtained from the BIPM, which retains full internationally protected copyright on the design and content of this document and on the BIPM's titles, slogans, or logos. The BIPM takes no responsibility for the accuracy or content of a reproduced document, as this is the responsibility of the person or organization making the reproduction. The only official version is the original version of the document published by the BIPM).

Some of this energy might be convertible to other forms of energy, such as through nuclear processes, most notably fission and fusion. All of the mass energy might be converted to a nonmaterial form of energy in the process of mutual annihilation of an elementary particle and its antiparticle when they collide. An electron and a positron, for instance, can undergo annihilation that results in a pair of gamma rays, which are photons, particle-like manifestations of electromagnetic energy. Alternatively, nonmaterial forms of energy might be converted to matter. A gamma ray can, under suitable conditions, transform into an electron-positron pair. In all such cases, Einstein's mass-energy formula is obeyed. Conversely, all energy possesses mass. So nonmaterial forms of energy, such as photons, exhibit inertia and both affect and are affected by the gravitational field. For example, light passing close to the Sun deviates from a straight path in the Sun's gravitational field.

As an example, the mass energy of one gram of matter is 9×10^{13} joules, which is the amount of energy that a 100-megawatt electric power plant supplies in about 10 days. Note, however, that this energy could be obtained and used in full only through the mutual annihilation of a half-gram of matter with a half-gram of corresponding antimatter, which is not feasible in practice.

Rest mass is an intrinsic property of an elementary particle, or of any body, and is defined as the particle's mass when the particle is at rest. A massless particle, such as a photon, is one that moves at the speed of light, only at that speed, and is never at rest. The special theory of relativity nominally assigns 0 rest mass to such a particle, although its energy still possesses a mass equivalent according to the previous mass-energy relation. Most references to mass relate to rest mass.

For the mass equivalent, m, of the total energy of a body, due both to the body's rest mass and to its motion, the special theory of relativity gives the expression

$$m = \frac{m_0}{\sqrt{1 - v^2/c^2}}$$

where m_0 denotes the body's rest mass and v is its speed in meters per second (m/s). For speeds that are sufficiently small compared to the speed of light, this relation takes the form

$$m = m_0 + \frac{1}{c^2}\left(\tfrac{1}{2}m_0 v^2\right) + \text{negligible terms}$$

This equation shows the mass equivalent of the total energy as a sum of the rest mass and the mass equiva-lent of the body's kinetic energy (in its low-speed form).

Masses of elementary particles are often given in terms of their energy equivalents, according to Einstein's mass-energy relation. The most common unit is the mass equivalent of one mega-electron-volt (MeV), which is $1 \text{ MeV}/c^2$, related to the kilogram by $1 \text{ MeV}/c^2 = 1.78266173 \times 10^{-30}$ kg, rounded to 1.78×10^{-30} kg.

Mass is all too often confused with weight, even by science writers and reporters. The weight of an object is the force with which the object is gravitationally attracted to an astronomical body, most commonly to Earth. Whereas the weight of an object is indeed related to the object's mass (more precisely to its passive gravitational mass), mass is an inherent property of an object and is the same no matter where the object might be located. An object's weight depends on not only its mass, but also the object's distance from an astronomical body and the mass and shape of the astronomical body. An object in outer space, far from any astronomical body, is truly weightless. The simple relation between the weight of an object and its mass is given by

$$W = mg$$

where W denotes the object's weight, the magnitude of the force with which it is attracted to, say, Earth, in newtons (N); m is the object's mass, in kilograms (kg); and g denotes the free-fall acceleration, the acceleration due to gravity, in meters per second per second (m/s^2), at the object's location. On the surface of Earth g has the approximate value 9.8 m/s^2 and depends on conditions such as altitude and geographic latitude. The value of g in general depends on the mass and shape of the astronomical body and on the distance from it.

As a numerical example, let the mass of a person be 60 kg. This person's weight on the surface of Earth is then about

$$W = (60 \text{ kg})(9.8 \text{ m/s}^2)$$
$$= 590 \text{ N}$$

On the Moon the free-fall acceleration is around 1.6 m/s^2. So this same 60-kg person, with no change in her mass, would have a weight of approximately

$$W = (60 \text{ kg})(1.6 \text{ m/s}^2)$$
$$= 96 \text{ N}$$

on the Moon, about a sixth of her Earth weight. And in the very unlikely case that this person could reach and survive on the surface of the Sun, where the free-fall acceleration is about 270 m/s², her weight there would be around

$$W = (60 \text{ kg})(270 \text{ m/s}^2)$$
$$= 1.6 \times 10^4 \text{ N}$$

One reason, perhaps, that weight and mass are often confused is that it is common to specify weight in mass units, in kilograms. Those are the units one reads when one is weighed on a metric scale. (In the United States the unit usually used for weight, the pound, is in fact a unit of force.) Nevertheless, when someone declares, for instance, that a newly discovered planet of some star is about 10 times *heavier* than Earth (rather than 10 times *more massive*), he is confusing the concepts. With respect to what astronomical body are the planet and Earth supposed to have weight?

See also ACCELERATION; EINSTEIN, ALBERT; ENERGY AND WORK; FISSION; FORCE; FUSION; GENERAL RELATIVITY; GRAVITY; MATTER AND ANTIMATTER; MOMENTUM AND COLLISIONS; MOTION; NEWTON, SIR ISAAC; NUCLEAR PHYSICS; SPECIAL RELATIVITY; SPEED AND VELOCITY.

FURTHER READING

Jammer, Max. *Concepts of Mass in Classical and Modern Physics*. Princeton, N.J.: Princeton University Press, 1999.

Serway, Raymond A., Clement J. Moses, and Curt A. Moyer. *Modern Physics*. 3rd ed. Belmont, Calif.: Thomson Brooks/Cole, 2004.

mass spectrometry *Mass spectrometry* and *mass spectroscopy* are two names for the same technique, which separates ionized compounds on the basis of the mass-to-charge ratio of the ions. This technique allows many types of compounds to be separated from one another and then detected or analyzed. Mass spectrometry in its simplest terms is an excellent means for analyzing samples for trace amounts of a certain chemical. Samples can be analyzed for identification of unknown chemicals, quantification of desired compounds, as well as structural information about molecules. Accurate detection can be achieved with sample quantities as low as 10^{-12} gram (one-trillionth of a gram). Many adaptations of the basic principle of mass spectrometry allow for the separation of diverse types of compounds. The future of this procedure in criminal investigations, drug development, and even forgeries makes it a very valuable tool. Mass spectrometry can even be coupled with other separation techniques to increase the amount of information that is gained from the techniques individually such as gas chromatography-mass spectrometry (GC-MS), in which the sample is first separated by gas chromatography and then subjected to mass spectrometry.

STRUCTURE OF MASS SPECTROMETER

Every mass spectrometer consists of three parts: the ionizer, the analyzer, and the detector. Many versions of each of these three parts allow for numerous variations in separation ability, making mass spectrometry a useful tool across a wide range of scientific fields.

The first step of mass spectrometry takes place in the ionizer, where the sample is ionized and separated, meaning any of the ionic bonds are broken, and the ions are present in the concentration at which they existed in the original sample. The analyzer sorts these ions and directs them to the detector, which totals each of the ions and represents their frequency in graphical form. Many times the signal requires amplification prior to interpretation because the substance analyzed is present in such trace amounts in the original sample. The multiple combinations of ionizer, analyzer, and detector help to make mass spectrometry one of the most flexible tools in science.

IONIZATION METHOD

Multiple types of ionizers exist to determine how the sample is separated and what types of samples can be ionized. Production of both positive and negative ions is likely during ionization. One must determine which of these to analyze and detect. The two most common types of ionization methods are electrospray ionization (ESI) and matrix-assisted laser desorption ionization (MALDI). These two techniques have helped to make mass spectrometry applicable to nearly all biological molecules.

ESI is useful for large, nonvolatile particles of a large range of masses that are often difficult to separate. The sample is inserted along with a solvent into an electrically charged metal capillary and ejected at a high rate of speed. The solvent is ionized either positively or negatively, and this repels the particles of the sample. This repulsion causes the ejection of the sample out of the capillary as individual particles. The solvent, being more volatile than the sample, evaporates. ESI is useful in separation of large biomolecules, such as oligonucleotides, fatty acids, carbohydrates, and proteins.

MALDI utilizes a laser to separate the sample into ions and is very useful for large biomolecules. The first stage is to dissolve the sample in an appropriate solvent. The addition of a matrix, a compound that absorbs ultraviolet (UV) light, is required in order to prevent the sample from being destroyed by the laser. The ratio of matrix to sample is very large,

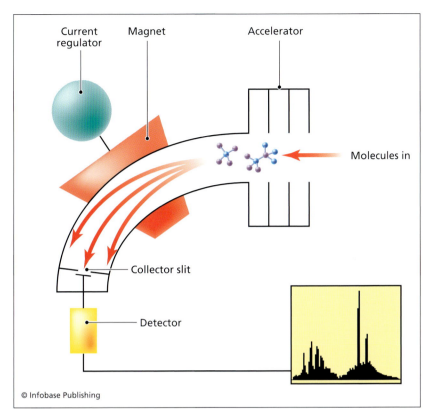

Current regulator   Magnet   Accelerator

Molecules in

Collector slit

Detector

© Infobase Publishing

A mass spectrometer separates samples on the basis of their mass-to-charge ratio as they travel through a magnetic field. The molecules are accelerated then passed through a magnetic field. Only those molecules with the appropriate mass-to-charge ratio will be bent to the appropriate angle to reach the detector.

netic field causes the ions to bend their path to the detector into an arc inversely proportional to the mass of the ion. Those ions with a mass/charge ratio that is too large will be bent too much to reach the detector, and those with a mass/charge ratio that is too light will not be bent far enough to reach the detector. Only the ions with a mass within a specifically set range will be bent with an appropriate arc toward the detector. When the ions reach the detector, the amount of charge is measured and represents the presence of the particular ion in the original sample. Controlling the intensity of the magnetic field controls which ions successfully reach the detector.

Quadrupole mass analyzers are made up of four parallel rods between which the ions travel. Subjecting the rods to an electrical current as well as a radio frequency allows for the passage of ions with the correct mass/charge ratio between the rods in order to reach the detector. The other ions will collide with the rods and not reach the detector. Adjusting the electrical current and/or the radio frequency determines which ions will reach the detector.

Time-of-flight (TOF) analyzers measure the amount of time the ions takes to move down a straight path to the detector. The lighter mass/charge ions move fastest and arrive first. Larger molecules move more slowly. This is a useful method for separating large biomolecules and is the fastest type of mass spectrometer analyzer.

FT-ICR analyzers function when ions are placed in a magnetic field then subjected to a perpendicular electric field. The frequency of the ions is determined, and scientists use a Fourier transform of the signal to calculate the mass of the substance. FT-ICR offers the highest mass resolution and accuracy of any type of mass spectrometer.

Some mass spectrometers have more than one analyzer set up in the device. This process uses tandem (MS/MS) mass spectrometers. The analyzers can be of the same type or different types to refine the separation and resolution of the sample. Quadrupole and FT-ICR can be combined to separate the sample by time. The quadrupole and time-of-flight analyzers can also be combined.

on the order of 10,000 to 1. The sample is placed in a chamber that is then evacuated, causing the solvent to evaporate and leaving the sample embedded in the matrix. The sample is subjected to a laser with a wavelength of approximately 300–360 nm, causing the matrix and the sample to absorb a large amount of energy. The interaction of the matrix and the sample causes the sample to ionize, and the energy absorbed causes the sample to vaporize, a rare occurrence for large molecules.

ANALYZERS

Once the samples are ionized, the next step involves sorting and resolution of the different ions based on their mass. Many types of analyzers specifically test for different components of mixtures, allowing mass spectrometry to be used for multiple applications. The most common types of analyzers include magnetic analyzers, quadrupole mass analyzers, time-of-flight analyzers, and Fourier transform ion cyclotron resonance (FT-ICR) analyzers.

Magnetic analyzers employ a magnetic field to control the trajectory of the stream of ions. The analyzer has a bent path to reach the detector. A mag-

Mass spectrometry is a valuable analytical chemistry technique. A Pacific Northwest National Laboratory scientist works at a Tesla Fourier Transform Ion Cyclotron (FT-ICR) mass spectrometer at the Hanford Site in Richland, Washington, August 31, 2001. *(DOE Photo)*

DETECTORS

The detector is responsible for collecting and detecting the ions that are separated by the analyzer. The detector turns this information into a signal that represents the mass of the ions and their relative abundance. Three common types of detectors are the electron multiplier, the photomultiplier, and the microchannel plate.

The electron multiplier uses a conversion dynode that changes all of the ions into electrons, regardless of their initial charge. A channeltron (a horn-shaped device that concentrates the electrons) creates a current from the electrons. The detector measures the current and converts it into a mass spectrograph.

A photomultiplier detector converts the ions to electrons using a conversion dynode, a series of successively more positive electrodes that undergo secondary emission after being struck by a charge. The electrons hit a phosphorescent substance that uses the energy of the electrons to emit a photon. This photon then proceeds to cause the release of an electron due to the photoelectric effect. These electrons are multiplied by passing through a series of dynodes that increases the number of electrons in the signal, thus amplifying weak signals. Microchannel plate

detectors have a plate with multiple small channels through it. Each microchannel is an individual dynode that multiplies the amount of electrons through to secondary emission. The signal is collected on the other side of the channels.

USES OF MASS SPECTROMETRY

Mass spectrometry has applications in more fields than nearly any other scientific procedure or device. The ability of mass spectrometry to detect and analyze trace quantities of material makes it ideal for such applications as detecting toxins, detecting trace quantities of materials in the blood, and determining the composition of unknown toxins or chemicals. Biotechnological applications include analysis of the composition of proteins, carbohydrates, nucleic acids, and other large polymers.

Archaeological research depends on the ability of mass spectrometry to identify components in a substance, thus allowing the composition to be known and compared to other samples. Forgeries have been discovered by using mass spectrometry based on the composition of antique substances versus newer components. By detecting the quantities of individual isotopes, mass spectrometry is essential for radioactive dating.

Pharmaceutical applications of mass spectrometry are numerous. Studying the rate of drug metabolism allows for the determination of appropriate doses. Identification and quantification of impurities in samples of medications are also accomplished using this technique. Drug testing using mass spectrometry detects even a trace amount of a substance in the blood or bodily fluids.

Environmental applications of mass spectrometry include analysis of air, water, and soil for pollution and contaminants. The effectiveness of water purification procedures is analyzed using mass spectrometry to determine whether there are trace amounts of harmful chemicals in the water. Mass spectrometry has become an indispensable analytical tool in all types of scientific applications.

HISTORY OF MASS SPECTROMETRY

Scientific discovery in the field of mass spectrometry led to six Nobel Prizes, awarded to Sir J. J. Thomson, Francis Aston, Wolfgang Pauli, Hans G. Dehmelt, John B. Fenn, and Koichi Tanaka.

Mass spectrometry began in the late 1800s with Sir J. J. Thomson's work with the cathode-ray tube. Thomson recognized that the ionization of a compound could prove to be a useful technique for scientific discovery. He discovered the electron using this method in 1897 and developed the beginnings of the first mass spectrometer (a parabola spectrograph) in the early 1900s. Thomson received the Nobel Prize in physics in 1906 for his work with electrical conductivity through gases.

The British chemist Francis W. Aston improved on Thomson's design and used his mass spectrometer for work on isotopes that led to his receiving the Nobel Prize in chemistry in 1922. Aston used both magnetic fields and electrostatic fields to focus onto photographic plates.

Wolfgang Pauli, a German physicist, developed the quadrupole and quadrupole-ion trap mass spectrometers. For his accomplishments in this field he received the Nobel Prize in physics in 1989. The German physicist Hans G. Dehmelt shared the prize that year for his development of the ion trap technique.

John B. Fenn, an American chemist, received the Nobel Prize in chemistry for his work with electrospray ionization mass spectrometry of biomolecules. ESI was proposed by the American chemist Malcolm Dole in the 1960s but was refined by Fenn.

The Japanese chemist Koichi Tanaka received the Nobel Prize in chemistry in 2002 for the development of MALDI. Innovations in the field of MALDI mass spectrometry were also contributed in the 1980s by the German scientists Franz Hillenkamp and Michael Karas.

Other notable contributions and advances include time-of-flight mass spectrometry, first proposed by the American physicist William H. Stephens in 1946. FT-ICR mass spectrometry was developed by Canadian chemists Melvin Comisarow and Alan Marshall in 1974.

See also ANALYTICAL CHEMISTRY; MASS; PHOTOELECTRIC EFFECT.

FURTHER READING
Gross, Jürgen H. *Mass Spectrometry—a Textbook*. Heidelberg, Germany: Springer, 2004.

materials science   Also known as materials engineering, materials science studies the properties and performance of matter, their relation to materials' microscopic composition and structure, and the application to science and engineering of the understanding obtained. Materials science is a highly interdisciplinary field. It overlaps with parts of physics, applied physics, chemistry, and various engineering fields, such as chemical, electrical, and mechanical engineering, and it includes nanotechnology.

Thermodynamics and statistical mechanics lie at the foundation of materials science, since together they are concerned with the fundamentals of relating macroscopic properties of matter to its microscopic properties. In practice, materials science seeks to achieve the ability to predict the properties and performance of materials from their chemical composi-

The successful design and production of an airplane, such as the one shown here, depend on the efforts of materials scientists. They create the materials needed to fulfill the various functions necessary for the airplane to perform as required. For example, the material the airplane's tires are made of must be able to withstand the forces produced by a rough landing, yet be as light as possible. *(Lars Christensen, 2008, used under license from Shutterstock, Inc.)*

tion and the processing they have undergone. Then, given certain desired properties of a material, one can determine the raw materials needed and how they are to be combined and processed in order to obtain those properties. So, materials science has immense relevance to industry. For instance, if it is known that the skin of a new kind of aircraft needs to possess certain elastic properties, have a certain strength, be able to withstand a certain temperature, yet have no greater than a certain density, then the materials scientist might devise a recipe for creating such a material. The result might be a metallic alloy or possibly a composite material, such as a network of carbon fibers embedded in a polymer, or perhaps something else altogether.

The first appearance of what one might call materials science was surely the art of ceramics and glass, whose development commenced in humanity's premetallic era, in the Stone Age. The Bronze Age introduced the art of metallurgy, which developed on through the Iron Age and into the 19th century. Then the arts of ceramics, glass, and metallurgy started to turn into sciences and by the 20th century became serious sciences and evolved into materials science as it is known today. Modern materials science is no longer confined to ceramics, glass, and metals, but includes all types of materials, such as plastics, semiconductors, polymers, and composites.

In order to give a more detailed idea of the field of materials science, the following is a list with brief descriptions of some of the field's subfields:

- *Metallurgy.* Still an important field, this is the study of metals, metallic compounds, and alloys of metals. It includes their extraction from ores and their processing into useful forms.
- *Crystallography.* The study of the structure of crystals in terms of the positions of their microscopic constituents—atoms, ions, and molecules—and of the defects of crystal structures. The latter include various classes of dislocation, which distort the regular lattice structure, and boundaries between crystal grains, when a crystal is composed of a conglomeration of smaller crystals possessing different orientations.
- *Vitreous state study.* The investigation of noncrystalline materials, including various kinds of glass and vitreous metals.
- *Materials characterization.* This is the study of the composition and microscopic structure of materials for the purpose of determining and understanding their properties. Methods used include chemical analysis, light and electron microscopy, various kinds of spectroscopy, chromatography, and diffraction of X-rays or particles.
- *Nanotechnology.* The creation and investigation of materials in particles whose dimensions are in the range of from about a nanometer (one-billionth of a meter, 10^{-9} m = 1 nm) to around 100 nanometers. Such nanoparticles can exhibit quantum effects, which do not appear in larger conglomerations of the same material. Nanotechnology is a "hot" field these days, since materials consisting of nanoparticles can possess useful properties, presently utilized for such applications as cosmetics and the delivery of drugs.
- *Biomaterials.* The production and study of materials that are derived from biological sources or are used in biological systems.

A major professional society devoted to materials science is the Minerals, Metals, & Materials Society (TMS). This is a United States–based international organization with nearly 10,000 members. Its declared mission is to promote the global science and engineering professions concerned with minerals, metals, and materials. TMS supports this mission by, among other activities, providing forums for the exchange of information (meetings and journals, in particular the monthly *JOM*) and promoting technology transfer and the education and development of materials science professionals. The society comprises five technical divisions:

- Electronic, Magnetic, & Photonic Materials Division
- Extraction & Processing Division
- Light Metals Division
- Materials Processing & Manufacturing Division
- Structural Materials Division

See also CHROMATOGRAPHY; ELASTICITY; ELECTRICAL ENGINEERING; HEAT AND THERMODYNAMICS; MECHANICAL ENGINEERING; NANOTECHNOLOGY; QUANTUM MECHANICS; STATISTICAL MECHANICS.

FURTHER READING

Askeland, Donald R., and Pradeep P. Fulay. *The Science and Engineering of Materials,* 5th ed. Boston, Mass.: CENGAGE Learning, 2005.

Gordon, J. E. *The New Science of Strong Materials or Why You Don't Fall through the Floor.* Princeton, N.J.: Princeton University Press, 2006.

"The Greatest Moments in Materials Science and Engineering." Available online. URL: http://www.materialmoments.org/top100.html. Accessed January 8, 2008.

The Minerals, Metals, & Materials Society (TMS) home page. Available online. URL: http://www.tms.org. Accessed January 8, 2008.

matter and antimatter   In its broadest sense, the term *matter* is generally understood to encompass everything in nature that possesses rest mass, that is, whatever can—at least in principle—be brought to rest. That excludes photons, which form the particle aspect of electromagnetic waves such as light, and gravitons, the putative particle aspect of gravitational waves. Photons and gravitons are sometimes referred to as "pure" energy. They travel at the speed of light (approximately 3.00×10^8 m/s) and only at that speed. Everything else, all the elementary particles and everything that is composed of elementary particles, comprises matter in the broad sense.

In elementary presentations, one often finds a definition of matter as anything that has mass and occupies space. Both parts of this definition are problematic. As for mass, even a photon, which is not generally considered matter, has mass in the sense of the mass equivalent of its energy, according to Albert Einstein's famous mass-energy relation

$$E = mc^2$$

or, perhaps better for our present purpose, the equivalent form

$$m = \frac{E}{c^2}$$

where E denotes the energy, of the photon in this case, in joules (J); m is the mass that is equivalent to that energy, in kilograms (kg); and c is the speed of light in a vacuum. However, a photon does not possess *rest mass* and cannot be brought to rest. Electrons present another problem with regard to occupying space. Electrons should clearly be considered matter, as they form an important component of all macroscopic matter (such as cars, coal, and cantaloupes). Yet, as far as is presently known, electrons seem to be point particles, which by definition means they do not occupy any space.

In a narrower sense, however, matter is opposed to antimatter and designates a class of elementary particles including the particles that make up ordinary matter, such as the electron, proton, and neutron, and other particles. For every matter particle type there exists an antimatter particle type, called its antiparticle, that has some properties in common with it and other properties opposed. A particle and its antiparticle both possess the same mass, spin, and magnitude of electric charge, for example. On the other hand, they have opposite signs of electric charge, as well as opposite signs of charges of other kinds, such as lepton number or baryon number. (The latter are numbers conveniently assigned to elementary particles and are conserved in certain interactions.) The antiproton, for example, has the same mass as the proton and has spin ½ and one elementary unit of electric charge, as does the proton, but its electric charge is negative, whereas that of the proton is positive. Also, its baryon number is –1, while that of the proton is +1. The situation for the electron is similar, but its antiparticle has its own name, the *positron*. Certain particles, all of whose various charges are 0, serve as their own antiparticles. Such are the neutral pion and the photon, for instance, although the photon is not strictly matter. In this narrower sense, conglomerations of matter particles are termed matter, while ensembles of antiparticles are called antimatter.

When a particle and its antiparticle collide with each other, they can undergo mutual annihilation, with the emission of a pair of photons. In this process, the energy of the pair is totally converted to electromagnetic energy. On the other hand, a sufficiently energetic photon (a gamma ray) passing through matter can convert to a particle-antiparticle pair, such as an electron and a positron, a process called pair production. Particle-antiparticle pairs feature in the picture of the vacuum that quantum physics describes. The vacuum is in a constant turmoil of activity, with particle-antiparticle pairs of all kinds ceaselessly and spontaneously materializing briefly and then dematerializing. The energy needed for such fleeting pair production

This bubble-chamber photograph caught the production of two particle-antiparticle pairs, here electron-positron pairs. A gamma ray, which did not leave a track, entered from the upper right, moving downward and slightly to the left, about south by southwest. It interacted with a hydrogen nucleus (a proton) and produced the first pair, while causing the proton to recoil to the lower left. The electron and positron made the two oppositely spiraling tracks emanating from the production point. The recoiling proton created the slightly curving track from the production point to the lower left. The bubble chamber was immersed in a magnetic field, which caused the charged particles' circular or curved motion. After creating the first electron-positron pair, the gamma ray continued in practically the same direction and, by interacting with another proton, created a second electron-positron pair, whose tracks are indicated by the inverted V in the center of the photograph. *(LBNL/Photo Researchers, Inc.)*

and annihilation is "on loan" for extremely short times, according to the quantum uncertainty principle of the 20th-century German physicist Werner Heisenberg.

As far as is presently known, the objects of the universe consist entirely of matter, with antimatter only being produced in various processes and annihilated soon after. The reason for that is still a mystery, especially in light of cosmological theories that matter and antimatter were created in nearly equal amounts at an early stage in the evolution of the universe. The matter and antimatter are theorized to have almost completely annihilated each other, leaving a very slight surplus of matter (rather than of antimatter), which is what we observe today. Why a surplus of matter rather than of antimatter and why any surplus at all are questions that are presently unanswered.

The notion of antimatter arose originally in the 1920s, when the British theoretical physicist Paul Dirac, in his attempt to combine quantum mechanics and the special theory of relativity, devised an elegant equation to describe an electrically charged point particle that has spin ½, such as the electron. However, the equation seemed to predict that negative-energy states would be available to such a free particle, which would preclude the existence of the particle as it is observed to exist. With negative-energy states available to it, the particle would spontaneously jump from its free, positive-energy state to a negative-energy one, while emitting a photon to carry away the energy difference between the two states. However, a reinterpretation of the equation was found that allowed the existence of both the electron and a kind of particle that was identical to the observed version except for its possessing opposite electric charge. Thus, Dirac's equation predicted the existence of the antielectron, a positively charged, electronlike particle to accompany the electron. Similarly, every type of particle should have an antiparticle counterpart. The American physicist Carl Anderson experimentally discovered the antielectron, renamed the *positron,* in 1932. The antiproton and antineutron were discovered in 1955. Since then, it has become clear that every type of particle indeed has an antiparticle counterpart, with certain particles, such as the photon and the neutral pion, serving as their own antiparticle.

See also ATOMIC STRUCTURE; BIG BANG THEORY; CONSERVATION LAWS; DUALITY OF NATURE; ELECTROMAGNETIC WAVES; GRAVITY; HEISENBERG, WERNER; MASS; NUCLEAR PHYSICS; PARTICLE PHYSICS; QUANTUM MECHANICS; SPECIAL RELATIVITY.

FURTHER READING

Serway, Raymond A., Clement J. Moses, and Curt A. Moyer. *Modern Physics,* 3rd ed. Belmont, Calif.: Thomson Brooks/Cole, 2004.

Young, Hugh D., and Roger A. Freedman. *University Physics.* 12th ed. San Francisco: Addison Wesley, 2007.

Maxwell, James Clerk (1831–1879) Scottish *Mathematician and Physicist* James Clerk Max-

well was a Scottish-born mathematician and physicist best known for his experimental and mathematical explanations of electromagnetism, including adaptations of the British physicist Michael Faraday's work on electromagnetism, his work on the kinetic theory of gases, and his calculations of the stability of the rings of Saturn. Maxwell published several important papers including "On the Nature of Saturn's Rings," covering his calculations on Saturn ring stability, and "On Faraday's Lines of Force," which used his mathematical mind to extend the presently known information about electricity and electromagnetism. Maxwell studied the motion and properties of gases and helped develop the kinetic molecular theory and the Maxwell-Boltzmann equations. James Clerk Maxwell was also responsible for the publication of the long neglected works of Henry Cavendish from the late 1700s.

James Clerk Maxwell was born to John Clerk Maxwell and Frances Maxwell on June 13, 1879, in Edinburgh, Scotland. He was educated at home until the death of his mother when he was eight years old and then transferred to the Edinburgh Academy, where he completed his schooling. Much of his time at Edinburgh Academy involved the study of Saturn's rings. Maxwell attended Trinity College at Cambridge and graduated in 1854 with a degree in mathematics. Despite his desire to remain at Cambridge, because his father became ill, in 1856 Maxwell took an appointment for his first teaching assignment at Marischal College in Aberdeen, where he was appointed chair. In 1857 Maxwell competed for the prestigious Adams Prize on the motion of Saturn's rings. Maxwell's research at Aberdeen was almost exclusively focused on that work. Maxwell's findings included the concept that the only way that Saturn's rings could exhibit their stability was if they were made up of a series of smaller particles. This concept was later demonstrated by the *Voyager* spacecraft visualizing Saturn's rings. Maxwell's work won him the prize. In 1859 Maxwell married the daughter of the head of Marischal College, Kathy Mary Dewar. The couple did not have any children.

SATURN'S RINGS

The rings of Saturn were first observed by Galileo Galilei with his development of the first telescope. He was not able to visualize their composition. In 1655 the Dutch mathematician Christiaan Huygens described the disks surrounding Saturn. The true understanding of the makeup of the rings remained a mystery. Maxwell worked on the nature of Saturn's rings. His paper "On the Nature of Saturn's Rings" contained nearly 60 pages of complicated mathematical calculations. Maxwell concluded first that it was not possible for Saturn's rings to be solid. The stability of a solid ring would depend on perfect balance. If there were any unsymmetrical mass distribution, one would easily be able to see the lopsided nature of the ring. No such unsymmetrical view of the ring had been witnessed.

Through his calculations, Maxwell determined that the rings must be made of particles that are independent of one another. They could be arranged as either concentric rings or multiple, independent particles that collide with one another. Maxwell's research determined that concentric rings would be too difficult to maintain, and therefore the most stable version of Saturn's rings was made of individual particles moving with their own speed.

ELECTROMAGNETISM

Maxwell began working at King's College in 1860. While working there, he completed his primary and thorough work on electromagnetism. Maxwell's equations presented the foundation for classic electromagnetism. In 1873 he published *A Treatise on Electricity and Magnetism,* which presented his theories on electromagnetism based on the foundational work of Michael Faraday. Maxwell's first contribution was the calculation that the speed of electromagnetic radiation was the same as that of light. From this work, he concluded that light is made up of electromagnetic radiation.

Prior to Maxwell's work, physicists commonly accepted the theory of "action at a distance" as standard. This theory stated that energy transferred between two objects separated by a distance occurred between the objects regardless of whether there were a medium between them or not. Electrical attraction or gravitational attraction required nothing more than the two objects; therefore, no information needed to be discerned about the material between them or the method of transmission. This theory created a problem in that all other physical transfers known at the time required contact between two types of medium.

Maxwell's work introduced the classical electromagnetic field theory, in which the electrical attraction or other interactions occur when the first object exerts a force on the other through an intervening medium when the two objects are separated by a distance. This concept is known as action by continuous contact when the contact is with a medium that contains an electromagnetic field that exists in the space between the two objects. Maxwell demonstrated that the electromagnetic field was the mediator between two charged objects.

Maxwell embraced Faraday's previous work on magnetic lines of force and was quoted in 1873 as saying, "Faraday saw a medium where they saw nothing

James Clerk Maxwell was a 19th-century Scottish mathematician and physicist best known for his work on electromagnetism and the kinetic theory of gases. *(Sheila Terry/Photo Researchers, Inc.)*

but distance. Faraday sought the seat of the phenomena in real actions going on in the medium, they were satisfied that they had found it in a power of action at a distance impressed on the electric fluids." The field lines that Faraday discovered represented the force acting between the two objects, supporting Maxwell's action by continuous contact theory. James Clerk Maxwell's contribution to the understanding of electromagnetism made him a founding father of physics.

MAXWELL-BOLTZMANN STATISTICS

Understanding the motion and properties of gases is an important topic in physical science. Gas is the state of matter with no definite shape and no definite volume. Gases assume the shape and volume of their container. Maxwell spent a large portion of his research career working on the motion of gas molecules. In 1867 he formulated the Boltzmann-Maxwell equation, a statistical relation for the kinetic energy of gas molecules. Maxwell determined this relationship independently of the Austrian physicist Ludwig Boltzmann, for whom the formula is named.

The equation that represents the speed of the molecules in a gas is given by the speed distribution function is as follows:

$$n(v) = \frac{4\pi N}{V} \left(\frac{m}{2\pi kT} \right)^{3/2} v^2 e^{-\frac{mv^2}{2kT}}$$

where the v is the speed in meters per second (m/s), T is the temperature in kelvins (K), N represents the total number of molecules, V is the volume of the gas in cubic meters (m^3), m is the mass of a single molecule in kilograms (kg), and k is the Boltzmann constant, whose value is rounded to 1.38×10^{-23} J/K.

The present-day kinetic molecular theory involves the following principles:

- The molecules of a gas are small relative to the distance between them.
- Molecules of a gas are in constant random motion.
- Collisions between molecules are frequent and random.

Maxwell's research helped to demonstrate the relationships among the motion of the particles, their kinetic energy, and the temperature of the gas. The faster the motion of the particles, the higher the kinetic energy they have, and the higher the kinetic energy, the higher the absolute temperature of the gas.

HENRY CAVENDISH PUBLICATIONS

Maxwell received the prestigious Cavendish Professor of Physics position at Cambridge in 1871. Here Maxwell designed the Cavendish Laboratory, which opened in 1874. Maxwell made major contributions to the field of electricity simply by saving work that had been completed nearly a century before his time. Henry Cavendish was an English physicist who made many significant discoveries in electricity without ever publishing them. Cavendish's reclusiveness and reluctance to share his work with others until he was completely confident of his research nearly resulted in the loss of his work forever. Maxwell prevented this loss, as much of his time at Cambridge was dedicated to the researching and reviving of Henry Cavendish's unpublished work and experiments. Maxwell repeated Cavendish's experiments and compiled them in his version of Cavendish's manuscripts as *The Electrical Researches of the Honourable Henry Cavendish* in 1879.

James Clerk Maxwell was a pioneer in physics. His work on the rings of Saturn and the fields of electromagnetism and electricity, his discoveries on the kinetic theory of gases, as well as his salvaging the work of Cavendish cemented his position in the history of physics. Maxwell died on November 5, 1879, in Cambridge.

See also CAVENDISH, HENRY; ELECTROMAGNETIC WAVES; ELECTROMAGNETISM; MAXWELL-BOLTZMANN STATISTICS; STATISTICAL MECHANICS.

FURTHER READING

Campbell, Lewis, and William Garnett. *The Life of James Clerk Maxwell: With Selections from His Correspondence and Occasional Writings.* Boston: Adamant Media, 2005.

Harman, James P. *The Natural Philosophies of James Clerk Maxwell.* New York: Cambridge University Press, 1998.

Maxwell-Boltzmann statistics Named for the 19th-century British physicist James Clerk Maxwell and 19th-century Austrian physicist Ludwig Boltzmann, Maxwell-Boltzmann statistics deals with collections of distinguishable particles in thermal equilibrium. One result is the expression for the distribution of speeds among the molecules of a gas in thermal equilibrium, called the Maxwell distribution. According to this, in such a gas at absolute temperature T, the number of molecules per unit volume that have speeds in the small range from v to $v + dv$ is given by $n(v)\ dv$, where $n(v)$, the speed distribution function, is

$$n(v) = \frac{4\pi N}{V} \left(\frac{m}{2\pi kT} \right)^{3/2} v^2 e^{-\frac{mv^2}{2kT}}$$

Here the speed v is in meters per second (m/s), the absolute temperature T is in kelvins (K), N denotes the total number of molecules; V is the volume of the gas, in cubic meters (m^3); m the mass of a single molecule, in kilograms (kg); and k is the Boltzmann constant, whose value is $1.3806503 \times 10^{-23}$ joule per kelvin (J/K), rounded to 1.38×10^{-23} J/K.

Expressed in terms of energy, the number of molecules per unit volume with kinetic energy in the small range from E to $E + dE$ is $n(E)\ dE$, where the energy distribution function, $n(E)$, is given by

$$n(E) = \frac{2\pi N}{V} \left(\frac{1}{\pi kT} \right)^{3/2} \sqrt{E}\ e^{-\frac{E}{kT}}$$

with the kinetic energy E in joules (J).

The root-mean-square speed of the molecules, v_{rms}, is the square root of the average of the molecules' squared speeds, $\sqrt{(v^2)_{av}}$. It is given by

$$v_{rms} = \sqrt{\frac{3kT}{m}}$$

For example, the mass of an oxygen molecule O_2 is 5.3×10^{-26} kg. So, at room temperature of 27°C, which is 300 K, the root-mean-square speed of an oxygen molecule in the air is

$$v_{rms} = \sqrt{\frac{3 \times (1.4 \times 10^{-23}) \times 300}{5.3 \times 10^{-26}}} \approx 490 \text{ m/s}$$

(which is about 1,800 kilometers per hour, or around 1,100 miles per hour).

The average kinetic energy of a molecule, $(E_k)_{av}$, is

$$(E_k)_{av} = \frac{1}{2} m (v^2)_{av}$$
$$= \frac{1}{2} m v_{rms}^2$$
$$= \frac{3}{2} kT$$

This is in accord with the principle of equipartition of energy, whereby the contribution of each degree of freedom to the total average energy of a molecule is $\frac{1}{2}kT$. In the present case, each molecule can move freely in three independent directions, giving three degrees of freedom, and thus three contributions of $\frac{1}{2}kT$ to the average kinetic energy of a molecule.

See also BOSE-EINSTEIN STATISTICS; ENERGY AND WORK; FERMI-DIRAC STATISTICS; HEAT AND THERMODYNAMICS; MAXWELL, JAMES CLERK; SPEED AND VELOCITY; STATISTICAL MECHANICS.

FURTHER READING

Serway, Raymond A., Clement J. Moses, and Curt A. Moyer. *Modern Physics,* 3rd ed. Belmont, Calif.: Thomson Brooks/Cole, 2004.

measurement To be of significance in science, a measurement must be objective and quantitative and thus produce a numerical outcome, which is the value of the physical quantity that is being measured. Outcomes of measurements must be expressed in terms of units, without which a statement of the value of a physical quantity is meaningless. For instance, if a time interval is specified simply as 203, the duration of the interval remains unknown. One also needs to know 203 of what, what it is that is being counted. Such a reference amount for a physical quantity is a unit of the quantity. A unit must be well defined in order for it to be useful. The unit for time, as an example, in the International System of Units is the second (s). The second is defined as the duration of 9,192,631,770 periods of microwave electromagnetic

The primary mission of the National Institute of Standards and Technology (NIST) is to promote U.S. economic growth by working with industry to develop and apply measurements, standards, and technology. This is a photograph of the NIST administration building in Gaithersburg, Maryland. *(Copyright Robert Rathe)*

radiation corresponding to the transition between two hyperfine levels in the ground state of the atom of the isotope cesium 133. Given that, it is clear what a time interval of 203 seconds is. The specification of the value of any physical quantity must include both a number and a unit. The only exceptions are certain dimensionless quantities whose values are independent of the system of units.

The International System of Units, or, in brief, SI (for Système International d'Unités, in French), is a unified, international system of units for physical quantities, which allows scientific communication that is free of the need for unit conversion. Most countries have adopted some variation of the SI as their national standard. The United States, however, holds the distinction of being the only country of major importance that has not done so, still generally using, for example, miles rather than kilometers, pounds rather than kilograms, and Fahrenheit degrees rather than Celsius.

At the foundation of the SI lies a set of seven base quantities, which are physical quantities that serve to define the seven base units. The base quantities are length, mass, time, electric current, temperature, amount of substance, and luminous intensity. The corresponding base units, which are defined directly from measurement and are not derived solely from other units, and their symbols are given in the table on page 402.

All other SI units are derived from the seven base units together with two supplementary units, based on the supplementary quantities of plane angle and

BASE QUANTITIES AND BASE UNITS FOR SI

Base Quantity	Base Unit	Symbol	Defined As. . .
length	meter	m	the length of the path traveled by light in vacuum during a time interval of 1/299,792,458 of a second
mass	kilogram	kg	the mass of a platinum-iridium prototype kept at the International Bureau of Weights and Measures (BIPM) in Sèvres, France
time	second	s	the duration of 9,192,631,770 periods of microwave electro-magnetic radiation corresponding to the transition between the two hyperfine levels in the ground state of the atom of the isotope cesium 133
electric current	ampere	A	the constant electric current that would produce a force of exactly 2×10^{-7} newtons per meter length, if it flowed in each of two thin straight parallel conductors of infinite length, separated by one meter in vacuum
temperature	kelvin	K	1/273.16 of the temperature interval between absolute zero and the triple point of water (the temperature at which ice, liquid water, and water vapor coexist)
amount of substance	mole	mol	the amount of substance that contains as many elementary entities (such as atoms, molecules, ions, electrons, etc., which must be specified) as there are atoms in exactly 0.012 kilogram of the isotope carbon 12
luminous intensity	candela	cd	the luminous intensity in a given direction of a source that emits electromagnetic radiation of frequency 5.40×10^{14} hertz at radiant intensity in that direction of 1/683 watts per steradian

solid angle. The corresponding supplementary units and their symbols are shown in the first table on page 403.

The second table on page 403 contains a list of some derived units that have names of their own, together with the corresponding physical quantities they are units of and their symbols.

The SI units can be used with multipliers—positive integer powers of 10, such as 100 or 1,000—and dividers—negative integer powers of 10, such as 1/10 or a 1/1,000,000. Multiples and submultiples are indicated by a prefix to the name of the unit and a symbol prefixed to the unit's symbol. As examples, a kilovolt, denoted kV, is 1,000 (10^3) volts, while a millisecond, ms, is one-thousandth (10^{-3}) of a second. (Note the exception, that the base unit for mass, the kilogram, is itself a multiple, a thousand grams. So a millionth [10^{-6}] of a kilogram, for instance, is called a milligram, denoted mg.) Multipliers and dividers that are used with SI units, listed with their corresponding prefixes and symbols, are shown in the third table on page 403.

Now consider some basic units in more detail. Mass is the measure of how much matter an object contains. The SI unit for mass is the kilogram (kg).

The English system of measurement is based on the mass unit of a pound (lb). Using the conversion factor of 2.2 lb = 1 kg allows one to convert measurements from one system to another. The measurements in grams of several common objects are shown.

1 dime $\cong$ 1 g

10 paper clips $\cong$ 10 g

two apples $\cong$ 1,000 g (1 kg)

Length is the measure of distance between two points. The SI unit of length is the meter (m). The English measurement system uses inches (in), feet (ft), and miles (mi). In order to convert between the English system and the SI system, one must use the conversion factor 2.54 cm = 1 inch (in), or 1,000 m (1 km) = 0.6 miles (mi). The measurements in meters of some common objects are shown.

width of a dime $\cong$ 0.01 m (1 cm)

height of a doorknob $\cong$ 1 m

distance of 15 average sized blocks $\cong$ 1,000 m (1 km)

SUPPLEMENTARY QUANTITIES AND UNITS FOR SI

Supplementary Quantity	Supplementary Unit	Symbol
plane angle	radian	rad
solid angle	steradian	sr

The measurement of time uses an SI unit that is familiar in both SI and the English system—the second (s). Some conversions between different units of time are as follows:

60 seconds = 1 minute

60 minutes = 1 hour

24 hours = 1 day

365.25 days = 1 year

A mole (mol) is the SI unit for measuring the amount of substance present in a sample in terms of the number of elementary entities (such as atoms, molecules, and ions) of that substance in the sample. The number of such entities for one mole of any substance is known as Avogadro's number and has the value

SOME DERIVED QUANTITIES AND UNITS FOR SI

Derived Quantity	Unit	Symbol
frequency	hertz	Hz
force	newton	N
pressure	pascal	Pa
energy	joule	J
power	watt	W
electric charge	coulomb	C
electric potential	volt	V
electric resistance	ohm	Ω
electric conductance	siemens	S
electric capacitance	farad	F
magnetic flux	weber	Wb
magnetic flux desity	tesla	T
inductance	henry	H
luminous flux	lumen	lm
illuminance	lux	lx

MULTIPLIERS AND DIVIDERS FOR USE WITH SI UNITS

Multiplier	Prefix	Symbol
10^1	deca	da
10^2	hecto	h
10^3	kilo	k
10^6	mega	M
10^9	giga	G
10^{12}	tera	T
10^{15}	peta	P
10^{18}	exa	E
10^{21}	zetta	Z
10^{24}	yotta	Y

Divider	Prefix	Symbol
10^{-1}	deci	d
10^{-2}	centi	c
10^{-3}	milli	m
10^{-6}	micro	μ
10^{-9}	nano	n
10^{-12}	pico	p
10^{-15}	femto	f
10^{-18}	atto	a
10^{-21}	zepto	z
10^{-24}	yocto	y

$6.02214199 \times 10^{23}$ particles per mole (mol^{-1}). Thus, for example, one mole of atomic sodium contains approximately 6.02×10^{23} atoms, while two moles of water consist of around 1.20×10^{24} molecules.

The SI unit for temperature is the kelvin (K). The absolute temperature scale is based on the same degree units as the Celsius scale with the 0 value for the Kelvin scale at $-273.15°C$. Absolute zero, 0 K, is the lower limit for temperatures. However, this temperature itself is unachievable according to the second law of thermodynamics, although one might approach it extremely closely. For any temperature, its values as expressed in the Kelvin and Celsius scales are related by the following equation:

$$K = °C + 273.15$$

Some units are derived units, meaning they are made by combining one or more base units. Volume is a measure of the amount of space an object takes up in three dimensions. The SI unit for volume is a derived unit, cubic meter, denoted m^3. When considering a solid rectangular object, the volume would be calculated using the formula

$$volume = length \times width \times height$$

Since the length, width, and height are each measured in the SI unit of meters, volume takes on the following derived unit

$$meter \times meter \times meter = m^3$$

Another unit that is often used for volume that is not the SI unit is the liter (L). The two systems are related by the following conversion factors:

$$1 \ m^3 = 1,000 \ L$$

$$1 \ L = 1,000 \ cm^3$$

$$1 \ cm^3 = 1 \ mL$$

Density is the measure of how much mass a substance contains in a unit volume. Density serves as another example of a derived unit. The formula for density is mass/volume. The unit for mass is kilogram and the unit for volume is cubic meter, so the derived unit for density is kg/m^3. The density of water is $1,000 \ kg/m^3$, also expressed as $1 \ g/cm^3$. When placed in water, objects that are denser than water will sink, and objects that are less dense than water will float.

Pressure is the amount of force acting perpendicularly on a unit area. Pressure is most commonly thought of using gas pressure. The SI unit for pressure is newton per square meter (N/m^2) and has its own name, the *pascal* (Pa). The English measurement of pressure is units of pound of force per square inch (psi). Pressure can also be measured in a non-SI unit known as the standard atmosphere (atm), where one atmosphere is defined as the pressure at sea level that supports a column of mercury 760 mm tall. The conversion between atmospheres and pascals is

$$1 \ atm = 1.013 \times 10^5 \ Pa$$

When scientists make measurements involving very large or very small numbers, or otherwise express such numbers, the numbers are written in scientific notation. Scientific notation requires a coefficient greater than or equal to 1 and less than 10. The second part of scientific notation is 10 raised to a power. For example,

$$3.93 \times 10^3 \ m = 3,930 \ m$$

or

$$3.14 \times 10^{-2} \ kg = 0.0314 \ kg$$

As illustrated in these examples, the exponent in this notation indicates how many places to move the decimal point in order to get rid of the power-of-10 part of the number:

positive exponent—move the decimal point to the right

negative exponent—move the decimal point to the left

Good science requires that measurements have the quality known as accuracy, an indication of how close a measurement is to the actual value. Another characteristic of a set of measurements is precision, which is a measure of how close to each other the measurements are. An accurate set of measurements of the same quantity should be both precise and unbiased, that is, not systematically overstating or understating the true value. Individual measurements can be inaccurate yet precise. The meanings of the terms *accuracy* and *precision* are demonstrated by the following example. When someone weighs himself on an improperly functioning bathroom scale that weighs 10 pounds light, he will consistently get the same result for his weight, for example, 150 pounds. This group of measurements are said to be precise. The measurements clearly are not accurate because the person actually weighs 160 pounds. This scale would be considered biased because it consistently understates the true value. If all of a group of three measurements are accurate, then it is also necessary by definition that they are precise, because if they are all close to the same number, then they must be close to each other and therefore be precise. Using the bathroom scale example, a set of measurements on a scale that is 10 pounds light and reads 140, 165, and 152 pounds is both inaccurate for each measurement and imprecise for the group.

The measure of accuracy in scientific measurement requires the use of significant figures. Significant figures are all of the measured digits in a measurement plus one estimated digit. Counting the number of significant figures requires the following rules of significant figures (often abbreviated to *sig figs* by teachers).

1. All nonzero digits *are* significant. The number 764 has three significant figures.
2. All trapped zeros *are* significant. Trapped zeros are zeros in between nonzero digits.

The number 4.5008 has five significant figures.

3. All leading zeros *are not* significant. These zeros simply serve as place holders. The measurement 0.000347 has three significant figures.

4. Trailing zeros are only significant if there is a decimal point in the measurement. The number 98.00 has four significant figures. The number 860 is ambiguous as to the number of significant figures and would better be represented by 8.6×10^2, if it has only two significant figures, or by 8.60×10^2, if it has three.

One must consider significant figures when performing calculations with measurements. When calculating answers with measurements, the answer can only be as specific as the least specific measurement. In addition and subtraction problems, the least specific measurement is given by the measurement with the least number of digits to the right of the decimal point. In multiplication and division problems, least specific measurement is given by the measurement that contains the least number of significant figures.

The calculated answers need to be rounded to the correct number of significant figures, or decimal places, in order to represent the correct measurement calculation. If two measurements differ greatly in their specificity, then the only way to utilize the specificity of the more specific measurement would be to remeasure the least specific measurement to the number of significant figures or decimal places.

Addition example:

$$(5.6309 \text{ m}) + (7.02 \text{ m}) = 12.6509 \text{ m}$$

Rounded answer = 12.65 m

In this example, the first measurement, 5.6309 m, contains four digits to the right of the decimal point, and the second measurement, 7.02 m, contains two digits to the right of the decimal point. When calculated, the answer originally has four places to the right of the decimal point. On the basis of the rule that the answer can only be as specific as the least specific measurement, and knowing that when adding and subtracting, the least specific measurement is the one with the fewest digits to the right of the decimal point, the answer must be rounded to two decimal places, as seen in the final answer of 12.65 m.

Multiplication example:

$$(9.6050 \text{ m}) \times (12.3 \text{ m}) = 118.1415 \text{ m}^2$$

Rounded answer = 118 m²

In this example, the measurement 9.6050 m has five significant figures. The 9, 6, and 5 are significant because they are nonzeros. The first 0 is significant because it is a trapped zero, and the last 0 is significant because it is a trailing zero with a decimal point in the number. The second measurement has three significant figures because they are all nonzero. When multiplying and dividing, the least specific measurement is the one with the fewest significant figures, and the answer must have that number of significant figures. The rounded answer, 118 m², contains only three significant figures.

See also ATOMIC STRUCTURE; ELECTRICITY; ELECTROMAGNETIC WAVES; ELECTROMAGNETISM; GRAPHING CALCULATORS; HEAT AND THERMODYNAMICS; MAGNETISM.

FURTHER READING
Wilbraham, Antony B., Dennis D. Staley, Michael S. Matta, and Edward L. Waterman. *Chemistry*. New York: Prentice Hall, 2005.

mechanical engineering This is the field of engineering that involves the application of the principles of physics to the design, analysis, testing, and production of mechanical systems. It is based mostly on the fundamental physics concepts of mechanics and thermodynamics, where mechanics includes, for example, kinematics, forces, momentum, and energy, and thermodynamics is concerned with the conversion of energy among its various forms while taking temperature into account. Mechanical engineers might, as examples, design or be involved in the production of automobiles, spacecraft, medical devices, or industrial plants.

The college or university education of a mechanical engineer starts with introductory courses in physics and chemistry and a heavy dose of mathematics that includes calculus and more advanced subjects. The engineering courses typically include statics, dynamics, strength of materials, solid mechanics, thermodynamics, fluid mechanics, and computer-aided design (CAD). The program requires many labs and often includes projects or some research. The mechanical engineering curriculum leads to a bachelor's degree.

The following are brief descriptions of several of the subdisciplines of mechanical engineering, some of them requiring postgraduate specialization:

● *Robotics.* The design and manufacture of devices that perform work that is dangerous or otherwise undesirable for humans. These include industrial robots, which are used, for example, in the manufacture of automobiles.

An automobile, such as the one shown, is a feat of mechanical engineering. A team of mechanical engineers labored to get everything to work properly and to work together smoothly within the constraints of the design. *(Kyodo/Landov)*

- *Marine engineering.* The design, construction, and maintenance of ships and of the various mechanical systems contributing to the operation of ships. Examples of the latter are propulsion systems, plumbing systems, and fire extinguishing systems.
- *Automotive engineering.* The design, manufacture, and maintenance of automotive vehicles, such as cars, trucks, and buses.
- *Aerospace engineering.* The design, production, and maintenance of craft that operate inside and outside Earth's atmosphere. These include civilian and military airplanes, missiles, spacecraft, and booster rockets.
- *Nanotechnology.* The design and fabrication of structures and devices whose sizes are on the scale of around one to 100 nanometers (where 1 nm = 10^{-9} m, which is one-billionth of a meter) or that operate at such a scale. An example of the former is a nanomotor, which might be utilized to propel a tissue repair device—itself a nanoscale device—that functions inside the body. The atomic force microscope serves as an example of a large-scale device that operates and manipulates at the nanoscale.
- *Heating, ventilation, and air conditioning (HVAC).* The design, manufacture, installation, and maintenance of equipment to control the quality of indoor air, including its temperature, humidity, and purity. HVAC systems are used in homes and are essential in office and industrial buildings.
- *Biomedical engineering.* The application of engineering principles and methods to medicine. This includes research and development in information, imaging, and materials, as examples. Specific applications include devices to replace human organs, such as artificial hearts, limbs, and joints; medication delivery devices, such as insulin pumps; and imaging and measurement devices, such as magnetic resonance imagers (MRI machines) and electroencephalography (EEG) and electrocardiography (EKG or ECG) systems.

A major professional organization for mechanical engineers is the American Society of Mechanical Engineers (ASME), with a membership of around 120,000. This organization focuses on technical, educational, and research issues of the engineering and technology community. It conducts one of the world's largest technical publishing operations, including the publication of journals such as *Mechanical Engineering, Journal of Engineering for Gas Turbines and Power,* and *Journal of Fuel Cell Science and Technology,* and holds numerous technical conferences worldwide. ASME offers hundreds of professional development courses each year and, very important, sets internationally recognized industrial and manufacturing codes and standards that enhance public safety.

See also ENERGY AND WORK; FORCE; HEAT AND THERMODYNAMICS; MOMENTUM AND COLLISIONS; MOTION; POWER.

FURTHER READING

American Society of Mechanical Engineers (ASME) home page. Available online. URL: http://www.asme.org. Accessed January 10, 2008.

Baine, Celeste. *Is There an Engineer Inside You? A Comprehensive Guide to Career Decisions in Engineering,* 3rd ed. Belmont, Calif.: Professional Publications, 2004.

Billington, David P. *The Innovators: The Engineering Pioneers Who Transformed America.* Hoboken, N.J.: Wiley, 1996.

Meitner, Lise (1878–1968) Austrian *Physicist*

A brilliant theoretician with an unassuming manner, Lise Meitner overcame gender bias in the late Victorian age and the racial discrimination of Nazi Germany to make key contributions to the emerging field of radioactivity. Work with the chemist Otto Hahn, her collaborator and peer for more than 30 years, led to the discovery of the element protactinium and, more significantly, the accurate description of the phenomenon of nuclear fission, the process whereby a large atom (like uranium) splits into two smaller atoms when bombarded with subatomic particles called neutrons. In addition to the creation of two

Lise Meitner was an Austrian physicist whose contributions to the field of radioactivity included the discovery of the element protactinium and discovery of the process known as nuclear fission, although the 1944 Nobel Prize in chemistry for this discovery went to her collaborator, Otto Hahn. *(© Bettmann/CORBIS)*

smaller atoms, nuclear fission frees neutrons that may bombard other surrounding large atoms, producing a chain reaction that releases a huge amount of energy. Although Meitner interpreted the results from Hahn's experiments that demonstrated the process of nuclear fission, only Hahn received the Nobel Prize in chemistry in 1944 for this research; even more troubling to Meitner, the Allied forces of World War II married nuclear fission to the development of nuclear weapons, harnessing the energy generated by the chain reactions to make an atomic bomb. Yet despite the trials and heartaches, Meitner persevered, and she continued her quest to understand the physical world until her death.

Born on November 7, 1878, Elise Meitner was the third of eight children born to Philip and Hedwig Meitner of Vienna, Austria. Both sets of grandparents practiced Judaism, yet, neither her father nor mother did. A free-thinking liberal, Philip became one of the first prosperous attorneys of Jewish descent in Vienna. He was also a minor chess master. In later years, Lise remembered the love and devotion that she felt in her home while growing up surrounded by books and music. A good student, Lise especially enjoyed classes in math and science, but she left school at 14 as all girls did during the end of the Victorian age. She grudgingly attended a teacher's preparatory school with the intent of becoming a French teacher. At the turn of the 20th century, universities across Europe began to accept women, and Meitner hired a private tutor to prepare herself for the rigor-

ous entrance exam. In 1901, at almost 23, Meitner enrolled at the University of Vienna, where she took a heavy course load in calculus, chemistry, and physics. After she was introduced to atomic theory (all matter is composed of atoms), physics particularly fascinated Meitner. Her physics professor, Dr. Ludwig Eduard Boltzman, believed that women deserved opportunities to prove themselves in the sciences and encouraged Meitner to pursue a career in the physical sciences. After the completion of her coursework in 1905, she prepared to defend her doctoral research on the conduction of heat in solids. Meitner passed her oral exam in 1906, becoming only the second woman to receive a physics doctorate from the University of Vienna.

After the completion of her graduate studies, Meitner could not find a job in physics so she began teaching at an area girls' school during the day, and she performed research in the evenings with Stefan Meyer, Boltzmann's successor at the University of Vienna. Trying to characterize the properties of alpha- (α-) and beta- (β-) rays, Meitner measured the absorption of these radioactive rays by different types of metals using a leaf electroscope. This instrument has a thin sheet of metal (gold or aluminum) attached to a metal rod. When an electric current passes through the rod, the sheet of metal is repelled; in the presence of radiation, the area around the rod becomes ionized, neutralizing the effect of the electricity, and the metal sheet reassociates with the metal rod. The reassociation occurs more quickly with higher levels of radioactivity. During this period, Meitner also observed that α-rays scatter when they hit a solid mass, and this scattering increases with increasing atomic mass of the atoms being bombarded with the radioactive rays.

Because radioactivity studies were more advanced in Berlin, Meitner asked her parents for money to go the University of Berlin to find a position. Upon her arrival, she was met with the same gender discrimination that she had encountered in Vienna and had to obtain special permission to attend the renowned physicist Max Planck's lectures. Although no one at the University of Berlin would give her a chance to prove herself as a physicist, Otto Hahn, a chemist also interested in the newly discovered radioactive elements, agreed to collaborate with her at the Chemistry Institute of Berlin. Women were forbidden to work in the laboratories at that time so the director of the institute, Emil Hermann Fischer, gave Meitner permission to set up a makeshift laboratory in the basement of the building. Initially, Meitner and Hahn tried to understand the phenomenon of radioactive decay. Over time, radioactive elements disintegrate (decay), releasing radioactive particles. Sometimes radioactive elements transform into

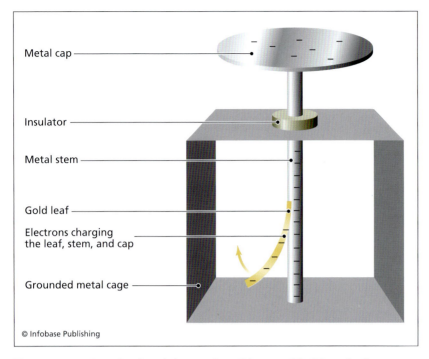

- Metal cap
- Insulator
- Metal stem
- Gold leaf
- Electrons charging the leaf, stem, and cap
- Grounded metal cage

© Infobase Publishing

The amount of static electricity produced is quantified by a leaf electroscope. A long, thin metal sheet, or leaf, hangs loosely along the metal stem of the device. When the electroscope is charged, the leaf is repelled from the stem. The measure of the displacement of the leaf is indicative of the amount of charge.

different elements as the makeup of their subatomic particles changes as they emit the radioactive particles; other types of radioactive elements decay into a related isotope, an element with the same number of protons but a different number of neutrons. These isotopes have the same atomic number as the parent atom (the original decaying element), yet they have a different atomic mass. The newly created daughter isotope subsequently also decays, giving rise to another daughter isotope, and these successive decays by the radioactive elements are called a series. Meitner and Hahn spent years identifying elements that form these radioactive series with Hahn using his chemistry background to analyze and identify the elements and Meitner using her physics background to identify the subatomic particles released during decay. In her analyses, Meitner observed a way to differentiate between parent and daughter isotopes. When exposed to α-particles, daughter isotopes recoil from these particles, whereas parent atoms do not. In 1912 Hahn joined the new Kaiser-Wilhelm Institute (KWI) for Chemistry in Dahlem (a city outside Berlin, Germany), and Meitner followed him to serve as his unpaid assistant. Fischer asked Meitner to join the institute as a scientific associate in 1913, and she and Hahn served as cochairs of the radioactivity division. Prior to the outbreak of World War I, Meitner also served as the first female assistant at the University of Berlin, grading papers for Max Planck at the Institute of Theoretical Physics. In 1914 the University of Prague offered Meitner a position; Fischer proposed a counteroffer from the KWI that Meitner could not refuse, raising her salary and creating a radiophysics department that she would lead.

Although Hahn stayed in the chemistry department, Meitner continued to collaborate with him during World War I. During this period, she asserted that another radioactive element must exist between uranium and actinium in that particular radioactive decay series. In 1918, with minimal help from Hahn, Meitner discovered a new element that she called protactinium (atomic number 91 on the periodic table), the first identified long-lived isotope of a radioactive element. Meitner and Hahn published their results in the *Physical Chemical Journal* in an article entitled "The Mother Substance of Actinium, a New Radioactive Element of Long Half-Life" listing Hahn as the primary author despite the fact that Meitner performed the majority of the work leading to the element's discovery. In 1922 Meitner became the first woman lecturer at the University of Berlin when she presented a seminar called "The Significance of Radioactivity in Cosmic Processes"; several reporters mistakenly told their readership that Meitner's topic was cosmetic physics instead. The humorous bungling of the press did not take away from Meitner's achievements, and the University of Berlin recognized her contributions to physics by appointing her as extraordinary professor in 1926, a position she held until Hitler came to power in the 1930s. Other accolades that she received prior to World War II included the silver Leibniz Prize Medal in 1924 from the Berlin Academy of Sciences, the Ignaz Lieben Prize in 1925 from the Vienna Academy of Sciences, and a yearly nomination for the Nobel Prize from 1924 through the mid-1930s.

Meitner's world changed when Adolf Hilter became chancellor of Germany in 1933. For more than 25 years in Germany, Meitner contended only with discrimination based on her gender. As Hitler's power grew, Meitner now faced discrimination based on her Jewish heritage as well. Briefly protected by her Austrian citizenship, Meitner came under the jurisdiction of Nazi laws when Germany annexed

Austria in 1938. Despite the fact that Meitner joined the Lutheran Church as a young woman, her Jewish ancestry relegated her to a subject of the state, meaning she lost all her rights as a citizen (her professorship, her civil rights, her passport, etc.). As Nazi policies increasingly impeded her ability to work and to travel, one of her peers, the Dutch physicist Dirk Coster, devised a plan of escape for Meitner using his own money and soliciting the aid of others in the scientific community. With Coster's help, she illegally crossed the border into Holland on July 13, 1938, carrying only a small suitcase filled with her personal belongings. Meitner made her way to Copenhagen, where she stayed with the renowned Nobel physicist Niels Bohr and his wife, who had learned of her plight from her nephew, Otto Robert Frisch, a graduate student working in Bohr's laboratory.

From Copenhagen, Meitner went to Sweden, where she accepted a position at the Nobel Institute for Experimental Physics in Stockholm and began a collaboration with the Swedish physicist Karl Manne Georg Siegbahn. Meitner also continued a long-distance collaboration with Otto Hahn and Fritz Strassman. In 1934 Hahn, Strassman, and Meitner began transuranic studies, bombarding uranium with newly discovered subatomic particles called neutrons in hopes of creating elements heavier than uranium. In late 1938, when Hahn and Strassman bombarded uranium with neutrons, they obtained radium, an element with a smaller atomic number. Hahn wrote

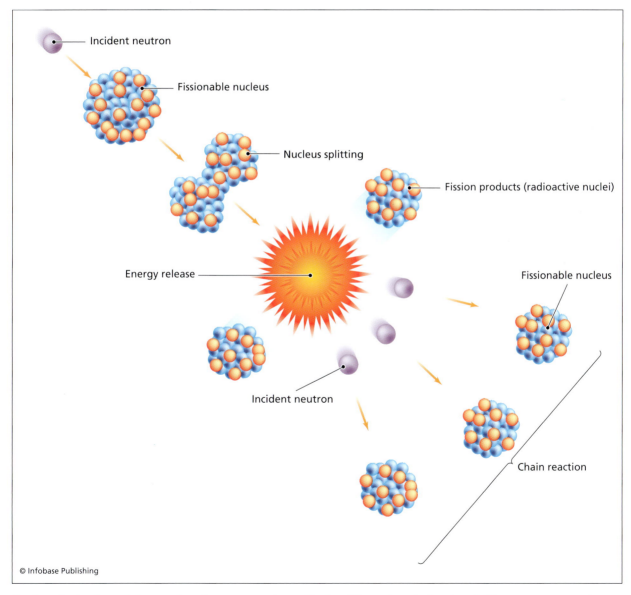

Nuclear fission is a chain reaction that occurs when a fissionable nucleus is bombarded by neutrons, causing the nucleus to split. This type of reaction releases a large amount of energy. Nuclear fission reactions are utilized in nuclear reactors.

to Meitner about these peculiar results, and Meitner suggested that the experiment needed to be repeated. Hahn and Strassman obtained similar results a second time, with neutron bombardment of uranium yielding products similar to radium but with the chemical properties of barium. Chemical analyses by Hahn confirmed that barium, with an atomic number of 56, was the product of their experiments. Barium was not a decay product of uranium, whose atomic number is 92, so Hahn was at a complete loss for an explanation of these results. Meitner suspected that bombarding uranium with neutrons resulted in the atom's being split into two smaller elements—barium with an atomic number of 56 and krypton with an atomic number of 36 (56 + 36 = 92). The idea of splitting atoms seemed almost ludicrous to Hahn, yet, mathematical calculations performed by Meitner and her nephew, Otto Frisch, proved its feasibility to the chemist. Meitner and Frisch borrowed a term from biology, *fission,* to name this phenomenon, and they proposed that a number of events occurred during nuclear fission. First, the uranium nucleus splits into two fission products, a barium nucleus and a krypton nucleus. When this fission occurs, the uranium nucleus releases a number of neutrons, and these neutrons are free to bombard surrounding nuclei, yielding more fission products. Thus, with a single nuclear fission event, a chain reaction of multiple nuclear fissions could be set off. Each one of these nuclear fissions released a large amount of energy as well, with about 200 million electron volts released per fission event. Hahn and Strassman published a paper in 1939 that described how they generated barium from uranium. One month later, Meitner and Frisch published in *Nature* the paper "Disintegration of Uranium by Neutrons: A New Type of Nuclear Reactions" describing the process of nuclear fission.

Even as the war raged around her, Meitner never anticipated that nuclear fission might be used to develop weapons of mass destruction, yet, the phenomenon that she described became the underlying principle that led to the creation of the atomic bomb. Scientists including Albert Einstein saw the potential ramifications of nuclear fission, and his letter to President Franklin D. Roosevelt explaining the magnitude of Meitner, Hahn, and Strassman's discovery spurred the inception of the Manhattan Project, the project that yielded the atomic bomb. In 1943 the United States extended an invitation to Meitner to help build the bomb, but Meitner refused. She was devastated when the bombs dropped on the Japanese cities of Hiroshima and Nagasaki in August 1945.

The hurt associated with the discovery of nuclear fission extended to her professional life as well. In 1944 Otto Hahn alone received the Nobel Prize in chemistry for the discovery of fission of heavy nuclei.

During the war, it made sense that Hahn made no mention of his collaboration with Meitner, yet he continued to deny her contribution to the discovery for the rest of his life. Hahn claimed that Meitner was merely one of his lab assistants, somehow forgetting that she had been named the head of the department of radiophysics in 1917 along with the countless other accolades that Meitner received in the years up to 1938. In Meitner's defense, Strassman boldly stated that Meitner, not Hahn, was the intellectual leader of the team. In any case, Hahn's denial hurt Meitner greatly, and she always felt that both she and Strassman deserved the Nobel, too. The wrong was righted to a certain extent when all three (Hahn, Meitner, and Strassman) received the Max Planck Medal in 1948 for the discovery of nuclear fission. This medal appeased Meitner greatly because she greatly admired Planck and considered him a good friend. Later, in 1966, the United States Atomic Energy Commission awarded the trio of scientists the Enrico Fermi Medal, with the additional honors of being the first non-American recipients and, in the case of Meitner, the first female recipient, too.

Physics was Lise Meitner's life, yet, she never lost her love for humanity. During World War I, Meitner served as an X-ray nurse administering aid to wounded soldiers in the Austrian army. As the Nazis gained power in Germany and later Austria, she hoped that the oppression that she endured because she was a woman and Jewish would be only temporary. Later, she refused to think the worst of people, even when Hahn denied her significant contributions to the work that led to his Nobel Prize. Over the course of her illustrious career, she published more than 150 articles. She continued working, driven by her unquenchable desire to understand the physical world, until she had a heart attack in 1960. Her peers acknowledged the quality of her work by electing her to the Royal Society of London in 1955 and to the American Academy of Sciences in 1960. A series of strokes in 1967 made it impossible for Meitner to speak clearly, and she died in her sleep at her nephew's home on October 27, 1968, two weeks shy of her 90th birthday. Fittingly, the Heavy Ion Research Laboratory in Darmstadt, Germany, named the artificial element with atomic number 109 (made by the nuclear fission of bismuth) *meitnerium* in her honor.

See also ATOMIC STRUCTURE; BOHR, NIELS; EINSTEIN, ALBERT; FERMI, ENRICO; FISSION; OPPENHEIMER, ROBERT; PLANCK, MAX; RADIOACTIVITY.

FURTHER READING

Bankston, John. *Lise Meitner and the Atomic Age.* Hockessin, Del.: Mitchell Lane, 2003.

Sime, Ruth Lewin. *Lise Meitner: A Life in Physics.* Los Angeles: University of California Press, 1996.

membrane potential The sum of electrical and chemical concentration gradients from one side of a cellular membrane to another is known as a membrane potential, electrochemical potential, or cellular membrane potential. The membranes within plant and animal cells are phospholipid bilayers, meaning they consist of two layers of phospholipids. The hydrophilic (water-loving) phosphate head groups of the phospholipids surround the outer and inner surfaces of the membrane while the hydrophobic (water-fearing) hydrocarbon tails make up the interior of the membrane. These membranes are selectively permeable and give the cell the ability tightly to control the movement of materials moving into and out of the cell. The chemistry of membrane potential plays an important role in many cellular processes. The generation and alteration of chemical concentration gradients and electrochemical gradients across membranes drive processes such as nerve impulse transmission, muscle contraction, photosynthesis, and cellular respiration.

Diffusion is the movement of substances from areas of high concentration to areas of low concentration across a semipermeable membrane. Large and small compounds alike easily move down their concentration gradient if given the opportunity. The ability of a substance to pass through a cell membrane is determined by how soluble the substance is in the membrane and whether there is an available energy source to move the molecule if necessary. The membranes within a cell do not allow every type of molecule to pass. Small, uncharged, nonpolar molecules can pass through the cell membrane on their own, but polar or charged molecules and ions cannot pass through the cell membrane without the help of protein molecules that act as carrier proteins, channels, or pumps.

Ions such as calcium (Ca^{2+}), sodium (Na^+), potassium (K^+), and protons (H^+) are not able to pass through the cellular membrane without assistance. These ions are of extreme importance to the cell, and they control many cellular signaling processes related to movement, nerve impulses, cellular respiration, photosynthesis, and cellular communication. The concentrations of these ions both inside and outside the cell or a cellular organelle are tightly regulated.

Establishing a concentration gradient requires an input of energy, and the quantity of that gradient for an uncharged molecule is determined by the following equation:

$$\Delta G = 2.303\, RT\log(c_2/c_1)$$

where ΔG is the amount of energy available to do work, R is the gas constant (8.31 kJ/mole), T is the temperature in kelvins, and c_1 and c_2 are the concentrations of the molecule on the starting side of the membrane and the ending side of the membrane, respectively. When the ΔG for this process is negative, the transport of the substance is considered passive and occurs without an input of energy. Transport where the ΔG is positive requires the input of energy and is an active process.

When the molecule being transported is charged, the process becomes more complicated. The amount of charge that the particle possesses contributes to the total amount of energy required to transport it across the membrane. The larger the charge of the particle, the harder it is to move across the hydrophobic membrane. The membrane potential must include this electrical component as well as the concentration gradient information in its calculation. The formula for the movement of a charged species is shown in the following:

$$\Delta G = 2.303\, RT\log(c_2/c_1) + ZF\Delta V$$

Z stands for the charge of the molecule being transported, F is the Faraday constant [96.5 kJ/(V·mol)], and ΔV is the voltage potential across the membrane.

SODIUM–POTASSIUM PUMP

The normal cellular concentration of sodium ions (Na^+) is 14 millimolar (mM) inside the cell and 143 mM outside the cell, representing an approximate 10-fold difference in the concentration outside compared with inside the cell. If sodium ions were able to pass through the hydrophobic membrane, there would be a mass diffusion of sodium ions from the outside to the inside. The high concentration of sodium outside the cell helps create a membrane potential (a difference between the charge inside and outside the cell) that can be utilized by opening up sodium ion channels in the cell membrane and allowing the ions to move down their concentration gradient and into the cell. No energy is required to move the sodium from the outside of the cell to the inside. The cellular concentration of potassium (K^+) is higher inside the cell than outside. Potassium would, if given the opportunity, move out of the cell down its concentration gradient.

The cell membrane contains passive channels called leak channels that allow potassium and sodium to leak back through the membrane down their concentration gradients. The membrane is more permeable to potassium than to sodium, so the cell loses potassium ions through diffusion more quickly than it gains sodium ions. This leads to the creation of an electrical gradient with the inside of the cell membrane having a slight negative charge relative to the outside, and the outside of the cell membrane having

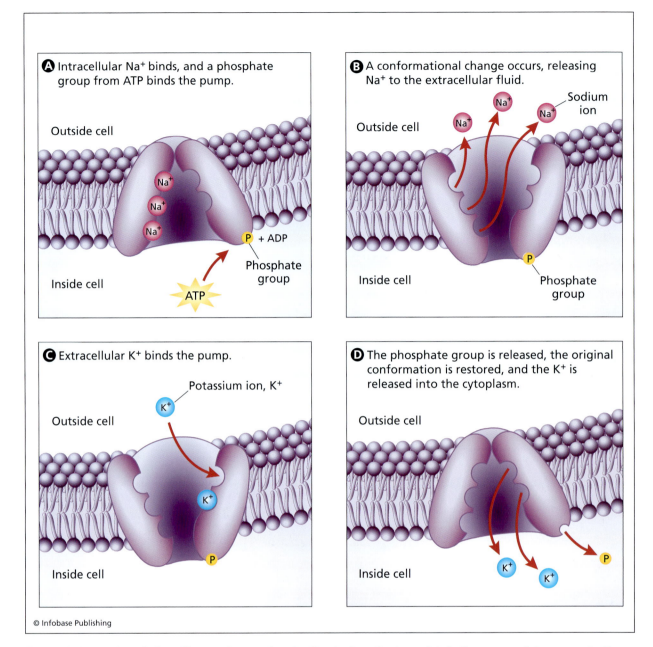

A Intracellular Na⁺ binds, and a phosphate group from ATP binds the pump.

Outside cell

Inside cell

Na⁺
Na⁺
Na⁺

P + ADP

Phosphate group

ATP

B A conformational change occurs, releasing Na⁺ to the extracellular fluid.

Outside cell

Inside cell

Na⁺
Na⁺
Na⁺

Sodium ion

P

Phosphate group

C Extracellular K⁺ binds the pump.

Potassium ion, K⁺

Outside cell

K⁺

K⁺

Inside cell

P

D The phosphate group is released, the original conformation is restored, and the K⁺ is released into the cytoplasm.

Outside cell

Inside cell

K⁺
K⁺

P

© Infobase Publishing

Cells tightly regulate their sodium and potassium ion levels. In order to maintain the appropriate concentration and electrochemical gradients, the sodium-potassium pump draws potassium into the cell and pumps sodium out of the cell. For every three sodium ions pumped out, two potassium ions are pumped in, creating a negative potential on the inside of the cell.

a slight positive charge relative to the inside. The net result of this charge movement is a potential difference that is measured in volts, or millivolts (mV). The combination of electrical and chemical gradients (electrochemical gradient, membrane potential) is a form of potential energy for the cell and is known as the resting potential.

Setting up and maintaining the gradient require the use of active transport that entails a membrane pump (known as the sodium-potassium pump) that forces the sodium out of the cell against its con-

centration gradient using energy released by the hydrolysis of adenosine triphosphate (ATP). This pump transfers three sodium ions out of the cell, with the concomitant transfer of two potassium ions into the cell.

MEMBRANE POTENTIAL IN NERVE IMPULSE TRANSMISSION

The resting potential due to the movement of potassium and sodium is responsible for the production of nerve impulses in neurons. The normal resting potential of a nerve cell is -70 mV. Changing this membrane potential by opening and closing gates and channels in the neuronal cell membrane produces a signal down the neuron known as an action potential (propagated changes in the membrane potential that move down the length of a neuron).

The action potential starts when a voltage-regulated sodium channel is opened in the membrane, allowing the positively charged sodium ions to move into the cell, thus decreasing the local membrane potential. The reduction of the membrane potential, known as depolarization, signals the closing of the sodium channels and opening of voltage-regulated potassium channels. These open potassium channels cause potassium to leave the cell, and the process of regenerating the membrane potential, called repolarization, begins. The change in charge sends the depolarizing signal down the neuron to an adjacent region that undergoes the same depolarization-repolarization process. The sodium-potassium pump is responsible for resetting the correct concentrations of sodium and potassium inside and outside the cell and thus regenerating the resting membrane potential.

MEMBRANE POTENTIAL IN MUSCLE CONTRACTION

The neuronal response reaches a muscle cell at a structure known as the neuromuscular junction. This is where the action potential caused by the depolarization-repolarization of the neuronal membrane meets the target structure, the muscle, to elicit a response. A muscle cell contracts as a result of an interaction between actin and myosin filaments. In a relaxed muscle cell, the binding site for myosin on an actin molecule is covered by a troponin/tropomyosin combination that does not allow actin and myosin to bind. When the neurotransmitter acetylcholine is released into the neuromuscular junction from the neuron it attaches to the acetylcholine receptors on the muscle cells. The response of the muscle cell is to open up the sodium channels, increasing the influx of sodium into the cell and decreasing the membrane potential. This depolarization is transferred to the sarcoplasmic reticulum (SR), a specialized organelle in a muscle cell similar to the smooth endoplasmic reticulum, causing a release of calcium from the SR into the cytoplasm. The released calcium ions bind to the troponin molecules, an event that changes the conformation of troponin and its interactions with tropomyosin and actin. This results in the exposure of myosin binding sites on actin, allowing for muscle contraction. Following contraction, the calcium levels return to resting levels, with a higher concentration inside the sarcoplasmic reticulum than in the cytoplasm.

MEMBRANE POTENTIAL IN CELLULAR RESPIRATION

Concentration and electrical gradients play a crucial role in the aerobic process known as cellular respiration. The breakdown of glucose to generate a usable form of cellular energy in the presence of oxygen occurs in three stages: glycolysis, the tricarboxylic cycle, and oxidative phosphorylation. The first two stages break down glucose and reduce electron carriers, NADH and FADH$_2$. These electron carriers pass the received electrons down the electron transport chain found in the inner mitochondrial membrane. As the electrons move to lower energy states progressively, the energy released in this process is used to pump protons from the mitochondrial matrix into the intermembrane space. This produces a concentration gradient across the membrane that serves as potential energy for the cell. The mitochondrial membrane contains multiple copies of the enzyme ATP synthase. This enzyme requires an input of energy to synthesize molecules of ATP, the energy source for the cell. ATP synthase uses the energy of the proton gradient to power ATP synthesis. Like water building up behind a dam, when ATP synthase opens, the protons rapidly diffuse down their concentration gradient back into the mitochondrial matrix, passing their energy to the synthase, which creates ATP.

The proton gradient is essential to the production of readily available cellular energy. Uncoupling the movement of protons into the matrix from the production of ATP produces lethal consequences. A class of compounds known as dinitrophenols (DNPs) uncouples this process, carrying protons across the membrane without the production of ATP. These chemicals are used in pesticides, explosives, and wood preservatives. In the 1930s, scientists discovered that DNP compounds helped in weight loss. By decreasing the amount of ATP produced per glucose molecule consumed, the body increases metabolism and thus increases weight loss. The only problem was that the administration of this chemical substance caused death. Interestingly, the lack of ATP produced was not the lethal factor: the increase in body temperature due to the uncoupling caused fevers that were deadly. A comparable situation occurs when revving a car engine in neutral; the eventual outcome is overheating.

MEMBRANE POTENTIAL IN PHOTOSYNTHESIS

Mitochondria and chloroplasts both produce ATP by the same method of chemiosmosis. In the mitochondria, the purpose of ATP production is to have a form of chemical energy that organisms can use to drive cellular processes. In the chloroplasts, the role of chemiosmosis is the production of ATP to drive the Calvin cycle, which produces carbohydrates. Photosynthesis in the chloroplast is an anabolic process during which atmospheric carbon dioxide is converted into carbohydrates and molecular oxygen. Photosynthesis occurs in two phases, the light-dependent reactions and the Calvin cycle. The light-dependent reactions occur in the thylakoid and function to generate the cellular energy required to run the Calvin cycle. The light-dependent reactions of photosynthesis involve two photosystems, known as photosystem I and photosystem II, that absorb light of different wavelengths. Photosystem II is responsible for the harvesting of light and the production of ATP using an electron transport chain. Photosystem I harvests solar energy and produces NADPH, also using energy provided by the movement of electrons down a transport chain. Photosystem II passes the energy of the photon through the electron transport chain with the simultaneous movement of protons from the stroma of the chloroplast into the thylakoid space. This high concentration of protons produces a membrane potential that is utilized by the chloroplast to power ATP synthase and produce ATP as the protons diffuse out of the thylakoid space back into the stroma. The ATP and NADPH produced in the thylakoid membrane diffuse into the stoma of the chloroplast and are used to fuel the Calvin cycle.

The processes of cellular respiration and photosynthesis use similar proton gradients, and the same type of enzyme (ATP synthase) uses the gradient to produce ATP. The main difference between electron transport chains in cellular respiration and those in photosynthesis is the primary source of the energy used to produce the gradient. In cellular respiration, the oxidation of glucose supplies the energy found in NADH and $FADH_2$ that runs the electron transport chain, which pumps protons into the intermembrane space from the matrix. In photosynthesis, the energy for running the electron transport chain and pumping protons into the thylakoid space is from sunlight.

See also BIOCHEMISTRY; ELECTRON TRANSPORT SYSTEM; EQUILIBRIUM; METABOLISM.

FURTHER READING

Saladin, Kenneth S. *Anatomy and Physiology.* 4th ed. Boston: McGraw-Hill, 2007.

Sten-Knudsen, Ove. *Biological Membranes: Theory of Transport, Potentials and Electrical Impulses.* New York: Cambridge University Press. 2002.

Mendeleyev, Dmitry (1834–1907) Russian Chemist

By the 19th century, chemists had identified more than 60 elements. The properties of these elements varied widely. Some were inactive metals like platinum; others were nonmetals like carbon; still others were poisonous gases like chlorine. Chemists found it difficult to remember the chemical and physical properties of this diverse group of elements, let alone the compounds they combined to form. Scientists started looking for patterns between atomic mass and similarities in chemical and physical properties to help organize the elements. While writing a textbook on inorganic chemistry, Dmitry Mendeleyev found that two groups of elements (the halogens, including chlorine, and the alkali metals, including sodium) with dissimilar properties shared similarities in the progressive nature with which the atomic weight of one member of the group varied from its neighbors. Expanding his analysis to other groups of elements, Mendeleyev discovered that the order of atomic weights could be used to arrange elements within groups and to arrange groups themselves. In 1871, he created the first periodic table of the elements, which systematically displayed a recurring pattern, or periodicity, of properties within groups of elements. Believing strongly in the validity of the periodic law that he had developed, Mendeleyev boldly predicted the locations and properties (both physical and chemical) of three yet unidentified elements. The scientific community gave little credence to Mendeleyev's work until the discoveries of the three predicted elements: gallium (1875), scandium (1879), and germanium (1886). Primarily recognized as the founder of the periodic law, Mendeleyev also made significant contributions to the scientific and technological communities through his undying interest in Westernizing his homeland, Russia.

One of the youngest children of the large family (between 13 and 17 siblings, depending on the source) of Ivan Pavlovich Mendeleyev, a gymnasium teacher, and Mariya Dmitriyevna Kornileva, Dmitry Ivanovich Mendeleyev was born in 1834 and grew up in the Siberian town of Tobolsk. His father became blind in the same year that Dmitry was born, making it necessary for his mother to support the family by operating a small glass factory owned by some of her family members. Shortly after the death of his father in 1847, the factory burned down, and his mother took him to St. Petersburg, Russia, to enroll him in the Main Pedagogical Institute. After graduating in 1855, Dmitry taught for a brief time in the Crimean Peninsula before he went back to school in St. Petersburg to obtain a master's degree in organic chemistry. Awarded a government fellowship, Mendeleyev studied abroad for two years at the University of Heidelberg during the tenure of such prominent chemists as Robert Bunsen, Emil Erlenmeyer, and August Kekulé, yet, rather than studying under the supervision of any of these renowned chemists, Mendeleyev set up a laboratory of his own in his apartment. In 1860, he attended the International Chemistry Congress in Karlsruhe, Germany, where chemists debated over the issues of atomic weights, chemical symbols, and chemical formulas. During the conference, Mendeleyev met and established contacts with many of Europe's leading chemists. Years later, he vividly remembered reading a paper written by Stanislao Cannizzaro that clarified his ideas about atomic weight. One year later, Mendeleyev went back to St. Petersburg, where he began work on his doctorate studying properties of water and alcohol. After he defended his dissertation in 1865, Mendeleyev initially accepted the position of professor of chemical technology at the University of St. Petersburg (now St. Petersburg State University). He assumed the role of professor of general chemistry in 1867 and remained in this position until 1890.

When he began to teach courses in inorganic chemistry, Mendeleyev found no textbook met his needs, so he decided to write one that would attempt to organize the extensive knowledge that already existed of the chemical and physical properties of the known elements. As early as 400 B.C.E., Greek philosophers proposed the atomic theory, stating that all matter is composed of atoms. Yet, chemists did not acquire experimental results that supported this theory until the late 18th century. Joseph Proust proposed the law of definite proportions, stating that the constituent elements in any particular compound are always present in a definite proportion by weight. The English scientist John Dalton went on to suggest that the reason elements combined in ratios of whole numbers as Proust observed was that the whole numbers represented individual atoms. Although the presence of atoms had been proposed centuries before, Dalton became the first scientist to suggest that each element is a distinct type of atom. Dalton then determined the relative mass of each of the known elements using hydrogen as the standard. Hydrogen proved to be a bad choice because it did not react with many of the elements. Since 1962 scientists have used one carbon 12 atom as the standard for atomic mass units, also referred to as daltons.

In 1865, the British chemist John Newlands took the known elements and arranged them according to their atomic mass. Then he gave each element a number corresponding to its assigned position. Newlands noticed that every eighth element shared common properties. Because the chemical and physical properties repeated every eight elements, he referred to this pattern as the law of octaves. For example, Newlands noticed that lithium, sodium, and potassium followed the law of octaves. He consequently grouped them together in the table he created. Today, these three elements belong to the very reactive Group I metals in the modern periodic table of the elements. Using the law of octaves as another means to order the elements, Newlands's work suggested that there were gaps in the table of elements. Unbeknown to

Newlands, these gaps represented undiscovered elements, and he eliminated these gaps in his table when he presented his research to his peers.

By assimilating the work of others while writing his book, Mendeleyev put all the pieces together to develop the first periodic table as we know it. For each of the known 63 elements, Mendeleyev created a card that included the element's physical and chemical properties as well as its relative atomic mass. On his laboratory walls, Mendeleyev arranged and rearranged these elements in various configurations until he settled upon arranging them in order of increasing atomic mass. He also took into account the law of octaves by placing elements in periodic groups arranged in horizontal rows in his table. Mendeleyev placed elements with similar properties in vertical columns, or groups. In 1871, he published a table similar to Newlands's, except he left gaps for the undiscovered elements. So confident was he in his arrangement of the elements that he even predicted the atomic mass and properties of these unknown elements using his table as a guide. For example, Mendeleyev left a gap between calcium and titanium. He predicted this element, which he called eka boron, would have properties similar to those of boron and an atomic weight of 44. Lars Fredrick Nilson and his team soon discovered eka boron (1879) and renamed it *scandium*, a transition element with properties similar to those of boron's and an atomic weight of 44.956. Successful predictions like this gave credence to Mendeleyev's periodic table of the elements. Mendeleyev developed a framework with his periodic table by which other scientists could discover minor discrepancies such as atomic mass differences and major deletions such as the entire bridge of transition elements between his Groups III and IV. Because his insight far outweighed his errors in the development of the modern periodic table of the elements, chemists credit Dmitry Mendeleyev with its establishment.

Mendeleyev thought that the atomic masses of the elements were the key to ordering them. The British scientist Henry Moseley x-rayed crystals of various elements and obtained data that led him to conclude that the real underlying order of the periodic table was not atomic mass, but the number of protons that each element possessed. Moseley might never have seen this pattern had Mendeleyev not previously organized the elements into the periodic table of the elements.

The modern periodic table of the elements groups elements into columns of families with similar physical and chemical properties and into rows of elements with the same outer electron shell. The atomic number of each element determines its position in the periodic table. The periodic table contains a great deal of information about individual elements. Using hydrogen as an example, the periodic table tells us that its atomic number is 1, that its symbol is *H*, that its atomic mass is 1.00794, and that its electron configuration consists of a single electron in its 1*s* orbital.

In the wake of the reputation he earned from the periodic table, Mendeleyev continued to teach and to do research. He studied the effects of isomorphism and specific volumes on various substances. In 1860, Mendeleyev defined the absolute point of ebullition, a property concerning gases and liquids, which is the point where the gas in a container turns to its liquid form simply by the exertion of pressure on the vessel; he also extensively studied the thermal expansion of liquids. As a theoretician, he tirelessly disputed the ideas of William Prout, a British chemist, who believed that all elements are from a primary source. Prout's idea, which is partially true with respect to the fact that all atoms consist of electrons, neutrons, and protons, continued to be the bane of Mendeleyev's existence as he vehemently argued that each element is unique.

Mendeleyev carried on many activities outside academic research and teaching. He became one of the founders of the Russian Chemical Society, now the Mendeleyev Russian Chemical Society. A prolific thinker and author, he wrote more than 400 articles and books over the course of his career. Mendeleyev started writing in order to earn money, authoring articles on popular science and technology for journals and encyclopedias beginning in 1859. His profound interest in spreading scientific and technological knowledge to a general audience never wavered throughout his career. Mendeleyev also yearned to integrate Russia into the Western world. Concerned with his country's national economy, he felt that Russia's agricultural and industrial resources needed to be developed. Traveling as far as Pennsylvania in order to learn about industries, he tried to encourage growth in Russia's coal and petroleum industry. Later, in 1893, the Russian government appointed Mendeleyev director of the Central Bureau of Weights and Measures, where he made significant contributions to the field of metrology, the science of measurement, including the introduction of the metric system to the Russian Empire.

In 1862, Dmitry Mendeleyev married Feozva Nikitichna Leshcheva, and they had a son and a daughter—Volodya and Olga. Mendeleyev divorced Leshcheva in 1881, and in early 1882 he married Anna Ivanova Popova, with whom he had four children: a son, Ivan; a pair of twins; and a daughter, Lyubov, who married the famous Russian poet Alexander Blok. Despite a legal divorce and subsequent legal marriage, the Russian Orthodox Church considered Mendeleyev a bigamist because the church

required a person to wait seven years before remarrying after a divorce. The divorce and the surrounding controversy contributed to Mendeleyev's failure ever to gain admission into the Russian Academy of Sciences. In 1907, Mendeleyev died in St. Petersburg of influenza.

Mendeleyev's contributions to the periodic table greatly enhanced the understanding of atomic structure and its relationship to periodic properties. Posthumously, the scientific community named a crater on the Moon (Mendeleyev's crater) and element number 101 (mendelevium) in his honor.

See also ATOMIC STRUCTURE; DALTON, JOHN; INORGANIC CHEMISTRY; PERIODIC TABLE OF THE ELEMENTS; PERIODIC TRENDS; PHYSICAL CHEMISTRY.

FURTHER READING

Gordin, Michael D. *A Well-Ordered Thing: Dmitrii Mendeleev and the Shadow of the Periodic Table.* New York: Basic Books, 2004.
Strathern, Paul. *Mendeleyev's Dream: The Quest for the Elements.* New York: St. Martins Press, 2001.

metabolism   Metabolism is the sum of all chemical reactions in an organism that involve the utilization and production of energy. Forms of metabolism include catabolism, breaking down food sources to obtain cellular energy, and anabolism, using cellular energy to build up compounds. The primary catabolic pathway is cellular respiration. Anabolic processes include such events as photosynthesis in plants and muscle building in humans. Catabolic and anabolic pathways involving the same products and reactants often exist as separate pathways, allowing for individual control.

ENERGETICS OF METABOLISM

Many metabolic reactions are endothermic, meaning they require an input of energy, but can be forced to happen by coupling them to a more favorable, exothermic reaction. Free energy (ΔG) is the amount of energy available to do work. In biological systems, the term *Gibbs free energy* was formerly used to honor Josiah Gibbs, who developed the relationship

$$\Delta G = \Delta H_{system} - T\,\Delta S_{system}$$

where ΔH_{system} represents the heat content of the system or enthalpy, T is the absolute temperature in kelvins, and ΔS_{system} represents the entropy (disorder) of the system. The free energy change is negative for exothermic reactions and is positive for endothermic ones. The overall free energy of two coupled reactions is the sum of the free energy of the individual reactions as shown in the following:

Reaction 1	A → B + C	$\Delta G^{\circ\prime}$ = +21 kJ/mol
Reaction 2	B → D	$\Delta G^{\circ\prime}$ = -34 kJ/mol
Overall coupled reaction	A → C + D	$\Delta G^{\circ\prime}$ = -13 kJ/mol

$\Delta G^{\circ\prime}$ stands for the standard free energy change at pH 7. Removal of the shared intermediate B drives the formation of products C and D from reactant A. Coupling these two reactions leads to a negative (spontaneous) overall free energy change, allowing both of the reactions to occur. The first reaction has a positive free energy change and would not occur spontaneously unless coupled to the second reaction.

In cellular processes, the commonly used energy-carrying molecule is adenosine triphosphate (ATP), composed of one unit of ribose, adenosine (a purine nucleotide), and three phosphate groups. The presence of ATP in an organism is essential for cellular processes such as motion, synthesis, signal transduction, nerve and heart impulses, and muscle contraction, to name a few. The bond between the second and third phosphates of ATP contains a high amount of energy, and the hydrolysis of this phosphate bond to form aldenosine diphosphate(ADP) and inorganic phosphate (P_i) is a thermodynamically favorable process. The hydrolysis of ATP to ADP has a $\Delta G^{\circ\prime}$ = -30.5 kJ/mol, and the further hydrolysis of ADP to AMP has a $\Delta G^{\circ\prime}$ = -45.6 kJ/mol. Many cellular processes that alone would be thermodynamically unfavorable are coupled to the hydrolysis of ATP to drive the reaction.

CELLULAR RESPIRATION

The ATP used in cellular processes of aerobic organisms is produced through the process known as cellular respiration, a series of organized, coupled reactions that convert food sources, primarily carbohydrates, into carbon dioxide, water, and energy. This process is the primary catabolic cellular pathway. The overall reaction for the oxidation of glucose by cellular respiration is shown:

$$C_6H_{12}O_6 + 6O_2 \rightarrow 6CO_2 + 6H_2O + energy$$

Glucose is oxidized with the final electron acceptor being oxygen.

Cellular respiration is broken into three stages known as glycolysis, the citric acid cycle (also called the Krebs cycle, tricarboxylic acid cycle, or TCA cycle), and oxidative phosphorylation. Each of these stages takes place in a different location inside the cell with glycolysis taking place in the cytoplasm, the Krebs cycle taking place in the mitochondrial matrix, and oxidative phosphorylation taking place in the membranes of the inner mitochondrial membrane. The final stage of cellular respiration (oxidative phosphorylation) requires molecular oxygen.

Each of the three stages is made up of multiple reactions that require multiple enzymes. Common types of chemical reactions reappear throughout all of the stages of respiration. These include hydrolysis and condensation reactions, oxidation-reduction reactions, isomerization reactions, and group transfer reactions.

Hydrolysis reactions are a type of addition reaction that involves adding water to a compound. If it is added across a double bond, the linkage is reduced to a single bond, and if it is added to a single bond, the bond can be broken and new products formed. Condensation reactions are essentially the

The primary energy currency of cellular reactions, adenosine triphosphate (ATP), consists of three phosphate groups, a ribose sugar, and an adenine base. When only two phosphate groups are present, the compound is known as adenosine diphosphate (ADP); when there is only one phosphate, the compound is adenosine monophosphate (AMP).

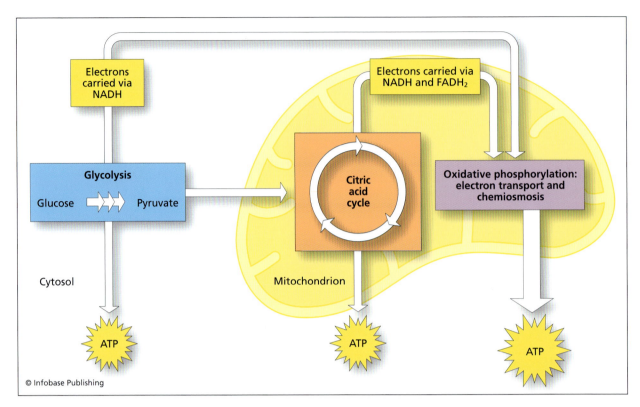

Cellular respiration occurs in three phases: glycolysis, citric acid cycle, and electron transport chain/oxidative phosphorylation. Glycolysis takes place in the cytosol, while the citric acid cycle and electron transport/oxidative phosphorylation take place in the mitochondrion of eukaryotic cells.

reverse of hydrolysis reactions, with water being removed and a new bond being formed. Oxidation and reduction reactions involve electron transfer between the reacting compounds. When a compound loses electrons, it is oxidized, and when electrons are gained, the compound is reduced. Isomerization reactions involve the rearrangement of a compound to form a new isomer, a compound with the same molecular formula but a different arrangement of atoms. The transfer of a functional group from one compound to another is known as group transfer. The different stages of cellular respiration utilize these reaction types in order to oxidize a carbohydrate completely into carbon dioxide and generate the ATP required for cellular functions. The more reduced the initial carbohydrate is (the fewer double bonds and more hydrogen atoms it possesses), the more energy that can be released upon its oxidation. This same principle is the reason that fats are a better fuel source than carbohydrates—because fats are far more reduced than carbohydrates and therefore release more energy upon oxidation.

While ATP serves as the energy source of the cell and is the end product of cellular respiration, often energy needs to be transferred from one cellular process to another. Electron carrier molecules perform this task. Examples of electron carriers important in cellular respiration are NADH and $FADH_2$ in their reduced form and NAD^+ and FAD in their oxidized form. They carry the reduction potential from glycolysis and the Krebs cycle to the electron transport chain, where it is used in oxidative phosphorylation to create ATP. Electron carriers can be thought of as the savings bond of cellular energy. While ATP is the cash that is used for cellular activities, savings bonds have all of the value of cash but cannot be utilized right away by the cell. The electron transport chain converts the energy carried by the electron carriers into ATP.

The energy created by the complete oxidation of one molecule of glucose by cellular respiration is summarized in the following table. The total value of 30 ATP produced varies from some published numbers of 36–38 ATP. The higher values are due to rounding of ATP produced (per NADH from 2.5 to 3 and per $FADH_2$ from 1.5 to 2).

GLYCOLYSIS

Glycolysis is the first stage in cellular respiration and is the oldest studied method of energy production. Since glycolysis takes place in the cytoplasm, it is not constrained to eukaryotic organisms and was present in ancient and extant prokaryotes. Glycolysis consists of 10 reactions that are tightly regulated

ENERGY PRODUCTION OF CELLULAR RESPIRATION

Stage	ATP Yield	NADH, FADH$_2$ Yield
Glycolysis	2	2 NADH (each will yield 1.5 molecules of ATP)
Pyruvate to acetyl CoA		2 NADH (each will yield 2.5 molecules of ATP)
Citric acid cycle	2 (GTP)	6 NADH (each will yield 2.5 molecules of ATP) 2 FADH$_2$ (each will yield 1.5 molecules of ATP)
Oxidative phosphorylation	26	10 NAD$^+$ regenerated 2 FAD$^+$ regenerated
Totals	30 ATP	

and controlled. The initial reactant for glycolysis is glucose, and the final products of glycolysis are two molecules of pyruvate that can be converted into acetyl coenzyme A (CoA) molecules that enter the Krebs cycle in aerobic organisms. In anaerobic organisms or in an aerobic organism under anaerobic conditions, these pyruvate molecules undergo fermentation rather than the Krebs cycle. Glycolysis produces four molecules of ATP per input glucose molecule but requires an investment of two ATP consumed for a net ATP production of two ATP molecules per glucose molecule. Glycolysis also generates one molecule of NADH per glucose molecule. In an aerobic eukaryotic cell, when molecular oxygen is present, this NADH is transported to the mitochondria and utilized in the production of ATP through oxidative phosphorylation by the electron transport chain. The energy from NADH is transferred from the cytoplasm to the mitochondria through one of two mechanisms. known as the glycerol 3-phosphate shuttle or the malate-aspartate shuttle. The resulting amount of energy transferred depends on the method of transfer, as the malate-aspartate shuttle maintains slightly more energy than the glycerol 3-phosphate shuttle.

THE CITRIC ACID CYCLE

The first stage of cellular respiration that is unique to aerobic organisms is the citric acid cycle. The starting material for this eight-step cycle is acetyl CoA. The pyruvate produced in glycolysis is transported to the mitochondria, where it undergoes a reaction to produce acetyl CoA (a two-carbon compound linked to a coenzyme A molecule), carbon dioxide, and two NADH molecules. Acetyl CoA enters the citric acid cycle by combining with one molecule of oxaloacetate, then proceeds through a series of reactions used to break down the carbon chain, producing carbon dioxide and releasing the energy of the bonds in the forms of ATP/GTP (guanosine triphosphate) and the electron carriers NADH and FADH$_2$. The final reaction of the Krebs cycle regenerates a molecule of oxaloacetate in order for the process to continue.

OXIDATIVE PHOSPHORYLATION

Oxidative phosphorylation takes place in the mitochondrial matrix and involves a series of membrane-bound molecules known as the electron transport chain. The electron transport chain converts the chemical energy present in the bonds of electron carrier molecules (NADH and FADH$_2$) from glycolysis and the citric acid cycle (Krebs cycle) to ATP. Electron transfer through the chain causes protons from NADH and FADH$_2$ to be pumped from the mitochondrial matrix to the intermembrane space, creating a proton gradient. As the protons are pumped, the electrons that were transported travel down the chain of the membrane-bound carriers through the mitochondrial matrix to molecular oxygen, which acts as the final electron acceptor, forming water in the process. Oxidative phosphorylation synthesizes ATP using an enzyme (ATP synthase) by harnessing the energy provided by this proton gradient across the inner mitochondrial membrane. The end result of oxidative phosphorylation by the electron transport chain is the regeneration of 10 NAD$^+$ and two FAD as well as the production of 26 molecules of ATP.

FERMENTATION

In anaerobic organisms or aerobic organisms under anaerobic conditions, the products of glycolysis do not enter the citric acid cycle and electron transport chain. In order to continue glycolysis, they need to regenerate the necessary NAD$^+$. (Glycolysis requires the oxidized form, NAD$^+$, as a reactant.) In these cases the cells undergo fermentation. Two common types of fermentation are lactic acid fermentation and alcohol fermentation, named for the end product of the process. The type of fermentation an organism carries out is dependent on the species.

Lactic acid fermentation occurs as the enzyme lactate dehydrogenase converts pyruvate to lactate and produces one molecule of NAD^+. This form of fermentation occurs in some bacteria, some fungi, and the muscles of animals. When an aerobic organism functions in an oxygen debt due to overworking muscles or poor physical conditioning, the organism undergoes lactic acid fermentation. The end product is lactic acid, which builds up in muscles in a crystalline form. This buildup leads to sore muscles and can be painful.

Alcohol fermentation is an alternative form of fermentation that creates alcohol, usually ethanol, as an end point. Pyruvate is first converted to acetylaldehyde by the enzyme pyruvate decarboxylase. Then alcohol dehydrogenase converts the acetaldehyde to ethanol. The second step also regenerates NAD^+ so that glycolysis can continue. Fermentation is essential for brewing and wine making processes.

METABOLISM OF OTHER ENERGY SOURCES

The discussion of catabolism generally focuses on the breakdown of carbohydrates, primarily glucose, through the processes discussed. Other food sources such as fats and protein also enter into cellular respiration and supply the body with energy as well as serving as building blocks, such as amino acids. Fats are compact storage molecules that supply much energy to the body and contain more energy per gram (nine kilocalories per gram) than carbohydrates or proteins (four kilocalories per gram). Structurally, fats (or triglycerides) are made of one molecule of glycerol as a backbone with three molecules of fatty acids (long hydrocarbon chains with a terminal carboxylate group) attached. In order for these molecules to enter cellular respiration, several reactions must first occur, including separation of the glycerol backbone from the fatty acid chains. The glycerol enters glycolysis in the form of dihydroxyacetone phosphate (DHAP), the product of the fourth reaction of glycolysis, or in the form of acetyl CoA. The fatty acid portion of the lipid is oxidized through a process known as β-oxidation, producing one NADH, one $FADH_2$, and one acetyl CoA molecule that enters the citric acid cycle.

Proteins are used less often as an energy source but are capable of entering the citric acid cycle as amino acids that are converted into acetyl CoA. More important, catabolism of proteins provides the cell with a necessary supply of free amino acids that can be used for the biosynthesis of new proteins.

ANABOLIC PROCESSES

Anabolic processes are cellular processes that involve the use of energy to build molecules to increase cell size or complexity. The production of proteins, nucleic acids, cell membranes, or polysaccharides provides examples of anabolic processes. The synthesis of carbohydrates through photosynthesis is also an example of anabolism; using the energy from the sun, a photosynthetic organism converts carbon dioxide and water into carbohydrates and oxygen. Photosynthesis converts the energy of the sun into a usable form and produces oxygen necessary for aerobic organisms to carry out cellular respiration. This energy is passed through the food chain when heterotrophic organisms eat the plants to obtain their chemical energy and use it to grow and develop.

Anabolic and catabolic pathways are linked. The breakdown of glucose by glycolysis is a catabolic process, while the buildup of glucose by gluconeogenesis is an anabolic process. While many of the steps between the two processes are conserved, there are two reactions that utilize different enzymes in order to maintain the independence of the two processes. These two processes are regulated separately and are not simply the reverse of one another. This separation of processes can also be controlled by location within the cell. Fatty acid catabolism takes place in the mitochondria of liver cells, while fatty acid anabolism takes place in the cytosol, keeping the two processes separated.

See also CARBOHYDRATES; CITRIC ACID CYCLE; ENZYMES; ELECTRON TRANSPORT SYSTEM; FATTY ACID METABOLISM; GLYCOGEN METABOLISM; GLYCOLYSIS; LIPIDS; PHOTOSYNTHESIS; PROTEINS.

FURTHER READING

Berg, Jeremy M., John L. Tymoczko, and Lubert Stryer. *Biochemistry,* 6th ed. New York: W. H. Freeman, 2007.

metallic compounds Metals are elements with characteristic properties such as being malleable and ductile and good conductors of electricity and heat. The periodic table is conveniently divided into two distinct regions, metals and nonmetals, separated by the stair-step line. The metals, found beneath the line, include all members of groups 1–12 as well as some elements of groups 13–15. Most of the known elements are metals. Except mercury, all metals are solid at room temperature.

METALLIC ELEMENTS

The periodic table of the elements is arranged in columns known as groups. The first group of representative elements, Group IA metals, is known as the alkali metals. This group consists of the elements lithium, sodium, potassium, rubidium, cesium, and francium. The elemental forms of the alkali metals are highly reactive. The alkali metals have a low density and can explode if they react with water. Each element in this group has one valence electron that is located in an *s* sublevel. These metals all contain one valence electron (s^1) and will become $^+1$ ions by losing this valence electron. The oxidation number of alkali metals is always $^+1$.

Group IIA elements, also known as the alkaline earth metals, include such elements as beryllium, magnesium, calcium, strontium, barium, and radium. They are gray-white in color, are soft solids, and have a low density like that of the alkali metals. The alkaline earth metals are very reactive, although they are less reactive than the alkali metals. Each element in this group has two valence electrons located in an *s* sublevel, and therefore forms $^+2$ cations by losing its two s^2 electrons.

Elements in Group IIIA have less metallic character than either the alkali metals or the alkaline earth metals. This group, often called the boron group, consists of boron, aluminum, gallium, indium, and thallium. Boron is not considered a metal: it is a metalloid. The other elements in this group exhibit a range of properties, and aluminum, gallium, indium, and thallium are considered poor metals. Aluminum, the most abundant element in the Earth's crust, is strong and, like all metals, is ductile and, therefore, very useful in a variety of manufacturing processes. Gallium is a semiconductor and has a melting point of 86°F (30°C) and a boiling point of 3,999°F (2,204°C). All of the elements in this group have three valence electrons located in the *s* and *p* sublevels. They usually form $^+3$ cations by losing their s^2 and p^1 electrons.

Transition metals are found in the center of the periodic table. They are malleable and ductile and are good conductors of electricity. Many elements in this block have multiple oxidation states. The transition metals contain a wide variety of elements including copper, gold, iron, silver, tungsten, zinc, mercury, nickel, and cobalt.

The inner transition metals are divided into two periods called the lanthanides and actinides. The lanthanides were once called the rare earth metals despite the fact that they are not rare. The lanthanides have high melting points and are silver in color. The lanthanides include elements with atomic numbers of 57–71: lanthanum, cerium, praseodymium, neodymium, promethium, samarium, europium, gadolinium, terbium, dysprosium, holmium, erbium, thulium, ytterbium, and lutetium. The actinides include elements with atomic numbers 89–103: actinium, thorium, protactinium, uranium, neptunium, plutonium, americium, curium, berkelium, californium, einsteinium, fermium, mendelevium, nobelium, and lawrencium. The actinides are all radioactive and have a wider range of physical properties than the lanthanides.

CHARACTERISTICS OF METALS

The characteristics of metals include low ionization energy, three or fewer valence electrons, luster, malleability, ductility, and are good conductors.

Ionization energy is the measure of how easily an atom will lose an electron. The higher the ionization energy, the more tightly the electrons are held. Metals generally have low ionization energies; that means that another atom can take an electron from a metal atom relatively easily. As a result, metal atoms are likely to donate electrons and become positively charged ions, or cations. Nonmetals, on the other hand, have high ionization energies, meaning that they hold on to their electrons tightly, and they also have a high electron affinity, meaning that they are attracted to the electrons of other atoms and tend to accept additional electrons. This is why when a metal and a nonmetal form a bond, the nonmetal usually steals the electron from the metal, which, compared to the nonmetal, has a low ionization energy and a low electron affinity.

The electron configuration of all elements determines the reactivity of those elements. In the case of metals, the outer energy level for their electrons always contains three or fewer electrons. To satisfy the octet rule, all atoms want eight valence electrons (with the exception of hydrogen and helium, which require two valence electrons). It is easier for a metal to lose its outer three valence electrons than to obtain five or more new electrons from another atom.

Conductivity is a property of metals that defines their ability to conduct an electrical current. While nonmetals are known as insulators and do not conduct a current, all metals are conductors to some extent. Some metals are better conductors than others. The table "Conductivity of Metals" on page 422 lists the conductivity capacity of several common metals. Silver has the highest conductivity of the metals and all of the metals have a higher conductivity than nonmetals. Sulfur, a nonmetal, is listed for the sake of comparison. The difference in conductivity between sulfur and silver is on the order of 10^{13} less.

Many metals exist in crystalline form. A characteristic of this crystal form is its luster, a measure of the amount of light that is reflected off a crystal of a metal. The type of luster is determined by the surface of the crystal and the index of refraction of the substance that is being evaluated. Many terms can be used to describe the luster of a metal, including *dull* (a nonreflective surface), *fibrous* (looks like fibers), *metallic* (looks like metal), *vitreous* (looks like glass), *earthy* (the look of dirt), *waxy, greasy,* and *adamantine* (brilliant). Metallic luster includes the bright shiny appearance, while all of the others are classified as nonmetallic.

Metals have the property of being malleable. This means that they can be bent and shaped by

CONDUCTIVITY OF METALS

Conductivity	Element (Symbol)
0.63×10^6/(cm × ohm)	Silver (Ag)
0.596×10^6/(cm × ohm)	Copper (Cu)
0.45×10^6/(cm × ohm)	Gold (Au)
0.377×10^6/(cm × ohm)	Aluminum (Al)
0.313×10^6/(cm × ohm)	Beryllium (Be)
0.298×10^6/(cm × ohm)	Calcium (Ca)
0.226×10^6/(cm × ohm)	Magnesium (Mg)
0.21×10^6/(cm × ohm)	Rhodium (Rh)
0.21×10^6/(cm × ohm)	Sodium (Na)
0.197×10^6/(cm × ohm)	Iridium (Ir)
5×10^{-18}/(cm × ohm)	Sulfur (S) (a nonmetal)

hitting them with a hammer or rolling them out by rollers. This allows metals to be formed into long sheets and makes them useful in the production of tools, automobile bodies, and other articles. Metals are also classified as ductile, meaning able to be drawn out thinly or into wires. Examples include copper wires that are used to carry electricity.

METALLIC BONDING

Metals can participate in two types of bonding—ionic and metallic bonds. An ionic bond results from the electrostatic attraction between positively charged and negatively charged ions. When a metal interacts with a nonmetal, the electronegativity difference between the two elements is great enough that the nonmetal will steal the outer valence electron(s) from the metal, creating a positively charged metal ion and a negatively charge nonmetal ion. The electrostatic interaction between the positive and negative ions creates a strong chemical bond.

A metallic bond is formed when a metal atom bonds with other like metal atoms. The ionization energy (energy required to remove an electron) for metals is quite low, and this causes the atoms to hold their electrons loosely. The stability of a metal increases when the metal shares its electrons with other atoms. The end result is that the electrons in a metal are not held tightly by any one atom; they exist as a "sea of electrons" surrounding each of the positive nuclei of the atoms. This type of bonding is responsible for the physical characteristics found in metals, such as malleability and ductility. Metals can be bent, hammered, and rolled without breaking

because the electrons can move with the substance as it moves. The "sea of electrons" concept also explains the conductivity of metals. An electrical current is simply the movement of electrons from one place to another. When the electrons have the freedom that they have in a metal, they are able to move and thus conduct an electrical current.

ALLOYS

Alloys are metals that are made up of more than one type of metallic element. Two of the most common alloys are steel and bronze. Steel is used in numerous manufacturing processes in the world today. Everything from food cans and razor blades to turbines and tools contains steel. Steel contains iron with varying concentrations of carbon and other metals added to it. The percentage of carbon is never greater than approximately 1.5 percent. Steel is classified as high-carbon steel, medium-carbon steel, or mild-carbon steel on the basis of the percentage of carbon in it. The higher the carbon content of the steel, the harder the type of steel. Heating also affects the properties of steel. Quick cooling molten steel creates a hard, brittle form of steel. Slower cooling creates softer steel.

In the early 20th century, inventors discovered that the addition of small amounts of chromium to the iron-carbon alloy known as steel created a shiny tarnish- and rust-resistant compound, which became known as stainless steel. This type of steel is now a common material for kitchen appliances.

Bronze, a mixture of copper and tin, and brass, a mixture of copper and zinc, are two common alloys. Items such as statues, bells, and silverware are made of bronze. Bronze is much harder an alloy than brass. Doorknobs and other decorative items are often made of brass, which has a shiny gold appearance.

METALLURGY

Metallurgy is the science of studying metals and removing them from their ores. The majority of metals on Earth are found in the form of ores, minerals that contain economically valuable metals. Minerals are naturally occurring and have a range of physical and chemical properties. Some of the most valuable metals found in the Earth's crust include aluminum, magnesium, calcium, iron, sodium, potassium, titanium, and manganese.

In order to separate the metals from their ores, the following three basic steps are required:

1. removal of the metals from the contaminants in the ore
2. reduction to produce the free metal
3. further purification of the metal to remove any impurities.

In order to prepare the ore to remove the metal, one must remove all of the other components, including such things as clay. These contaminants can be removed by several methods, including flotation or using a magnet. Many types of metals can be floated on the surface of a water mixture containing oil and detergent. The metals are collected by the oil and remain on the top of the mixture, where they can be collected. A magnet easily separates highly magnetic metals from contaminants in the ore.

Reduction of the metals from the ores to release the free metal is a necessary step. Very few of the metals exist in nature in their free metal form. Metals are quite reactive, meaning they quickly react with other substances to create metallic compounds. The metals always form the positive ion in an ionic compound and have a positive oxidation number. Releasing the free metal will lower its oxidation number to 0, thus requiring a reduction reaction. This reduction generally takes place at high temperatures in a process known as pyrometallurgy.

Once the free metals are reduced, they often require further purification by methods such as distillation, which allows for the separation of substances based on different boiling points. As a mixture is heated, the substance with the lowest boiling point vaporizes first and can be removed from the mixture. Metals with low boiling points such as mercury and zinc are often separated in this way.

The isolation, purification, and production of metals and metallic compounds compose a multibillion-dollar industry. One unique method of metal purification from ores, biohydrometallurgy, was initially discovered in the 1940s and commercially applied in the 1970s. In this method, microbes are used to release metals such as copper, lead, and zinc by oxidizing the sulfur that holds the metals in the ore. This process is more environmentally friendly than the traditional method known as smelting. Scientists are researching new and improved methods of producing stronger and more durable metallic products. Recycling of metals such as steel and aluminum is of growing environmental importance.

See also COVALENT COMPOUNDS; IONIC COMPOUNDS; PERIODIC TABLE OF THE ELEMENTS.

FURTHER READING
Zumdahl, Stephen, and Susan Zumdahl. *Chemistry,* 7th ed. Boston: Houghton Mifflin, 2007.

Molina, Mario (1943–) *Mexican Chemist*
Mario Molina is a chemist who has spent his life studying the physical chemistry behind chemicals such as chlorofluorocarbons and their effect on the atmosphere. In 1995, Molina became the first Mexican scientist to receive the Nobel Prize in chemistry, which he shared with the American chemist F. Sherwood Rowland and the Dutch chemist Paul J. Crutzen, for their work illustrating the direct link between chlorofluorocarbons (CFCs) and destruction of the ozone layer.

EARLY EDUCATION AND POSITIONS
Mario Molina was born on March 19, 1943, in Mexico City, Mexico. His father, Roberto Molina Pasquel, was a lawyer, politician, and professor at Universidad Nacional Autonoma de Mexico (UNAM). Mario's mother was Leonor Henriquez de Molina.

Molina was fascinated by science from an early age. His interests were focused in the area of physical chemistry. Molina even set up his own chemistry lab in his childhood home with the help of his aunt, who was a chemist. He received his bachelor of science in chemical engineering from UNAM in 1965. Although he was more interested in physical

Mario Molina is a Mexican chemist renowned for the discovery of the link between chlorofluorocarbons (CFCs) and destruction of the ozone layer, October 11, 1995. *(AP Images)*

chemistry, that degree was not available to him at the time, but a chemical engineering major provided him the opportunity to take more difficult math, chemistry, and physics courses.

In 1967, Molina received an advanced degree from the University of Freiburg in West Germany. This gave him the opportunity to spend more time working on the mathematical areas about which he wanted to learn more. Molina then returned to his alma mater in Mexico City to serve as an associate professor in 1967–68. After finding substitutes at other universities, Molina still truly wanted to study physical chemistry. He finally traveled to the United States and received his Ph.D. in chemistry from the University of California at Berkeley in 1972. There, he studied under American chemist George C. Pimentel, the inventor of the chemical laser. Molina's graduate work involved the internal energy of chemical and photochemical reaction products utilizing Pimentel's lasers. In Pimentel's lab, Molina met his future wife, Luisa. Together, the Molinas had one son, Felipe.

CFCS AND THE OZONE LAYER

After his work at Berkeley, Molina traveled to the University of California at Irvine as a postdoctoral student and worked with F. Sherwood Rowland on the relationship between CFCs and the ozone layer.

CFCs are synthetic compounds originally developed in the 1920s and 1930s as a safer alternative to refrigerants and coolants of the time. The market for CFCs expanded rapidly up until the 1980s as they were nontoxic, nonreactive, and nonflammable. CFCs were utilized in refrigeration (both home and commercial) under the common name *Freon,* aerosol propellants, insulation (foam-type), and Styrofoam.

An ozone molecule (O_3) is made up of three covalently bonded oxygen atoms. Ozone at ground level is a major component of smog, especially surrounding large cities. The presence of ozone in the lower portions of the stratosphere, the second level of Earth's atmosphere, plays an important role in protection from ultraviolet radiation reaching Earth's surface. Ozone is formed when molecular oxygen (O_2) is struck with solar energy according to the following reactions:

$$O_2 + sunlight \rightarrow O + O$$

$$O + O_2 \rightarrow O_3$$

When Molina chose his research project, scientists still believed that CFCs were essentially inert substances with no harmful effects. Molina set out to determine what happened to these molecules once they were released. This project afforded Molina the opportunity to apply his physical chemistry knowledge to the new field of atmospheric chemistry. Molina achieved very few positive results when he attempted reacting CFCs with compounds in the troposphere. Only after he began to explore the possibility of their passing into the stratosphere did he begin to see a potential downside of CFCs.

Molina and Rowland were able to show that the reaction between chlorofluorocarbons and ozone caused the destruction of the ozone molecule. The method of destruction of the ozone layer by CFCs does not involve the entire CFC molecule. Upon being subjected to intense solar energy at the higher levels of the atmosphere, the chlorofluorocarbons break down, leaving primarily reactive chlorine (Cl). The chlorine reacts with a molecule of ozone and atomic oxygen (O) to produce two molecules of molecular oxygen according to the following reactions:

$$Cl + O_3 \rightarrow ClO + O_2$$

$$ClO + O \rightarrow Cl + O_2$$

As is shown in these reactions, chlorine reacts with ozone to create one molecule of molecular oxygen and one molecule of chlorine monoxide (ClO), which then reacts with an atom of oxygen to reproduce the chlorine atom and form an additional oxygen molecule. The net result of these reactions is that O_3 and O are converted into two molecules of O_2. Chlorine is not consumed in the reaction, but ozone is.

The loss of the ozone layer or even the depletion of the ozone layer has dramatic effects on the human population. The ozone layer acts as a protective blanket that eliminates the majority of the Sun's ultraviolet rays. Without it, life on Earth would cease to exist. Destruction of the ozone layer could lead to increases in skin cancer and damage to crops, wildlife, and habitats. CFCs also impact the environment by trapping solar energy in Earth's atmosphere, contributing to the greenhouse effect. Greenhouse gases in the atmosphere prevent solar energy reflected from Earth's surface from escaping, leading to a rise in global temperature. CFCs are greenhouse gases and are more effective greenhouses gases than the more commonly discussed carbon dioxide.

Molina and Rowland published a paper in *Nature* in 1974, relating the destruction of the ozone layer to CFCs. The initial reaction to their findings was skepticism. Scientists understood the information, but not much was done to reduce the use of CFCs, because many believed that it would never have a dramatic effect. Business and political interests did not believe that such a critical component of the economy as CFCs could be eliminated. Others had difficulty believing that a compound with such a safe history could produce such a dramatic result.

Several countries did begin eliminating the production of CFCs in aerosols in 1978. Serious environmental implications of the reaction between CFC and ozone began to appear in 1985, when the English scientist Joseph Farman identified a large hole in the ozone layer over Antarctica. The hole encompassed an area larger than the size of the United States. This dramatic finding coupled with the solid understanding that CFCs were directly linked to the production of this hole led to action. In 1985, an international treaty known as the Montreal Protocol was passed to reduce the number of CFCs produced starting in 1985 and attempt to phase out their production by 1996. Developing countries were given extended deadlines. Modifications to the phaseout programs led the Environmental Protection Agency to regulate the phase out of CFCs in the United States by 2000. CFCs have already proven to be an environmental hazard by their destruction of the ozone layer of Earth's atmosphere. CFCs have been taken off the market, and the release of CFCs from older appliances into the atmosphere is unlawful.

According to a 2006 report by the U.S. Department of State on the condition of the ozone layer, the destruction appears to be slowing. The CFCs that have already been released have an incredibly long life span in the stratosphere and can continue to damage the ozone layer for nearly 100 years after they have been released. Despite this, the elimination of CFCs will eventually have a significant effect on preventing further loss of the ozone layer.

Mario Molina continues to use his knowledge in atmospheric and physical chemistry to impact the condition of Earth. From 1982 to 1989, Molina worked at California Institute of Technology in Pasadena in the jet propulsion lab. After that work, he accepted a professorship at the Massachusetts Institute of Technology (MIT) in Cambridge, with a joint appointment in the department of Earth, atmospheric and planetary sciences and in the department of chemistry. The focus of Molina's lab at MIT included three areas of research: gas phase chemical kinetics and photochemistry, chemistry of atmospheric aerosols, and an integrated program on urban, regional, and global air pollution—a Mexico City case study.

In 2005, Molina left MIT for work in both Mexico City and the University of California at San Diego (UCSD). He started the Center for Strategic Studies in Energy and the Environment, where he works on air pollution in large cities.

See also AIR POLLUTION (OUTDOOR/INDOOR); ATMOSPHERIC AND ENVIRONMENTAL CHEMISTRY; CRUTZEN, PAUL; GREEN CHEMISTRY; GREENHOUSE EFFECT; ROWLAND, F. SHERWOOD.

FURTHER READING

Gribbin, John. *The Hole in the Sky.* London: Corgi Books, 1998.
The Nobel Foundation. "The Nobel Prize in Chemistry 1995." Available online. URL: http://nobelprize.org/nobel_prizes/chemistry/laureates/1995. Accessed July 25, 2008.
Rowland, F. S., and Mario Molina. "Ozone Depletion: Twenty Years after the Alarm." *Chemical and Engineering News* 72 (1994): 8–13.

momentum and collisions Momentum is a property of a moving body that takes into account both the body's mass and its velocity. As a physical quantity, momentum is the product of the mass and the velocity. Its importance lies in its relation to force, via Newton's second law of motion; in the conservation law that applies to it; and, consequently, in collisions.

MOMENTUM

The linear momentum $\mathbf{p}$ of a particle of mass m and velocity $\mathbf{v}$ is

$$\mathbf{p} = m\mathbf{v}$$

where $\mathbf{p}$ is in kilogram-meters per second (kg·m/s), m in kilograms (kg), and $\mathbf{v}$ in meters per second (m/s). Linear momentum is a vector quantity. The adjective *linear* is used to distinguish linear momentum from angular momentum and can be dropped when the situation is clearly one of linear motion. The total momentum of a collection of particles is the vector sum of the particles' individual momenta.

Newton's second law of motion is often formulated as "force equals mass times acceleration":

$$\mathbf{F} = m\mathbf{a}$$

$$= m\frac{d\mathbf{v}}{dt}$$

where $\mathbf{F}$ denotes the force, in newtons (N); $\mathbf{a}$ the acceleration, in meters per second per second (m/s^2); and t the time, in seconds (s). It is more correctly expressed as "force equals time-rate-of-change of the linear momentum":

$$\mathbf{F} = \frac{d\mathbf{p}}{dt}$$

$$= \frac{d(m\mathbf{v})}{dt}$$

Einstein's special theory of relativity uses the law in the latter form. The former expression is an approximation to the latter for speeds that are sufficiently low compared to the speed of light. In that case, the mass, which in general depends on speed according to special relativity, can be considered constant, giving

$$\mathbf{F} = \frac{d(m\mathbf{v})}{dt}$$
$$= m\frac{d\mathbf{v}}{dt}$$
$$= m\mathbf{a}$$

Linear momentum obeys a conservation law, which states that in the absence of a net external force acting on a system, the system's total linear momentum remains constant in time. This conservation law is particularly useful in the analysis of collisions. Since momentum is a vector quantity, its conservation implies the separate conservation of each of its components. Thus, for a collision in which the net external force vanishes, the x-, y-, and z-components of the total momentum are separately conserved. Even if the net external force does not vanish, if its component in any direction does vanish, then the component of the total momentum in that direction is conserved.

COLLISIONS

The term *collision* has very much the same meaning as in everyday use, with the extension that colliding objects do not actually have to have contact with each other but instead can interact in other ways. Two electrons might collide by passing each other closely while interacting through the electromagnetic and weak interactions. In another example, the maneuver in which a spacecraft swings closely around a planet and thereby gains speed is a collision between the spacecraft and the planet through the gravitational interaction. On the other hand, colliding automobiles, for instance, *do* affect each other through direct contact. In the aftermath of a collision, the colliding objects might or might not emerge unscathed, and additional objects might be produced. After the flyby, the spacecraft and the planet obviously still retain their identities, but the results of collisions of atomic nuclei, say, might involve free protons, neutrons, and various lighter

The games of billiards and pool demonstrate almost-elastic collisions, when the balls collide with each other. Legend has it that these games served as motivation for developing the mechanics of collisions. The photograph shows a backhand shot at a pool table. *(Dorothy Ka-Wai Lam, 2008, used under license from Shutterstock, Inc.)*

nuclei. Or colliding nuclei might merge and fuse into heavier nuclei. Physicists investigate the properties of the fundamental forces by causing elementary particles to collide and studying the particles that the collisions produce.

For collisions in isolated systems, certain conservation laws are always valid. In other words, for systems that are not influenced by their surroundings, certain physical quantities retain the same total value after a collision that they had before the collision. Energy, linear momentum, angular momentum, and electric charge are the most commonly known quantities that are always conserved in isolated systems. Other quantities might also be conserved, depending on the interactions involved.

Although total energy is always conserved in collisions in isolated systems, the total *kinetic* energy of the collision results might not equal that of the colliding objects. In a head-on car crash, for example, almost all the initial kinetic energy of the colliding cars is converted to energy in the form of heat, sound, and energy of metal deformation. Those collisions in which kinetic energy *is* conserved are called elastic collisions. One example is a planetary flyby, as described previously. If the spacecraft gains speed, and thus kinetic energy, from such an encounter, the planet loses kinetic energy. Collisions in which kinetic energy is not conserved are called inelastic collisions. When the colliding objects stick together and do not separate after collision, maximal loss of kinetic energy occurs, a phenomenon termed completely inelastic collision. In the extreme, for instance, when two bodies of equal mass and speed collide head on and stick together, the bodies stop dead, and all their initial kinetic energy is lost.

As an example of the use of momentum conservation for analyzing collisions, let us confine our considerations to motion along a line, say, along the *x*-axis, and let two otherwise free bodies, A and B, collide. Denote their masses m_A and m_B; their initial velocities, immediately before collision, v_{Ai} and v_{Bi}; and their final velocities, their velocities immediately after collision, v_{Af} and v_{Bf}. Conservation of the *x*-component of total momentum gives the equation

$$m_A v_{Ai} + m_B v_{Bi} = m_A v_{Af} + m_B v_{Bf}$$

This relation can be used, for instance, to solve for the final velocity of one of the particles, given the final velocity of the other particle, the particles' masses, and their initial velocities. If B is the particle whose final velocity is to be found, we have

$$v_{Bf} = \frac{1}{m_B}\Big[m_A(v_{Ai} - v_{Af}) + m_B v_{Bi}\Big]$$

Alternatively, all the velocities might be measured and the mass of A known. Then the mass of B can be calculated:

$$m_B = -m_A \frac{v_{Af} - v_{Ai}}{v_{Bf} - v_{Bi}}$$

For another example, let the two bodies stick together when they collide in a completely inelastic collision. Denote the final velocity of the combined body by v_f. The conservation of momentum equation now becomes

$$m_A v_{Ai} + m_B v_{Bi} = (m_A + m_B)v_f$$

giving for the final velocity of the combined body in terms of the masses and the initial velocities

$$v_f = \frac{m_A v_{Ai} + m_B v_{Bi}}{m_A + m_B}$$

As a numerical example, let a body of mass 1.0 kg moving to the right at 2.2 m/s collide with a 3.5-kg body moving to the left at 1.5 m/s. If the bodies stick together, what is their common velocity immediately after the collision? With motion to the right taken as positive, the data are $m_A = 1.0$ kg, $v_{Ai} = 2.2$ m/s, $m_B = 3.5$ kg, $v_{Bi} = -1.5$ m/s. By the previous equation, the final velocity is

$$v_f = \frac{(1.0 \text{ kg})(2.2 \text{ m/s}) + (3.5 \text{ kg})(-1.5 \text{ m/s})}{(1.0 \text{ kg}) + (3.5 \text{ kg})}$$

$$= -0.68 \text{ m/s}$$

So the combined body moves at 0.68 m/s to the left immediately after the collision.

For the final example, let the bodies collide elastically. Then, in addition to conservation of the *x*-component of total momentum,

$$m_A v_{Ai} + m_B v_{Bi} = m_A v_{Af} + m_B v_{Bf}$$

the collision also conserves kinetic energy,

$$\tfrac{1}{2} m_A v_{Ai}^2 + \tfrac{1}{2} m_B v_{Bi}^2 = \tfrac{1}{2} m_A v_{Af}^2 + \tfrac{1}{2} m_B v_{Bf}^2$$

These two simultaneous equations allow finding two quantities, given all the others. For instance, given the masses and initial velocities, the final velocities of both bodies can be calculated. Because of the quadratic nature of the second equation, the solution in practice is extremely messy and invariably turns out wrong. However, by appropriate algebraic manipulation of the two equations, the following linear relation emerges:

$$v_{Af} - v_{Bf} = -(v_{Ai} - v_{Bi})$$

This relation shows that as a result of the elastic collision the relative velocity of one body with respect to the other simply reverses direction (changes sign). This equation, rather than the kinetic-energy conservation equation, should be used together with the momentum conservation equation for solving for any two quantities.

See also ACCELERATION; CONSERVATION LAWS; FORCE; EINSTEIN, ALBERT; MASS; MOTION; NEWTON, SIR ISAAC; ROTATIONAL MOTION; SPECIAL RELATIVITY; SPEED AND VELOCITY; VECTORS AND SCALARS.

FURTHER READING

Young, Hugh D., and Roger A. Freedman. *University Physics*, 12th ed. San Francisco: Addison Wesley, 2007.

Mössbauer effect   Named after the 20th–21st-century German physicist Rudolph Mössbauer, the effect is related to the frequencies of electromagnetic radiation whose photons (gamma rays) are absorbed and emitted by nuclei as they make transitions between the same pair of energy levels. For energy difference ΔE, in joules (J), between two states of a nucleus, one might expect, on the basis of conservation of energy, that photons of radiation having exactly the following frequency

$$f = \frac{\Delta E}{h}$$

will excite such resting nuclei from the lower-energy state to the higher, where f is in hertz (Hz) and h is the Planck constant, whose value is $6.62606876 \times 10^{-34}$ joule-second (J·s). When such resting nuclei deexcite from the higher-energy state to the lower, one might similarly expect that the radiation whose photons are emitted will possess the same frequency. As a result of conservation of linear momentum, however, the nuclei undergo recoil upon absorption and emission of photons. So, in order to excite a resting nucleus, the photon must have a higher energy than ΔE, and consequently its radiation must possess a higher frequency, since some fraction of the photon's energy subsequently appears as kinetic energy of the recoiling excited nucleus. In the inverse process of deexcitation, the amount of energy released by a resting nucleus, ΔE, is divided between the photon and the recoiling nucleus (as its kinetic energy), so that the photon's energy is less than ΔE and the frequency of its radiation less than $\Delta E/h$. When the nuclei are initially in motion, the Doppler effect, too, affects the radiation's frequencies of absorption and emission. Therefore, such nuclei in a sample of material, because of their recoil and thermal motion, neither absorb nor emit photons of radiation of a single, well-defined frequency.

Mössbauer discovered in 1958 that the recoil can be eliminated by having the nuclei firmly attached to a crystal structure, so that it is the sample as a whole that recoils. Since the mass of the sample is enormously greater than that of a single nucleus, the linear momentum transferred to the sample rather than to the nucleus results in negligible recoil motion. This is the Mössbauer effect. If, in addition, one cools the sample considerably in order to reduce thermal motion, then the absorption and emission energies become very well defined and equal to each other. These methods allow production of photons of very precisely defined energy (and, accordingly, radiation of precisely defined frequency) as well as high-resolution measurement of photon energy. As to the latter, if the photon energy does not quite match the nuclear excitation energy, one can precisely measure the difference by utilizing the Doppler effect and moving the sample toward or away from the arriving photons at such a speed that the radiation's frequency, and the photons' energy, relative to the sample are shifted to that required for excitation.

One important use of the Mössbauer effect was to confirm Albert Einstein's prediction that photons lose or gain energy, with a corresponding decrease or increase of frequency of their radiation, when they move upward or downward (i.e., against a gravitational field or with it), respectively, according to the general theory of relativity. The effect has found applications beyond the confines of physics, such as in chemistry and in industry. In 1961 Mössbauer was awarded half the Nobel Prize in physics "for his researches concerning the resonance absorption of gamma radiation and his discovery in this connection of the effect which bears his name."

See also DOPPLER EFFECT; DUALITY OF NATURE; EINSTEIN, ALBERT; ELECTROMAGNETIC WAVES; ENERGY AND WORK; GENERAL RELATIVITY; GRAVITY; HEAT AND THERMODYNAMICS; MOMENTUM AND COLLISIONS; MOTION; WAVES.

FURTHER READING

Eyges, Leonard. "Physics of the Mössbauer Effect." *American Journal of Physics* 33 (1965): 790–802.

Hesse, J. "Simple Arrangement for Educational Mössbauer-Effect Measurements." *American Journal of Physics* 41 (1973): 127–129.

Lustig, Harry. "The Mössbauer Effect." *American Journal of Physics* 29 (1961): 1–8.

Vandergrift, Guy, and Brent Fultz. "The Mössbauer Effect Explained." *American Journal of Physics* 66 (1998): 593–596.

motion   A change in spatial position or orientation that occurs over time is called motion. An ideal point particle, a hypothetical object with no size, can only change position, while an object that is endowed with size (think of an apple) can change both its position and its orientation. Change of position is linear motion, or translation, or translational motion, while change of orientation is rotation. For

High-speed computers analyze the motion of the spacecraft that is being launched here (Space Shuttle Orbiter *Endeavour,* carrying a part of the *International Space Station*), in order to predict its course and examine the results of possible course corrections. *(NASA Marshall Space Flight Center)*

a body, it is convenient and common to consider separately the motion of its center of mass and the body's rotation about its center of mass. If the center of mass moves, then the body is undergoing translational motion. The most general motion of a body involves simultaneous motion of its center of mass and rotation about the center of mass.

As an example, consider the motion of the Moon, which is revolving around Earth while keeping its same side facing Earth. So, as its center of mass performs nearly circular translational motion around Earth with a period of about a month, the Moon is simultaneously rotating about its center of mass at exactly the same rate, as it happens. This article deals only with linear motion. Rotational motion is discussed in a separate entry.

The study of motion without regard to its causes is kinematics. Dynamics studies the causes of motion and of changes in motion. Classical, nonrelativistic kinematics and dynamics are based on Sir Isaac Newton's laws of motion (discussed later). Albert Einstein generalized those laws in his special theory of relativity and further generalized them in his general theory of relativity to include gravitation.

Free motion, or inertial motion, is motion in the absence of forces. According to Newton's first law of motion (see later discussion), such motion is one of constant velocity (i.e., motion at constant speed and in a straight line, which includes rest [zero speed]). However, in Einstein's general theory of relativity, free motion is generally neither at constant speed nor in a straight line. This motion is termed geodesic motion. It is the generalization of constant-velocity motion as appropriate to the general theory of relativity. The path of a photon passing near the Sun on its way from a distant star to Earth, for example, is not a straight line: it bends toward the Sun. According to the general theory of relativity, the photon is in free motion and is following the trajectory that is closest to a straight line in the space-time geometry determined by the nearby presence of the Sun.

KINEMATICS

As mentioned earlier, kinematics is the study of the motion of bodies without regard to the causes of the motion. Here we consider a body's translational motion, which is related to a body's position as a function of time. This can be expressed in terms of a body's coordinates (i.e., the coordinates of its center of mass or its actual coordinates, if it is a point particle) with respect to some reference frame, or coordinate system, say, $(x(t), y(t), z(t))$, or in terms of its position vector, $\mathbf{r}(t)$, which is a vector pointing from the coordinate origin to the body's position and whose x-, y-, and z-components are the coordinates, $(x(t), y(t), z(t))$.

A body's instantaneous velocity, $\mathbf{v}(t)$, a vector, is the time rate of change of its position (i.e., the derivative of its position vector with respect to time):

$$\mathbf{v}(t) = \frac{d\mathbf{r}(t)}{dt}$$

The x-, y-, and z- components of the velocity vector, $(v_x(t), v_y(t), v_z(t))$, are the time derivatives of the body's x-, y-, and z-coordinates, respectively:

$$v_x(t) = \frac{dx(t)}{dt}$$

$$v_y(t) = \frac{dy(t)}{dt}$$

$$v_z(t) = \frac{dz(t)}{dt}$$

The body's instantaneous speed, $v(t)$, a scalar, is the magnitude of its instantaneous velocity:

$$\mathrm{v}(t) = |\mathbf{v}(t)| = \sqrt{v_x^2 + v_y^2 + v_z^2}$$

As an example, let a body move in a single direction only, say, along the x-axis, such that its position is given by

$$x(t) = -3t^2 + 2t - 6$$

By taking the time derivative of this expression, one obtains for the x-component of the instantaneous velocity

$$v_x(t) = -6t + 2$$

These expressions allow the calculation of the body's position and instantaneous velocity for any time. At $t = 3$ time units, for instance, the body's location is at $x(3) = -25$ on the x-axis, while its instantaneous velocity is $v_x(3) = -16$ units of velocity, meaning 16 units of velocity in the direction of the negative x-axis. The instantaneous speed of the body is $v(3) = 16$ units of velocity and has no direction.

The average velocity for the interval from time t_1 to time t_2 is defined as the displacement of the body during the interval divided by the time duration:

$$\mathbf{v}_{av} = \frac{\mathbf{r}(t_2) - \mathbf{r}(t_1)}{t_2 - t_1}$$

This can also be written as

$$\mathbf{v}_{av} = \frac{\Delta \mathbf{r}}{\Delta t}$$

where $\Delta\mathbf{r}$ denotes the body's displacement during time interval Δt. Since the time derivative is the limit of the finite displacement divided by the finite time interval as the time interval approaches zero, the instantaneous velocity is the limit of the average velocity for smaller and smaller time intervals.

The earlier example can be extended. The body's position at time $t = 1$ time unit is $x(1) = -7$ on the x-axis. So, its displacement (which is along the x-axis only) during the time interval from $t_1 = 1$ to $t_2 = 3$ is

$$\Delta x = x(t_2) - x(t_1) = -25 - (-7) = -18$$

During this time interval, the body moves 18 space units in the direction of the negative x-axis. The time interval itself is

$$\Delta t = t_2 - t_1 = 3 - 1 = 2$$

time units. So the average velocity for this time interval is

$$v_{x\,av} = \frac{\Delta x}{\Delta t} = \frac{-18}{2} = -9$$

meaning 9 velocity units in the negative x-direction. So no matter where the body was between the times of $t = 1$ and $t = 3$ and no matter what its instantaneous velocities were during those times, its net displacement during this two-second time interval was 18 space units in the negative x-direction, giving the average velocity just obtained.

The instantaneous acceleration, $\mathbf{a}(t)$, is the derivative of the instantaneous velocity with respect to time:

$$\mathbf{a}(t) = \frac{d\mathbf{v}(t)}{dt}$$

Its x-, y-, and z- components, $(a_x(t), a_y(t), a_z(t))$, are the time derivatives of the respective velocity components:

$$a_x(t) = \frac{dv_x(t)}{dt}$$

$$a_y(t) = \frac{dv_y(t)}{dt}$$

$$a_z(t) = \frac{dv_z(t)}{dt}$$

Accordingly, the instantaneous acceleration is the second time derivative of the position vector:

$$\mathbf{a}(t) = \frac{d^2\mathbf{r}(t)}{dt^2}$$

and its components are the second time derivatives of the respective coordinates:

$$a_x = \frac{d^2x(t)}{dt^2}$$

$$a_y = \frac{d^2y(t)}{dt^2}$$

$$a_z = \frac{d^2z(t)}{dt^2}$$

Continuing the earlier example of motion only in the x-direction, the time derivative of the instantaneous velocity gives $a_x = -6$ acceleration units for all times. In other words, for this example, the acceleration is a constant 6 units of acceleration in the negative x-direction.

The average acceleration for the interval between times t_1 and t_2 is defined as

$$\mathbf{a}_{av} = \frac{\mathbf{v}(t_2) - \mathbf{v}(t_1)}{t_2 - t_1}$$

or also

$$\mathbf{a}_{av} = \frac{\Delta\mathbf{v}}{\Delta t}$$

where $\Delta\mathbf{v}$ denotes the change of the body's velocity during time interval Δt.

The SI units for the quantities are time in seconds (s), coordinates in meters (m), velocity and speed in meters per second (m/s), and acceleration in meters per second per second (m/s^2). Higher-order derivatives can be useful in certain circumstances.

SIMPLE TRANSLATIONAL MOTIONS

Although translational motion can, in general, involve an arbitrary dependence of the body's position on time, there are some particularly simple motions that both have importance and are easy to treat mathematically. The simplest is rest, the situation of constant, time-independent position,

$$\mathbf{r}(t) = \mathbf{r}_0$$

where $\mathbf{r}_0$ denotes the body's constant position vector, whose components are (x_0, y_0, z_0). In terms of coordinates:

$$x(t) = x_0$$
$$y(t) = y_0$$
$$z(t) = z_0$$

This is motion at constant zero velocity

$$\mathbf{v}(t) = 0$$

Example: A body whose coordinates are (4, –2, 7) and are unchanging is at rest.

The next simplest motion is motion at constant nonzero velocity:

$$\mathbf{v}(t) = \mathbf{v}_0$$

where $\mathbf{v}_0 \neq 0$ and represents the body's constant velocity, whose components are (v_{x0}, v_{y0}, v_{z0}). This is straight-line motion at constant speed, motion with constant zero acceleration:

$$\mathbf{a}(t) = 0$$

The position as a function of time is then given by

$$\mathbf{r}(t) = \mathbf{r}_0 + \mathbf{v}_0 t$$

where $\mathbf{r}_0$ now denotes the position at time $t = 0$. In terms of components and coordinates,

$$v_x(t) = v_{x0}$$
$$v_y(t) = v_{y0}$$
$$v_z(t) = v_{z0}$$

and

$$x(t) = x_0 + v_{x0}t$$
$$y(t) = y_0 + v_{y0}t$$
$$z(t) = z_0 + v_{z0}t$$

For example, take the position at time $t = 0$ to be $\mathbf{r}_0 = (4, -2, 7)$ and let the constant velocity be $\mathbf{v}_0 = (0, 6, -3)$. Then the coordinates as functions of time are

$$x(t) = 4$$
$$y(t) = -2 + 6t$$
$$z(t) = 7 - 3t$$

The graph of any coordinate against time has the form of a straight line whose slope equals the respective component of the velocity. Note that the state of rest is a special case of constant-velocity motion, when the velocity is constantly zero. Constant-velocity motion is, according to Newton's first law of motion (discussed later), inertial motion (i.e., motion in the absence of forces).

The next simplest motion after constant-velocity motion is motion at constant nonzero acceleration

$$\mathbf{a}(t) = \mathbf{a}_0$$

where $\mathbf{a}_0 \neq 0$ and is the constant acceleration, whose components are (a_{x0}, a_{y0}, a_{z0}). It follows that the time dependence of the velocity is

$$\mathbf{v}(t) = \mathbf{v}_0 + \mathbf{a}_0 t$$

where $\mathbf{v}_0$ denotes the instantaneous velocity at time $t = 0$. The position is given as a function of time by

$$\mathbf{r}(t) = \mathbf{r}_0 + \mathbf{v}_0 t + \mathbf{a}_0 t^2/2$$

In terms of components and coordinates, these relations take the form

$$a_x(t) = a_{x0}$$
$$a_y(t) = a_{y0}$$
$$a_z(t) = a_{z0}$$

for the acceleration,

$$v_x(t) = v_{x0} + a_{x0}t$$
$$v_y(t) = v_{y0} + a_{y0}t$$
$$v_z(t) = v_{z0} + a_{z0}t$$

for the velocity, and for the coordinates

$$x(t) = x_0 + v_{x0}t + a_{x0}t^2/2$$
$$y(t) = y_0 + v_{y0}t + a_{y0}t^2/2$$
$$z(t) = z_0 + v_{z0}t + a_{z0}t^2/2$$

As an example, take the position and instantaneous velocity at time $t = 0$ to be $\mathbf{r}_0 = (4, -2, 7)$ and $\mathbf{v}_0 = (0, 6, -3)$, respectively, and let the constant acceleration be $\mathbf{a}_0 = (6, 0, -10)$. Then the coordinates as functions of time are

$$x(t) = 4 + 3t^2$$
$$y(t) = -2 + 6t$$
$$z(t) = 7 - 3t - 5t^2$$

The graph of any velocity component against time shows a straight line whose slope is the respective component of the acceleration, while the graph of any coordinate against time has the form of a parabola. Motion at constant velocity is a special case of constant-acceleration motion, that with constant zero acceleration. Motion at constant acceleration is, according to Newton's second law of motion (see later discussion), the result of a constant force acting on a body. Such is the situation, say, near the surface of Earth, where the acceleration of free fall is, indeed, constant and has the same value for all bodies falling in vacuum at the same location, about 9.8 meters per second per second (m/s^2) toward the center of Earth.

DYNAMICS

In mechanics, dynamics is the study of the causes of motion and of its change, that is, the study of the effect of forces on motion. Classical mechanical dynamics is based on Newton's laws of motion, especially on his second law of motion. This law relates an applied force, as a cause, to a change in momentum, or to an acceleration, as an effect.

NEWTON'S LAWS OF MOTION

Named for their discoverer, the British physicist and mathematician Sir Isaac Newton, Newton's laws of motion comprise three laws. These are found to govern classical mechanics (the mechanics of the macroscopic world) as long as the speeds involved are not too great (compared to the speed of light), the densities not too high, and the distances between bodies not too small, but less than cosmological distances.

Newton's First Law of Motion

In the absence of forces acting on it, a body will remain at rest or in motion at constant velocity (i.e., at constant speed in a straight line).

This law is also referred to as the law of inertia, since it describes a body's inertial motion (i.e., its motion in the absence of forces). The law states that the inertial motion of a body is motion with no acceleration. The effect of a force, then, is to bring about noninertial motion, which is accelerated motion. As an example, an object in outer space, far from any astronomical body, has negligible forces acting on it, so it moves at constant speed in a straight line. Near Earth, however, because of gravitational attraction, a body's motion is different. As evidence, toss a pillow or eraser across the room.

Another view of this law is as a definition of what are known as inertial reference frames, which are reference frames in which the second law of motion is valid. One assumes that the presence of forces acting on a body is readily identifiable independently of their effect on the body. This might be done by noting the body's contact with another body or the presence of nearby bodies affecting the body gravitationally or electromagnetically. Then, when forces are known to be absent and a body's motion is unaccelerated, the reference frame in use is an inertial one. But if in the absence of forces a body's motion is accelerated, it is being observed in a noninertial reference frame. Such accelerations are sometimes "explained" by fictitious forces.

For instance, when a bus or car makes a sharp turn, the passengers appear to be pushed toward the outside of the turn, while no physical outward-pushing force is actually operating. But such apparent acceleration (with respect to the vehicle) is sometimes "explained" by a fictitious outward-pushing force, given the name *centrifugal force*. In an inertial reference frame, the passengers are tending to move in a straight line, while it is the vehicle that has a sideways acceleration.

Newton's Second Law of Motion

The effect of a force on a body is to cause it to accelerate, where the direction of the acceleration is the same as the direction of the force and the magnitude of the acceleration is proportional to the magnitude of the force and inversely proportional to the body's mass.

In a formula

$$\mathbf{F} = m\mathbf{a}$$

where $\mathbf{F}$ denotes the force, a vector, in newtons (N); m the body's mass in kilograms (kg); and $\mathbf{a}$ the body's resulting acceleration, also a vector, in meters per second per second (m/s^2).

For example, a force in the upward vertical direction produces upward vertical accelerations only. If the magnitude of the force is 100 N, its direction upward, and it is acting on a 50-kg body, then Newton's second law says that the body's acceleration is 2 m/s^2 in the upward direction.

An alternative formulation of the second law of motion, one that is more generally valid, is this. The effect of a force on a body is a change in the body's linear momentum, where the direction of the change is the same as the direction of the force and the magnitude of the rate of change over time is proportional to the magnitude of the force. Expressed in a formula

$$\mathbf{F} = \frac{d\mathbf{p}}{dt}$$

where $\mathbf{p}$ denotes the body's momentum, a vector, in kilogram-meters per second (kg·m/s) and t the time in seconds (s). The momentum is

$$\mathbf{p} = m\mathbf{v}$$

with $\mathbf{v}$ the body's velocity, also a vector, in meters per second (m/s).

The second law of motion is valid only in inertial reference frames, those in which the first law of motion holds. In noninertial reference frames, such as in an accelerating automobile or in a bus rounding a curve, bodies appear to undergo accelerations with no real forces acting on them. For instance, in an accelerating car or airplane, passengers appear to be thrown backward, although there is no force pushing them backward. Passengers in a turning vehicle seem to be thrown outward, away from the direction the vehicle is turning, again with no real force pushing them outward, as in the earlier example. In all such cases, the passengers are simply tending to obey Newton's first law of motion, to continue moving at constant speed in a straight line, while the vehicle is changing its speed or direction of motion. Such apparent accelerations are sometimes attributed to fictitious forces. In a rotating reference frame, such as in a carousel or turning car (see earlier example), the fictitious force, the apparent outward-pushing force, is called centrifugal force.

Newton's Third Law of Motion

When a body exerts a force on another body, the second body exerts on the first a force of equal magnitude and of opposite direction.

This law is also referred to as the law of action and reaction. What it is telling us is that forces occur in pairs, where one is the action force and the other the reaction force. Note that the forces of an action-reaction pair act on different bodies. Either of such a pair can be considered the action force, although one usually thinks of the force we have more control over as the action, with the other serving as its reaction. If you push on a wall, for instance, the wall pushes back on you with equal magnitude and in opposite direction. We usually think of your push as the action and the wall's push as its reaction. But the reaction force is no less real than the action force. If you are standing on a skateboard, for example, and push on a wall, you will be set into motion away from the wall by the wall's force on you, according to the second law of motion.

While the first and third laws of motion appear to be generally valid, the second law is correct only for speeds that are small compared to the speed of light. The second law, in momentum form, was generalized for any speed by Albert Einstein's special theory of relativity. Accordingly, Newtonian mechanics is an approximation to Einsteinian mechanics that is valid for sufficiently low speeds.

HISTORY

In the fourth century B.C.E. the Greek philosopher Aristotle thought that an object would only move in the direction of a straight line. He believed, for instance, that a cannon ball would move in a straight line until it fell straight down to Earth. Aristotle also believed that objects fell at a rate that was proportional to their masses, meaning that a heavy object would fall more quickly than a lighter object. As was common at the time, Aristotle had a geocentric view of Earth and the planets. This model presents Earth at the center of the universe, with all

the planets revolving around it. In the late 1500s, the Italian scientist Galileo Galilei, along with the Polish astronomer and clergyman Nicolas Copernicus and the German astronomer and mathematician Johannes Kepler, developed the heliocentric model of planetary motion accepted today, which places the Sun at the center of the solar system.

Galileo also contributed significantly to the field of physics and is considered one of its founders for his experiments on the laws of motion. Galileo developed the concept of friction, which he described as the force that resists motion. He reportedly dropped objects off the Leaning Tower of Pisa in order to investigate the speeds of falling objects. He discovered that objects of the same size and shape but of different masses fell at exactly the same rate. The mass did not determine how fast an object fell, disproving Aristotle's statement that heavier objects fall faster than lighter objects. Galileo did observe that if the objects had distinctly different shapes, then one would fall faster than another. The more compact the shape of the object, the more quickly the object would fall. Galileo discovered that the objects with compact shapes experienced less air resistance than objects that had their mass spread over a large area. Therefore, since friction is a force that slows objects, the more air resistance an object experiences, the more slowly it will fall. (This is easily seen by dropping two similar sheets of paper, one intact, for which the air resistance is larger, and the other wadded tightly into a ball, giving it noticeably less air resistance.)

Newton formalized the rules of moving objects in his magnum opus, *Principia*, published in 1687. Newton discussed the three laws of motion and the law of universal gravitation. Newton was a revolutionary scientist who not only developed the laws of motion and explained gravitational forces on objects, but also invented the field of mathematics known as calculus.

See also ACCELERATION; CLASSICAL PHYSICS; EINSTEIN, ALBERT; FORCE; GALILEI, GALILEO; GENERAL RELATIVITY; GRAVITY; KEPLER, JOHANNES; MASS; NEWTON, SIR ISAAC; ROTATIONAL MOTION; SPECIAL RELATIVITY; SPEED AND VELOCITY; VECTORS AND SCALARS.

FURTHER READING

Cohen, I. Bernard, and George E. Smith. *The Cambridge Companion to Newton.* New York: Cambridge University Press, 2002.

Newton, Isaac. *The Principia.* Translated by Andrew Motte. Amherst, N.Y.: Prometheus Books, 1995.

Mullis, Kary (1944–) *American* *Biochemist*
While driving down Pacific Coast Highway 1 in 1983, Kary Mullis conceptualized the revolutionary

technique of polymerase chain reaction (PCR), a molecular biology technique that replicates (makes copies of) DNA (deoxyribonucleic acid) with reagents in a test tube rather than using a living organism such as bacteria or yeast. Beginning with a single molecule of DNA, PCR can generate 100 billion copies of that same molecule in a single afternoon. Because the technique requires only a few components and a heat source, whole new venues of study in molecular biology and genetics opened up with the introduction of PCR. In a short time, this influential technique also found its way into practical applications in medicine, forensics, and paleontology. In 1993, Mullis received the Nobel Prize in chemistry for the development of PCR.

Named after his maternal great-grandfather Cary, who was from the North Carolina city of the same name, Kary Banks Mullis was born in the small rural community of Lenoir, North Carolina, at the base of the Blue Ridge Mountains on December 28, 1944, to Cecil Banks Mullis and Bernice Alberta Barker Mullis. For the first five years of his life, Mullis lived near his maternal grandparents' dairy farm, where he and his cousins watched their mothers shell peas, string beans, and peel apples and peaches on the screened back porch and played in spider-infested attics and haylofts. His family moved to Columbia, South Carolina, where he attended public school. After graduating from Dreher High School, Mullis enrolled in the Georgia Institute of Technology (Georgia Tech) taking courses in chemistry, physics, and math—referring to the information he gleaned from these studies as the technical stuff that he would need for later work; he graduated in 1966 with a bachelor of science in chemistry six years after he first enrolled.

Moving from the South to the West Coast, Mullis pursued his doctoral degree in biochemistry at the University of California, Berkeley, in the laboratory of Dr. J. B. Neilands, obtaining a Ph.D. in 1972. After a one-year lecturer position at the University of California, he accepted a postdoctoral fellowship in pediatric cardiology at the University of Kansas Medical School with an emphasis in angiotensin (an oligopeptide in the blood that affects circulation by constricting arteries and veins) and pulmonary vascular physiology. Mullis returned to California in 1973 to begin a second postdoctoral fellowship in pharmaceutical chemistry at the University of California in San Francisco. In 1979, Cetus Corporation, one of the first biotechnology companies, hired Mullis as a DNA chemist responsible for making oligonucleotides, short chains of DNA, for other scientists working at their facilities in Emeryville, California.

In the early 1980s, Cetus wanted to develop methods for detecting gene mutations that could be

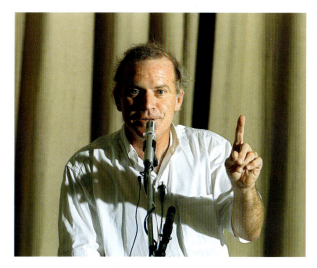

Kary Mullis is an American biochemist credited with the development of the technique known as polymerase chain reaction (PCR). Mullis is shown speaking at a meeting sponsored by HEAL (Health, Education, AIDS Liaison) in Hollywood, California, October 25, 1995. (Vince Bucci/AFP/Getty Images)

useful in genetic screening of common diseases. To develop these protocols, millions of copies of mutated DNA had to be readily available. When Mullis started working at Cetus, the task of amplifying DNA fragments was laborious and time-consuming—biochemists had to insert the DNA into a living organism that reproduced quickly (i.e., bacteria or yeast), extract the DNA from the organism, and analyze the DNA to make sure the organism had not altered the nucleotide sequence in any way before using the fragment for additional studies. Mullis needed to come up with a way to make lots of copies of DNA in a less arduous fashion. While brainstorming about new ways to tweak existing biochemical techniques of amplifying DNA using oligonucleotides and hybridizing them to target DNA sequences, Mullis first conceived the notion of polymerase chain reaction (PCR) while driving to his cabin one night in 1983. With PCR, the amplification procedure occurs in a test tube and requires a small number of components: a DNA template, two primers (short synthetically made oligonucleotides typically 18–25 nucleotides long), a heat-resistant DNA polymerase, deoxynucleotide triphosphates (dNTPs), a buffer solution that stabilizes the pH, divalent cations like magnesium and manganese, and monovalent cations like potassium.

The DNA template serves as a guide for the multiple copies of DNA to be made. Because the procedure occurs in vitro (in a test tube), there are fewer restrictions on the form of DNA used as a template. Extensively modified DNA can be amplified with PCR in a manner that is not feasible with the in vivo methods. The DNA to be amplified can be an entire gene, part of a gene, or a section of a noncoding region of the genomic DNA under study. PCR amplifies shorter DNA fragments (up to 10 kilobases, or 10,000 nucleotides) with fewer mistakes and in greater amounts than in vivo methods. Researchers have successfully amplified DNA fragments up to 25,000 base pairs using modified versions of the original PCR method proposed by Mullis. Fragments of this size are small when compared to the size of most eukaryotic genomes (the total DNA found in each cell), with more than 3 billion base pairs of DNA in a single human cell.

Short DNA fragments called primers mark the beginning and end of the DNA fragment to be amplified by forming complementary base pairs with the sequence at those boundaries. Primer length is an important consideration, as using primers that are too short results in an increased number of nonspecific, random bindings and the generation of unwanted fragments. Use of primers that are too long results in primers that require too much time to melt (dissociate from the template DNA), thereby compromising the integrity of the DNA polymerase by keeping it at an elevated temperature for an extended period.

The base composition of the primers also plays a role in determining the effectiveness of a primer. The number of hydrogen bonds between the base pairs determines how strongly those bases are connected to the template DNA. If the primer is attached too tightly, removing it from the template will be difficult in future cycles. Primers with weak interactions are easily removed from the template strand. Base pairs of guanine-cytosine are held together by three hydrogen bonds, while the base pair made of adenine-thymine is held together by two hydrogen bonds, making G-C base pairs stronger than A-T pairs. Primer design strives for 40–60 percent guanine-cytosine base pairs as the goal when choosing a primer.

The factors of primer length and base composition, along with others, contribute to a primer's melting temperature, T_m, at which half of the primer binding sites are not occupied. Primers that are used in the same PCR are designed to have comparable melting temperatures, ideally between 55°C and 65°C (131–149°F). Sometimes biochemists only know the protein sequence of an area of interest, then they must design degenerate primers, a pool of primers with different nucleotide sequences that still encode the same amino acid sequence. This is necessary because several amino acids are encoded by more than one set of triplet nucleotides. Software programs that help design primers that meet all of these various parameters are available.

The other remaining components—the polymerase, the free dNTPs, buffer, and the salts—make the amplification steps possible. In his original PCR reactions, Mullis needed to add more DNA polymerase to the reactions between rounds because the DNA polymerase was not stable at high temperatures needed to denature the DNA. In 1986 *Taq* DNA polymerase was purified at Cetus. *Taq* DNA polymerase comes from a heat-resistant species of bacterium called *Thermus aquaticus* (a bacterium discovered in 1969 thriving in geysers in the Great Fountain region of Yellowstone National Park in the United States). Because the bacteria survive in temperatures that range from 50° to 80°C (122–176°F), *Taq* DNA polymerase became an ideal candidate enzyme to carry out PCR, as it could withstand temperatures high enough to denature the double-stranded DNA to permit multiple replication events without requiring the addition of more enzyme at the beginning of each new cycle. The major drawback of the polymerase is its propensity for replication errors. The enzyme has no proofreading mechanism (a function normally relegated to a $3' \rightarrow 5'$ exonuclease) so accidental mismatches in newly synthesized DNA remain. Specifically, commercially sold *Taq* polymerase has an error rate of one in every 10,000 base pairs. Biochemists do not use *Taq* DNA polymerase in cases that demand error-free amplification; instead, they use enzymes with more faithful replication, including *KOD* DNA polymerase (a recombinant, or synthetic form, of an enzyme derived from *Thermoccus kodakaraensis KOD1*), *Vent* polymerase (extracted directly from *Thermococcus litoralis*), *Pfu* DNA polymerase (extracted from *Pyrococcus furiosus*), or *Pwo* DNA polymerase (extracted from *Pyrococcus woesii*). All of these polymerases work best when the DNA to be amplified is less than 3,000 base pairs in length. When the DNA fragment exceeds this length, the polymerase loses its processivity (going along replicating the DNA until the end) and starts falling off the fragment. Biochemists have created a new type of DNA polymerase that has a DNA binding protein fused to it in the hope of enhancing the processivity by making the polymerase stick to the DNA longer.

Mullis originally performed the PCR procedure by placing the reactions in a series of heated water baths, but now most biotechnology labs have instruments called thermocyclers, machines that heat and cool reactions within them to the precise temperature required for each step of the reaction. To minimize condensation inside the test tubes, these thermocyclers often have heated lids; to minimize evaporation, a layer of oil is sometimes placed on top of each reaction since the reaction volumes are typically only 15–100 microliters (10^{-6} liters) at the outset. The PCR process usually proceeds for 20 to 35 cycles consisting of three steps each: a denaturation step, an annealing step, and an elongation step. In the denaturation step, the thermocycler heats the double-stranded DNA to 94–96°C (201–205°F), breaking the hydrogen bonds that tie the two strands together. This step might be five to 10 minutes in length during the first cycle, then one to five minutes in the subsequent cycles. In the annealing step, primers attach to single-stranded DNA signaling the start and stop sites to the DNA polymerase. In this second step, the annealing temperature depends on the primers being used, typically about 5°C (9°F) below the lowest melting temperature of the two primers. When the annealing temperature is too low or too high, either the primers bind at random DNA sequences or do not bind to the template DNA at all, respectively. The annealing cycles usually range from 30 seconds to two minutes in length. The final step in PCR is the elongation step; in this step, the heat-resistant DNA polymerase makes copies of the DNA fragment flanked by the annealed primers. The DNA polymerase starts at the annealed primer then works its way along the DNA strand until it reaches the end of the template, which is determined by the position of the second primer. The elongation temperature depends on the DNA polymerase being used. For example, *Taq* DNA polymerase elongates optimally at 72°C (162°F). The time for this step depends both on the DNA polymerase and on the length of DNA being amplified. A general rule of thumb is that the elongation step should be at least one minute per every 1,000 base pairs being amplified. Sometimes the final elongation step is extended to 10 to 15 minutes in length to ensure that all single-stranded DNA fragments are completely copied. The entire process is very sensitive as it amplifies fragments exponentially; therefore, one must take measures to prevent DNA contamination from other sources (such as bacteria, yeast, or laboratory workers). For every set of PCR reactions, a control reaction (an inner control) with no template should be performed to confirm the absence of contamination or primer problems.

The importance of polymerase chain reaction cannot be overemphasized, as this new technology has revolutionized the fields of molecular biology, genetics, medicine, and forensics. As Mullis and his new method of amplifying DNA became known, controversies developed that centered around the originality of the idea that Mullis claimed as his own and the subsequent rights to its licensing for others to use. Some scientists argue that Mullis borrowed the idea of PCR from Kjell Kleppe, who first proposed using the intrinsic repair mechanism of DNA polymerases to reproduce multiple copies of the same fragment of DNA with primers as guides in 1971. The method described by Kleppe only quadrupled the DNA fragment of interest, versus the 100-fold

and greater amplification that Mullis achieved, and Kleppe did not envision the use of a heat-resistant DNA polymerase to permit more than 20 cycles of amplification before needing additional enzyme. The second controversy involved the role that Cetus played in the development of PCR and how the company's part in its invention would be recognized. Many at the company questioned why Mullis received the Nobel Prize in 1993 rather than a team of peers including David Gelfand and Randy Saiki, who also worked on the development of PCR. Finally, a dispute arose concerning who had a right to the $300 million that Cetus received when its PCR patent was sold to Roche Molecular Systems, another biochemical firm.

While controversies stirred around him, Kary Mullis reveled in all of the attention as he unabashedly went about his professional and personal life. Mullis left Cetus Corporation in 1986 to become the director of molecular biology at Xytronyx Incorporated, a biotechnology firm that permitted Mullis to concentrate on DNA technology and photochemistry. He has served as a consultant in nucleic acid chemistry for various companies, including Angenics, Cytometrics, Eastman Kodak, and Abbott Laboratories. In 1991, he received the Research and Design Scientist of the Year award; two years later, he received the Thomas A. Edison Award for his role in the invention of PCR. He also received a prestigious international award, the Japan Prize, in 1993, as many of his peers proclaimed PCR to be the technique of the century. Most recently, Mullis formed a new company, Altermune LLC, to develop a means to induce a rapid immune response. The process occurs by cross-linking known antigens to pathogens or toxins that the body has never seen, thus diverting existing antibodies and in turn evoking a faster immune response against those pathogens or toxins. Mullis served as distinguished researcher at Children's Hospital and Research Institute in Oakland, California; in addition, he sits on the board of scientific advisers of several companies providing input regarding DNA technologies. Combining his love of travel and talking about science, Mullis frequently lectures on college campuses, corporations, and scientific meetings. Sometimes scorned by others in the scientific community for his brazen, often erratic behavior, he became known to a wider public during the infamous O. J. Simpson trial as the defense questioned the reliability of the state's DNA fingerprinting evidence. Mullis has even tried his hand at writing. His autobiography *Dancing Naked in the Mind Field* was published in 1998 and gave a humorous look at the scientific process. Married four times, Mullis has one daughter and two sons. He lives with his fourth wife, Nancy Cosgrove Mullis, and they split their time between homes in Newport Beach, California, and Anderson Valley, California.

See also BIOCHEMISTRY; DNA FINGERPRINTING; DNA REPLICATION AND REPAIR; NUCLEIC ACIDS.

FURTHER READING

Mullis, Kary B. *Dancing Naked in the Mind Field.* New York: Pantheon Books, 1998.

———. "The Unusual Origin of PCR." *Scientific American* 262 (1990): 56–65.

The Nobel Foundation. "The Nobel Prize in Chemistry 1993." Available online. URL: http://nobelprize.org/nobel_prizes/chemistry/laureates/1993/. Accessed July 25, 2008.

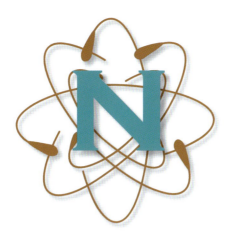

nanotechnology Nanotechnology involves the creation and control of materials that are of an extremely small size, often billionths of a meter in size. Many of the advances in computer storage and processing are the direct results of applied nanotechnology. Computer mainframes that once filled entire rooms now fit in exponentially smaller spaces. The creation of targeted medicines to treat such diseases as cancer and the development of robotics to perform a variety of functions are also examples of applications of nanotechnology.

Nanotechnology was simply a concept of science fiction for most of the 20th century. In 1959, the Nobel laureate Richard Feynman predicted the future of nanotechnology. He gave a lecture at the American Physical Society meeting at the California Institute of Technology in which he foreshadowed the problems and potential of creating things on an extremely small scale. During his often-quoted speech, Dr. Feynman laid the groundwork for this new field of research and discussed how "you could decrease the size of things in a practical way," the essential focus of nanotechnology.

A PROBLEM OF SCALE

The scale of much of the work in nanotechnology is awkward and is one of the main problems in this emerging field. Devices in nanotechnology are obviously smaller than those measured on a macroscopic scale; however, they are distinguishably larger than the scale of individual atoms. The mesoscale, the scale employed for nanotechnology, is at a level of aggregates of atoms or billionths of meters in dimension, intermediate to the atomic and macroscopic scales. Nanotechnology can be thought of as the science of the mesoscale; many of the devices do not

behave in a fashion characteristic of those produced according to the atomic or macroscopic scales. Many problems are inherent to a scale of this size, and they are inhibiting progress in the field.

Production of nanotechnologic devices highlights the problem of scale. Two methods of production are a top-down production method and a bottom-up method. The top-down method essentially uses larger devices and strips them of components to produce a smaller device. It uses properties that are well defined macroscopically, but removing some of the components might also remove the properties of that device that are desirable. The bottom-up method builds from individual atoms to create a device measured on the mesoscale. Again, properties present at a smaller scale may or may not be replicated as more elemental components are introduced. These design methods are borrowed from manufacturing strategies. Both manufacturing and nanotechnology involve the end goal of creating a device or product. Nanotechnology has additional problems that manufacturing does not in that the characteristics of the device may be drastically affected by the buildup or removal of components. A ripple effect may completely change the stability and operating characteristics of the device as changes occur in its production.

ELECTRICAL CONDUCTANCE AND HEAT TRANSFER

Although many of the properties of devices produced on the mesoscale are unknown, some founding principles have been developed and more are currently being discovered. One principle is that the nanotechnologic device must be chemically stable. Instability will prevent the device from performing in any predictable way as its components will not be able to integrate into a functioning whole. Once chemical stability has been established, two quantifiable characteristics must be determined, namely, electrical conductance and heat flow. *Electrical conductance* is the ability of electric energy to flow from input device to output drain of the device. *Heat flow* similarly refers to the ability of heat to pass from one component to another.

Electrical conductance is important in regard to nanotechnology because components that are linked together can have considerable resistance. Electrical energy must flow from an input source to an output drain for the device to function. If too much resistance is encountered, then the electrical energy will dissipate and the components of the device will be deprived of the electrical energy needed to function. Feynman anticipated this problem in 1959, stating, "I have thought about some of the problems of building electric currents on a small scale, and the problem of resistance is serious." In 1987, Hank van Houten at Philips Research Laboratories, Bert J. van Wees at Delft University, and other collaborators made an important discovery regarding the conductance of mesoscale devices. They determined that the amount of electricity that flows through the components of the device is predictable. This quantification of electrical conductance allows resistance to be predicted and can be used to ensure that energy sufficiently passes from input to output without excessive dissipation. In short, the devices can be created to behave according to electrical principles that are governed by predictable quantum mechanics.

In 1999, the American physicist Keith Schwab determined that heat flow could similarly be quantified much as electronic conductance can. Mesoscopic components that can transfer electrical energy also can have the propensity to overheat. When components overheat, their ability to function may diminish, potentially diminishing the ability of the device to function properly. Schwab formulated an equation for determining the maximal rate by which heat can be transported from the input to the output of the device. It is very similar to the governing principles for the transportation of electrical energy from input to output. The mesoscopic components can be arranged similarly, with heat dissipation, a major problem with microprocessors, in mind in order to make a better-functioning device.

ADVANCES AND APPLICATIONS

Some of the more recent advances in nanotechnology have related to the creation of very precise weighing apparatuses that can measure the "weight" of individual atoms. Transistors are being designed that, according to projections by the International Technology Roadmap for Semiconductors, will measure only 20 nanometers by the year 2014. Another development in nanotechnology is the creation of extremely sensitive devices for measuring vibrations such as sound waves. Many of these newer developments reflect the problems encountered by nanotechnology in general. Devices used to measure vibrations have a large surface area to capture the vibrations but in contrast have a relatively small volume. An example of this type of device are the sensors that measure the eye movement of quadriplegics and allow them to move their own wheelchairs or use computers. The physical properties of structures with both large surface area and volume are well defined by classical physics; however the behavior of high-surface area and low-volume structures is relatively unknown. Many of the problems with nanotechnology are captured in microcosm in the inherent duality of making a device sensitive to vibration or weight while maintaining insensitivity to background vibrations and interferences to weight. For example, the behavior of gases can drastically interfere with the accuracy of the results when measuring the weight of a single atom.

In 2007, Albert Fert and Peter Grunberg were jointly awarded the Nobel Prize in physics for their usage of nanotechnology in reading information from hard disks. Hard disks have been increasingly able to store larger amounts of information in smaller spaces. Fert and Grunberg independently discovered a new physical effect called giant magnetoresistance (GMR), in which small magnetic changes result in very large changes in electrical resistance. Hard disks store information using small areas magnetized in various directions. Read-out heads use the GMR effect discovered by Fert and Grunberg to decipher where and what information is stored in the hard disk space. As smaller and smaller areas record and store more and more information, the read-out heads used to decipher the changes in magnetism must be extremely sensitive. Research in nanotechnology has been applied to create these increasingly sensitive read-out heads that are capable of detecting extremely small magnetic changes.

The world of science fiction has focused on the future possibilities of nanotechnology as opposed to the difficulties encountered in the reality of this relatively uncharted discipline. For example, in Michael Crichton's novel *Prey*, nanotechnology was used to create artificial, self-replicating killer devices that were used to asphyxiate victims. More recently science-fiction novels have focused on the usage of nanotechnology in creating the illusion of invisibility, called cloaking, through diffusion of energy by nanotechnologic devices. Other topics in science fiction follow a much closer-to-fact premise.

Future research in nanotechnology is extremely diverse, and the difference between science fiction and future reality is becoming increasingly blurred. Scientists have been studying the behavior of smaller colonies of animals, such as ants, and replicating their behavior in a bottom-up building technique for creating smaller, similarly behaving devices. Some of the research in this area is leading toward the development of small devices that could ingest pollution or excess oil from spills. Robotics and artificial intelligence also employ advances in nanotechnology to create smarter and more lifelike robots that process information more quickly. Another future application of nanotechnology is in the targeting of specific cells for the delivery of medicine. Cancer tumors or individual cells can be specifically targeted instead of using radiation or chemotherapy.

See also NUCLEAR MAGNETIC RESONANCE (NMR); X-RAY CRYSTALLOGRAPHY.

FURTHER READING

Rogers, Ben, Sumita Pennathur, and Jesse Adams. *Nanotechnology: Understanding Small Systems.* Boca Raton, Fla.: CRC Press, 2007.

Scientific American, eds. *Understanding Nanotechnology.* Foreword by Michael L. Roukes. New York: Warner Books, 2002.

Newton, Sir Isaac (1642–1727) English *Physicist and Mathematician*   Natural laws allow for the prediction of the effects of forces such as gravity and processes such as motion. While humans have been able to predict such consequences and even use natural laws to their advantage for thousands of years, Sir Isaac Newton was the first to formulate such laws precisely and use mathematics to prove them rigorously. Considered one of the greatest scientists of all time, he was able to make a connection between the fall of an apple and the operation of the universe. Scientists and engineers, as well as athletes and artists, still use principles that he delineated in his two major works, *Principia* and *Opticks*. Newton also invented the branch of mathematics called calculus and made significant contributions to many other areas of mathematics. In addition, he designed the first reflecting telescope, which has allowed the human race to explore farther into space than ever before.

CHILDHOOD YEARS

Sir Isaac Newton was born prematurely on December 25, 1642, at Woolsthorpe Manor, in Lincolnshire, England. His father, Isaac, was a farmer and had died three months earlier. Three years later, his mother, Hannah, was remarried to the Reverend Barnabas Smith, who was much older than she and forced her to move to a nearby village, leaving young Isaac with his grandmother. He seems to have had a lonely childhood, perhaps by his own choosing, as he has been described as a serious and somber lad. He began his education at the local school, where he learned to read and write. His free time was spent making models of sundials, windmills, kites, and other such mechanical devices. When he was 10 years old, his stepfather died, and Hannah returned to Woolsthorpe Manor to live with Isaac's three younger half siblings, Mary, Benjamin, and Hannah.

At age 12, Isaac moved seven miles (11 km) away to Grantham and boarded with the Clark family so he could attend the King's Grammar School. Mr. Clark was a pharmacist, and Isaac learned the basics of chemistry while working in his shop. Isaac performed satisfactorily at school but gave no early sign of his hidden genius. His study habits were rumored to have improved after a fight with the school's bully, named Arthur Storer, who made very high marks. Isaac was very competitive, and though he won the physical fight, he was determined to beat out Arthur in the intellectual realm as well. (Interestingly, Isaac and Arthur corresponded later in life about astronomy.) His schoolmaster began to

Sir Isaac Newton, who lived in the 17th and 18th centuries, was one of the most famous scientists of all time and is best known for his laws of motion and gravitation and the invention of calculus. *(Art Resource, NY)*

recognize that Isaac had abilities that could take him beyond his father's farm.

When he was age 16, Isaac's mother pulled him out of school to run the farm left by his father, but Isaac had no aptitude for farming. Instead of going to market or watching the animals, he would hide out and read or lie under a tree and stare into space. The family servants thought he was lazy and dumb, but Isaac's uncle and the schoolmaster saw potential in the young man and persuaded Hannah to send him back to school to prepare for college. He reentered school and one year later was admitted to Trinity College at Cambridge University.

CAMBRIDGE UNIVERSITY

During his first few years of college, Isaac was a sizar, a servant to older students and to faculty in return for reduced fees. The position, in addition to the extra time Isaac had to spend working to earn his stay at Trinity, set him apart from the other students, and he had few friends in college. He eventually found a roommate with whom he shared a similar lifestyle and spent the next four years buried in his books.

One year Isaac went to a nearby fair, where he proceeded to purchase a triangular glass prism that was meant to be a child's toy. He also purchased some books for pleasure reading, including a text on astronomy. To help him understand this, he also purchased a text on Euclidean geometry. After beginning to read it, however, he thought it was silly and bought a more modern text of Descartes's geometry (René Descartes was a famous 17th-century French philosopher). This was much more challenging to comprehend, yet he trudged through it on his own and in the process neglected his regular studies, failing to impress his teachers. He reportedly even failed an exam on Euclidean geometry. During this time, he developed the binomial theorem, a shortcut for solving problems that involve multiplying a binomial (an expression with two variables) by itself many times over, $(a + b)^n$. Of this, the faculty took notice. He began to develop a relationship with Dr. Isaac Barrow, a professor of mathematics. Isaac Newton received his bachelor's degree at age 22, in 1665.

THE MIRACLE YEAR

Shortly afterward, the bubonic plague took over London. This horrific infectious disease killed about 17,440 people out of London's total population of 93,000. The university was forced to close, and Newton retreated to Woolsthorpe Manor. The next 18 months turned out to be the most intellectually productive of his life. This period is often referred to by scholars as *annus mirabilis*, or the "miracle year."

During this time, Newton continued playing with prisms. He explored the relationship between light and color. Prisms are intriguing toys because when light hits them, they create a rainbow of colors. Scientists thought this might occur because the light that entered them was dirtied or darkened as it passed through the glass itself. Newton wondered about this and the nature of light, which was a popular topic at the time. To explore light, he darkened his room, letting only a small beam of light through the window. This light was aimed at the prism, which had a black screen behind it. The light spread out onto the screen in the expected rainbow of colors. Newton wondered why the shape of the light emitted onto the screen was oblong, whereas the light hitting it was just a small circle. He then set a second prism upside down between the first prism and the screen, not knowing what to expect. Surprisingly, the rainbow disappeared, and white light was restored. He also isolated single colors of light from the spectrum by using a board with a tiny hole in it and sent those colored beams of light through a second prism onto the screen. When individual colors were sent through the second prism, the same single color shone on the screen. He concluded that each individual color was a component part of white light—white light was due to the presence of all the colors. This was contrary to the popular belief that white light was due to the absence of any color. What was happening was that the individual colors were refracted, or bent, by the prism to different degrees; the red was bent the least, and the blue light was bent the most. Thus a circle of white light entered the prism and exited as an elongated spectrum of the individual colors.

Newton was also intrigued by the principles and mathematics of motion. There was no method for solving certain equations concerning motion because variables changed as rates of speed changed. Thus, he developed the foundation of a new type of mathematics he called the method of fluxions. The modern name of this branch of mathematics is *calculus*, and it is concerned with solving problems that have ever-changing variables. He used these new methods in many of his mathematical proofs.

The German astronomer Johannes Kepler had proposed that planets orbited elliptically around the Sun, that their speed changed along their orbits, and that the length of time it took to complete one orbit was related to the distance the planet was located from the Sun. Newton was learned in Kepler's laws concerning planetary motion, but he wondered what kept the planets and the Moon in their orbits. He spent a lot of time pondering this question, and legend has it that one day as he sat under an apple tree, with his mind primed to discover the law of universal gravitation, an apple fell to the ground. (Another version of legend has the apple falling on Newton's head.) Newton contemplated why the apple fell downward. Was Earth pulling the apple toward itself?

He proposed the existence of an attractive force that acted between all pieces of matter. This force of gravity depended on the masses of the bodies involved (for example, of the apple and of the Earth) and on the distances between them. Perhaps, he mused, the same force attracts the Moon, but then why does the Moon not fall to Earth as the apple does? The Moon is kept balanced by its attraction to Earth and its inertia. It is being pulled downward while it is moving sideways, a process that bends its path and keeps it in orbit; the Moon neither falls to Earth nor flies away from it.

Newton started working out mathematical calculations of Earth's pull on the Moon as well as the elliptical orbits of the planets around the Sun, but his estimate for the radius of Earth's orbit was incorrect. In addition, he was not sure whether to use Earth's surface or center as the point of gravity's pull. Thus, he was unable to finish this proof completely. Frustrated, he set it aside, but he never forgot about it. As time would have it, 18 months after his productive holiday began, Cambridge University reopened after the Great Fire of London, which ended the plague by killing off all the rats that were spreading it.

Newton returned to London and, at age 25, began working on a master's degree, which he completed in 1668. The lifestyle of a research fellow suited him. He had plenty of time to study independently and could live at the university. Two years later, Professor Barrow resigned, and Newton replaced him as a professor of mathematics at Trinity College.

Newton continued his studies of optics. One problem with the common reflecting telescopes of the time was chromatic aberration. This means that the observed images were surrounded by rings of color that blurred the image. He devised a telescope that replaced a second convex lens with an angled mirror to focus the light on the eyepiece to reduce the effects of chromatic aberration. In addition, he put the eyepiece on the side of the telescope rather than the end, making it more convenient to use. He showed this reflecting telescope to Professor Barrow, who took it to London in 1671 to show it off. He took one to the premier academic organization, the Royal Society of London, whose members were so impressed that the next year they elected Newton a member. Modern astronomers still use reflecting telescopes to gather astronomical data.

Excited finally to have an audience for his scientific musings, Newton presented his experiments on light that led to his invention of the reflecting telescope to the Royal Society. At the time, Robert Hooke was the curator of experiments for the society. Hooke was an eminent English scientist who contributed to a broad range of developing fields, including physics, astronomy, and microbiology. He had previously performed similar experiments with light, though less extensive and not as explanatory as Newton's. Hooke took offense that some of the comments Newton made in his paper disagreed with some of Hooke's own beliefs, and a lifelong bitterness between Newton and Hooke ensued. Newton was furious at having to defend the validity of his conclusions and vowed never again to publish his findings. The two men bickered for decades. Their disdain for one another was fueled further when Newton published his theory of universal gravitation in 1686, and Hooke claimed priority credit for the inverse square law, the fact that the attraction between the Sun and Earth varied inversely as the square of the distance between them. The truth is Hooke's theory was incomplete, and Hooke was unable to prove it mathematically whereas Newton did.

ELLIPTICAL ORBITS

Newton continued living his solitary and studious lifestyle. As a professor at Cambridge, he was required to give weekly lectures. He was not a popular speaker, however, and his lectures were not well attended. Much of his time was spent privately studying alchemy, a shady business concerned with the alteration of ordinary metals into gold and the preparation of a magic elixir that would extend life. While it was an early form of chemistry, at the time, most regarded alchemy as wizardry. Newton's interest in alchemy was probably based on his interest in the nature of matter and of life. He probably would have been embarrassed if his contemporaries found out about his secret studies. This was not a problem, though, since he did not collaborate much and was a loner.

As a member of the Royal Society, Newton corresponded regularly with Hooke, who now was secretary and in charge of keeping abreast of the members' activities. Through this correspondence, Newton was reminded of his earlier calculations concerning elliptical orbits of the planets. He now had access to the correct value for Earth's radius and was able to complete the solution to this problem successfully. Because Newton recognized the significance of this accomplishment, he was extremely excited. Rumor had it that he had to have an assistant finish writing down the calculations because he was too anxious to do that himself. However, he was still hesitant to share his work with others, since he had been publicly harassed by Hooke years before. Thus he shoved his amazing proof in a drawer.

LAWS: THERE IS ORDER IN THE WORLD

by Joe Rosen, Ph.D.

In science generally, and in physics particularly, a law is an expression of a pattern or some order that exists among natural phenomena. Consider Charles Augustin de Coulomb's law, as an example. This law states that the magnitude of the electrostatic force of attraction or repulsion between two electric point charges is proportional to the magnitude of each charge and inversely proportional to the square of the distance between them. The pattern and order that the law expresses are, first of all, that the magnitude of the force is not random but depends on physical quantities that can be measured and controlled. In this case, it depends on three quantities: the magnitude of each charge and the distance between the charges. Second, the dependence on these quantities is not wild and useless, but tame and serviceable. Because it can be expressed mathematically, one can apply the law and make use of it. In the present example, the dependence involves the mathematical relations of proportionality, inverse proportionality, and second power.

Scientists discover laws by searching for and finding patterns in experimental and observational data. Take Coulomb's law again. In order to discover the law, Coulomb performed many measurements of the force with which two small, charged spheres attract or repel each other. He repeated his experiments using different magnitudes of charges and different separation distances. He might have noticed, for instance, that whenever he doubled the charge on one sphere, the magnitude of the force doubled as well. He noticed that when he decreased the separation to a half of its value, the magnitude of force quadrupled. In such a manner, he arrived at his law for the electrostatic force:

$$F = k\,\frac{|q_1|\,|q_2|}{r^2}$$

Here F denotes the magnitude of the electrostatic force, q_1 and q_2 are the values of the charges, $|q_1|$ and $|q_2|$ their respective absolute values, and r represents the distance between the point charges. The value of the proportionality constant k depends on the units used for the various quantities. This relation summarized in a succinct mathematical expression all the data that Coulomb had collected.

Laws are based on a finite amount of data. The amount might be very large, but it is always limited. The mathematical relation that expresses a law forms a summary of the relevant data, but it offers more than that—it allows the prediction of the results of new experiments or observations. Thus the law can be tested and confirmed, or tested and contradicted. In the example of Coulomb's law, one can take a set of values for q_1, q_2, and r that Coulomb never tried, calculate the resulting force, run an experiment with those values, measure the force, and test the law.

When a law has been confirmed by overwhelming evidence for its validity, that is, by many successful results of predictions, it becomes accepted as correct and can serve as a useful, predictive tool. Such is the case now with Coulomb's law. It is so well confirmed that no physicist feels the need to reconfirm it, and it serves as a standard tool in the toolbox of physics. At least in principle, there always exists the chance, however

---

Years later, in 1684, while Newton remained hidden away in his lab, three other members of the Royal Society, Hooke, Christopher Wren, and Edmond Halley, were discussing the problem of elliptical orbits while drinking coffee together. They had figured out that the force needed to move planets in a circle around the Sun obeyed the inverse square law. This meant that a planet twice as far from the Sun would only require one-fourth the force exerted by the Sun to keep it in orbit. But these three extremely intelligent men could not figure out why they traveled in elliptical orbits. After months without success, Halley set out to visit Newton. He was shocked when he presented the problem to Newton only to hear Newton remark that he had solved that problem 15 years ago. This was an amazing accomplishment, and Halley could hardly believe Newton had kept it to himself. Newton could not locate the calculations immediately, however, and he promised to redo them. Several months later, Newton mailed his proof to Halley. By 1686, Newton had fleshed out the nine-page proof into his most famous work, which described the workings of the universe using mathematics.

PRINCIPIA

The full title of this work is *The Mathematical Principles of Natural Philosophy*, though it is most often referred to by the abbreviation of its Latin name, *Principia*. Because of shortness of funds, the Royal Society almost did not publish it. Halley himself arranged the financial support for the publication of this work. In addition, Hooke again was angered, claiming that Newton stole his ideas. Newton balked, but Halley smoothed matters over, and the world is forever in his debt.

The first volume of *Principia* described what are known today as the three laws of motion. The

tiny, that the law might fail for some as yet untried combination of q_1, q_2, and r. Every law, through its use, is continually being retested and reconfirmed.

Every law possesses a limited known range of validity. What this means is that every law is known to be correct only for certain ranges of values of the physical quantities that enter into it. For Coulomb's law, the known range of validity covers charges that are not too large and distances that are neither too large nor too small. Within its known range of validity, a law serves as a tool. Under conditions that exceed its known range of validity, a law is being tested: its envelope is being stretched, so to speak.

Every known law of physics is expected to fail under sufficiently extreme conditions. For some laws, the actual range of validity has been found. For others, it is surmised. Here are some examples. Because of quantum effects, Coulomb's law is expected to fail for sufficiently large charges and for sufficiently small separations. For large distances, its range of validity is not known. The ideal gas law fails for gas densities that are not sufficiently low. Sir Isaac Newton's law of gravitation fails for strong gravitational fields and for high speeds. Albert Einstein's general theory of relativity fixes that but is expected to fail at very small

distances, since it is incompatible with quantum physics.

Consider another example of the discovery of a law. Galileo Galilei performed many experiments on the behavior of uniform spheres rolling down straight, inclined tracks. He released spheres from rest and measured the distances d they roll from rest position during various elapsed times t. In this way, Galileo collected a lot of experimental data in the form of d-t pairs. He studied the numbers, performed calculations on them, and perhaps plotted them in various ways, all in search of pattern and order. What he found was, for a fixed angle of incline, the distance a rolling sphere covers from rest position is proportional to the square of the elapsed time from its release,

$$d = bt^2$$

The value of the proportionality coefficient b depends on the incline angle and on the units used. It does not depend on the size, mass, or material of the sphere.

This mathematical relation summarizes Galileo's experimental results and forms a law. Although it is based on only a finite number of d-t pairs, the relation is able to predict the distance covered for *any* elapsed time, whether Galileo actually had that d-t pair among his data or not. Thus, it predicts the results of

new experiments. Perhaps Galileo never measured the distance for an elapsed time of exactly 2.5903 seconds, for instance. This relation predicts what that will be. The predictive power of the law allows it to be tested, and it has been tested and confirmed sufficiently to be considered correct. The law is valid as long as the mass of the rolling sphere is not too small nor its speed too high (i.e., as long as the force of gravity acting on the sphere is sufficiently greater than the retarding force on it due to the viscosity of air).

Although physicists are always happy to discover laws, laws themselves do not fulfill the need of physicists to understand nature. Laws can be useful for physics applications. They might serve engineers and chemists for their purposes. But to reach the understanding that physicists are striving for, laws are only a step in the right direction. The next step is to develop theories to explain the laws. That is the subject of another essay.

FURTHER READING

Feynman, Richard. *The Character of Physical Law.* Cambridge, Mass.: M.I.T. Press, 1965.

Hatton, John, and Paul B. Plouffe. *Science and Its Ways of Knowing.* Upper Saddle River, N.J.: Prentice Hall, 1997.

first law summarizes inertia, the tendency originally articulated by the Italian astronomer and mathematician Galileo Galilei—unless acted upon by an outside force, an object in motion remains in motion at a constant speed in a straight line and an object at rest remains at rest. Because air resistance and friction act upon most earthly motions, Newton's first law is not apparent. If you kick a ball, it will not roll forever at constant speed: it will eventually stop. However, in space, where there is no air and no friction, the first law is manifest. The second law of motion states that force is equal to the product of mass and acceleration. This law explains why it is harder to throw a bowling ball than a tennis ball. The third law maintains that when one object exerts a force on another object, the second object exerts an equal but opposite force on the first. (This is often expressed imprecisely in the form: For every action there is an equal and opposite reaction.) For example, when someone pushes off the

ground to jump vertically, the force her feet exert on the ground is equal in magnitude to the force that the ground simultaneously exerts back on her feet.

Using these laws, Newton calculated the gravitational force between Earth and the Moon, proving that it followed the inverse square law: the force was directly proportional to the product of the two masses (the mass of Earth times the mass of the Moon) and inversely proportional to the square of the distance between the centers of Earth and the Moon. Previously, it had been assumed that the universe and Earth followed different sets of natural laws. Amazingly, Newton went on to prove that his predictions concerning gravitational forces could be applied throughout the universe. Thus, Newton was able to explain mathematically Kepler's laws of planetary motion and demonstrate why the planets orbited the Sun in elliptical orbits (rather than in circles, as had earlier been assumed).

In the second volume of *Principia,* Newton disproved Descartes's explanation of planetary motion. Descartes had proposed that the universe was filled with a fluid that was whirlpooling the planets and the stars around the universe by its vortex motion. While many people accepted this explanation, Newton disproved it mathematically.

Before the third and last volume of *Principia* was published, Newton became annoyed and sulky as a result of Hooke's claims that he deserved credit for discovering the inverse square law. Newton almost refused to continue. Halley panicked, worried that the first two volumes would not sell as well without the third volume completed and that society would never receive this wonderful contribution Newton was capable of making. Luckily, he was again able to smooth matters over. The third volume applied Newton's new theories to the Moon, planets, and comets. It contained predictions using the laws of motion and gravity, which he formulated. One prediction he made was that gravity should cause Earth to be a perfect sphere, but the rotation of Earth about its axis should cause a bulge at the equator. Newton predicted the size of this bulge; it has since been proven correct to within 1 percent accuracy. He also predicted that many comets would follow elliptical paths just as planets do, but that they would have more elongated paths. Halley was quite excited by the realization that the motion of comets could be predicted using Newton's laws. Credited for discovering the comet named for him, he was able to use Newton's laws and methods to predict the return of this comet every 76 years. The comet has indeed reappeared every 76 years since.

LIFE OUTSIDE ACADEMICS

Later in his life, Newton became more involved in both university and national politics. In 1689, he was elected to Parliament. This was shortly after the "Glorious Revolution," when King James II fled to France and Prince William of Holland took over. During this year, Newton voted to make William and Mary the king and queen of England; in favor of the Bill of Rights, which gave more power to the citizens of England; and for the Toleration Act, which allowed more religious freedom.

In 1693, Isaac Newton suffered a mental breakdown. He started writing strange letters to his friends and accused them of plotting against him and making odd threats. These episodes might have resulted from overwork, as people who knew him recalled that he often forgot to eat or sleep when he was in the midst of a scientific breakthrough. Another possibility was that his antagonistic personality and the stress from the continual controversy surrounding the intellectual ownership of his work broke him down. A more recent explanation suggests that he was suffering mercury poisoning from his experiments in alchemy.

By 1696, Newton seemed completely recovered. He was offered an administrative position as warden of the Royal Mint, which was in charge of producing the currency. When he took this position, it was supposed to be a reward, mostly in title, but Newton was a hard worker. The nation was having a coin crisis. The coins in circulation were being clipped and the precious metal melted down for other uses. Moreover, the design was easy to duplicate, making counterfeiting too easy. Under Newton's commission, all the old coins were recalled and replaced with a newer coin with a more intricate design and a less precious composition. In 1700, he was named master of the Mint, and in 1701 he resigned his position as professor at Cambridge.

Around this time, he moved out of the warden's house at the Mint and into his own home with his niece, Catherine Barton, who served as his hostess. Newton attended Royal Society meetings only occasionally, as he hated facing Hooke. After Hooke's death in 1703, Newton was elected president of the Royal Society, a position to which he was reelected every year until his death 23 years later. During his tenure, he gained the society financial security and offered encouragement for many new scientists.

With the Mint running smoothly and his archenemy gone forever, he began composing another masterpiece, *Opticks.* This book described his earlier experiments on light and color and formed the basis of the field of spectroscopy, the study of the colors composing light. He also began writing updated revisions of *Principia.* To accomplish this, he corresponded with the astronomer John Flamsteed to obtain current astronomical data. Newton was not generous in his explanation for needing these data; thus, Flamsteed was wary of sharing all this information. In addition, some of Flamsteed's figures were incorrect, irritating Newton, who had to repeat his calculations.

Hooke and Flamsteed were not Newton's only foes. The German mathematician Gottfried Wilhelm von Leibniz argued with Newton over who invented calculus first. Leibniz published his results before Newton, but Newton may have arrived at them first while formulating his theory of universal gravitation. Leibniz was no match for Newton's powerful position as president of the Royal Society and his international reputation, so Newton received credit for this achievement.

Queen Anne knighted Newton in 1705. He was the first scientist to receive such an honor. In January 1725, his health began to deteriorate. He moved out to Kensington (which used to be outside London) in the hope of healing a cough. He died on March 20, 1727, at age 84 and was buried at Westminster Abbey.

Remarkable as he was, Newton was not always right about everything. For example, he thought that light was composed of bunches of corpuscles that were ejected from the light source. By the 1800s, most scientists believed in the wave theory of light, as did Robert Hooke. However, in the 20th century, when the theoretical physicist Albert Einstein described photons as particles of light, many similarities between them and Newton's corpuscles became apparent. Newton also carried out experiments in 1717 that he believed demonstrated the existence of ether. Ether was a mysterious invisible substance believed to fill all space. Its existence was disproved in 1887 by Albert Michelson and Edward Morley.

The verse inscribed at the bottom of a monument erected in his honor reads, "Let mortals rejoice that there has existed so great an Ornament to the Human Race." And the poet Alexander Pope wrote, "God said, Let Newton be! and all was Light." Very few scientists have been so successful at unlocking nature's secrets. Not only was Newton an amazing puzzle solver, but he was also methodical and meticulous in his calculations and record keeping, and he recognized the significance of his numerous discoveries. While not many people are capable of understanding his original works, everyone should appreciate the huge impact he has had on the maturation of science.

His research formed the pinnacle of the scientific revolution. Before Newton, scientists relied heavily on conjecture and assumptions made by Greek philosophers hundreds of years prior. After Newton, scientists started doubting claims; they began to believe only that which they observed. They began to test hypotheses, many times over, before drawing conclusions. Newton was able to explain in mathematical language, very logically and concretely, that which had amazed and awed generations for thousands of years—the workings of the universe. He may not have had many friends among his contemporaries, but he was respected by all, feared by some, and inspirational to the upcoming scientists. And that which he was the first to explain controls them all.

See also ACCELERATION; FORCE; GALILEI, GALILEO; GRAVITY; KEPLER, JOHANNES; MASS; MOTION; OPTICS.

FURTHER READING

Cropper, William H. *Great Physicists: The Life and Times of Leading Physicists from Galileo to Hawking.* New York: Oxford University Press, 2001.

Fauvel, John, Raymond Flood, Michael Shorthand, and Robin Wilson, eds. *Let Newton Be!* New York: Oxford University Press, 1988.

Gleick, James. *Isaac Newton.* New York: Vintage, 2004.

Newton, Isaac. *Opticks.* Amherst, N.Y.: Prometheus, 2003.

———. *The Principia: Mathematical Principles of Natural Philosophy.* Berkeley, Calif.: University of California Press, 1999.

Spangenburg, Ray, and Diane K. Moser. *The Birth of Science: Ancient Times to 1699.* New York: Facts On File, 2004.

Nobel, Alfred (1833–1896) Swedish *Industrialist, Scientist, Inventor, Philanthropist* The Swedish-born industrialist and scientist Alfred Nobel was responsible for the development and production of dynamite. Upon his death, Nobel bequeathed most of his estate for the generation of yearly international prizes (Nobel Prizes) in the fields of chemistry, physics, physiology or medicine, literature, and peace.

Alfred Nobel was a 19th-century Swedish industrialist and scientist. His legacy was the establishment after his death of the Nobel Prize, awarded every year to individuals who have made significant contributions to chemistry, physics, physiology or medicine, peace, and literature. *(HIP/Art Resource, NY)*

Alfred Nobel was born on October 21, 1833, to Immanuel Nobel and Andriette Ahlsell Nobel in Stockholm, Sweden. Nobel's father was an entrepreneur in the bridge-building and construction industries, and his career varied from being very successful to resulting in bankruptcy, which he encountered twice. When times were good for the family, Alfred studied with well-trained tutors and visited several countries. He loved literature and science and became fluent in five languages. After his father's first bankruptcy, the Nobel family hit on very hard times. Immanuel was forced to leave his country and move to Russia, leaving Alfred and the rest of his family behind. In Russia, he started a new business and produced mines from gunpowder for the Russian navy. His business was highly successful during the Crimean War but led to bankruptcy once again after the war. Nobel's family history and his father's work influenced him greatly. Armed with his life experience in business and explosives, Nobel began his own career as an inventor and industrialist in 1860. By the end of his life, Alfred Nobel had received 355 patents.

Nobel began his period of invention by studying under the French scientist J. T. Pelouze with Ascanio Sabrero, an Italian chemist who had previously invented nitroglycerine, a highly explosive and unpredictable chemical substance. Nitroglycerine is a colorless, oily, slightly toxic liquid also known as glyceryl trinitrate. The International Union of Pure and Applied Chemistry (IUPAC) name for nitroglycerine is propane-1,2,3-triyl trinitrate. Prior to nitroglycerine, the only explosive available for blasting roads, bridges, and other development projects was gunpowder or black powder. Nobel saw the potential of nitroglycerine as a more effective means of completing projects such as these. In 1863, Nobel received his first patent for formulations of nitroglycerine that he commercially called blasting oil. Nitroglycerine is an incredibly dangerous substance, and sometimes it exploded unexpectedly, while at other times it would not explode when desired. Also in 1863, Nobel received a patent for his development of the blasting cap. Previous blasting detonators for nitroglycerine involved heat to detonate the blast. Nobel's detonator employed a strong shock rather than heat to detonate. Mercury fulminate and potassium chlorate mixtures were used to explode with a strong shock wave. This type of detonator was safer than traditional methods and was used unchanged for many decades.

In 1864, Alfred Nobel's own life was struck by tragedy when his younger brother, Emil, was killed in an industrial accident along with four other workers in a plant in Heleneborg, Stockholm, while working with nitroglycerine. Despite the tragedy and the danger of the product, Nobel became even more convinced that nitroglycerine had to be made available on the production market. In 1864, Nobel opened the world's first nitroglycerine manufacturing plant, Nitroglycerine Aktiebolaget, with the financial backing of his family and Swedish businessmen. This company continued to produce nitroglycerine for decades. Previous industrial accidents led the city of Stockholm to outlaw nitroglycerine plants within the city limits, so the first order of business for the new company was to find a location for the plant. Initially, they worked on a barge floating in a lake, but eventually Stockholm lifted the ban and Nobel was able to move inland to a town known as Vinterviken. Nitroglycerine production at Vinterviken continued until 1921, when Nitroglycerine Aktiebolaget purchased their largest competitor and moved production to central Sweden away from the crowded areas of Stockholm.

Nobel started his first foreign company in Germany in 1865—Alfred Nobel and Company (later part of Nobel Dynamite Trust Company)—and began production in a factory near Krümmel, which was approximately 19 miles (30 km) from Hamburg and surrounded by sand dunes. Here Nobel began searching for a mechanism to stabilize nitroglycerine to increase its safety. After repeated unsuccessful attempts, in the sand hills surrounding Krümmel, Nobel stumbled across a substance known as kieselguhr that was composed of diatomaceous earth. When mixed with nitroglycerine, it created a malleable stable substance that could be shaped and manipulated. This product, dynamite, was patented in 1867. The citizens of Krümmel understood the dangers of dynamite and even in hard times were reluctant to work in the nitroglycerine factory. Many of Nobel's employees were Swedish workers who relocated with him. Despite these difficulties and the continued dangers of dynamite production, Nobel's company expanded to include factories around Europe and distribution around the world. The factory at Krümmel was destroyed near the end of World War II and was not rebuilt.

ESTABLISHMENT OF THE NOBEL PRIZE

Despite his industrial and business success, the primary reason most people know the name *Alfred Nobel* is a unique bequest upon his death. Nobel died in 1896 at his home in San Remo, Italy, and his last will and testament included an important one-page provision stating that the preponderance of his estate was to go into a trust for the development and distribution of awards for people who were doing pioneering and significant work in the fields that Nobel himself found intriguing; physics, chemistry, physiology or medicine, literature, and peace. The

Nitroglycerine, developed by Nobel, is an important tool in the construction of roads and bridges as well as demolition. In this photo concrete counterweights that helped open and close the old John Ringling Causeway Bridge are demolished using dynamite in Sarasota, Florida. *(AP Images)*

physics and chemistry prizes were to be awarded to scientists who contributed the most new and innovative or significant work in their field. The award in physics or medicine was for the scientist(s) who contributed the most significant work toward the betterment of understanding human health. The prize in literature stemmed from Nobel's childhood fascination with literature. The Nobel Prize in peace was to be awarded to the individual who best represented attempts to prevent war and increase world peace. This award stemmed from the relationship Nobel had with his friend, onetime assistant, and longtime correspondent, an Austrian named Bertha von Suttner. She went on to become a peace activist, and Nobel had many discussions with her about the responsibility of every individual to the peace of the world. She was awarded the Nobel Prize in peace in 1905. An intriguing condition of the will established that the national origin of the individual could not be considered in selection of the prize recipients. This caused many to suggest that Nobel's will was unpatriotic.

During his lifetime, Alfred Nobel was one of the wealthiest individuals in Europe, and his business interests had become international ventures. He lived in San Remo, Italy, at the time of his death, but he considered himself a citizen of the world. Nobel was quoted, "Home is where I work, and I work everywhere." This philosophy later played a role in the interpretation of his will. At the time of his death, Nobel had residences in San Remo, Italy; Paris, France; Krummel, Germany; Laurleston, Scotland; and Bjürkborn, Sweden. Determining which country was his legal residence determined who was to interpret his will. Nobel drafted his will without the help of a lawyer, and if it had been strictly challenged, the results may not have been the same. The deciding factor of Nobel's permanent residency was based on the acquisition of a stable of horses in Sweden, demonstrating that he considered Sweden his primary residence.

The contents of Nobel's will and testament were a complete surprise to his family. They were significantly disappointed in their portion of the inheritance and fought the development of the Nobel Foundation. The executors of the estate were two prior assistants, Ragnar Sohlmann and Rudolf Lilljequist. Sohlmann was a chemical engineer who worked with

Nobel in his laboratory in Italy as Nobel's personal assistant. Lilljequist was a civil engineer who had also worked with Nobel. The executors fought the Nobel family, and the institutions named in the will over the course of many years before the will was finally verified and supported. Sohlmann even went so far as to sneak all of Nobel's papers, valuables, and money out of Paris in a carriage then ship them by boat to Stockholm before the governmental agencies of Paris or the Nobel heirs knew what was happening. Sohlmann worked for Nobel's companies the remainder of his life. He was the chief executive officer of one of Nobel's dynamite companies and served as director of the Nobel Foundation from 1929 to 1946.

Other countries found the establishment of Nobel's primary residence as Sweden to be politically charged and unjust. They believed that the establishment of a peace prize by a committee from the Swedish parliament was unfair to the rest of the nations that had benefited from Nobel's influence. The executors of Nobel's estate finally appealed to the Swedish parliament for assistance in having the will approved. Although it was not common practice for the government to become involved in areas of personal finances, since Nobel was a Swedish resident, they listened to the executors' statements. In 1897, the attorney general declared the will to be legally valid.

DEVELOPMENT OF THE NOBEL FOUNDATION

Alfred Nobel's will and testament named four different bodies to oversee the distribution of the Nobel Prizes. Even these institutions did not easily accept their roles in the process. It took several years after Nobel's death for each of the appointed agencies to approve the will. Nobel gave the Storting (Norwegian parliament) the responsibility of selecting and administering the peace prize. On April 26, 1897, the Norwegian Nobel Committee of the Storting (known since 1977 as the Norwegian Nobel Committee) was formed, consisting of five members who were responsible for the selection. After much debate about the validity of the will, the three other named bodies approved the will and founded selection committees. The Karolinska Institute approved the will on June 7, 1898; the Swedish Academy on June 9, 1898; and the Royal Swedish Academy of Sciences on June 11, 1898.

The Norwegian Nobel Committee is responsible for selection and administration of the Nobel Prize in peace. The Swedish Academy oversees the selection and administration of the Nobel Prize in literature. The Royal Swedish Academy of Science distributes the Nobel Prize in physics and the Nobel Prize in chemistry. The Karolinska Institute is responsible for the Nobel Prize in physiology or medicine.

The first Nobel Prizes were distributed in 1901. The winners that year included six individuals who had made significant contributions to their fields. The first Nobel Prize in chemistry was presented to the Netherlands-born Jakobus van't Hoff for his discoveries in solution chemistry. He demonstrated that thermodynamic laws apply to dilute solutions. The van't Hoff factor is used in colligative property and solution calculations. Wilhelm Conrad Röntgen, a German physicist, received the first Nobel Prize in physics for the discovery of X-rays, once called Röntgen rays. Frederic Passy from Switzerland and Henry Dunant from France split the first Nobel Prize in peace. Passy was responsible for the formation of the Geneva Convention and was the founder of the International Committee of the Red Cross. Dunant founded and served as the first president of the original French Peace Society. The 1901 Nobel Prize in physiology or medicine was awarded to the German Emil von Behring for his work on immunizations and antitoxins for diphtheria and tuberculosis. The first Nobel Prize in literature was presented to a French poet, Sully Prudhomme.

In 1968, the Swedish National Bank created an additional prize that is coordinated through the Nobel Foundation for the field of economic sciences. The Royal Swedish Academy of Science is responsible for its administration and the selection of the prizewinners.

To date, 766 individuals and 19 organizations have received the Nobel Prize, for a total of 785 prizes. The prize consists of a medal emblazoned with Alfred Nobel's likeness, a diploma, and a cash award, which in 2007 was 10,000,000 kronor ($1,531,393). When joint prizes are awarded, the money is split.

See also CHEMICAL REACTIONS.

FURTHER READING

The Nobel Foundation. *Alfred Nobel—the Man behind the Nobel Prize.* Available online. URL: http://nobelprize.org/alfred_nobel. Accessed July 25, 2008.

nondestructive evaluation (NDE) *Nondestructive evaluation* (NDE) and *nondestructive testing* (NDT) both refer to the application of methods and technology to provide the public safety from unexpected structural failures. Nondestructive testing methods are also used to maintain products in safe operating condition. A large number of NDT methods contribute to the maintenance of both public safety and low-cost public services in a variety of areas. Though NDT is applicable at the individual level, the clear picture of the benefits of NDT can be seen at the community level.

SAFETY AND THE DESIGN PROCESS

NDT is located in the middle of the "product safety" road map, a process followed by all manufacturers with the goal of protecting consumers. The development of millions of products made globally each year shares a nearly identical "safe design" strategy. The safe product road map begins with attention to safety during the product design stage. Whether governed by governmental standards, manufacturing societies, or a conscientious manufacturer, the selection of materials and construction procedures is the initial step in the process. Once properly designed, consumer protection extends to the manufacturing process. Quality control of the materials and manufacturing procedures results in a safe product that is placed into service.

Once in service, the product is out of "design control." The environmental exposure of two identical products can be radically different. For example, consider two identical cell phones. One is infrequently used by a motorist and is kept in the automobile's glove box. The other cell phone is owned by an energetic teenager, and the cell phone's keypad experiences the daily wear and tear of near-constant text messaging. There is a large difference in the number of force cycles applied to the two keypads. Eventually the two cell phone keypads will both fail, but when they fail depends on their usage.

This simple example highlights the issue associated with all products once they are placed into their intended service environment. For the case of the cell phone, the eventual keypad failure will be followed by the rapid replacement of the cell phone. In fact, for many consumer products, the replacement of old products has more to do with trends and fashion than failures from usage beyond the product's original design life. Nondestructive testing does not play much of a role in the manufacture of these types of consumer products.

When a product has the potential for serious risk to an individual or to the population at large, the situation is different. Bridges, airplanes, gas lines, and thousands of other products that society deems essential require verification that the product has safe operational life remaining. During the in-service life of major products, NDT is the primary inspection and maintenance method.

NONDESTRUCTIVE TESTING VERSUS DESTRUCTIVE TESTING

NDT is favored by the population service provider because of the "nondestructive" nature of the inspection. The counterpart to NDT is DT (destructive testing). DT is absolute and unambiguous. A DT examination of a commercial passenger airliner jet engine compressor shaft could result in the conclusion that the shaft would continue to provide safe engine operation for the next 25,000 flight hours. In the process of making this decision, however, the engine shaft would be destructively examined and must now be replaced. This type of testing provides accurate results but is very expensive.

The challenge that faces all users of nondestructive testing methods is to approach or equal the accuracy of DT without needing to replace the tested component. Consider a highway bridge. The bridge has been properly designed for a certain traffic rate, for the climatic conditions of the locality where it is built, and with materials to allow for safe traffic flow for 50 years. After construction, however, the traffic patterns may differ from what the designer originally imagined. Because of this, the regularity and quality of the bridge maintenance significantly impact the bridge's safe operation. By applying NDT methods, the community's engineers can ensure that the bridge remains safe.

Despite the obvious economic benefits, NDT must accurately provide component safety assessments. The development of NDT methods is time-consuming and detailed work. NDT methods developed for the example bridge may be totally inappropriate for the inspection of the jet engine shaft mentioned earlier. Thus NDT methods must be qualified, or tailored, for a specific application.

NDT CERTIFICATION

In addition to having a qualified NDT method, the test personnel must be certified. NDT methods are implemented by using approved procedures executed by certified operators. Both the test procedure and operator certifications are governed by specific agencies. In the United States, several organizations are involved in the governance of NDT. The three major organizations that oversee the application of NDT methods are as follows:

- American Society of Nondestructive Testing (ASNT)
- American Society of Testing Methods (ASTM)
- American Society of Mechanical Engineering (ASME)

All NDT inspection procedures follow standardized protocols to identify defective components and have the flaw remedied, avoiding an unsafe condition.

NDT TECHNOLOGIES

Four types of technology exist for nondestructive testing:

- liquid penetration inspection
- X-ray

- ultrasonic
- eddy current

The oldest NDT method is liquid penetrant inspection (LPI), in which a substance that can penetrate cracks is added, making the cracks easier to visualize. LPI was developed to enhance the ability of the NDT inspector to identify a surface crack during a visual component inspection. A visual examination of a cleaned component by a skilled inspector with 20/20 vision can resolve a surface crack that is 0.003 inch (0.0076 cm) in width. The limit of human vision is such that smaller surface cracks are undetectable to the unaided human eye. Magnification allows for the detection of smaller surface cracks, but the use of magnification requires additional training and requires longer inspection times, thus increasing the cost of testing.

The advantage of LPI over unaided visual inspection is that LPI produces a higher level of crack contrast and an enlargement of the crack due to component surface staining when the penetrant reverse flows from the surface crack. LPI can be used on metals, ceramics, glass, and plastic. Among its many advantages are ease of use and the ability to estimate crack volume. The cost of an LPI inspection is low and LPI has a high rate of inspection (square meters/hour). The most significant disadvantage of LPI is that it can only detect cracks that have broken the component surface. Even a significant flaw that is below the surface cannot be detected by an LPI inspection.

SUBSURFACE DEFECT DETECTION

As a result of LPI's significant subsurface defect detection limitation, numerous NDT technologies have been developed to allow for subsurface defect detection. None is as simple and cost effective as LPI, but the desirability of defect detection early in its development, well before a surface crack develops, provides a level of safety that LPI cannot provide. Subsurface defect detection techniques include X-ray detection, ultrasonic acoustics, and eddy current detection. Each of these three methods can identify and locate a material defect prior to a surface crack indication. This ability significantly increases the safety level and provides a material assessment of comparable accuracy to destructive testing analysis.

X-RAY DETECTION

X-ray detection involves a series of X-rays that pass through the component being tested. The X-ray energy at the source is the same for each source path. Thus the level of X-ray energy at the detector is a function of the component's "attenuation path" through the material. Since defective material has

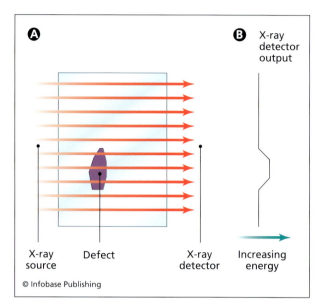

© Infobase Publishing

(a) X-rays from an X-ray source are directed through the material to be tested. The X-ray energy passing through the test material is measured by an X-ray detector. If there is a defect or flaw in the test material, the energy of the X-rays passing through the flaw will differ from that of the X-rays passing through the rest of the material. (b) The X-rays that reach the detector after passing through the defect possess higher energy than those traversing the intact material. In this manner the X-ray detector reveals the location of the defect.

different X-ray attenuation than unflawed material, the X-ray detector output provides a profile of the subsurface flaw. X-rays provide a sensitive method for the location and size of subsurface defects within components. The shorter the wavelength of the X-rays, the higher the energy and the thicker is the material that can be inspected.

X-ray NDT inspection machines are large and expensive and require a significant amount of electrical power. Another disadvantage is the requirement to protect test personnel from the X-rays while performing the inspection. While most X-rays pass directly through the test piece during the inspection, some X-rays follow unpredictable scattering paths. Testing personnel require protection from inadvertent X-ray exposure.

Despite the need for extensive radiological safety training and exposure history documentation, X-ray NDT is the most reliable method for the detection of flaws in thick-walled metal constructions. X-ray NDT can inspect complex geometries without a concern of a missed inspection volume. Submarine hull plates and nuclear reactor vessels are both inspected with X-ray systems capable of detecting small defects within the walls of 10–14-inch (25–36-cm) steel plates.

ULTRASONIC (UT) NDT TECHNOLOGY

Ultrasonic NDT technology uses high-frequency sound energy that is inaudible to human hearing. Ultrasonic testing measures the time required for a transmitted UT wave to be returned and received. The higher density of the material (kg/m^3) the higher the speed of sound in the material. Knowing the speed at which sound waves travel through a material is crucial to the ability of ultrasonic NDT to determine the depth of any flaw below the surface of the component under test.

Ultrasonic testing (UT) utilizes piezoelectric materials, which are capable of converting an electrical voltage to a pressure and thus producing ultrasound. A piezoelectric crystal both transmits and receives the UT sound. When transmitting, the UT crystal converts a voltage spike to a UT pulse. In the receiving mode, the crystal converts a UT pulse back into a voltage. Measuring the amount of time the UT sound takes to travel through the substance demonstrates whether any flaws are present. In order to convert the time into depth, an accurate measure of the speed of sound in the material under test is required. Once that speed is known, then the depth of the flaw can also be established.

Ultrasonic testing is a simpler NDT method than X-ray inspection. UT has the same advantage as X-ray detection in its ability to locate subsurface defects, and it shares the advantages of simplicity and low cost with liquid penetrant inspection. The only significant deficiency with UT inspection is that it is difficult to direct the UT wave into all component areas for inspection.

EDDY CURRENT TESTING

Eddy current testing (ECT) is most appropriate for detecting subsurface defects when the inspection must be made without contact between the component and the NDT sensor. By virtue of being a noncontact NDT method, ECT can be applied while a machine is in operation. For example, the shaft of a large shaft-driven machine can be monitored for defects while the shaft is rotating.

Eddy currents were predicted by James C. Maxwell when he derived the physical laws that govern electrical and magnetic interactions. Maxwell's equations can be stated as a pair of related laws:

A time-varying electrical field will give rise to a magnetic field.

A time-varying magnetic field will give rise to an electrical field.

Eddy currents are produced in the remote target component as the result of the ECT sensor's generating a time-varying electrical field, which first generates a magnetic field in the remote component. The varying magnetic field produces a time-varying electrical field and eddy current, which the remote ECT sensor detects. Despite the complicated explanation, the usage of eddy current NDT for remote inspection is common. In particular, eddy current NDT is utilized when thin-walled tubing is required. Thin-walled components pose problems for UT inspection because there is such a short time between UT wave launch to back wall reflection. Eddy current NDT avoids the time-of-flight issue and is therefore the preferred NDT method for thin wall inspections.

Eddy current exploits the physics of electromagnetic theory to locate a subsurface defect within a component without the requirement that the NDT sensor be in direct contact with the component under test. ECT is used whenever NDT inspection is needed while the equipment is in operation. ECT is more effective than X-ray or UT when the component under test is thin wall tubing.

APPLICATIONS OF NDT

NDT is used in every sector of industrial and consumer manufacturing, including the electrical power industry, production of consumer goods such as vehicles, and purification and production of energy sources. Nuclear power plants, coal-fired boilers, and gas turbines all have components that are regularly inspected using NDT in order to maintain public safety and provide a reliable source of electricity. Even environmentally "green" power generation such as photovoltaic (PV) solar panels, wind turbines, and hydroelectric plants all utilize NDT methods.

For example, a large hydroelectric dam has a series of required scheduled NDT inspections. The dam's concrete and steel structure provides the elevation by which the stored lake water has its mechanical energy converted into electricity. NDT methods are applied to the dam structure, the hydroturbine, and the electrical generator.

NDT plays an important role in the production of low-cost, reliable consumer goods. Paper products, electronics, and clothing are all produced in factories where manufacturing costs are kept low through the application of NDT methods to factory machines. NDT keeps production lines running reliably and provides incoming inspection of raw materials to ensure that the product quality is sufficiently high to maintain customer satisfaction.

Even the family car relies on NDT methods in its manufacture and operation. For instance, today's high-tech automobiles utilize an NDT method to monitor tread depth on tires and brake system component thickness above minimal safety levels. The gasoline that fuels the family automobile is transported from the oil field to the refinery by tanker ships and oil pipelines. Both of these transportation methods require scheduled NDT inspections to prevent accidental spills and protect the environment.

NDT provides an inspection process whereby the operational condition of equipment can be assessed without destroying the component under test. NDT has spread into virtually every area of industry. NDT can produce flaw detection capability equivalent to that of destructive testing (DT) without the consequence of replacement of the component tested. With NDT, healthy equipment can be inspected and recertified and then returned to operational service. Collectively, the NDT technologies described here add a significant layer of population safety while lowering the risk of an unexpected failure. NDT is preferred to destructive testing methods because it offers major cost savings when NDT inspection establishes a component to be free of flaws and eligible for return to service.

See also MATERIALS SCIENCE; MAXWELL, JAMES CLERK.

FURTHER READING

American Society for Nondestructive Testing. "Introduction to Nondestructive Testing." Available online. URL: http://www.asnt.org/ndt/primer1.htm. Accessed April 20, 2008.

American Society of Testing Methods (ASTM) home page. Available online. URL: http://www.astm.org/. Accessed April 20, 2008.

nuclear chemistry Nuclear chemistry is a branch of chemistry concerned with reactions that occur within the nucleus of atoms (nuclear reactions) and the processes that involve nuclear reactions. Nuclear energy is the energy found within the attraction between protons and neutrons, the subatomic particles that make up an atomic nucleus. Atomic nuclei that become unstable spontaneously decay and emit radioactive particles. These reactions are dramatically different from chemical reactions that involve the making and breaking of bonds between atoms. All nuclides (forms of an atom with a specific number of protons) with an atomic number greater than 84 are naturally radioactive. Nuclear reactions occur within an atom at the subatomic level. Three main types of radiation exist: alpha (α) radiation, beta (β) radiation, and gamma (γ) radiation. A better understanding of nuclear chemistry has led to applications in many fields including medicine, archaeology, the environment, and nuclear energy sources.

RADIOACTIVE DISCOVERIES

Wilhelm Conrad Röntgen, a German physicist, first discovered radioactivity in 1895. While performing experiments using a cathode-ray tube, Röntgen discovered a new form of radiation originally referred to as Röntgen rays, and now commonly known as X-rays.

The next major development in the field of nuclear chemistry occurred when a French physicist, Henri Becquerel, discovered spontaneous radioactivity in 1896. He received the Nobel Prize in physics for his work in 1903.

Marie and Pierre Curie, a French scientific couple, also made significant contributions to the field of radioactivity. The Curies examined pitchblende, the unrefined ore from which uranium was obtained. They observed that pitchblende gave off hundreds of times more radioactivity than pure uranium. This led to their refinement of the pitchblende and the discovery of two new radioactive elements that they named polonium (Po) and radium (Ra). Marie Curie determined that the decay of a radioactive element could be measured and predicted, setting the stage for the determination of half-lives in later studies. The Curies received a joint Nobel Prize in physics in 1903 for their work with radioactivity. After Pierre's untimely death, Marie also received an individual Nobel Prize in chemistry in 1911 for the discoveries of radium and polonium.

TYPES OF RADIOACTIVITY

Several types of radioactivity exist, each with distinct properties: alpha radiation, beta radiation, and gamma radiation. Alpha radiation occurs when alpha particles are released from the nucleus of an atom. An alpha particle contains two protons and two neutrons, the equivalent of a helium nucleus. New Zealand–born Sir Ernest Rutherford utilized alpha particles in his classic gold foil experiment that demonstrated that atoms consisted mostly of empty space, with a small, centrally located positive nucleus. Rutherford shot alpha particles at a sheet of gold foil and watched the deflection of the particles, using this information to build his model for atomic structure. When an alpha particle is released, the radioactive element transmutates; that is, it turns into a new element. The loss of two protons reduces the atomic number of the original element by 2, and the loss of two additional neutrons reduces the mass number by a total of 4. A classic example of alpha radiation occurs with uranium 238.

$$^{238}_{92}\text{U} \rightarrow \,^4_2\text{He} + \,^{234}_{90}\text{Th}$$

The symbols in the formula have superscripts that represent the mass number and subscripts that represent the atomic number of the element. The reaction illustrates the reduction of atomic mass by 4 and the reduction of atomic number by 2. Alpha radiation is the least penetrating of all radioactivities and can be blocked by a single sheet of paper.

Beta radiation has higher penetrating power than alpha radiation but can be stopped by a single sheet of aluminum. This form of radiation causes the most damage when it is ingested. In close proximity to cells, it may cause irreparable damage within a short period. Beta radiation can be classified as one of two types: positive beta emission and negative beta emission (beta particle emission). Positive beta emission releases a positron with a neutrino while converting a proton to a neutron, thus decreasing the atomic number. Negative beta emission releases an electron with an antineutrino while converting a neutron to a proton, thus increasing the atomic number.

Gamma radiation differs from alpha and beta radiation in that gamma rays do not give off a particle and do not change the mass, atomic number, or mass number of the original element. Gamma rays are a form of electromagnetic radiation with a very high frequency. Release of gamma radiation by an unstable nucleus often accompanies alpha or beta emission in an attempt for the isotope to regain stability. Gamma radiation is able to penetrate nearly everything. A lead sheet several inches (several centimeters) thick is necessary to block the penetrating effects of gamma radiation.

RADIOACTIVE HALF-LIFE

The practical uses of radioactivity as well as the damaging effects and disposal issues related to radioactivity depend upon the amount of time the isotope lasts. The measure of radioactive decay is known as half-life, the time it takes for half of the amount of original radioactive material to decay. After one half-life, half of the original material will be present; after two half-lives, half of the material that existed at the end of the first half-life will persist. For example, if the value of the original material was 10 units, five units would remain after one half-life, and after two half-lives 2.5 units would be left, then 1.25 units, and so on. Simply put, after two half-lives, one-fourth of the true original starting material would still be available.

Half-lives of radioactive isotopes range from fractions of a second to millions of years. Knowing the half-lives for certain radioactive isotopes allows for the determination of the date or age range of artifacts and archaeological finds. This method, known as radiometric dating, is valuable in many branches of science. One performs radiometric dating by comparing the amount of parent isotope (the original, unstable, radioactive isotope) with the production of the daughter isotope (the isotope created by radioactive decay). This type of dating is more effective when the daughter isotope is unique or rare. When the daughter isotope is a common element on Earth, ratios of the stable form of an atom to the radioactive isotope form can be used to determine age.

In the dating of biological specimens, the most common isotope combinations are carbon 14 and carbon 12. Carbon 12 is the stable form of carbon; carbon 14 contains two additional neutrons and is radioactive. All living organisms take in carbon dioxide at a fixed ratio of carbon 12 to carbon 14. Once the organism has died, it no longer takes in carbon dioxide and the carbon 14 within its body begins to decay. The parent isotope, carbon 14, decays into the daughter isotope, nitrogen 14, which is normal nitrogen in the atmosphere and is therefore undetectable from normal atmospheric nitrogen through negative beta decay according to the following formula:

$$^{14}_{6}C \rightarrow {}^{14}_{7}N + electron + antineutrino$$

Scientists use the ratio of carbon 14 to carbon 12 to determine the age of the sample. The half-life of carbon 14 is 5,730 years. If the ratio of carbon 14 to carbon 12 is half of what it is naturally, then one knows that 5,730 years have passed. Carbon 14 dating only works for previously living samples. The useful technique of radiocarbon dating was developed by the American chemist William Libby in 1949. Libby received the Nobel Prize in chemistry in 1960 for his discovery.

Nonliving materials such as rocks, metals, and glass cannot be dated using carbon dating techniques. Other elements can be used to help determine age of inorganic matter. An example is radium 226, which decays into the unique and easily identifiable radon 222. If a sample is tested and contains a ratio of radon 222 to radium 226, the age of the object is determined using the half-life of 1,600 years. Successful radioactive dating requires choosing an isotope with an appropriate half-life for the sample. After 10 half-lives, the detectable amount of radioactive sample may be too low to measure. Scientists determine the possible age range of their sample and then select an isotope accordingly. For example, potassium 40 decays into argon 40 with a half-life of 1.3 billion years, making it an excellent choice for dating samples older than 100,000 years.

Half-life calculations are completed using the following formula:

$$t_{1/2} = 0.693/k$$

where $t_{1/2}$ is the half-life of an isotope, and k is the rate constant for decay of the specific nuclide. If the half-life is known, one can determine the rate constant using the equation for half-life. For example, a useful radioactive tracer used in medicine is technetium 99. The rate constant is known to be 1.16×10^{-1}/hour. One can calculate half-life as follows:

$$t_{1/2} = 0.693/(1.16 \times 10^{-1}/\text{hour})$$
$$t_{1/2} = 6.0 \text{ hours}$$

NUCLEAR MEDICINE

Nuclear medicine is an application of nuclear chemistry to the health and well-being of the body. In the medical field, radioactivity is used to treat cancers, and iodine 131 is used to detect and treat thyroid disorders. Many nuclides, such as thallium 201 and phosphorus 32, can be used as tracers that facilitate visualization of internal structures in order to diagnose and treat human diseases.

FUSION

Nuclear fusion is a method of producing energy by combining two smaller atomic nuclei into one larger nucleus. The fusion of any nuclei with a mass lower than iron (with an atomic mass of 26) releases energy, as long as the mass of the resulting nucleus is less than the sum of the two individual nuclei prior to fusion. The law of conservation of mass states that mass can be neither created nor destroyed. This nuclear reaction appears to disobey this law by losing mass. Albert Einstein determined the relationship between mass and energy that explains the apparent disappearance of the additional mass. Einstein's theory of relativity demonstrates the conversion of mass to energy by the following formula:

$$E = mc^2$$

where E stands for the amount of energy released, m is the mass, and c is the speed of light (3×10^8 m/s). Thus, the loss of mass translates into the release of large quantities of energy in fusion reactions.

Fusion requires extremely high temperatures to overcome the natural repulsion between the positively charged nuclei, temperatures that naturally occur on the surface of the Sun, catalyzing the fusion of hydrogen and helium nuclei with the concomitant production of energy.

NUCLEAR FISSION

Nuclear fission is a process that releases energy by splitting the nucleus of an atom such that smaller atoms are produced. In order for fission to occur, the nucleus of the atom being broken apart (the fissionable material) must be unstable. The fissionable material is bombarded with particles to break

it apart. One can visualize the process by thinking of billiard balls on a pool table. When the balls are racked, they are packed closely together. When they are struck by the cue ball, the collision with the particles causes them to bounce into each other and be broken apart from the group. When nuclear fission occurs in a controlled environment, the energy from the collisions can be harnessed and used to create electricity, to propel submarines and ships, or to be applied to research purposes. An atomic bomb is the result of uncontained nuclear fission.

NUCLEAR ENERGY

Nuclear power plants are an alternative energy source that currently supplies 20 percent of the U.S. energy demand. Nuclear fission reactors power these plants, and the energy obtained from the fission reaction is harnessed to run generators and produce electricity. The most common type of fuel cell for a nuclear fission reactor is uranium 235. Nuclear reactors are inspected and regulated by the Nuclear Regulatory Commission.

NUCLEAR ACCIDENTS AND NUCLEAR WASTE

Serious concerns in the field of nuclear chemistry are the potential for nuclear accidents and the production of vast quantities of nuclear waste. Nuclear energy has potential for leaks and accidents. In 1979, a problem in the reactor at Three Mile Island in Pennsylvania led to a meltdown. Although well contained and no true risk to the community, this accident caused widespread public concern about the safety of nuclear reactors. A more serious accident occurred in 1986 at the Chernobyl power plant near Kiev, USSR. The widespread release of radioactivity had drastic consequences for the environment and for those living in the area. Disposal of radioactive isotopes has created a complicated environmental issue. Isotopes with long half-lives will exist for generations. A true solution for the maintenance, storage, and disposal of these radioactive isotopes has not yet been found.

See also ALTERNATIVE ENERGY SOURCES; CONSERVATION LAWS; FISSION; FUSION; ISOTOPES; NUCLEAR PHYSICS; RADIOACTIVITY.

FURTHER READING
Loveland, Walter, David Morrissey, and Glenn Seaborg. *Modern Nuclear Chemistry.* Hoboken, N.J.: John Wiley & Sons, 2006.

nuclear magnetic resonance (NMR) Nuclear magnetic resonance (NMR) is a technique that reveals the location and arrangements of atoms in a molecule on the basis of their spin in a magnetic field when they are bombarded with radio waves. Every atom is made up of protons (positively charged particles), neutrons (neutral particles), and electrons (negatively charged particles). The protons and neutrons are arranged in a central nucleus, while the electrons rotate in an electron cloud outside the nucleus. When placed in a magnetic field, each of these particles can align with the magnetic field in a north-south-north-south (NSNS) arrangement often called "spin up." Alternatively, they may be opposed to the magnetic field in a north-north-south-south (NNSS) arrangement often called "spin down."

The atoms spin according to the way they line up in the magnetic field. Slightly more than half usually line up spinning in the same direction as the magnetic field, and the rest line up opposite to the magnetic field. Each of these particles will spin, and as they

spin, they experience a wobble known as precession, much like the wobble of a spinning top. Once the atoms reach equilibrium in the magnetic field, the samples are subjected to radio waves of particular frequencies. When the radio frequency matches the frequency of the precession of the atom, the atom changes its direction of spin. Energy will be absorbed from the radio waves if the atom goes from the low-energy state to the high-energy state. Energy will be given off by the atom to the radio waves if the change is from high-energy state to low-energy state. NMR measures how much energy must be gained or lost in order to change the direction of spin of the nucleus. The peaks present in the NMR display demonstrate the frequency of energy given off or taken up by the atom and thus information about the location of an atom and its neighbors.

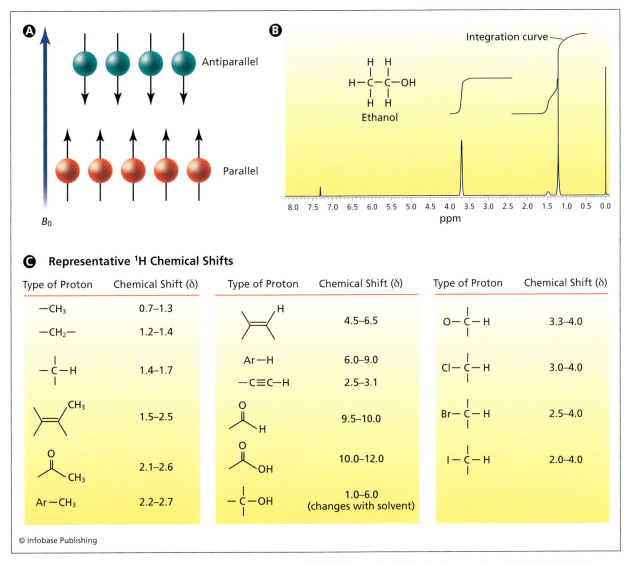

© Infobase Publishing

The antiparallel/parallel motion created by a magnetic field (a) can be detected in a proton NMR (^{1}H-NMR) spectrum (b) and interpreted on the basis of the representative ^{1}H chemical shifts (c).

The frequency of energy given off depends on the atom type and its neighboring atoms, as well as the strength of the applied magnetic field. In order to compare results, scientists describe a chemical shift as the difference in spin frequency of an atom and a standard, usually tetramethylsilane [$(SiCH_3)_4$]. This allows for the comparison of NMR spectra. In the original form of one-dimensional (1-D) NMR, the area under the peaks is proportional to the number of atoms at that position, making it a very informative experiment. The disadvantage of 1-D NMR involves the interaction of adjacent atoms in the molecule. The presence of an atom that is less than three bond lengths away and that is spinning in the opposite direction will influence the frequency of the atom being studied. Scientists term this phenomenon *spin-spin coupling,* and it compromises some of the resolution of the NMR. The peak could split and reduce the strength of a signal. The development of decoupled 1-D NMR helped remove this problem by using a radio frequency equal to the frequency of the neighboring atoms. This increases the signal of the atom being studied; however, the area under the curve is no longer proportional to the number of atoms at that location.

The most common element used for NMR study is hydrogen, and the form of NMR is known as 1H-NMR, or proton NMR. The hydrogen nucleus contains one proton and the atom also has one electron spinning on the outside. Hydrogen makes up a large portion of bodily fluids and tissues including water and fat. Estimates place the human body at greater than 60 percent hydrogen. The elements ^{13}C, ^{15}N, ^{19}F, and ^{31}P can also be used in NMR, but they are not as abundant as hydrogen and therefore give a weaker signal.

DEVELOPMENT OF NMR

The discovery, refinement, and applications of NMR have led to six Nobel Prizes in the fields of chemistry, physics, and physiology or medicine. The first person to perform an NMR experiment successfully was Isidor Rabi, a physicist working at Columbia University in the 1930s. Rabi received the 1944 Nobel Prize in physics for his development of a type of NMR called molecular beam magnetic resonance. He first subjected a beam of lithium chloride to radio waves and then subjected them to varying magnetic fields. When the frequency of the precession of an atom matched the frequency of the radio wave, the atom either absorbed energy from the radio waves and jumped up an energy level or released energy as it fell back down an energy level. These flips were recorded on detectors. Since the frequency of the magnetic moment of each of the atoms is relative to the type of atom and its neighbors, knowing the

frequency of each atom gave Rabi much information about the location and surroundings of the atoms in his molecular beam. Rabi later determined that if the magnetic field remained constant, then the radio waves could be varied. Future developments and medical applications of MRI built on Rabi's discoveries.

Edward Purcell, an American physicist, and Felix Bloch, a Swiss-born physicist, shared the Nobel Prize in physics in 1952 for their work on NMR performed in the 1940s. They applied nuclear magnetic resonance to solids and liquids instead of a beam, as Rabi did. Both Purcell and Bloch used hydrogen as their atom to study. Purcell used solid paraffin as a hydrogen source, while Bloch used water. They developed much of the present-day procedure for performing NMR. They applied a magnetic field, allowed the atoms to reach spin equilibrium with the magnetic field, applied radio waves, and measured the amount of energy taken or given off. Their method was called nuclear magnetic resonance in condensed matter.

In 1948 Nicolaas Bloembergen, Purcell's graduate student, first accurately measured relaxation times, represented by T_1 and T_2. T_1 represents the time a nucleus takes to attain equilibrium with the magnetic field. T_2 represents the duration of time in which the sample gives off a signal. Bloembergen, Purcell, and the American physicist Robert Pound were able to determine that these relaxation times took place in seconds or less, knowledge that became incredibly important in the development of biologically useful testing methods with NMR.

A significant discovery was that of two independent American physicists, Henry Torrey and Erwin Hahn. Rather than subjecting samples to continuous radio waves while samples were under a magnetic field, they were able to pulse the radio waves of known wavelengths. This pulsed NMR technique expanded the potential applications. While performing his experiments, Hahn also recognized and characterized a spin echo that followed a series of radio wave pulses. This spin echo gave him even more information about the type, location, and shape of molecules. In order for magnetic resonance imaging (MRI) to become a functional clinical diagnostic tool, perfection of the concepts of pulsed NMR and spin echo was necessary.

Despite all of these discoveries, the technique of NMR would not have become mainstream without the work of Russell Varian in the 1950s. Having known Bloch from Stanford, Varian started a company, Varian Associates, with a primary goal of creating user-friendly functioning NMR machines. The first NMR spectrometer, used in Texas in 1952, was known as the Varian HR-30. Varian Associates created the first superconducting magnet for NMR. A superconducting magnet is cooled so that the magnet becomes a superconductor and requires much less electrical energy to generate the magnetic field. This allowed stronger magnets to be used in NMR experiments. Varian was also involved in the development of Fourier transform NMR (FT-NMR), which uses the signal versus time data received from the NMR experiment and converts the data into a discernible picture with a mathematical formula known as a Fourier transform.

Richard R. Ernst, a Swiss chemist, received the Nobel Prize in chemistry in 1991 for his significant contributions to FT-NMR and 2-D NMR in the 1960s. By using Fourier transformation, a range of radio frequencies could be run at the same time, and the computer would be able to sort out all of the individual frequency results, whereas in standard NMR, they would be scrambled together.

In 1971, Raymond Damadian, an American physician and scientist, demonstrated the principle that

Richard R. Ernst is a Swiss chemist who received the 1991 Nobel Prize in chemistry for his contributions to high-resolution nuclear magnetic resonance (NMR) spectroscopy. *(AP Images)*

tumors and regular tissues showed different relaxation times, and he received the first patent for MRI. This presented the possibility of using NMR to diagnose disease and distinguish between healthy and damaged tissue, thus allowing physicians to determine such phenomena as tumor boundaries, neural damage due to stroke, or blood vessel integrity. Damadian was also responsible for the development of field-focusing NMR in 1977.

The 1970s brought dramatic improvement and the application of NMR to medicine. In 1973, Paul Lauterbur, an American chemist, demonstrated that two-dimensional images (2-D NMR) could be produced using gradients in a magnetic field. At the same time, Peter Mansfield, a British physicist, established that the use of gradients gave a signal that could be turned into an image. Mansfield developed echo-planar imaging, which used quick gradient changes. All of these NMR improvements led to distinct changes in NMR use for medical diagnostics. For their work and discoveries, Lauterbur and Mansfield received the Nobel Prize in physiology or medicine in 2003.

In 1992, an NMR advancement involving the development of functional MRI (fMRI) occurred. This technology allowed for the study of the physiology and function of tissues, whereas MRI normally just shows the anatomy of the tissues. This discovery has significantly improved brain studies.

APPLICATIONS OF NMR

NMR has multiple applications in both the lab and the medical field. Chemists are able to use proton NMR and ^{13}C NMR to study the chemical structure, atom arrangement, and bonding patterns of a vast array of compounds. The advances in NMR even allow scientists to study the structure of proteins and subunit associations as well as perform reaction rate studies.

The best-known application of NMR is in magnetic resonance imaging (MRI), a technique that applies the principles of NMR to diagnosis of biological injuries, diseases, and conditions. Rather than X-rays, which can be damaging, MRI uses radio waves, which have no known side effects. MRI also differs from X-rays by visualizing soft tissue rather than bones. This makes MRI irreplaceable in diagnosis of brain and spinal cord diseases and injuries, cardiovascular issues, aneurysms, strokes, and multiple sclerosis, as well as sports injuries and other soft tissue damage.

When having an MRI, the patient lies on a table that slides into a big tube encircled by the magnet. When the magnetic field is applied, the patient hears pounding noises that indicate radio waves are being pulsed through the chamber. MRI scans last from

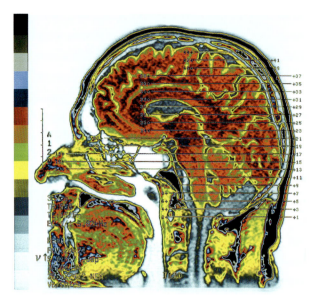

Molecular resonance imaging (MRI) is a form of nuclear magnetic resonance. MRI allows visualization of human tissues. Shown here is an MRI image of the human head. *(Eye of Science/Photo Researchers, Inc.)*

approximately 20 minutes to one and a half hours, depending on the areas of the body being tested. Patients must remove all metal objects prior to the MRI. The intense magnetic field used during the test could be very dangerous. In 2001, a flying oxygen canister that was inadvertently left in the room with a magnet killed a six-year-old child during an MRI. Patients with pacemakers, metal plates, shrapnel, or bullet fragments should not be tested by MRI because of the risk these objects will move during the test, causing potentially fatal damage. During the MRI, the technician maintains communication with the patient via headphones. The subject must remain perfectly still during the course of the test to prevent blurred images.

See also ATOMIC STRUCTURE; MAGNETISM; MASS SPECTROMETRY.

FURTHER READING

Conlan, Roberta, with the assistance of Richard Ernst, Erwin L. Hahn, Daniel Kleppner, Alfred G. Redfield, Charles Slichter, Robert G. Shulman, and Sir Peter Mansfield. "A Lifesaving Window on the Mind and Body: The Development of Magnetic Resonance Imaging." In *Beyond Discovery: The Path from Research to Human Benefit*. The National Academy of Sciences. Available online. URL: http://www.beyonddiscovery. org/content/view.page.asp?I=133. Last updated August 16, 2002.

Hornak, Joseph P. "The Basics of NMR." Available online. URL: http://www.cis.rit.edu/htbooks/nmr/inside.htm. Accessed July 25, 2008.

nuclear physics   This is the field of physics that studies atomic nuclei. A review of the nature of the nucleus is relevant to the discussion of this field.

The compact central component of an atom, called the nucleus, contains practically all of the mass of the atom, while its size is of the order 10^{-14} meter (m), about one ten-thousandth the size of an atom. Nuclei are in general composed of both protons, which carry one elementary unit of positive electric charge each, and neutrons, which are electrically neutral. (The sole exception is the hydrogen nucleus, consisting of only a single proton.) Both kinds of particles are called nucleons. The strong force, which in its internucleon version is mediated by pions, binds together the nucleons in a nucleus.

The number of protons in a nucleus is its atomic number, which determines the nucleus's electric charge and its atom's chemical properties and identity (i.e., its corresponding chemical element). The total number of nucleons in a nucleus is its mass number (or atomic mass number). A form of matter whose nuclei possess both a definite atomic number and a definite mass number is an isotope of the corresponding chemical element. Naturally occurring elements generally consist of a mixture of isotopes. The term *nuclide* is used for a type of nucleus characterized by both a particular atomic number and a particular mass number. Thus the nuclei of an isotope are all of the same nuclide.

The electric charge of a nucleus, *q*, is proportional to its atomic number, *Z*, and has the value

$$q = Ze$$

where *q* is in coulombs (C) and *e* is the magnitude of the elementary unit of electric charge, with the value $1.602176462 \times 10^{-19}$ C.

The nucleons in a nucleus are closely packed, so all nuclei possess practically the same density, which is uniform throughout most of a nucleus's volume and has the value of around 2.3×10^{17} kilograms per cubic meter (kg/m^3). So the volume of a nucleus is proportional to its mass number, giving the relation between the radius, *R*, of a nucleus and its mass number, *A*:

$$R = R_0 A^{1/3}$$

Here *R* and R_0 are in meters (m), and R_0 has the value of about 1.2×10^{-15} m.

If the masses of the individual nucleons composing a nucleus are added together, the result is greater than the mass of the nucleus. The mass difference is called the mass defect of the nucleus. Denote by *N* the number of neutrons in the nucleus, where $N = A - Z$, then

$$\text{Mass defect} = Zm_p + Nm_n - (\text{mass of nucleus})$$

where m_p is the mass of the proton, which is $1.67262158 \times 10^{-27}$ kg, and m_n is the neutron's mass, $1.67492716 \times 10^{-27}$ kg. Now convert the mass defect into its energy equivalent, according to Einstein's relativistic mass-energy relation, and the result is called the binding energy of the nucleus:

$$\text{Binding energy} = (\text{mass defect})c^2$$

where the binding energy is in joules (J), the mass defect is in kilograms (kg), and c denotes the speed of light in vacuum and equals 2.99792458×10^8 m/s. The binding energy is the amount of energy required to decompose the nucleus completely into its component nucleons. The binding energy per nucleon, given by

$$\text{Binding energy per nucleon} = \frac{\text{Binding energy}}{A}$$

serves as an indication of the stability of the nucleus: the larger the value, the more stable the nucleus.

Unstable nuclides correspond to radioactive isotopes. They undergo decay to daughter nuclides in a number of possible ways, including alpha decay, beta decay, inverse beta decay, and gamma decay. The specific decay process depends on the nuclide.

Nuclear physics studies the properties of nuclei, their structure, and their reactions, as well as applications of all those. While the strong interaction is at its strongest when it binds together the triplets of quarks that form nucleons, it "leaks" from the nucleons in a weakened version that holds nucleons together as nuclei, against the repulsive electric force among the protons. So the study of nuclei is a way of gaining understanding of the strong interaction. The particle accelerator is a major tool for the study of nuclei. One of the important applications of nuclear physics is nuclear power, the production of energy from nuclear reactions. Controlled production is performed exclusively in nuclear reactors, in which energy is obtained from controlled fission reactions, in which heavy nuclei are split into lighter nuclei with concomitant release of energy. Uncontrolled

The nuclear power station at Dukovany, Czech Republic, contains a nuclear reactor that produces heat, which boils water. The resulting high-pressure steam causes a turbine to rotate, turning an electric generator. Those processes take place in the buildings shown left of center. Electric energy is carried to consumers over high-tension power lines. The eight similar tall structures—four on the left and four on the right—are cooling towers. *(Tatonka, 2008, used under license from Shutterstock, Inc.)*

nuclear fission is the basis of nuclear-fission bombs, also called atom bombs. Nuclear-fusion bombs, often referred to as hydrogen bombs, are based on uncontrolled fusion reactions, in which light nuclei merge to form heavier nuclei and, in doing so, release energy. Such applications are the domain of nuclear engineering. Nuclear physicists are actively pursuing the goal of obtaining controlled nuclear power from fusion reactions. Another important range of application of nuclear physics is in medicine. Medical uses of nuclear physics form a field called nuclear medicine. These uses include radiation therapy for treating cancer and positron emission tomography (PET) for imaging.

See also ACCELERATORS; ATOMIC STRUCTURE; EINSTEIN, ALBERT; ELECTRICITY; ENERGY AND WORK; FISSION; FUSION; ISOTOPES; MASS; PARTICLE PHYSICS; RADIOACTIVITY; SPECIAL RELATIVITY.

FURTHER READING

Mackintosh, Ray, Jim Al-Khalili, Björn Jonson, and Teresa Peña. *Nucleus: A Trip into the Heart of Matter.* Baltimore: Johns Hopkins University Press, 2001.

Serway, Raymond A., Clement J. Moses, and Curt A. Moyer. *Modern Physics,* 3rd ed. Belmont, Calif.: Thomson Brooks/Cole, 2004.

nucleic acids In 1868, the Swiss biochemist Friedrich Miescher isolated a class of nitrogen-containing compounds from white blood cells found in the pus of infected wounds. Miescher originally called these molecules nuclein, but biochemists changed the name of these complex phosphoric compounds, acidic in nature, to *nucleic acids.* All living cells contain these long threadlike molecules because they play a biological role in the transmission of genetic material from one generation to the next. The base and sugar content distinguish two types of nucleic acids: deoxyribonucleic acid (DNA) and ribonucleic acid (RNA). For most organisms (including many viruses), DNA stores the genetic information to make the necessary proteins for a particular cell. Each monomer, or nucleotide, of DNA consists of the pentose (five-carbon) sugar deoxyribose, a phosphate group, and one of four nitrogenous bases. Two of these bases, adenine and guanine, are double-ringed, aromatic structures called purines, and two of these bases, thymine and cytosine, are single-ringed, aromatic structures called pyrimidines. The sugar and phosphate groups play a structural role in DNA, creating a backbone from which the nitrogenous bases protrude. The nitrogenous bases carry the information needed to make requisite proteins. DNA exists as a double helix with two linear polynucleotide strands running antiparallel (in opposite directions)

with respect to the orientation of the deoxyribose. Hydrogen bonds between nitrogenous base pairs, adenine interacting with thymine via two hydrogen bonds and guanine interacting with cytosine via three hydrogen bonds, link the two strands of DNA together. The specific complementary base pairing of the nitrogenous structures provides a means for the faithful replication of the DNA molecule. Biochemists have identified three forms of DNA: A-form, B-form, and Z-form. All forms have two strands that twist around one another, but B-form is the physiological configuration found in vivo.

Although it carries genetic information in some plant viruses and animal tumor viruses, the second type of nucleic acid, RNA, normally acts as a mediator between DNA and the protein-synthesizing machinery of cells. RNA nucleotides consist of a ribose sugar rather than a deoxyribose sugar, and the nitrogenous base uracil replaces thymine as the complementary base for adenine. RNA exists as a single-stranded entity with multiple functions. The three major types of RNA are messenger RNA (mRNA), transfer RNA (tRNA), and ribosomal RNA (rRNA). Determination of the unique sequences of both DNA and RNA helps elucidate their respective biological roles, and sequencing occurs through chemical, enzymatic, and automated methods. The monomers of these polynucleotides also play important physiological roles. The ribonucleotide adenosine triphosphate acts as a universal carrier of chemical energy. Coenzymes, such as nicotinamide adenine dinucleotide (NAD^+), flavin adenine dinucleotide (FAD), and coenzyme A (coA), that are necessary in metabolic pathways consist of adenine nucleotides. Signaling molecules such as cyclic adenosine monophosphate (cAMP) are also derivatives of nucleotides.

NUCLEOTIDES

The building blocks, or nucleotides, of DNA and RNA consist of three basic components: a pentose sugar, a phosphate group, and a nitrogenous base. A nucleoside has only the base and the sugar. As with polysaccharides and polypeptides, condensation reactions link nucleotides to one another. A specific type of ester bond (a link between an alcohol and an acid) called a phosphodiester bond occurs between the phosphate group of one nucleotide, acting as the acid, and the hydroxyl group of the adjacent sugar, behaving as an alcohol. The linkage is a diester bond because the phosphate group interacts with the sugar molecules on each side of it. As the bond forms between the adjacent nucleotides, the sugar releases a hydroxyl group and the phosphate group releases a hydrogen atom, or, in other words, the reaction results in the elimination of a water molecule (H_2O). Depending on the organism, the number of nucleotides in a polynucleotide may reach or even surpass 1,000,000 residues in a single polymer. With more structural variability and diversity than polypeptides, polynucleotides (the sugar-phosphate groups) have a backbone with six potential sites of rotation versus two for polypeptides. Puckering of the pentose ring and the orientation of the base and the sugar with respect to the glycosidic bond (*trans* occurs when the base and sugar are on opposite sides of the glycosidic bond, and *syn* occurs when

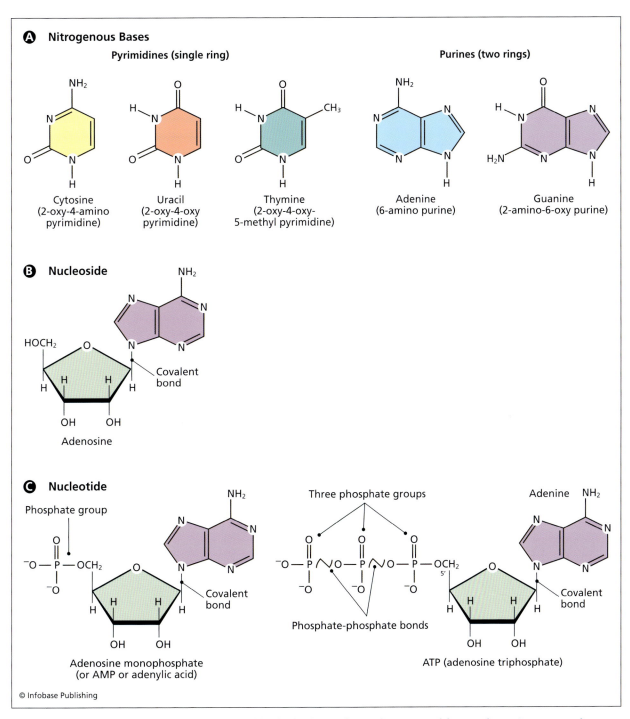

A Nitrogenous Bases

Pyrimidines (single ring)

Cytosine (2-oxy-4-amino pyrimidine)

Uracil (2-oxy-4-oxy pyrimidine)

Thymine (2-oxy-4-oxy-5-methyl pyrimidine)

Purines (two rings)

Adenine (6-amino purine)

Guanine (2-amino-6-oxy purine)

B Nucleoside

Covalent bond

Adenosine

C Nucleotide

Phosphate group

Covalent bond

Adenosine monophosphate (or AMP or adenylic acid)

Three phosphate groups

Phosphate-phosphate bonds

Adenine

Covalent bond

ATP (adenosine triphosphate)

© Infobase Publishing

(a) Nitrogenous bases may be pyrimidines with single rings of two nitrogen and four carbon atoms or purines with double rings of four nitrogen and five carbon atoms. (b) Nucleosides are nitrogenous bases linked to a sugar; shown here is an adenine linked to a ribose. (c) Nucleotides consist of a nitrogenous base, a sugar, and a phosphate group. Adenosine monophosphate (AMP) has a single attached phosphate group, and adenosine triphosphate (ATP, the energy currency of all cells) is a nucleotide consisting of an adenine, a ribose, and three phosphate groups.

the base and the sugar are on the same side of the glycosidic bond) affect the overall shape of the macromolecule and local intermolecular and intramolecular events. Nascent, or newly made, DNA and RNA form long linear chains that fold into specific three-dimensional shapes. DNA assumes the double helix with its two polynucleotide chains running in opposite directions held together by hydrogen bonds, and RNA exists as a single polynucleotide chain with unequal proportions of the four bases since no base pairing occurs, as well as the occasional unusual base, such as pseudouridine, incorporated in its length. As with polysaccharides and polypeptides, hydrolysis of polynucleotides back to single nucleotides occurs in vivo through interactions with specific enzymes, or nucleases, that cleave the phosphodiester bond, releasing nucleotides. Some nucleases cut only at specific regions within a DNA or RNA strand, and some nucleases cut at sequence-specific sites along the nucleic acid strands. For example, ribonuclease A (RNase A), a nuclease made in the mammalian pancreas and secreted into the intestine, breaks RNA into smaller fragments by cutting after cytosine and uracil residues along the polynucleotide.

In addition to serving as building blocks for DNA and RNA, nucleotides play significant roles in metabolic and signaling processes within the cell. Nucleotides act as short-term carriers of chemical energy. The ribonucleotide adenosine triphosphate (ATP) participates in the transfer of energy in hundreds of cell reactions. Every ATP molecule has three phosphates linked directly to each other by two phosphoanhydride bonds, and the hydrolysis of these two bonds, particularly the most terminal bond, releases large amounts of energy that the cell couples to many energy-requiring biosynthetic reactions. In addition to its role as a universal currency of energy, the presence of ATP alters the activities of enzymes such as glycogen synthase (involved in polysaccharide synthesis) and glutamine synthetase (involved in the synthesis of the amino acid glutamine). Adenine nucleotides are precursors to ubiquitous signaling molecules such as cyclic adenosine monophosphate (cAMP) and three major coenzymes: nicotinamide adenine dinucleotide (NAD$^+$), flavin adenine dinucleotide (FAD), and coenzyme A (CoA). Another nucleotide, guanine triphosphate (GTP), powers movement of macromolecules within the cell, moving new polypeptide chains on ribosomes. Activated intermediates in many biosynthetic pathways contain nucleotides, such as the precursors of glycogen and phosphoglycerides—uracil diphosphate (UDP)-glucose and cytosine diphosphate (CDP)-diacylglycerol.

DEOXYRIBONUCLEIC ACID

In 1953, Francis Crick and James Watson elucidated the three-dimensional structure of DNA to be two polynucleotide chains of deoxyribonucleotides wrapped in a right-handed spiral shape around each other. The hydrophilic backbone of these chains consists of alternating deoxyribose and phosphate residues with one of the four nitrogenous bases (adenine, thymine, guanine, or cytosine) linked to each sugar along the chain. The bases with their large hydrophobic surfaces stack themselves in groups of two, perpendicular to the backbones, resembling rungs on a ladder. Measurements from X-ray crystallography data indicated that the distance between the chains required one of the double-ringed aromatic bases (A or G) on one strand to interact with a single-ringed aromatic base (T or C) on the opposing strand. Studies by Erwin Chargaff in the early 1950s determined that the amount of adenine always equaled the amount of thymine in a specific DNA molecule, and the amount of guanine always equaled the amount of cytosine in the same DNA molecule. Model building showed that these ratios reflected the formation of specific hydrogen bonds between the respective pairs, adenine with thymine and guanine with cytosine. Two hydrogen bonds form between each adenine and thymine, and three hydrogen bonds form between guanine and cytosine. The combination of intermolecular interactions through these hydrogen bonds and intramolecular interactions through base stacking forces between neighboring bases on the same chain yields a highly stable double helix. Denaturation, or separating the two strands of a DNA molecule, occurs either at a critical temperature (melting point) or by chemical treatment. The base composition of the DNA molecule dictates the melting point. Because G:C base pairs have three hydrogen bonds to separate rather than two with A:T pairs, a greater fraction of G:C base pairs results in a higher melting point. If denatured DNA cools slowly from its melting point, renaturation (hydrogen bonds reform between base pairs along both strands) can occur.

DNA has more potential structural variability within each of its monomers than polypeptides. Whereas only two bonds rotate within an amino acid, six bonds freely move along the DNA backbone. Three types of helical configurations exist in DNA: A-form, B-form, and Z-form. DNA assumes these different configurations by changing the shape (puckering) of the sugar residue, by the angle in which one nucleotide is related to neighboring residues varying from 28 to 42 degrees, and by the orientation of the base and sugar with respect to the glycosidic bond (*trans* versus *syn*). The nitrogenous bases also assume different positions through phenomena called propeller twisting and base rolling. In propeller twisting, bases twist with respect to their base pairs as blades of a propeller do, and base rolling involves the base pairs tilting toward and away from neighboring residues. The degree of propeller twisting and base rolling depends on the sequence of the DNA molecule.

Watson and Crick proposed a model for the structure of B-form DNA using X-ray diffraction patterns collected from DNA fibers, which provided an average set of characteristics of all the DNA molecules in the sample. The development of in vitro DNA synthesis procedures permitted more detailed structural analyses of DNA through X-ray studies of DNA crystals with defined sequences shedding light on differences among the three helical configurations. The configuration found most often in vivo is B-form DNA. Longer and narrower than A-form but shorter and wider than Z-form, B-form has an overall shape intermediate between the other two. Unlike A-form and Z-form DNA, B-form DNA smoothly bends into an arc without kinking. The ability to bend and to twist about itself (supercoil) permits B-form DNA to associate more easily with proteins so that it can be compacted to fit inside a cell. The two chains of DNA in A-form and B-form DNA wrap around each other to the right, meaning that the helix has a right-handed screw sense; in contrast, the strands of DNA in Z-form wrap around each other to the left, making this form have a left-handed screw sense. The glycosidic bond between the deoxyribose and the nitrogenous base dictates screw sense in all three forms. Both A-form and B-form DNA have glycosidic bonds in which the sugar and the base assume the anticonfigura-

tion in all four nucleotides, with the sugar and base being on opposite sides of the bond between them. In Z-form DNA, the sugar and nitrogenous bases of cytosine and thymine nucleotides assume this anti-orientation with respect to the glycosidic bond, but the purines (A and G) assume a *syn* configuration, with the sugar and base closer together, leading the backbone to take on a zigzag shape, twisting to the left. Because the phosphate groups along the Z-form repel each other as like charges in general repel one another, the conformation is thermodynamically unfavorable except for short distances. In B-form DNA, each turn of the double helix contains about 10 base pairs separated by a rise of 3.4 nm. A-form DNA has 11 base pairs per turn of the helix, and Z-form DNA has 12. The base pairs lie almost perpendicular to the backbone axis in B-form DNA but tilt with respect to the backbone in both A-form and Z-form DNA. Indentations, or grooves, exist along the external portion of all three configurations as the two individual strands twist around each other. B-form DNA has a major groove that is wide (12 nm) and deep and a minor groove that is narrow (6 nm) and deep. Hydrogen bond donors and acceptors protrude from the surface of these grooves, permitting DNA-protein interactions to occur. Because the grooves of A-form and Z-form DNA have fewer available donors and acceptors, less water interacts with these two conformations, making them less water-soluble overall.

The sequence of purine and pyrimidine bases within the coding region of a piece of DNA contains the information necessary to yield a specific polynucleotide sequence. Every three DNA nucleotides within these coding regions (codons) represent a

COMPARISON OF A-, B-, AND Z-FORM DNA

	A-Form DNA	B-Form DNA	Z-Form DNA
Width	Widest	Intermediate	Thinnest
Rise per base pair	2.3 nm	3.4 nm	3.8 nm
Helix diameter	25.5 nm	23.7 nm	18.4 nm
Screw sense	Right-handed	Right-handed	Left-handed
Glycosididic bond	Anti	Anti	Anti for C, T Syn for A, G
Base pairs per turn of helix	11	10	12
Tilt of base pairs with respect to backbone	19 degrees	1 degree	9 degrees
Major groove	Narrow and very deep	Wide and quite deep	Flat
Minor groove	Very broad and shallow	Narrow and quite deep	Narrow and deep

specific amino acid to be incorporated into a protein. For example, the DNA sequence G-G-A represents the amino acid glycine to the protein-synthesizing machinery of the cell. Because the sequence of DNA directly correlates to proteins made by a cell, biochemists often identify the nucleotide sequence of genes as a means to study the structure and function of the encoded proteins. Biochemists use automated, chemical, and enzymatic DNA sequencing techniques to determine the order of nucleotides in a particular nucleic acid fragment. In general, sequencing involves four steps. Biochemists first purify nucleic acid from a starting tissue, then, using restriction enzymes (nucleases that cut at specific nucleotide sequences), cut the nucleic acid into smaller fragments. For each generated fragment, biochemists determine the sequence of residues, starting from one end and working to the other. Finally, they fit together the sequences of the individual fragments to ascertain the sequence of the entire gene of interest. Frederick Sanger developed a specific sequencing method, now called the Sanger method, to determine the 5,386-base-pair sequence of the Φ×174 genome. Sanger isolated DNA from the bacteriophage (a virus that infects bacteria), and he generated different lengths of radioactively labeled fragments from the genomic DNA. He separated the fragments according to size (length), using polyacrylamide gel electrophoresis. In this separation procedure, negatively charged DNA fragments move through the porous matrix of the gel toward a positive anode after electrical charge is applied. The smallest fragments travel fastest; thus, smaller fragments tend to be at the bottom of the gel. Larger fragments are caught in the pores of the gel, making them move more slowly through the substance, causing the larger pieces of DNA to stay toward the top of the gel. After exposing the gel containing the radioactively labeled fragments to X-ray film, a ladderlike pattern appears on the film. Each nucleotide has its own lane, so A, G, C, and T lanes exist, and one determines the sequence of the DNA by reading one nucleotide at a time from bottom to top. Genes responsible for many human diseases have been sequenced in this manner by the medical research community. The Human Genome Project used automated DNA sequencing as the workhorse to progress to its goal of determining the entire nucleotide sequence of the human genome.

RIBONUCLEIC ACID

RNA is a long, unbranched molecule consisting of ribonucleotides joined by $3' \rightarrow 5'$ phosphodiester bonds. As the name implies, the ribonucleotides consist of the pentose sugar ribose, a phosphate group, and a nitrogenous base. RNA and DNA consist of three of the four same bases: adenine, cytosine, and guanine. RNA molecules replace the fourth nitrogenous base found in DNA, thymine, with uracil (U). The number of nucleotides in a molecule of RNA ranges from as few as 75 to many thousands. RNA molecules are usually single-stranded, except in some viruses; consequently, an RNA molecule need not have complementary base pair ratios (equal amounts of A:U and G:C).

Most DNA molecules share a common genetic function and structural configuration. The different types of RNA molecules have different functions in the cell, and their structural properties vary accordingly. The cellular process of transcription, whereby one of two strands of DNA serves as a template for the synthesis of a complementary strand RNA, generates all of these molecules. Messenger RNA (mRNA) serves as a temporary copy of genetic instructions that can be used by the protein-synthesizing machinery in the cytoplasm, and its single-stranded structure lacks defined shape. The composition of an mRNA molecule corresponds to the DNA template, except that mRNA replaces thymine residues in the DNA with uracil residues. Transfer RNA (tRNA) acts as an adapter molecule that attaches to a specific amino acid and pairs it to the appropriate codon in the mRNA. Molecules of tRNA assume a cloverleaf, or T-shaped, conformation that permits the region that binds to an amino acid and the region that binds to the codon to be well separated although the RNA is single-stranded. Ribosomal RNA (rRNA) combines with specific proteins to form ribosomes, the physical sites of protein synthesis; thus, the conformation of rRNA is intimately associated with its interactions with proteins in the ribosomal unit. Some viruses, such as the plant virus tobacco mosaic virus (TMV), have long chains of RNA that serve as the genome wrapping around the internal core of the virus.

In most types of RNA, no sharp melting point exists because the molecule is single-stranded with no crucial number of hydrogen bonds between base pairs to break. Exceptions, such as the RNA of reovirus, assume a double-helical structure; consequently, these particular RNA molecules possess a defined melting point. Although most RNA molecules do not form a double helix, the molecules occasionally fold back on themselves in small regions, permitting the formation of hydrogen bonds in these small looped regions. Small tRNAs fold in this manner to form the cloverleaf structure, creating areas of base pairing with hairpinlike helical sections.

See also biochemistry; carbohydrates; Crick, Francis; DNA replication and repair; enzymes; Franklin, Rosalind; lipids; proteins; Watson, James; Wilkins, Maurice; X-ray crystallography.

FURTHER READING

Bloomfield, Victor A., Donald M. Crothers, Ignacio Tinoco, John E. Hearst, David E. Wemmer, Peter A. Killman, and Douglas H. Turner. *Nucleic Acids: Structures, Properties, and Functions.* Sausalito, Calif.: University Science Books, 2000.

Calladine, Chris R., Horace Drew, Ben Luisi, and Andrew Travers. *Understanding DNA: The Molecule and How It Works,* 3rd ed. San Diego, Calif.: Elsevier Academic Press, 2004.

Neidle, Stephen. *Nucleic Acid Structure and Recognition.* New York: Oxford University Press, 2002.

nutrient cycling   The constant use and regeneration of chemical elements through geological and life processes is known as nutrient cycling. Examples of nutrient cycles include the water cycle, the carbon cycle, the nitrogen cycle, and the phosphorus cycle. Each of these cycles has important biological, chemical, geological, and ecological implications.

WATER CYCLE

The water cycle involves the changes of state of the water molecule from a solid to a liquid to a vapor. Water makes up more than two-thirds of the human body and three-fourths of the surface of the planet Earth. The phase changes that water undergoes during cycling include solid to liquid, known as melting; liquid to vapor, known as evaporation; vapor to liquid, known as condensation; and liquid back to solid, known as solidification, or freezing, in the case of water and ice. The cycling of water on the planet is a continual process with no beginning or end. The evaporation of water from the oceans supplies 90 percent of all atmospheric water. Ocean evaporation occurs at different speeds, depending on the temperature of the water, the temperature of the air, the humidity in the air, and other factors. The process of transpiration, the loss of water through the leaves of plants, returns most of the remaining 10 percent of water to the atmosphere. A small fraction of 1

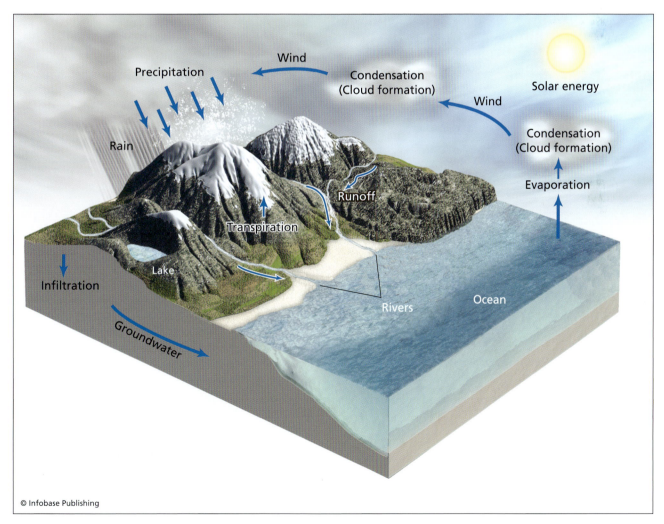

Water in the oceans, on Earth's surface, and within plants and animals evaporates from a liquid to a vapor. The water vapor then condenses in clouds and falls to Earth in the form of precipitation.

percent of the water in the atmosphere results from sublimation of snow and ice in the mountains, the direct change from a solid into a vapor. This occurs when ice or snow experiences a rapid increase in temperature that causes it to skip the intermediate liquid phase. Once the water evaporates, becoming water vapor, and reaches the atmosphere, the water vapor condenses because of lower temperature characteristic of the decreased air pressure at the higher altitudes. A cloud forms when this condensation occurs around particles in the air. When the water particles become large enough, they fall from the clouds in the form of precipitation (snow, ice, rain, or sleet). Once the water hits the ground, it either enters the ground in a process known as infiltration or stays on the surface of the ground as surface runoff. The water that infiltrates can become part of freshwater storage systems such as aquifers or springs. Gravity forces all of the surface runoff water to find the lowest point, which is the oceans; thus, all water ultimately returns to the oceans. Surface runoff collects in natural channels such as creeks, streams, and rivers, which serve a major function in returning the water to the ocean in order to continue the cycle.

The total amount of global fresh water represents only 2 to 3 percent of the total amount of all global water. The majority of the water exists in the oceans in the form of saltwater. People utilize the evaporation of water from the oceans in the reclamation of such compounds and elements as salt, magnesium, and bromine.

NITROGEN CYCLE

Another nutrient that is cycled through Earth is nitrogen. Because nitrogen is a critical component of the biomolecules proteins and nucleic acids, all living organisms require this nutrient. Nitrogen makes up 78 percent of the air. This atmospheric form of nitrogen is diatomic (N_2) and is bonded together by a triple bond between the nitrogen atoms. While this form of nitrogen is readily abundant, most organisms are unable to utilize the nitrogen in this form. In order to be useful for most life-forms, it first must undergo a process known as nitrogen fixation, carried out by certain types of bacteria that live in the soil or are catalyzed by lightning. Bacteria perform the majority of this work, fixing an estimated 140 million metric tons of nitrogen each year. Nitro-

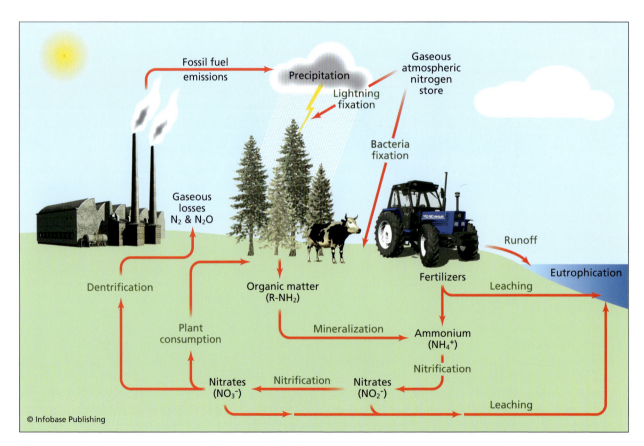

Nitrogen exists in the atmosphere in the form of N_2. Plants require nitrogen but cannot use it in the gaseous form. Bacteria and lightning perform a process known as nitrogen fixation, which results in the creation of forms of nitrogen, such as nitrates, that are usable for plants. After animals eat the plants, creating other organic matter, the nitrogen returns to earth. Denitrification can also occur, resulting in the return of nitrogen to the atmosphere.

gen-fixing bacteria often participate in mutualistic relationships (symbiotic relationships in which both partners benefit) with the roots of legumes, such as peas, soybeans, clover, alfalfa, and beans. Nitrogen fixation occurs in the bacteria by way of nitrogenase, the enzyme that catalyzes the reduction of N_2 to ammonia (NH_3), which readily ionizes into ammonium ions (NH_4^+).

The journey for the nitrogen atoms does not stop there. While ammonium is usable by plants, it can become toxic at high concentrations. Other types of bacteria can convert NH_4^+ to nitrite (NO_2^-), and then still others can turn it into nitrate (NO_3^-) through a series of oxidation reactions known as nitrification. Most plants and algae and other organisms can utilize NH_4^+ and NO_3^-. Once the plants incorporate the nitrogen (from ammonium or nitrate) into their tissues, primary consumers or decomposers can obtain nitrogen by ingesting the plant. Once the organism containing the nitrogen compounds dies, decomposers break down the nitrogenous organic material and release NH_4^+, which can be absorbed through plant roots or can cycle back into the nitrification process.

The loss of nitrogen compounds from the soil occurs through denitrification, a process carried out by some bacteria. The denitrification process is a reduction of nitrate (NO_3^-) to nitrogen (N_2) or dinitrogen oxide (nitrous oxide, N_2O), gases that are released back into the atmosphere, completing the nitrogen cycle.

Human activities significantly impact the nitrogen cycle, especially in the area of fossil fuel emissions, sewage leaching, and agricultural fertilization. Nitrous oxide compounds can be produced by the burning of fossil fuels and are important contributors to acid rain and the acidity of the soil. As the population continues to grow and the world continues to depend on fossil fuels as a primary source of energy, the amount of nitrogen compounds being released into the environment rises. Increased use of fertilizers by farmers raises the nitrate content in the soil, and, as a result, the amount of runoff in groundwater or streams and lakes increases. The same is true for leaching (percolation through the soil) of water-soluble nitrates from sewage into waterways. When the nitrate content in a lake, river, or slow-moving stream increases, the population of algae grows (causing algal blooms). This depletes the oxygen dissolved in the water, a condition that can eventually lead to eutrophication of the lake and death of most of its inhabitants. Responsible farming practices are necessary in order to prevent environmental damage from nitrogen-based fertilizers.

PHOSPHORUS CYCLE

The phosphorus cycle differs from the water and nitrogen cycles in that it does not include a gas phase. All living things require phosphorus, as it makes up the backbone of deoxyribonucleic acid and ribonucleic acid and is part of the primary energy source of the cell, that is, ATP. Phosphorus is second only to calcium in abundance in human bones. The elemental form of phosphorus is rare; most phosphorus is found in the form of phosphates (PO_4^{3-}), either inorganic or organic (associated with a carbon-based molecule). Animals can use phosphates in the form of either organic or inorganic phosphates, but plants can only use the inorganic phosphate form. Soil and rocks contain much of Earth's phosphorus, and when these rocks break down through erosion, the phosphorus enters the soil in the form of phosphates. Plants can absorb phosphates from the soil, and then primary consumers can eat the plants as their source of phosphorus. Secondary consumers

Natural weathering and erosion contribute to the recycling of rocks and minerals. Erosion can also be destructive, as shown by this chalet falling off a cliff eroded by rising tides in Norfolk, East Anglia, United Kingdom. *(Photofusion Picture Library/Alamy)*

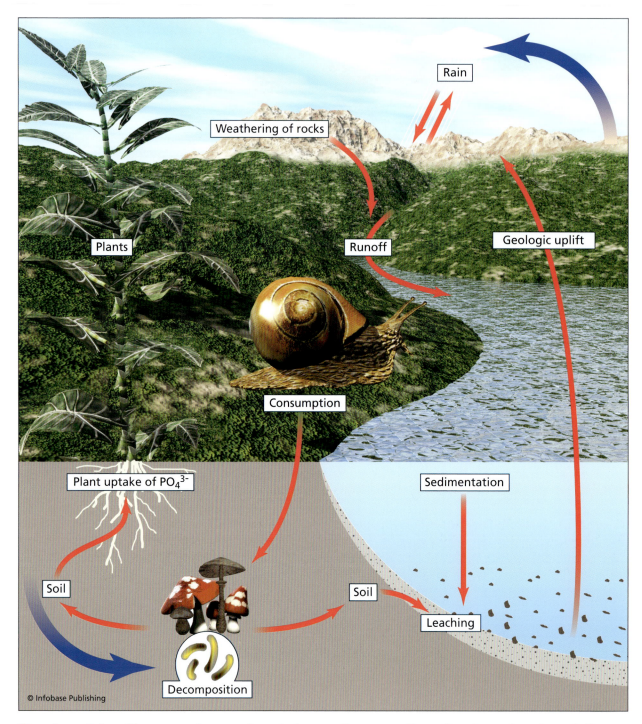

Phosphorus is found in many rocks on earth, and when erosion occurs, the rocks release phosphorus in the form of phosphates into the water. Dissolved phosphates in the water are used by plants and animals. Fertilizers contribute to soil phosphate levels. Plants and animals take up these phosphates and release them back into the soil and water by animal waste and decomposition.

eat the primary consumers, and the phosphates travel up the food chain. As the animals produce wastes, phosphates can be returned to the soil for use by the plants or enter waterways, where they can be dissolved and taken up by aquatic plants, which are then ingested by aquatic animals. When these organisms decompose, the phosphates become part of the sediment that makes its way into the rocks and soil, thus completing the phosphorus cycle.

Phosphorus is an important nutrient for plants and is a component of most fertilizers. In an average year, U.S. corn, wheat, and alfalfa crops remove from the soil 1.5 billion pounds of phosphorus that needs to be replenished in order to continue growing crops. However, as with any other chemical, the overproduction and overuse of phosphates by humans in the environment can have a large ecological impact. The overuse of fertilizer can lead to excessive phosphates in surface runoff because phosphates do not bind to the soil well and run off easily. This surface runoff makes its way to the lakes, rivers, and streams, where it can lead to the overgrowth of weedy plant life and algae that can lead to depletion of the dissolved oxygen, killing aquatic life through eutrophication, as happens with excess nitrogen. Human impact on the excess amount of phosphorus in the form of phosphates is due to fertilizer runoff, sewage leachate, as well as the use of phosphate-based detergents in washing of clothes. These phosphate-based detergents have been nearly eliminated today because of their negative environmental impact.

CARBON CYCLE

Carbon is the fourth most abundant element on the Earth and is the basis for all organic life forms, including humans. The versatile carbon atom is able to form many intricate compounds composed of straight chains, branched chains, and rings. Carbon can also have multiple substituents attached because of its ability to bond covalently with up to four partners. The most significant form of carbon in the atmosphere is carbon dioxide (CO_2). Carbon cycles through living systems in the form of carbon dioxide and carbohydrates. Photosynthetic organisms such as plants take up carbon dioxide from the atmosphere and reduce it through the process of the light-dependent reactions and the light-independent reactions (the Calvin cycle) of photosynthesis. The end products of photosynthesis are oxygen gas and carbohydrate molecules, also known as sugars. Both of these chemicals are necessary for plant and animal life. Animals and other life-forms take up the carbohydrates for use as an energy source and oxygen as a reactant in cellular respiration. During cellular respiration, cells oxidize glucose (the primary carbohydrate for cellular respiration) through a three-stage process.

The first stage, glycolysis, involves the breakdown of glucose into two three-carbon molecules known as pyruvate. The second stage is the Krebs cycle, during which the three carbon molecules of pyruvate are oxidized further to produce carbon dioxide as well as some ATP and electron carriers such as NADH and $FADH_2$. These molecules then make their way to the electron transport chain, where the oxygen that was produced via photosynthesis in plants is reduced to water, accompanied by the simultaneous production of 34–36 molecules of ATP. The carbon dioxide that is produced in cellular respiration is released back into the atmosphere and can be taken up by plants to begin the cycle again.

The release of CO_2 into the atmosphere by animals is not the sole contributor of CO_2 in the atmosphere. The burning of fossil fuels and other organic fuel sources creates carbon dioxide as a waste product. This carbon dioxide is being produced at an alarming rate and contributing to the dramatic environmental phenomenon known as global warming. As a greenhouse gas, carbon dioxide helps to trap solar radiation in the atmosphere and keep the planet warm and livable. The rapid rise in the amount of carbon dioxide released into the atmosphere means that more solar radiation is being trapped, and the average global temperature of the planet is slowly but significantly rising. The control of carbon emissions is a monumental environmental and political task that faces this generation.

Water, nitrogen, phosphorus, and carbon are examples of chemicals that naturally cycle through the environment. The availability of each of these chemicals in all of the forms in the cycle is necessary in order for the cycle to continue. The human impact on each of these cycles can have a dramatic effect on each of these natural cycles.

See also ACID RAIN; ATMOSPHERIC AND ENVIRONMENTAL CHEMISTRY; BIOCHEMISTRY; GREENHOUSE EFFECT.

FURTHER READING
Campbell, Neil A., and Jane Reece. *Biology,* 8th ed. San Francisco: Pearson Benjamin Cummings, 2007.

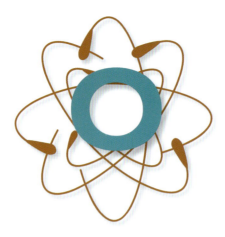

Oppenheimer, J. Robert (1904–1967) American *Theoretical Physicist*

J. Robert Oppenheimer was a pioneer in the field of quantum mechanics, the study of matter and radiation at the atomic level. As head of the Manhattan Project, he established one of the first world-class theoretical research centers, at Los Alamos, New Mexico, where during World War II, he led a team of 1,500 scientists and technicians in the design and construction of the first atomic bomb. The bombs they constructed ended World War II and propelled science into a new era. Oppenheimer was an effective, charismatic leader, whose contributions during and after the war advanced theoretical physics.

CHILDHOOD AND EDUCATION

J. Robert Oppenheimer was born to an affluent New York couple, Julius and Ella Freedman Oppenheimer, on April 22, 1904. His father was a German immigrant, who made his fortune in the textile import business, and his mother was a talented painter, who taught art. Robert had one younger brother, named Frank, and they grew up comfortably with a maid, a butler, and a chauffeur. The family frequently traveled, and on one trip to Germany, Robert's grandfather presented him with a set of rocks, sparking an interest in rock and mineral collecting. At age 11, Robert was elected the youngest member of the New York Mineralogical Society, and he presented his first paper when he was only 12 years old.

Robert attended a private school, the Ethical Culture School in New York, and graduated with honors in 1921. He planned to enroll at Harvard University that fall, but a bout with dysentery contracted during a trip to Europe prevented such plans. After a year of recovery, Robert's parents suggested he travel out west with his high school English teacher, Herbert Smith. They visited a ranch in the Pecos River Valley of New Mexico, where Robert learned to ride horses and fell in love with the geography of the region. He returned to this area later in life.

In autumn 1922, Oppenheimer enrolled at Harvard University, where he studied everything from classical literature to mathematics. He majored in chemistry but fell in love with physics. When he was not in the library, Oppenheimer was in the laboratory working with his physics professor, Percy Bridgman. In only three years, Oppenheimer graduated summa cum laude and went to England to study at the Cavendish Laboratory of Cambridge University. He hoped to work with the Nobel laureate Sir Ernest Rutherford (chemistry, 1908), who had discovered the atomic nucleus, but Rutherford did not consider Oppenheimer qualified. Another Nobel laureate (physics, 1906), Sir Joseph John Thomson, who had discovered the electron in 1897, accepted Oppenheimer into his laboratory. As intelligent as Oppenheimer was, he was clumsy in the laboratory, and his ineptness frustrated him. Being at a world-class research institution, however, he had contact with several other brilliant physicists. When he was exposed to theoretical physics, he realized that he had found his niche.

QUANTUM PHYSICS

Some physicists, called experimentalists, perform experiments to test hypotheses. They then compare the observed results with predictions that are based on theories put forth by theoretical physicists, who use mathematics to explain natural phenomena. Quantum mechanics, based on quantum theory, which helps scientists to understand the behavior of atoms and subatomic particles such as electrons, was just emerging as an exciting new field of physics. The dawn of modern physics brought with it the tools to solve phenomena that were inexplicable by classical methods. When Oppenheimer was invited to work with Max Born, who was crafting the mathematical basis for the new quantum theory at the University of Göttingen, he immediately accepted.

Before arriving in Göttingen in 1926, Oppenheimer had published two papers in the *Proceedings of the Cambridge Philosophical Society:* "On the Quantum Theory of Vibration-Rotation Bands" and "On the Quantum Theory of the Problem of the Two Bodies." The papers demonstrated his command of the new ideas and established Oppenheimer as a gifted theoretical physicist. With Born, he cowrote "Quantum Theory of Molecules," which introduced the Born-Oppenheimer approximation to explain the behavior of molecules using quantum mechanics. The method that they proposed for partitioning the energy of a molecule became standard in quantum physics. Oppenheimer also demonstrated that an electric field could be used to extract electrons from the surface of a metal. The scanning tunneling microscope, which allows scientists to observe atomic scale images of metal surfaces, is based on this discovery. Oppenheimer earned a doctorate in theoretical physics in spring 1927, and from 1928 to 1929 he continued performing research on quantum mechanics in Leiden, Holland, and Zürich, Switzerland. Oppenheimer was one of the first to apply quantum mechanics to solve problems that classical physics could not solve. Others recognized and admired his ability to grasp problems quickly and to ask meaningful probing questions.

THEORETICAL PHYSICS IN THE UNITED STATES

In 1929, Oppenheimer accepted simultaneous assistant professorships at the University of California in Berkeley and the California Institute of Technology (Caltech) in Pasadena. Before his arrival in California, Europe was the core for research in theoretical physics, but Oppenheimer established a world-class theoretical research center at Berkeley. Although he was an unpolished lecturer at first, students and colleagues recognized his underlying brilliance and were willing to ignore his intellectual snobbery to learn from and with him. His students and postdoctoral fellows followed him to and from Berkeley and Pasadena each season.

Under Oppenheimer's direction, Berkeley scientists focused on problems of relativistic quantum mechanics (quantum mechanical systems in which the number of particles can change) and the theory of electromagnetic fields. Research performed by Oppenheimer in 1930 dispelled the belief that protons are

The American physicist J. Robert Oppenheimer directed the design and construction of the atomic bomb as the head of the Manhattan Project. *(Courtesy of the J. Robert Oppenheimer Memorial Committee)*

the antimatter equivalent of electrons. He essentially predicted the existence of the positron, a positively charged particle with a mass equal to that of a negatively charged electron. In 1932, the American physicist Carl David Anderson of Caltech discovered the predicted positron, earning him the Nobel Prize in physics in 1936. In 1939, Oppenheimer predicted the existence of black holes. He discovered a phenomenon called the cascade process, in which cosmic ray particles could break down into another generation of particles. He also advanced research in particle physics (the branch of physics concerned with the behavior and properties of elementary particles) by showing that deuterons (particles made up of a proton and a neutron) could be accelerated to a greater energy than neutrons by themselves and, therefore, used to bombard positively charged nuclei at high energies.

During his 13 years at Berkeley, Oppenheimer associated closely with members of the Communist Party. Many liberal students and professors sought answers to the nation's problems by embracing ideals shared with communists, such as racial integration, an end to unemployment, and fair pay for employees. Oppenheimer's own brother and sister-in-law were members, as was a woman he seriously dated for three years. The repressive actions of Joseph Stalin later repelled Oppenheimer and turned him off communism, but these associations would haunt him later in life.

In 1940, Oppenheimer married Katherine (Kitty) Puening Harrison, who had had three previous husbands. They had their first child, a son named Peter, in 1941, and a daughter, Katherine (Toni), in 1944.

THE MANHATTAN PROJECT

Physicists began working on the theory of nuclear fission in the 1930s. Nuclear fission is the process of splitting a large nucleus into two or more smaller nuclei, thus changing the atom of one element into atoms of different elements. German scientists discovered that the bombardment of a uranium atom (atomic number 92) with a neutron caused it to break into two smaller nuclei, converting a tiny bit of matter to an enormous amount of released energy. In addition to catalyzing fission, neutrons are released during nuclear fission. The newly released neutrons bombard other adjacent nuclei, causing them to split, and a chain reaction results. This happens millions of times per second. Scientists realized this phenomenon could be exploited to create an atomic bomb more powerful than any other form of weaponry.

World War II in Europe began on September 1, 1939, when Germany invaded Poland, and the United States was drawn into the war when Japanese naval air forces bombed Pearl Harbor on December 7, 1941. American scientists raced to create an atomic bomb, a feat they believed would win the war. The scientists of the Radiation Laboratory at Berkeley decided uranium was the best choice of chemical element to use in a nuclear fission reaction. The idea was simple in theory—start a chain reaction by hitting a chunk of uranium metal with neutrons—but there were many technical problems that had to be overcome. The director of the Radiation Laboratory, Ernest Lawrence, worked furiously to try to separate uranium 235 from its chemically indistinguishable isotope, uranium 238. Isotopes are atoms of the same chemical element that differ in the amount of matter that they contain. Nuclei of uranium 235 will undergo nuclear fission, and so it was the desired form of the element; unfortunately, it was more than 100 times less abundant in nature than uranium 238. Elsewhere, scientists debated the feasibility of a nuclear reactor and examined other aspects of bomb production, such as how best to initiate the chain reaction and how to assemble the bomb. The lack of

a coordinated effort coupled with the necessity for extreme secrecy prevented real progress.

In 1942, the U.S. government organized a collective effort, code-named the *Manhattan Project,* with the objective of developing an atomic bomb. General Leslie Groves showed tremendous insight when he selected Oppenheimer, who had no previous administrative experience, to direct the project. While Oppenheimer generally was opposed to war, the repulsive Nazi regime forced him to fight back in the only way he knew—with his brain. Oppenheimer suggested that the goal would be met more easily if the scientists all worked together in one lab, where they could share ideas regularly and communicate freely without worrying about security. Groves agreed. Oppenheimer's first task was to recruit top-notch scientists to work on the secret project and convince them to move to a remote desert plateau in Los Alamos, New Mexico, where they would be housed in a dormitory of a former boys' boarding school. Military forces quickly assembled the makings of a little town to house 1,500 scientists and staff and their 3,000 accompanying family members, all of whom had to pass stringent security clearance testing.

Oppenheimer was an excellent choice for director. He was able to attract the nation's top minds and encourage them to work together toward a common goal. He defused situations caused by overinflated egos, set up a democratic system of leadership, organized weekly progress discussions and brainstorming sessions for group leaders, and established a regular seminar series for all scientists on staff. Four main divisions were created: experimental physics, theoretical physics, chemistry and metallurgy, and ordnance. Oppenheimer fully comprehended the technical information of each area. The workers described their leader as being omnipresent and having the uncanny ability of being able to enter a meeting, immediately grasp what needed to be accomplished, and jumpstart the group into moving in the right direction to complete its objectives. He was respected and admired, yet the enormous responsibility he shouldered took its toll.

Under intense pressure, his six-foot (1.8-m) frame wasted away to 110 pounds (50 kg). The government pushed for a deadline to test the bomb, which still only existed in the imagination. One pressing practical problem was the lack of an abundant supply of radioactive material for experimentation, forcing the team to rely heavily on theoretical calculations. Factories in Oak Ridge, Tennessee, and Hanover, Washington, were working around the clock to obtain enough fissionable material (uranium and plutonium). The uranium 235 was difficult to purify, and the production of plutonium was lagging.

Another difficulty was figuring out how to control the initiation of the fission reaction that would create the intensely powerful explosion. A condition called critical mass must be achieved, in which the nuclear reactions occurring inside the fissionable material compensate for the neutrons leaving the material. Oppenheimer calculated the critical mass that was necessary to maintain a chain reaction of fission and struggled over how the critical mass should be assembled. To detonate the bomb, two masses containing less than the calculated critical mass had to join together, so that the combined mass would exceed the critical mass. This had to be accomplished extremely rapidly so that the energy released by the initial fission would not propel the masses apart and terminate the chain reaction. One possible method for quickly merging the subcritical masses was the gun assembly prototype, which involved firing a slug of fissionable material into a target mass, as when a bullet is shot from a gun. A second method was the implosion method, in which the fissionable material was shaped like a hollow ball encircled by chemical explosives. When the explosives were detonated, the force would instantaneously crush the fissionable material, increasing its density and causing it to reach critical mass. (Critical mass is not only dependent on total mass, but also on concentration.) The implosion method required evenly distributing the explosives around the core of fissionable material, an engineering task that proved to be extremely difficult. Oppenheimer had scientists working on both methods and anxiously awaited an optimistic report from either team. The teams determined that the uranium bomb would need to use the simpler but more rugged gun method to initiate the chain reaction, whereas the plutonium bomb would employ the more efficient implosion method. In the end, the Manhattan Project created two plutonium "Fat Man" bombs and one uranium "Little Boy" bomb.

The world's first nuclear weapon was detonated on July 16, 1945, over the Alamogordo desert sands in New Mexico. Observers described the early morning sky as being illuminated more brightly than 1,000 Suns. Words from Hindu scripture came to Oppenheimer's mind, "I am become death, shatterer of worlds." The Los Alamos community was elated by accomplishing a seemingly insurmountable task in such a short period. The proud efforts of open collaboration among scientific geniuses under the effective leadership of one of their own were celebrated, though only briefly, as awe was replaced by recognition of the terrible reality.

Germany surrendered on May 7, 1945, but sadly, Japan would not. The government asked a committee of four scientists (Oppenheimer, Lawrence, Enrico Fermi, and Arthur Compton) for their opinions on the use of an atomic bomb against Japan. Millions had already died in the war, and it was certain a million more would die if the Allies (United States, Soviet Union, United Kingdom, France, and China) proceeded with a military invasion. On the basis of this assumption, the panel recommended using the bomb. President Harry S. Truman gave the order, and on August 6, 1945, the bomber *Enola Gay* dropped a uranium Little Boy bomb on Hiroshima, demolishing five square miles (13 km²) from the city's center and ending between 78,000 and 200,000 lives. When Japan refused to surrender, on August 9, a Fat Man was dropped on Nagasaki, killing 40,000 people. Many more died later of the effects of radiation. Japan surrendered to the Allies on August 14, 1945, ending the war, which had cost 17 million military deaths and millions more civilian deaths of war-related causes. Oppenheimer felt as if he had done something terribly wrong. He admitted to Truman that he felt as if he had blood on his hands and resigned from the project. For his efforts, he received a certificate of gratitude from General Groves and the Presidential Medal of Merit in 1946, and he then returned to teaching at Caltech.

OPPENHEIMER'S LOYALTY QUESTIONED

During the war, the Soviet Union and the United States grew into superpowers. After the war, the amicable union deteriorated, ushering in the cold war. The United States feared that the Soviets would expand communism across the world and became suspicious of anyone with ties to the Communist Party. Oppenheimer became very concerned about potential future uses of atomic weaponry and did what was in his own power to halt the nuclear arms race. As a scientific adviser to the highest levels of government, he helped compose the Acheson-Lilienthal Report, which called for stringent international controls on the development of atomic energy to ensure world peace. The plan, which was altered slightly and proposed to the United Nations, was vetoed by the Soviet Union. In 1947, Oppenheimer was appointed the chair of the General Advisory Committee of the Atomic Energy Commission. In this position until 1952, he advocated placing the development of nuclear power in the control of civilians.

Oppenheimer became director of the Institute for Advanced Study at Princeton University in 1947. The institute was an intellectual think tank, and several prominent scientists, including Albert Einstein, worked there. Under Oppenheimer's leadership, the reputation of the center as a mecca for theoretical physics research as well as humanities and social science studies escalated; unfortunately, he rarely had time to perform his own research anymore. He was a frequent lecturer on atomic energy controls and the politics of science.

President Truman was pushing for the speedy development of a hydrogen bomb, which is hundreds of times more powerful than the bombs developed under Oppenheimer's leadership. Oppenheimer, whose public opinions greatly influenced the minds of Americans, openly opposed developing the hydrogen bomb, angering many top officials. In December 1953, under suspicion of disloyalty, Oppenheimer's security clearance was withdrawn, meaning he no longer had access to any classified information. He chose to withstand a hearing, which more closely resembled a mock trial, to regain his security clearance. He admitted to having relationships with members of the Communist Party during his years at Berkeley. He had little opportunity to explain his relationships or the reasons for his opposition to the creation of the hydrogen bomb, the circumstances that aroused suspicion of his loyalties. His former ties with members of the Communist Party, interestingly, had been known by the government at the time he was appointed director of the Manhattan Project. Neither Oppenheimer nor his lawyer had access to any of the documentation that was being presented as evidence in his hearing, including memos Oppenheimer himself had written years before. His phones were illegally tapped, and the government had access to his private conversations with his lawyer. The decision to revoke Oppenheimer's security clearance appeared to have been made before the three-week-long hearing even began. Oppenheimer's years of dedication to the nation's military efforts and their success seemed forgotten. In the end, no evidence of sharing information with Soviets was found, but his clearance was revoked in April 1954 on the basis of supposed defects in his character.

Downhearted, Oppenheimer returned to Princeton to serve as director of the Institute of Advanced Study, where he found tremendous support from his colleagues, who were convinced of his loyalty. He continued to lecture frequently and wrote about the relationship between science and culture, including an influential book titled *Science and the Common Understanding*, in 1953. In an act of reconciliation, in 1963, President Lyndon B. Johnson presented Oppenheimer the Enrico Fermi Award from the Atomic Energy Commission. Though it was not a direct apology from the government, it was a public acknowledgment of the many significant contributions and sacrifices he had made for his country.

In 1966, Oppenheimer was diagnosed with throat cancer, and he retired from the institute. He died on February 18, 1967, at the age of 62, in Princeton, New Jersey.

Oppenheimer quickly mastered quantum theory and guided a whole generation of young theorists in the new physics. As a researcher, he made significant advances in the field of particle physics and contributed to many other fields. As director for the design and construction of the atomic bomb, he used his talents to create the world's most terrifying weapon and, simultaneously, to end the world's most murderous war. After becoming involved in the Manhattan Project, his personal physics research was diminished, but as a public figure and a scientific consultant to the highest levels of government, he continued to have an enormous influence on physics research and science policy for the remainder of his life.

See also FERMI, ENRICO; ISOTOPES; NUCLEAR PHYSICS; QUANTUM MECHANICS; RUTHERFORD, SIR ERNEST; THOMSON, SIR J. J.

FURTHER READING

Bethe, H. A. "J. Robert Oppenheimer (April 22, 1905–February 18, 1967)." In *Biographical Memoirs: National Academy of Sciences,* Vol. 71, 175–219. Washington, D.C.: National Academy of Sciences, 1997.

Cassidy, David C. *Oppenheimer and the American Century.* New York: Pearson Education, 2005.

Herken, Gregg. *Brotherhood of the Bomb: The Tangled Lives and Loyalties of Robert Oppenheimer, Ernest Lawrence, and Edward Teller.* New York: Henry Holt, 2002.

Smith, Alice Kimball, and Charles Weiner, eds. *Robert Oppenheimer, Letters and Recollections.* Stanford, Calif.: Stanford University Press, 1995.

optics   The study of the behavior of light, in particular, and of electromagnetic and other waves and radiation, in general, is called optics. Ray optics, or geometric optics, deals with situations in which light propagates as rays in straight lines and its wave nature can be ignored. Major topics in ray optics include reflection and refraction, which relate to mirrors, lenses, and prisms. The entry *waves* includes a discussion of reflection and refraction. Mirrors, lenses, and prisms are discussed in detail in the following. Wave optics, or physical optics, takes into account the wave nature of light and involves effects such as diffraction and interference. The entry *waves* addresses such matters. A particular application of interference, Young's experiment, is addressed in the following, as is the matter of resolving power, which is a diffraction issue.

MIRRORS

An optical device that modifies, by means of reflection and in a controlled manner, the direction of light rays impinging on it is called a mirror. (The concept is suitably generalized for other types of waves, such as other electromagnetic radiation and sound waves.) A mirror for light is made of polished metal or of a metallic coating on a polished supporting substrate. (For waves other than light waves, a mirror might be constructed of other materials.)

The most familiar type of mirror is a plane mirror. Its action is to reflect the light rays striking it so that they seem to be emanating from sources behind the mirror. The apparent sources are the *virtual images* of the actual sources, which are termed the *objects*. The designation *virtual* indicates that the image appears to be located where no light rays actually reach. Although a virtual image can be viewed, it cannot be caught on a screen. In the case of a plane mirror, the location of an object's image appears to be the same distance behind the mirror as the object is in front. The image and object have the same size, and they both possess the same orientation; the image of an upright object is upright as well.

A mirror in the shape of a concave rotational paraboloid has the useful property that all rays that strike the mirror's surface parallel to the mirror's axis are reflected to the same point on the axis, called the focal point of the mirror. This property enables a microphone to pick up sounds only from a particular distant source, ignoring other sounds. The micro-

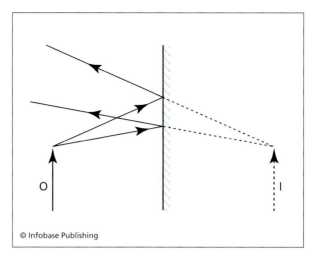

The action of a plane mirror is such that the virtual image I of object O appears to be located on the other side of the mirror from the object and at the same distance from the mirror as the object. It possesses the same size and orientation as does the object. Two rays from the object to the viewer are shown.

phone is placed at the focal point of a paraboloidal (commonly called "parabolic") reflector, which is aimed at the sound source.

Paraboloidal mirrors can often be reasonably approximated by spherical mirrors, mirrors whose shape is that of part of a surface of a sphere, as long as their depth is small compared to their diameter. Since paraboloidal mirrors for light are relatively expensive to manufacture, spherical mirrors are used when extreme precision is not necessary. The following discussion deals only with spherical mirrors. A spherical mirror is characterized by its radius of curvature, which is the radius of the sphere of which the surface forms a part. Note that a plane mirror is a limiting case of a spherical mirror, one of infinite radius of curvature. It is also assumed that the

Two mirrors—a larger concave mirror (a shaving or makeup mirror) and a smaller plane mirror—create virtual images of the ruler. *(Leslie Garland Picture Library/Alamy)*

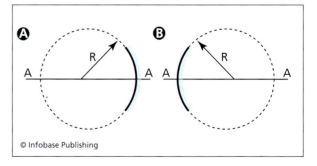

The figure shows the radius of curvature R of (a) a concave mirror and (b) a convex mirror. R is taken to be negative for a concave mirror and positive for a convex one. The mirrors' axes are denoted A-A.

light rays impinging on a mirror are almost parallel to each other and are almost perpendicular to the mirror's surface where they strike it (i.e., are almost parallel to the radius of the surface there).

FOCAL LENGTH OF MIRROR

Consider the effect of a mirror on rays that are parallel to each other and to the radius of the mirror's surface in the center of the area where they strike. After reflection, such rays are either made to converge to a point, the focal point, or made to diverge as if emanating from a point behind the mirror,

called the virtual focal point. One says that the rays are brought, or focused, to a real or virtual focus. In the former case, the mirror is a converging mirror; in the latter, a diverging mirror. A converging mirror is always concave, and a diverging mirror convex. The distance between the mirror's surface and its real or virtual focal point is the lens's focal length, denoted f. A positive value for the focal length f indicates a converging mirror, while a diverging mirror has a negative value of f. Since the focal length is a length, the SI unit of f is the meter (m). Let us denote the radius of curvature of a mirror by R in meters (m), taking R positive for a convex mirror and negative for concave. Then this relation holds:

$$f = -\frac{R}{2}$$

The focal length of a spherical mirror equals half its radius of curvature and has the opposite sign.

MIRROR EQUATION

For the focusing effect of a mirror on nonparallel rays, we assume, for the purpose of presentation, that the mirror is positioned so that its central radius is horizontal, extending left-right in front of us, with the reflecting surface on the left. The light source, or object, is located to the left of the mirror, so the light diverges from the source and strikes the mirror on its reflecting surface. Denote the distance of the object from the mirror by d_o in meters (m) and give it a positive sign. The light rays from the object might converge as they are reflected from the mirror, in which case they converge to an image, called a real image, on the left side of the mirror. Denote the distance of the image from the mirror by d_i in meters (m) and assign it a positive value. If the reflected light rays diverge, they will diverge as if they are from a virtual image located on the right side of the mirror. The distance of a virtual image from the reflecting surface is denoted by d_i as well, but with

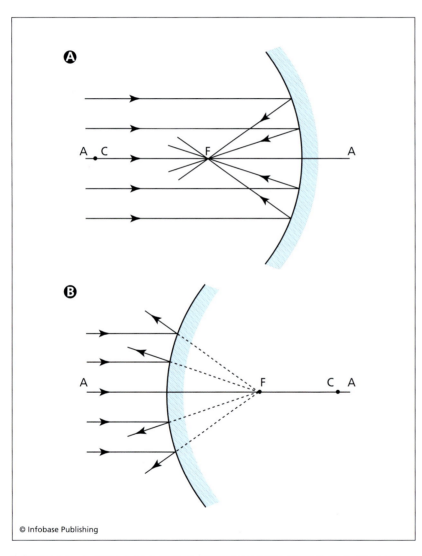

© Infobase Publishing

(a) When rays strike a converging mirror parallel to its axis A-A, they are made to converge to a point F on the axis, the focal point of the mirror. The distance of this point from the surface of the mirror along the axis is the mirror's focal length f, a positive number. (b) Similar rays that are incident on a diverging mirror are made to diverge as if they emanate from a point F on the mirror's axis A-A, a virtual focal point. This point's distance from the surface of the mirror along the axis is the mirror's focal length f, taken as a negative number. In both cases, the focal point is located midway between the mirror's center of curvature, C, and its surface.

CONVERGING MIRROR (*f* > 0)

Object Distance (*d*₀)	Image Distance (*d*ᵢ)	Image Type
∞	*f*	Real
Decreasing from ∞ to 2*f*	Increasing from *f* to 2*f*	Real
2*f*	2*f*	Real
Decreasing from 2*f* to *f*	Increasing from 2*f* to ∞	Real
f	$\pm\infty$	
Decreasing from *f* to *f*/2	Increasing from -∞ to -*f* (while decreasing in absolute value)	Virtual
f/2	-*f*	Virtual
Decreasing from *f*/2 to 0	Increasing from -*f* to 0 (while decreasing in absolute value)	Virtual

a negative sign. In brief, a positive or negative value for d_i indicates a real image (on the left) or a virtual image (on the right), respectively. If the object is located at the focal point, the rays are reflected from the mirror parallel to the radius through the focal point, neither converging nor diverging. In that case, the image distance is infinite and its sign is not important.

A real image can be caught by placing a screen, such as a movie screen, at its location. A virtual image cannot be caught in that way. Rather, a virtual image can be seen or photographed by aiming the eye or camera toward it, such as in the case of a bathroom or automobile mirror.

The rays impinging on the mirror from the left might converge rather than diverge. This occurs when the mirror intercepts the converging rays emerging from, say, another mirror or from a lens. In the absence of the mirror under consideration, the rays would converge to a real image, located to the right of the mirror. That real image serves as a virtual object for our mirror. Denote the distance of the virtual object from the surface of our mirror by d_o, and give it a negative sign. A positive or negative sign for d_o indicates a real object (on the

left) or a virtual object (on the right), respectively. If a virtual object is located at the virtual focal point of a diverging mirror, the rays reflected from the mirror are parallel to the axis and the image distance is infinite.

The object and image distances and the focal length of a mirror are related by the mirror equation:

$$\frac{1}{d_o} + \frac{1}{d_i} = \frac{1}{f}$$

This relation implies the behavior of d_i as a function of d_o for fixed f and real object (positive d_o) that is described in the tables on this page.

MIRROR MAGNIFICATION

Denote the size, or height, of an object in a direction perpendicular to the light rays by h_o and that of the corresponding image by h_i. The SI unit for both quantities is the meter (m). A positive value for h_o indicates that the object is upright in relation to some standard for orientation, while a negative value shows it is inverted. Similarly, positive h_i indicates an upright image, and a negative value an

DIVERGING MIRROR (*f* < 0)

Object Distance (*d*₀)	Image Distance (*d*ᵢ) - All Images Are Virtual
∞	-|*f*|
Decreasing from ∞ to |*f*|	Increasing to from -|*f*| to -|*f*|/2 (while decreasing in absolute value)
|*f*|	-|*f*|/2
Decreasing from |*f*| to 0	Increasing from -|*f*|/2 to 0 (while decreasing in absolute value)

inverted image. The object and image sizes are proportional to the object and image distances, respectively, through the relation

$$\frac{h_i}{h_o} = -\frac{d_i}{d_o}$$

It follows immediately that the real image of a real object is always oriented oppositely to the object, while its virtual image always has the same orientation. On the other hand, a virtual object and its real image both have the same orientation, and a virtual image of a virtual object is oppositely oriented. For a given object size, the farther the image is from the mirror, the larger it is.

The magnification is defined as the ratio of image size to object size,

$$M = \frac{h_i}{h_o}$$

which, by the preceding relation, is also

$$M = -\frac{d_i}{d_o}$$

Magnification is dimensionless and has no unit. An absolute value of M greater than 1 ($|d_i| > |d_o|$) indicates strict magnification, that the image is larger than the object. When $|M|$ equals 1 ($|d_i| = |d_o|$), the image is the same size as the object, while for $|M|$ less than 1 ($|d_i| < |d_o|$), we have reduction. Positive M (d_i, d_o of opposite signs) shows that the image and the object have the same orientation, while negative M (d_i, d_o of the same sign) indicates they are oriented oppositely. For a series of magnifications, where the image produced by the first mirror (or other optical device) serves as the object for the second, the image formed by the second serves as the object for the third, and so on, the

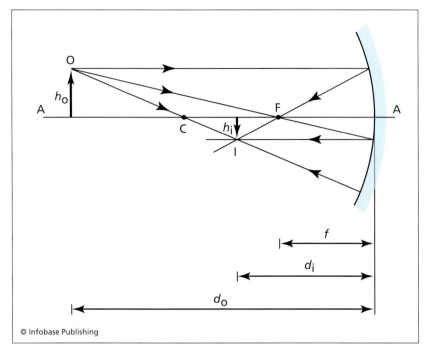

The figure shows an object O and its real image I as formed by a converging mirror of focal length f (> 0). The mirror's focal point is denoted F, its center of curvature C, and its axis A-A. The object and image distances are denoted d_o and d_i, respectively. Both are positive. The heights of the object and image are denoted, respectively, h_o and h_i. The former is positive and the latter negative. Three rays from the object to the image are shown. A converging mirror forms a real image whenever the object is farther from the mirror than is the focal point, as in the figure.

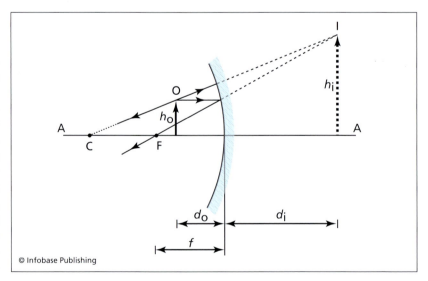

The figure shows the formation of a virtual image I of an object O by a converging mirror of focal length f (< 0). The mirror's focal point is denoted F, its center of curvature C, and its axis A-A. The object and image distances are denoted d_o and d_i, respectively, where d_o is positive and d_i is negative. The heights of the object and image are denoted, respectively, h_o and h_i. Both are positive. Two rays from the object to the viewer are shown. As in the figure, whenever the object is located between a converging mirror and its focal point, the image is virtual.

SIGN CONVENTIONS FOR MIRRORS

Quantity	Positive	Negative
R	Surface is convex	Surface is concave
f	Converging mirror	Diverging mirror
d_o	Real object	Virtual object
d_i	Real image	Virtual image
h_o	Upright object	Inverted object
h_i	Upright image	Inverted image
M	Object and image in same orientation	Object and image oppositely oriented

total magnification is the product of the individual magnifications,

$$M = M_1 M_2 \ldots$$

The table above summarizes the sign conventions used.

Real mirrors in real situations are subject to various kinds of defects, which are termed aberrations. They result in images that are distorted in various ways and are thus less than optimal in quality.

LENSES

A lens is an optical device that modifies, by means of refraction and in a controlled manner, the direction of light rays passing through it. (The concept is suitably generalized for other types of waves, such as other electromagnetic radiation and sound waves.) A lens is made of a transparent material, often glass or plastic, and typically causes light rays either to converge or to diverge.

This discussion deals with the simplest situation and assumes the following:

- The lens material is homogeneous. It is characterized by its index of refraction n, which is defined as $n = c/v$, where c is the speed of light in vacuum, with approximate value 3.00×10^8 m/s, and v denotes the speed of light in the lens material.

- The lens is immersed in air. To sufficient accuracy, the index of refraction of air is taken equal to 1: that is, the speed of light in air is taken as equal to the speed of light in a vacuum.
- Both lens surfaces have the form of spherical surfaces. So, each lens surface possesses a radius of curvature, which is the radius of the sphere of whose surface it is part. A plane surface is a special case of a spherical surface, one with infinite radius of curvature.
- The lens possesses an axis. This is an imaginary line through both surfaces, such that the

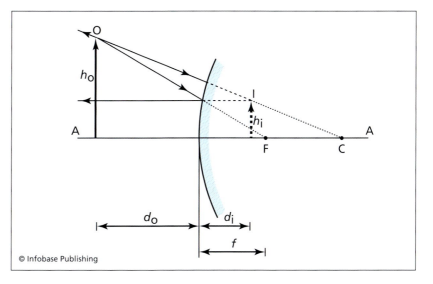

© Infobase Publishing

A diverging mirror always forms a virtual image of a real object, labeled I and O, respectively. The mirror's focal length is f (< 0). Its virtual focal point is denoted F, its center of curvature C, and its axis A-A. Object and image distances are denoted d_o and d_i, respectively, where d_o is positive and d_i negative. The heights of the object and image are denoted, respectively, h_o and h_i. Both are positive. Two rays from the object to the viewer are shown.

A collection of lenses. Converging lenses are thicker at the center than at the edge, while diverging lenses are thicker at the edge than at the center. *(Photo courtesy Jerry Bachur, Heussner Optics, Inc.)*

- surfaces are perpendicular to the line at their respective points of intersection with it.
- The lens is thin. This means that its thickness is much smaller than its other dimensions. The center of the lens is at the midpoint of the segment of the lens's axis that is between the lens surfaces.
- The light rays are close to the axis and almost parallel with it. Such rays are called paraxial rays.

FOCAL LENGTH OF LENS

Consider the effect of a lens on rays that are parallel to its axis. Such rays are either made to converge to a point, termed a focal point, or made to diverge as if emanating from a point on the other side of the lens, called a virtual focal point. One says that the rays are brought, or focused, to a real or virtual focus. In the former case, the lens is a converging lens, in the latter a diverging lens. This effect is independent of the direction the light passes through the lens. A converging or diverging lens is always thicker or thinner, respectively, at its center than at its edge. The distance between the center of the lens and its real or virtual focal point is the lens's *focal length*, and is denoted f. The focal length is the same for either direction of light passage. A positive value for the focal length f indicates a converging lens, while a diverging lens has a negative value. Since the focal length is a length, the SI unit of f is the meter (m).

A mathematical relation relates the focal length of a lens to its material and shape. Consider any one of the lens's surfaces that is not planar. Denote

its radius of curvature by R_1 in meters (m). If the surface bulges away from the lens (i.e., if the surface is convex), make R_1 a positive number. If the surface is concave, bulging in toward the lens, make R_1 negative. Denote the radius of curvature of the other surface by R_2, also in meters (m). If this surface is convex, assign a negative value to R_2. If the surface is concave, make R_2 positive. (Note that the sign convention for R_2 is opposite to that for R_1.) If the second surface is planar, then $R_2 =$ (infinity), and the sign is not important. For a lens material possessing index of refraction n, the lens maker's formula is

$$\frac{1}{f} = (n-1)\left(\frac{1}{R_1} - \frac{1}{R_2}\right)$$

The inverse of the focal length is called the optical power of the lens, D:

$$D = \frac{1}{f}$$

Its SI unit is the inverse meter (m^{-1} or $1/m$), also referred to as diopter (D). The significance of this is that the closer the focal point is to the center of the lens, and the smaller the focal length in absolute value, the greater the convergent or divergent

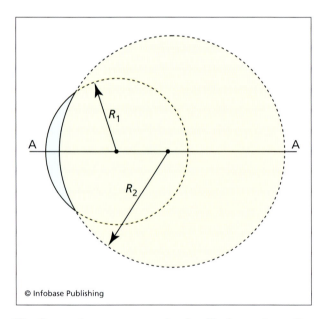

© Infobase Publishing

The figure shows an example of radii of curvature of lens surfaces. A-A denotes the lens axis. The radius of curvature of the surface on the left is R_1, and that of the right surface is R_2. According to the sign convention, both R_1 and R_2 are positive in this example.

effect of the lens on the light passing through it. The optical power expresses that numerically. As it is for focal length, a positive or negative value for D indicates a converging or diverging lens, respectively.

Combinations of lenses can be formed by having them touching at their centers with a common axis. The focal length f of such a combination of lenses having focal lengths $f_1, f_2, \ldots$, is given by

$$\frac{1}{f} = \frac{1}{f_1} + \frac{1}{f_2} + \ldots$$

Thus the optical power of the combination D equals the sum of the optical powers of the components:

$$D = D_1 + D_2 + \ldots$$

LENS EQUATION

For the focusing effect of a lens on nonparallel rays, the discussion assumes that the light source, or object, is located to the left of the lens, so the light diverges from the source and passes through the lens from left to right. Denote the distance of the object from the lens by d_o in meters (m). The light rays from the object might converge as they emerge from the lens, in which case they converge to an image, called a real image, on the right side of the lens. Denote the distance of the image from the lens by d_i in meters (m) and assign it a positive value. If the emerging light rays diverge, they will diverge as if they are from a virtual image located on the left side of the lens. The distance of a virtual image from the lens is denoted by d_i as well, but given a negative sign. In brief, a positive or negative value for d_i indicates a real image (on the right) or a virtual image (on the left), respectively. If the object is located at the focal point, the rays emerge from the lens parallel to the axis, neither converging nor diverging. In this case, the image distance is infinite, and its sign is not important.

A real image can be caught by placing a screen, such as a movie screen, at its location. A virtual image cannot be caught in that way. Rather, a

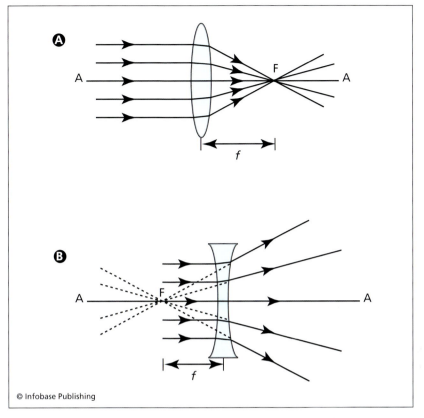

© Infobase Publishing

(a) When rays strike a converging lens parallel to its axis A-A, they are made to converge to a point F on the axis, a focal point of the lens. The distance of this point from the center of the lens is the lens's focal length f, a positive number. The lens possesses another focal point, at the same distance from the lens on the other side. (b) Similar rays that are incident on a diverging lens are made to diverge as if they emanate from a point F on the lens's axis A-A, a virtual focal point. This point's distance from the center of the lens is the lens's focal length f, taken as a negative number. The lens's other virtual focal point is located on its other side at the same distance from the lens.

virtual image can be seen or photographed by aiming the eye or camera toward it through the lens, possibly with the help of a magnifying glass or binoculars.

It might happen that the rays entering the lens from the left are converging rather than diverging. This occurs when the lens intercepts the converging rays emerging from, say, another lens. In the absence of our lens, the rays would converge to a real image, located to the right of our lens. That real image serves as a virtual object for our lens. Denote the distance of the virtual object from our lens by d_o and give it a negative sign. So a positive or negative sign for d_o indicates a real object (on the left) or a virtual object (on the right), respectively. If a virtual object is located at the focal point, the rays emerging from a diverging lens will be parallel to the axis and the image distance will be infinite.

CONVERGING LENS ($f > 0$)

Object Distance (d_o)	Image Distance (d_i)	Image Type
∞	f	Real
Decreasing from ∞ to $2f$	Increasing from f to $2f$	Real
$2f$	$2f$	Real
Decreasing from $2f$ to f	Increasing from $2f$ to ∞	Real
f	$\pm\infty$	
Decreasing from f to $f/2$	Increasing from -∞ to $-f$ (while decreasing in absolute value)	Virtual
$f/2$	$-f$	Virtual
Decreasing from $f/2$ to 0	Increasing from $-f$ to 0 (while decreasing in absolute value)	Virtual

The object and image distances and the focal length of a lens are related by the lens equation

$$\frac{1}{d_o} + \frac{1}{d_i} = \frac{1}{f}$$

This relation implies the behavior of d_i as a function of d_o for fixed f and real object (positive d_o) that is summarized in the tables above and below.

LENS MAGNIFICATION

Denote the size, or height, of an object in a direction perpendicular to the axis by h_o and that of the corresponding image by h_i. The SI unit for both quantities is the meter (m). A positive value for h_o indicates that the object is upright in relation to some standard for orientation, while a negative value shows it is inverted. Similarly, positive h_i indicates an upright image, and a negative value an inverted image. The object and image sizes are proportional to the object and image distances, respectively, through the relation

$$\frac{h_i}{h_o} = -\frac{d_i}{d_o}$$

It follows immediately that the real image of a real object is always oriented oppositely to the object, while its virtual image always has the same orientation. On the other hand, a virtual object and its real image both have the same orientation, and a virtual image of a virtual object is oppositely oriented. For a given object size, the farther the image is from the lens, the larger it is.

The magnification is defined as the ratio of image size to object size,

$$M = \frac{h_i}{h_o}$$

DIVERGING LENS ($f < 0$)

Object Distance (d_o)	Image Distance (d_i)—All Images Are Virtual
∞	$-|f|$
Decreasing from ∞ to $|f|$	Increasing to from $-|f|$ to $-|f|/2$ (while decreasing in absolute value)
$|f|$	$-|f|/2$
Decreasing from $|f|$ to 0	Increasing from $-|f|/2$ to 0 (while decreasing in absolute value)

which, by the preceding relation, is also

$$M = -\frac{d_i}{d_o}$$

Magnification is dimensionless and has no unit. An absolute value of M greater than 1 ($|d_i| > |d_o|$) indicates strict magnification, that the image is larger than the object. When $|M|$ equals 1 ($|d_i| = |d_o|$), the image is the same size as the object, while for $|M|$ less than 1 ($|d_i| < |d_o|$), we have reduction. Positive M (d_i, d_o of opposite signs) shows that the image and the object have the same orientation, while negative M (d_i, d_o of the same sign) indicates they are oriented oppositely. For a series of magnifications, where the image produced by the first lens (or other optical device) serves as the object for the second, the image formed by the second serves as object for the third, and so on, the total magnification is the product of the individual magnifications,

$$M = M_1 M_2 \ldots$$

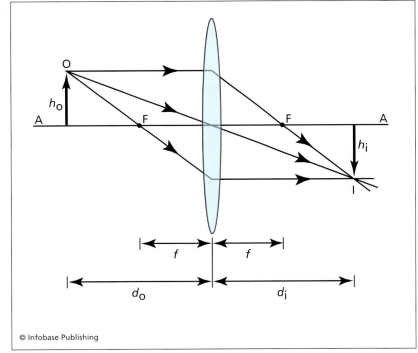

© Infobase Publishing

The figure shows an object O and its real image I as formed by a converging lens of focal length f (> 0). The lens's focal points are denoted F and its axis A-A. The object and image distances are denoted d_o and d_i, respectively. Both are positive. The heights of the object and image are denoted, respectively, h_o and h_i. The former is positive and the latter negative. Three rays from the object to the image are shown. A converging lens forms a real image whenever the object is farther from the lens than is the focal point, as in the figure.

The table on page 487 summarizes the sign conventions used.

Real lenses in real situations are subject to various kinds of defects, which are termed aberrations. They result in images that are distorted in various ways and are thus less than optimal in quality.

PRISMS

A prism is an optical device made of a block of uniform transparent material with polished planar faces. Prisms are made in different sizes and of diverse shapes, according to their purpose. One use of a prism is to change the direction of light rays, with or without inverting the image that is being transmitted. To this end, the effect of total internal reflection is exploited, whereby the light entering the prism is reflected once or more within the volume of the device before exiting. Another use of a prism is to separate light into its component colors (or, equivalently, frequencies, or wavelengths), into a spectrum. This makes use of the effect of dispersion, whereby the speed of light in the material from which the prism is formed depends on the frequency of the light. As a result, the extent of bending of light rays passing through the prism, their deviation, which results from their refraction at the prism's surfaces, also depends on the light's frequency.

YOUNG'S EXPERIMENT

Named for Thomas Young, an 18–19th-century English scientist, this experiment, carried out in the early 19th century, was pivotal in determining the wave nature of light. In this and similar experiments, light from a single source illuminates a screen, but only after the light first passes through a pair of holes close together or parallel slits in a barrier. As a result, patterns of variously colored and illuminated areas appear on the screen. They are very different from those produced by illumination through a single opening. Wave behavior explains the effect. The light passing through the openings spreads out as a result of diffraction (which is a characteristic wave effect), and so the screen has an area where illumination from the two openings overlaps. The rays from the two sources interfere at each point in the area of overlap, depending on the wavelength (i.e., the color) of the light and on the difference of distances between the point and each of the openings. In the

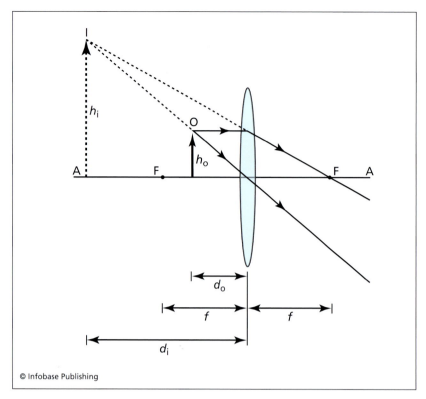

The figure shows the formation of a virtual image I of an object O by a converging lens of focal length f (> 0), which is acting as a magnifier. The lens's focal points are denoted F and its axis A-A. The object and image distances are denoted d_o and d_i, respectively, where d_o is positive and d_i is negative. The heights of the object and image are denoted, respectively, h_o and h_i. Both are positive. Two rays from the object to the viewer are shown. As in the figure, whenever the object is located between a converging lens and its focal point, the image is virtual.

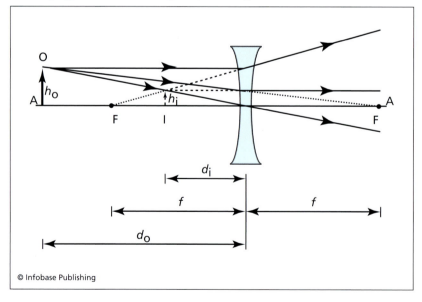

A diverging lens always forms a virtual image of a real object, labeled I and O, respectively. The lens's focal length is f (< 0), its virtual focal points are denoted F, and its axis is A-A. Object and image distances are denoted d_o and d_i, respectively, where d_o is positive and d_i negative. The heights of the object and image are denoted, respectively, h_o and h_i. Both are positive. Three rays from the object to the viewer are shown.

extreme, constructive interference produces maximal intensity, while destructive interference produces minimal intensity and even cancellation. Intermediate degrees of interference correspondingly result in intermediate intensity. A pattern of colors and of brightness and darkness appears on the screen.

When it is demonstrated today, an experimenter passes light of a single wavelength, called monochromatic light, through two parallel slits. The resulting pattern on the screen is a series of light bands, called fringes, separated by dark bands (with the single-slit diffraction pattern superposed on it). In the usual case, when the distance D between the centers of the slits is much smaller than the distance L from the slits in the screen, the spacing between adjacent fringes near the center of the interference pattern is given in meters (m) by $L\lambda/D$, where λ denotes the wavelength of the monochromatic light, and L, λ, and D are all in meters (m).

RESOLVING POWER

The ability of an optical imaging system to exhibit fine detail that exists in the object being imaged is called the resolving power of the system. The component parts, such as lenses and mirrors, affect the quality of the image. Even when the components are as nearly perfect as possible, diffraction effects limit the resolving power. The underlying factor is that the image of a point source does not appear as a point image, but rather as a blob surrounded by concentric rings. So the images of nearby points on an object might blend together and not appear separated, if the points are too close to each other.

The Rayleigh criterion for minimal image resolution of two

SIGN CONVENTIONS FOR LENSES

Quantity	Positive	Negative
R_1	First surface is convex	First surface is concave
R_2	Second surface is concave	Second surface is convex
f	Converging lens	Diverging lens
d_o	Real object	Virtual object
d_i	Real image	Virtual image
h_o	Upright object	Inverted object
h_i	Upright image	Inverted image
M	Object and image in same orientation	Object and image oppositely oriented

point light sources, named for Lord Rayleigh (John William Strutt), a 19–20th-century British physicist, is that the center of the image of one source falls on the first diffraction ring of the second. What this turns out to mean is that for two point sources to be at least marginally resolvable by an imaging system, the angle between them as viewed by the system must not be less than a certain minimal angle, called the angular limit of resolution, and given in radians (rad) by

$$\text{Angular limit of resolution} = \frac{1.22\lambda}{D}$$

Here λ denotes the wavelength, in meters (m), of the radiation used for imaging, and D is the diameter of the imaging system's aperture (opening), also in meters (m). The resolving power of the system, in the diffraction limit, is just the inverse of the angular limit of resolution,

$$\text{Resolving power} = \frac{D}{1.22\lambda}$$

This expression shows that resolving power can be increased both by increasing the aperture and by decreasing the wavelength. As for the former, satellite-borne spy cameras, which, it is said, can take pictures of the license plate numbers of motor vehicles, are constructed with enormously wide lenses. And birds of prey that search for their food from up high have especially wide pupils. Wavelength dependence is utilized through electron microscopes, for

Prisms can take any of a large variety of shapes and sizes, according to their purpose. (© Mark Karrass/ Corbis)

White light entering the prism from the left has its color components refracted at different angles at each face, causing the ray to separate into a spectrum, a continuous spread of colors. (© Matthias Kulka/zefa/Corbis)

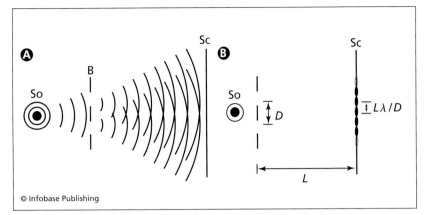

© Infobase Publishing

This experiment demonstrates interference of light and thus light's wave nature. (a) Monochromatic light from a single source (So) falls on two narrow, close, parallel slits in the barrier (B). The light passing through the slits spreads out, as a result of diffraction, and illuminates the screen (Sc). (b) As a result of interference of light from the two slits, there appears on the screen a pattern of light bands, called fringes, separated by dark bands. The distance between adjacent fringes near the center of the screen is $L\lambda/D$, where λ denotes the wavelength of the light, L is the distance from the barrier to the screen, and D is the distance between the centers of the slits.

example, which can image molecular structures that cannot be resolved by light microscopes. The wavelengths of electron beams, according to wave-particle duality, can be considerably smaller than those of visible light.

See also ACOUSTICS; DUALITY OF NATURE; ELECTROMAGNETIC WAVES; QUANTUM MECHANICS; WAVES.

FURTHER READING

Pedrotti, Frank L., Leno M. Pedrotti, and Leon S. Pedrotti. *Introduction to Optics*, 3rd ed. San Francisco: Addison Wesley, 2007.

Serway, Raymond A., Jerry S. Faughn, Chris Vuille, and Charles A. Bennet. *College Physics*, 7th ed. Belmont, Calif.: Thomson Brooks/Cole, 2006.

Young, Hugh D., and Roger A. Freedman. *University Physics*, 12th ed. San Francisco: Addison Wesley, 2007.

organic chemistry The subdiscipline of organic chemistry is the study of the chemistry of carbon. Organic chemistry involves an understanding of carbon-containing molecules: their structure, properties, reactions, composition, and synthesis. Organic compounds contain carbon and hydrogen in addition to a variety of other nonmetal elements including oxygen, nitrogen, sulfur, phosphorus, and the halogens.

The term *organic* means "derived from living systems." Originally, organic chemistry was limited to the study of compounds derived and extracted from living systems; today, this field of organic chemistry is called natural product isolation. At the beginning of the 19th century, chemists generally thought that molecules from living organisms were structurally too complicated to be synthesized in a laboratory. In 1828, a German chemist, Friedrich Wöhler, first manufactured the organic chemical urea, a constituent of urine, from inorganic ammonium cyanate (NH_4OCN), in what is now called the Wöhler synthesis. A few decades later, in 1856, William Henry Perkin, while trying to manufacture quinine, accidentally manufactured the first organic dye, now called Perkin's mauve. These two events ushered in the modern discipline of organic chemistry. The development of the field of organic chemistry continued with the discovery of petroleum and the separation of its components into various fractions by distillation. Use of these fractions enabled the petroleum industry to manufacture artificial rubbers, organic adhesives, the property modifying petroleum additives, and plastics. The production of many common plastic items, from baby bottles to disposable contact lenses to floor covering and roofing systems, has changed everyday life. The pharmaceutical industry originated in Germany during the last decade of the 19th century, when Bayer started manufacturiung acetylsalicylic acid, the active ingredient in aspirin.

Carbon atoms have unique properties that enable them to form many different compounds with numerous reactive or functional groups. Carbon's atomic number of 6 indicates that two electrons fill the first energy level, and four electrons exist in the valence shell; thus, carbon can form covalent bonds with four other nonmetal atoms simultaneously to obtain an outer shell of eight electrons (a full octet). Carbon is unique in that it can share electrons to form single, double, and triple bonds. Single bonds contain two electrons, double bonds contain four electrons, and triple bonds contain six electrons. These functional groups include alkanes, which contain C–C single bonds; alkenes, which contain C=C double bonds; and alkynes, which contain C≡C triple bonds. Other important functional groups include alcohols, alkyl halides, amines, nitriles, ethers, thiols, sulfides, sulfoxides, and sulfones. Many of the more reactive carbon groups contain a carbon-oxygen double bond called a carbonyl (C=O). These groups include aldehydes, ketones, carboxylic acids, esters, amides, acid chlorides, and acid anhydrides. Compounds containing all single bonds are saturated (alkanes), and compounds that contain double and triple bonds are classified as unsaturated (alkenes and carbonyls). This broad range of functional groups means that organic compounds form the basis of numerous products, including plastics, drugs, paints, foods, and materials used in nanotechnology.

Organic compounds have complex names, determined by using a standard set of rules that were developed by the International Union of Pure and Applied Chemistry (IUPAC). The IUPAC system works consistently with all the different functional groups of organic chemistry from alcohols to carboxylic acids. For convenience, organic chemists depend on numerous abbreviations and acronyms for many of the chemicals, reagents, and conditions they use when performing reactions. For example, the solvent dimethyl sulfoxide is referred to as *DMSO,* and *Me* is used to indicate a methyl group, which consists of one carbon atom with three hydrogen atoms attached. At first sight, this may appear very confusing, but it allows for precise communication by chemists when discussing complex organic molecules.

Most of the drugs used to treat symptoms of disease or infection, such as aspirin and ibuprofen, which are taken to reduce pain and fever, are organic compounds. Other organic chemical compounds that are used to treat bacterial infections, cancer, and a whole spectrum of other ailments and disorders are large, complex molecules containing six or seven different functional groups. These include Taxol®, which is used to treat ovarian cancer, and antibiotics such as tetracycline and Ciprofloxin®, which are used to combat bacterial infections.

Organic chemistry encompasses many subdisciplines including organometallic chemistry, the study of compounds that contain both carbon and metals, such as aluminium or magnesium, bonded together. Organometallic compounds are varied and can include compounds with alkali metals, alkali earth metals, and transition metals. Examples include Grignard reagents (MeMgBr) and organolithium reagents (MeLi), versatile reagents for many organic reactions.

The field of plastics and polymers is another specialized application of organic chemistry. Polymers are large molecules composed of small repeating units called monomers bonded together. Monomers are generally alkenes containing at least one carbon-carbon double bond such as ethylene, the simplest alkene, and vinyl chloride, which has a chlorine atom attached to the double bond. These are used to make polyethylene and polyvinyl chloride, respectively. Such polymers are versatile enough to be molded into the numerous plastic goods we see today—plastic wraps, foam insulation, and plastic bottles, to name a few. Modern society uses plastics in almost everything from clothing to car fenders, and from artificial hearts to sports equipment.

Newer specialties in carbon research are fullerenes and carbon nanotubes. Fullerenes are a family of carbon allotropes, different molecular configurations or forms of an element. The fullerenes were named after Richard Buckminster Fuller, an American architect, who was known for the geodesic dome. Carbon nanotubes are cylindrically shaped fullerenes with a closed end. Carbon nanotubes are composed of a lattice of carbon atoms in which the atoms are covalently bonded to three other carbon atoms. Three different types of nanotubes are produced by different orientations of the lattices at the closed ends. Nanotubes are quite strong, are able to conduct heat and electricity as metals do, and can be either single-walled (SWNTs) or multiwalled nanotubes (MWNTs). While the nanotube field is still developing, nanotubes can be used in electronics, lightweight building materials, and microscale machinery.

The related chemistry discipline of biochemistry involves the study of organic compounds in living systems, including the important carbon compounds of amino acids, proteins, carbohydrates, lipids, and vitamins. Understanding the ways these biomolecules react in laboratory experiments helps reveal the function and importance of these molecules in living systems, and the importance of maintaining a healthy diet.

Chiral chemistry is an important field involving the synthesis of chiral compounds with specific configurations, (R) or (S). Chiral carbon compounds contain a saturated carbon atom bonded to four different atoms or groups, and they exist in pairs called enantiomers. Many medications are presently made as racemic mixtures, containing equal amounts of both enantiomers. Chiral chemistry allows for the synthesis and isolation of only the active isomer of the chemical, thus increasing effectiveness and reducing side effects of some medications. Green chemistry initiatives use developing technologies and methods to synthesize organic molecules more efficiently with less waste, require either fewer or no organic solvents, and use far fewer hazardous reagents.

Organic chemistry is an important field of study for all medical doctors, dentists, and veterinarians, because they must know and understand how biochemicals in a living body interact with certain drugs and how to prevent unfavorable interactions.

Many professionals in the sciences must have a core understanding of the reactions and functional groups relevant to organic chemistry. These include endocrinologists, molecular biologists, material scientists, toxicologists, environmental scientists, and food technologists. Other professionals in related health fields include pharmacists, who dispense medications composed of organic chemicals; dietitians, who need to understand how organic molecules contribute to a healthy diet; health officers, who evaluate the potential hazards posed by some organic molecules; and crime investigators, who need knowledge of how different materials react with one another.

Organic chemists often describe the reactions they observe in terms of mechanisms. This approach helps them to formulate an understanding of what is actually happening in the reaction mixture as the reactants are combined. The reactants are the chemicals and organic compounds that are used in the reaction. Nucleophiles, electron-rich atoms or centers in molecules, always react with electrophiles, electron-deficient atoms or centers. A nucleophilic attack on an electrophile will form a new bond and, therefore, produce a new compound. There are four major reactions observed in organic chemistry: addition reactions, in which atoms and unsaturated molecules combine to produce new saturated compounds; elimination reactions, in which atoms or groups are lost from saturated molecules to yield unsaturated molecules; substitution reactions, in which one atom or group replaces another within a molecule; and rearrangement reactions, which involve the relocation of atoms on a molecule to produce a new functional group.

The advent of supercomputers has led to the development and application of quantum mechanical calculations of small organic molecules to model their macromolecular structure, dynamics and thermodynamics, and structure-function relationships. Using these molecular modeling strategies, organic chemists can reasonably predict the outcomes of reactions and the stability of organic compounds.

The majority of the knowledge about the field of organic chemistry has been gained by experimentation in the laboratory. Numerous reactions are performed in specialized glassware, and the products are isolated using chromatography and then analyzed for purity and composition. Thin layer (TLC), gas (GC), and liquid chromatography (LC) are all methods used by organic chemists to isolate their products from the reaction solutions. The products are then evaluated as potential candidates for pharmaceutical activity or other use. Organic chemists traditionally perform numerous reactions using various reaction conditions and reagents to produce

a plethora of compounds that can be screened for potential applications, such as therapeutic drugs. Reaction conditions include the temperature at which a reaction is performed, the length of time a reaction proceeds, or the type of solvent in which a reaction is carried out. Solvents are the liquids in which the reactions are performed. The essential requirement is that the reactants must be soluble in the solvents used. Examples of versatile organic solvents include ether, dichloromethane, and DMSO. Often these compounds are prepared on a small scale using microscale glassware.

Another significant change in the way chemists perform organic experiments has been facilitated by the availability of computer-driven instrumentation and software programs. Organic chemists are now able to prepare hundreds of molecules simultaneously by using high-throughput screening techniques. Traditionally, chemists prepared one compound at a time from one reaction by varying the reaction conditions and reagents until a successful product was obtained. This was then repeated with the next reactant. Using these new methods, organic chemists are able to prepare multiple compounds during one reaction. Multiple reactants with the same functional group can react in one flask with a specific reagent. Using unique apparatuses, labeling techniques, and multiple reagents enables all of the different products to be isolated and the successful ones identified. This has significantly changed the way drugs and target molecules are prepared and screened.

Once the compounds are prepared, they are analyzed for purity, composition, and structure using a variety of techniques, including infrared (IR) spectroscopy to determine the types of bonds and functional groups present in a molecule, mass spectrometry to identify the molecular mass and component fragments of the molecule, and nuclear magnetic resonance (NMR), which can determine the overall structure and connectivity of the atoms in the molecule.

Many of the Nobel Prizes in chemistry have been awarded to pioneering organic chemists. Robert B. Woodward received the Nobel Prize in 1965 for his outstanding achievements in the art of organic synthesis. Through his work, the field of organic chemistry gained a great deal of stature and importance in the later part of the 20th century. Other recent winners include Elias J. Corey for his development of the theory and methodology of organic synthesis in 1990; Robert F. Curl Jr., Sir Harold W. Kroto, and Richard E. Smalley for their discovery of fullerenes in 1996; and K. Barry Sharpless for his work on chirally catalyzed oxidation reactions in 2001.

In direct applications of the knowledge of organic chemistry, chemists can simulate and predict reactions seen in living systems. They can develop specific drugs that interact with selected tissues and receptors in the body to treat known ailments and conditions.

One current hot research topic is the synthesis of natural products using automation. This involves the use of automated synthesis systems, "robot chemists," to generate molecules. These systems reduce the chemist's workload and can run continuously and have cartridges loaded with reagents and microreactors rather than using traditional glassware. The advantage of this technique is that chemists no longer have to perform repetitive simple reactions; their time can be spent better planning and thinking about the science. Other significant trends are more interdisciplinary in nature, involving work with structural and molecular biology researchers to develop a better understanding of biological systems. Chemists also work with material scientists to develop organic compounds with the properties of metals and alloys.

See also BIOCHEMISTRY; COVALENT COMPOUNDS; FULLERENES; GREEN CHEMISTRY; ISOMERS; MATERIALS SCIENCE; ORGANIC NOMENCLATURE.

FURTHER READING

Carey, Francis A. *Organic Chemistry,* 7th ed. New York: McGraw Hill, 2007.

Smith, Janice G. *Organic Chemistry.* New York: McGraw Hill, 2006.

organic nomenclature Organic compounds are covalent compounds that contain carbon, a versatile atom containing six protons and six electrons. The electrons in carbon are distributed in the energy levels with the following electron configuration:

$$1s^2 2s^2 2p^2$$

meaning carbon has four valence electrons (electrons in its outer energy level). Valence electrons are critical in the behavior of an atom because they are the electrons that form bonds. Carbon atoms covalently bond with four partners, including other carbons, creating multiple structures of straight chains, branched chains, and rings.

HYDROCARBONS

One important class of organic compounds, the hydrocarbons, contain only the elements carbon and hydrogen. Hydrocarbons fall into two categories: aliphatic hydrocarbons and aromatic hydrocarbons. Aliphatic hydrocarbons have straight-chain and branched-chain configurations. Aromatic hydrocar-

bons, also called arenes, have carbons arranged in a ring structure.

Naming organic compounds follows the International Union of Pure and Applied Chemistry (IUPAC) rules for organic compounds. The names of both aliphatic and aromatic hydrocarbons depend on the number of carbon atoms present in the compound, the structure of the compound, and the location of any double or triple bonds. The number of carbons in the compound is represented by a set of prefixes shown in the table.

The type of bonds a compound possesses determines the class of organic molecules to which it belongs—alkanes, alkenes, or alkynes—depending on whether the compound contains all single carbon-carbon bonds, one or more double bonds, or triple bonds.

Writing out every carbon, hydrogen, and functional group symbol in a large organic compound becomes quite cumbersome; thus, organic chemists often use a shorthand method with lines to represent the carbon chain and the symbols of the functional groups attached to the chain. This method removes some of the confusion that occurs when all of the carbon and hydrogen atoms are shown and facilitates the identification of the functional groups on the organic molecule.

ALKANES

Alkanes are organic compounds made up entirely of single carbon-carbon bonds. The alkanes can be straight chains, branched chains, or ring structures, as long as only single bonds are present. Hydrocarbons that are alkanes are termed saturated because they contain the maximal amount of hydrogen atoms possible in a covalent compound with that number of carbons. Saturated hydrocarbons play an important role in many biological components, such as the cell membrane, where their fluid nature allows for dynamic interaction of membrane components. The straight-chain fatty acid tails of the phospholip-

PREFIXES FOR ORGANIC COMPOUNDS

Prefix	Number of Carbons	Prefix	Number of Carbons
meth-	1	hexa-	6
eth-	2	hepta-	7
prop-	3	octa-	8
but-	4	nona-	9
pent-	5	dec-	10

ids found in the lipid bilayer of the cell membrane are extremely hydrophobic and function to exclude water from the internal environment of the cell. The rotation around the single bonds is not restricted. Saturated fats found in food, such as butter and animal fat, stay solid at room temperature since the straight-chain hydrocarbons stack neatly together to form a solid.

Naming straight-chain alkanes involves adding the suffix -*ane* to the prefix for the appropriate number of carbons.

$$CH_4$$

is named *methane* and makes up 75–85 percent of natural gas.

$$CH_3-CH_3$$

is named *ethane* and is the second most abundant gas present in natural gas. The dashes representing single covalent bonds may be omitted; thus ethane can also be written

$$CH_3CH_3$$

A common fuel source in many homes, camping equipment, and outdoor grills is propane, written as follows:

$$CH_3CH_2CH_3$$

Butane fuel is found in cigarette lighters and has the formula C_4H_{10}.

$$CH_3CH_2CH_2CH_3$$

Octane is a component in automobile fuel. Gasoline grades are rated by the percentage of octane; the higher the octane, the higher the grade, and generally the higher the cost. The rest of the straight-chain parent alkanes are similarly named, including *pentane*, *hexane*, *heptane*, *nonane*, and *decane*.

ALKENES

Alkenes are organic compounds that have one or more carbon-carbon double bonds. The alkenes can form straight-chain, branched-chain, or ring structures. Hydrocarbons that are alkenes are unsaturated because they contain fewer hydrogen atoms than are theoretically possible. Rotation around carbon-carbon double bonds is not possible; therefore, the structure of alkenes contains a much more rigid backbone than that of alkanes, where there is free rotation around the carbon-carbon single bond. Functional groups located on the double-bonded carbon atoms are therefore fixed in place, and two different compounds can be formed by the very same chemical formula on the basis of whether the functional groups are located on the same side of the double bond (*cis*) or are located on opposite sides of the double bond (*trans*). The alkenes are named using the prefix system based on the number of carbons just as the alkanes are; however, they use the ending -*ene* to represent the presence of double bonds. Single-carbon compounds cannot exist because of the need for at least two carbons in order to make a carbon-carbon double bond. The lowest chain length possible would contain two carbons and be known as ethene. The rest of the parent alkene straight-chain compound names would be *propene*, *butene*, *pentene*, *hexene*, *heptene*, *octene*, *nonene*, and *decene*.

When double or triple bonds are present in the carbon chain, the number location (locant) of the double or triple bond needs to be included in the name, separated from the name with a dash.

In the name 1-hexene, the prefix *hex-* shows that the compound has six carbons. The suffix -*ene* indicates that there is a double bond. The number 1 shows that the double bond is between carbons one and two. The chemical formula for 1-hexene is

$$CH_2=CHCH_2CH_2CH_2CH_3$$

In the name *2-pentene*, the prefix *pent-* shows that the compound has five carbons. The suffix -*ene* represents the double bond, and the number 2 indicates that the double bond is between the second and the third carbon. The chemical formula for 2-pentene is

$$CH_3CH=CHCH_2CH_3$$

Alkenes can also exist in branched chains and ring structures. A prefix system indicates where the branches occur and ring structures are named by placing the prefix *cyclo-* in front of the compound's name as in the compound *cyclohexene*. This is a six-membered ring that contains a double bond. If there is not a number specified in the name of an aromatic hydrocarbon, then the functional group is found at the first carbon in the ring.

ALKYNES

Alkynes are carbon compounds that contain at least one carbon-carbon triple bond. Naming alkynes follows the same pattern as for alkanes and alkenes with the simple change of the suffix to -*yne* to represent the presence of triple bonds. As with double bonds, there is no rotation around a triple bond, making the structure rigid at those points. The terms *cis* and *trans* apply to functional groups on alkynes as well as alkenes to represent the functional group location relative to each other and the double bond. The straight-chain hydrocarbons beginning with a two-carbon chain have parent names of *ethyne, propyne, butyne, pentyne, hexyne, heptyne, octyne, nonyne,* and *decyne*. The locant must also be given to the parent name to represent the location of the triple bond. For example,

$$CH_3CH_2CH_2CH_2CH_2C \equiv CH$$

is named *1-heptyne*. The *hept-* indicates that there are seven carbons present in the compound. The *-yne* indicates that there is a triple bond in the compound, and the 1 indicates that the triple bond is located between carbon one and carbon two.

FUNCTIONAL GROUPS

Each hydrocarbon can be changed by adding a functional group, an atom, or a covalently bonded group of atoms that gives the hydrocarbon its chemical properties. The smallest functional group is a hydrogen atom. Though it is not very reactive, hydrogen serves as a functional group for alkanes, alkenes, alkynes, and arenes. Alkenes, alkynes, and arenes each have their double bonds, triple bonds, and rings, respectively, that act as functional groups as well. An alkene or alkyne that has functional groups located on the doubly or triply bonded carbon atoms encounters a new type of naming system. The terms *cis* and *trans* must precede the name of the compound. The term is not included when alphabetizing but is italicized and separated from the rest of the name by a dash.

Alkyl groups such as methyl ($-CH_3$), ethyl ($-CH_2CH_3$), propyl ($-CH_2CH_3CH_3$), and vinyl ($-CH=CH_2$) are considered functional groups when they branch from a hydrocarbon chain. Branched hydrocarbons represent a particular challenge in organic nomenclature. They often make it difficult to locate where the longest straight chain is in the compound. After the chain is identified, the attached alkyl groups are named as all the other functional groups, including carbonyl groups, carboxyl groups, amines, hydroxyl, and sulfhydryl groups, are.

The carbonyl group, which contains a carbon atom double-bonded to an oxygen atom, is used in two classes of organic molecules. Compounds that have a carbonyl group at the end of the molecule so the carbon is bonded to a hydrogen atom are called aldehydes. Ketones are organic compounds that have a carbonyl group attached to one of the middle carbon atoms.

The carboxyl group contains an oxygen group double-bonded to a central carbon that also has an attached hydroxyl group. The hydroxyl hydrogen of the carboxyl group confers an acidic characteristic to the hydrocarbon. Organic compounds that have carboxylic functional groups are carboxylic acids and include such molecules as the fatty acid linoleic acid and the acid present in vinegar, acetic acid.

Amines confer basic characteristics to the compounds to which they are attached. The most familiar amine compound is the fundamental building block of proteins known as an amino acid. The formula for an amino acid is

$$NH_2-CH-COOH$$

The hydroxyl and sulfhydryl groups both confer acidic properties to their compounds. Compounds that contain hydroxyl groups are polar and are called alcohols. Sulfhydryl groups are also polar and serve an important role aiding proteins during folding into their active conformation after polypeptide synthesis. Sulfhydryl groups cross-link different regions of protein secondary structure together in a covalent bond called a disulfide bridge.

NAMING COMPOUNDS WITH FUNCTIONAL GROUPS

Naming compounds containing functional groups requires numbering the longest carbon chain in the compound beginning with the carbon closest to the functional group. The functional group should always be given the lowest locant possible in the name of the compound. The ending of the organic compound name is often changed to the ending of the functional group name. In the compound

$$CH_3CH_2CH_2CH_2CH_2OH$$

the functional group is a hydroxyl group and the compound is called an alcohol. The numbering system begins at the carbon containing the hydroxyl group. This compound is called 1-pentanol, not 5-pentanol. The number 1 indicates that the alcohol group is found on carbon-1, the *penta-* prefix indicates that there are five carbons in the compound, and the ending *-ol* indicates the hydroxyl group.

When a compound contains more than one of the same functional group, the numbers are written consecutively, separated by commas, and separated from the words with a dash. Prefixes such as *di-* or *tri-* appear before the parent name. For example,

$$CH_3CH_2CHClCH_2CH_2CH_2Cl$$

is named *1,4-dichlorohexane*.

A compound that contains more than one type of functional group is numbered on the basis of a hierarchy of the functional groups and named by listing the functional groups in alphabetical order (ignoring prefixes such as *di-* or *tri-*). Double bonds have

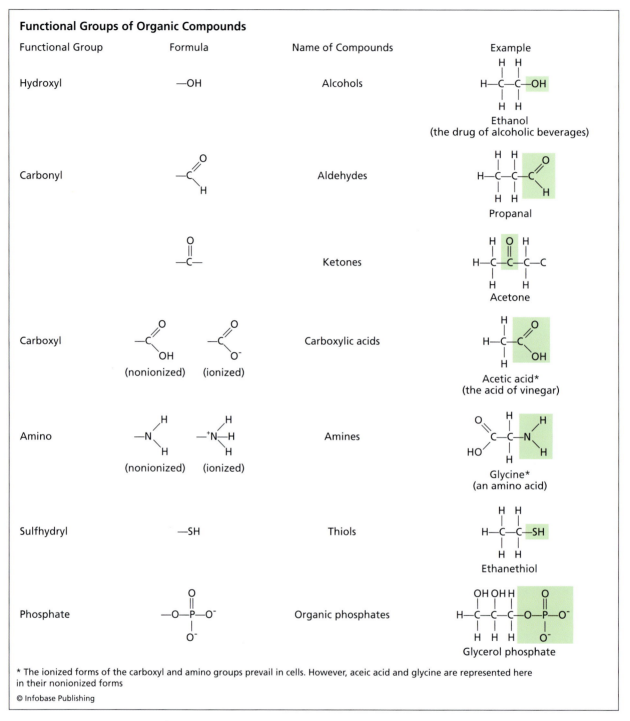

Functional Groups of Organic Compounds

Functional Group	Formula	Name of Compounds	Example

* The ionized forms of the carboxyl and amino groups prevail in cells. However, aceic acid and glycine are represented here in their nonionized forms

© Infobase Publishing

Biologically significant functional groups are important to living things.

precedence over single bonds and hydrogens. When double bonds and triple bonds appear together, the double bond has numbering precedence. The locants of the double or triple bonds occur prior to the *-ene* or *-yne* in the compound name. Alcohols have precedence over halogens and alkyl groups, while halogens and alkyl groups are considered equal, and the compound is numbered to assign the lower number to whichever of the functional groups is closer to the end of the chain. The locants of the functional groups are placed before the functional group in the compound name.

In organic nomenclature, the name includes every piece of information about the atoms in the compound. Knowing the name of the compound allows one to write the chemical formula; for example, 2,3-

dichloro-1-butene would be written as the following compound:

$$CH_3CHClCHCl=CH_2$$

The prefix *but-* explains that there are four carbons, *-ene* indicates there is a double bond present between carbons one and two, and the numbers 2 and 3 separated by commas indicate the location of the chloride groups.

Proper identification of the longest consecutive straight chain and proper numbering of the chain based on its being closest to the most important functional group are critical to successful naming of organic compounds. The following list summarizes the rules of organic compound nomenclature.

1. Locate the longest, consecutive, straight-carbon chain in the compound.
2. Locate and identify all of the functional groups attached to the straight chain.
3. Identify the most powerful functional group on the chain and begin numbering the carbon chain there, at the end closest to this functional group.
4. Include the locant for every functional group and every bond type (-yne or -ene).
5. All numbers must be separated from each other by commas.
6. All numbers must be separated from words by dashes.
7. The geometric configuration of the functional groups around a double or triple carbon bond is represented using the *cis-trans* nomenclature.
8. Double check that each functional group located on the carbon chain is represented in the name of the compound by locating it in the compound using your name.

See also BIOCHEMISTRY; COVALENT COMPOUNDS; ELECTRON CONFIGURATIONS; INORGANIC NOMENCLATURE; ORGANIC CHEMISTRY.

FURTHER READING

Carey, Francis A. *Organic Chemistry,* 5th ed. New York: McGraw-Hill, 2003.
International Union of Pure and Applied Chemistry: Commission on Nomenclature of Organic Chemistry. *Nomenclature of Organic Chemistry,* 4th ed. New York: Pergamon Press, 1979.
Smith, Janice G. *Organic Chemistry.* New York: McGraw-Hill, 2006.

oxidation-reduction reactions Reactions in which electrons are transferred between reactants are known as oxidation–reduction (or redox) reactions. *Oxidation* refers to the loss of electrons by a substance during a reaction, and *reduction* refers to the gain of electrons by a substance during a reaction. A simple acronym that simplifies the memorization of the process in redox reactions is *OIL RIG,* which stands for **o**xidation **i**s **l**oss, **r**eduction **i**s **g**ain. Oxidation and reduction reactions occur in pairs; while one reactant is oxidized, the other is reduced, and vice versa.

The scientific understanding of oxidation-reduction reactions began in the 18th century with the work of Antoine-Laurent Lavoisier. He worked to disprove the long-held theory of phlogiston, which stated that many chemical reactions resulted in the release or input of a mysterious substance known as phlogiston. Lavoisier demonstrated that these reactions were losing or gaining atoms, a demonstration that led to his formulation of the law of conservation of mass. At the time of Lavoisier, oxidation was considered the reaction of any substance with elemental oxygen (O_2), hence the name *oxidation.* Lavoisier's classic experiment that helped solidify the law of conservation of mass demonstrated that magnesium burned in air gained mass because it reacted with oxygen to form magnesium oxide according to the following reaction:

$$2Mg(s) + O_2(g) \rightarrow 2MgO(s)$$

Chemists did not discover the true basis for oxidation and reduction reactions until the 20th century. They determined that oxidation was related to the loss of electrons. Considering the oxidation of magnesium in these new terms, the reaction of magnesium and oxygen gas can be described in this way. The magnesium begins with no charge but, during the reaction, becomes incorporated into the compound of magnesium oxide, in which the magnesium has a +2 charge. The oxygen molecule also begins with no charge but ends up with a -2 charge when it forms the compound magnesium oxide. The electron configuration of the magnesium atom consists of two valence electrons in the outer shell. The octet rule states that the most stable electron configuration occurs when there is a complete outer shell of electrons, in other words, a configuration having a total of eight electrons (or two electrons, in the case of the first electron shell). Rather than gaining six electrons to fill the outer electron shell, magnesium loses its two valence electrons to the more electronegative oxygen atom.

Oxygen, on the other hand, has an electron configuration in which the outer shell has six valence electrons. To achieve stability, the oxygen atom gains two additional electrons to complete its octet rather than lose the six electrons in its valence shell. Therefore, the oxygen atom becomes a -2 ion. The magnesium atom loses its electrons to the oxygen atom, which gains them.

$$Mg + O_2 \rightarrow Mg^{2+} + O^{2-}$$

The products of the reaction subsequently combine to form MgO. The magnesium is said to be oxidized as it loses electrons, and the oxygen atom is said to be reduced as it gains the electrons. When referring to the components in an oxidation-reduction reaction, the substance that is oxidized is known as the reducing agent (in this case, magnesium), and the substance that is reduced is known as the oxidizing agent (in this case, oxygen).

Another classic reaction that demonstrates oxidation-reduction is the reaction of copper with silver ions in the following reaction:

$$Cu + Ag^+ \rightarrow Cu^{2+} + Ag$$

Copper atoms are losing electrons to the silver atom while the silver atoms gain electrons from the copper to become neutral. The copper is oxidized and acts as the reducing agent, while the silver is being reduced and acts as the oxidizing agent.

OXIDATION NUMBER

In order to understand oxidation-reduction reactions fully, one must first understand the concept of oxidation number, or the effective charge of an atom if it were to become an ion. Even if the compound is not ionic, the oxidation number is based on the charges the atoms would assume if the compound were ionic.

Oxidation numbers are used to keep track of electron transfers during reactions. The basis for oxidation numbers lies in the electron configurations of the element involved. Metals, which contain fewer than three valence electrons, will become positive ions, losing their valence electrons. Nonmetals tend to gain valence electrons, becoming negative. The rules for assigning oxidation numbers are as follows:

1. If the atom is part of a compound consisting of only one type of atom, then the oxidation number is 0. For example, atoms found in F_2, O_2, Cl_2 will all have oxidation numbers of 0.
2. The alkali metals (Group IA) have one valence electron and have an oxidation number of +1.
3. The alkaline earth metals (Group IIA) have two valence electrons and have an oxidation number of +2.
4. Group IIIA elements contain three valence electrons and have an oxidation number of +3.
5. Group IVA elements contain four valence electrons. The elements in this group include tin (Sn) and lead (Sb), which can have multiple oxidation numbers (+2 or +4). The rest of the elements in this group do not generally form ions but are given an oxidation number of +4.
6. The nitrogen family elements (Group VA) contain five valence electrons and have an oxidation number of –3 as they will gain three electrons to complete their valence shell.
7. The oxygen family elements (Group VIA) contain six valence electrons and have an oxidation number of -2 due to completing their valence shell by gaining two electrons. Oxygen is assumed to have an oxidation number of –2.
8. The halogens (Group VIIA) contain seven valence electrons and complete their octet by gaining one additional electron; thus they have an oxidation number of –1.
9. The noble gases (Group VIIIA) have a full octet and therefore have an oxidation number of 0.
10. Transition metals and inner transition metals generally have two valence electrons, so, on the basis of traditional electron configurations, they would have an oxidation number of +2. Several transition metals have multiple oxidation numbers, however.
11. Hydrogen will have an oxidation number of either -1 or +1, depending on its partner. Hydrogen only needs two electrons to complete its valence shell. It can either gain or lose its lone electron. If it reacts with a more electronegative element, it will give away its electron and become +1. Reacting with a less electronegative element, it will accept an electron and become -1.
12. All of the oxidation numbers in a compound must equal the total charge of the molecule.
13. When comparing the oxidation numbers of two atoms in a covalent compound that would not normally have a charge, the element with the higher electronegativity will be given the negative charge.
14. Oxidation numbers can be used to determine the charge of polyatomic ions as well. Sulfate (SO_4^{2-}) has a –2 charge because the sulfur atom has a +6 charge and oxygen has an oxidation of –2 for each atom with a total charge of the oxygen atoms being –8.

This gives an overall charge for the polyatomic ion of

$$+6 - (-8) = -2$$

In order to recognize an oxidation-reduction reaction, one simply needs to determine whether any of the species in the reaction have had a change in oxidation number. If they have not, the reaction is known as a metathesis reaction. To illustrate this, consider the following single-replacement reaction and the process by which the substances being reduced and oxidized are identified:

$$Cu + Li_2SO_4 \rightarrow CuSO_4 + Li$$

Copper (Cu) begins the reaction with an oxidation number of 0 based on rule 1. Elemental forms of the atom have an oxidation number of 0. When copper is ionically bonded to the sulfate ion, it has an oxidation number of +2. The oxidation state for copper changes to become more positive, indicating a loss of negatively charged electrons; thus the copper in this reaction was oxidized (oxidation **is l**oss).

Lithium (Li) begins the reaction with an oxidation number of +1, according to rule 2, as lithium is an alkali metal with only one valence electron. When lithium is in its elemental form on the product side of this reaction, it has an oxidation number of 0, indicating that it has gained an electron. This change in oxidation state demonstrates that lithium was reduced in this reaction (reduction **is g**ain).

Sulfate begins the reaction with an overall oxidation state of -2 and completes the reaction with the same overall oxidation number, meaning it is not oxidized or reduced. Copper serves as the oxidizing agent in the reaction, while lithium is the reducing agent.

BALANCING REDOX REACTIONS

When studying redox reactions, it is necessary first to divide the reaction into separate oxidation and reduction half-reactions and then to study the individual half-reactions. When writing the half-reactions, the electrons being transferred are written as free electrons (e^-). The half-reactions of the following copper/silver reaction are shown in the following:

$$Cu + Ag^+ \rightarrow Cu^{2+} + Ag$$

$$Cu \rightarrow Cu^{2+} + 2e^- \text{ (oxidation half–reaction)}$$

$$Ag^+ + e^- \rightarrow Ag \text{ (reduction half–reaction)}$$

In order to balance these equations, the number of atoms must be balanced as well as the charge. The appropriate half-reaction must be multiplied by a value that makes the charges of the two half-reactions balance. This is generally the number of electrons that were lost. Multiplying the silver reduction half-reaction by 2 balances the charge of the two half-reactions, giving the following:

$$Cu + 2Ag^+ + 2e^- \rightarrow Cu^{2+} + 2Ag + 2e^-$$

Removing the common electrons from both sides of the reactions leaves the simplified balanced reaction:

$$Cu + 2Ag^+ \rightarrow Cu^{2+} + 2Ag$$

Copper gave up electrons, so it was oxidized, while silver gained the electrons, so it was reduced. The copper is known as the reducing agent, as it caused the silver to be reduced, while silver is the oxidizing agent.

Some examples of oxidation-reductions reactions do not show a direct transfer of electrons, and yet these reactions are still considered to be redox reactions. The ownership of electrons is partially shifted, resulting in a change in oxidation number. An example of this is the reaction between carbon dioxide and hydrogen gas. This reaction is shown next.

$$CO_2 + H_2 \rightarrow H_2O + CO$$

In this reaction, the effective oxidation number of the carbon atom in CO_2 is +4. This is calculated by using the oxidation number of each oxygen atom as -2 (based on rule 7) and two oxygen atoms in the molecule. The oxidation number of carbon in CO would be +2, calculated by using the oxidation number of oxygen as -2 again and having one oxygen atom in the molecule. This is a decrease in oxidation number, meaning the carbon was reduced.

The oxidation number of the hydrogen atoms in H_2 is 0, while the oxidation number of the hydrogen atom in H_2O is +1. Again, this is calculated by using the oxidation number of oxygen as -2. This is an increase in oxidation number for hydrogen, and, therefore, the hydrogen atom is oxidized.

Although no direct electron transfer occurred, this reaction is still considered an oxidation-reduction reaction. This led to the development of a new definition for redox reactions: oxidation results in an increase in oxidation number, while reduction reactions result from a decrease in oxidation number.

ELECTRIC POTENTIAL OF REDOX REACTIONS

The electric potential (emf) of half-reactions under specific conditions known as standard state can be looked up in standard potential tables. Standard state conditions are a concentration of 1 mole/liter, gas pressures of 1 atm, and a standard room temperature of 25°C. These values are useful for comparing the amount of electrical potential between different types of half-reactions.

When a reaction occurs under conditions other than standard conditions, a calculation must be made to determine the electrical potential of the half-reactions. The Nernst equation is used to perform this calculation.

$$E = E° - (RT/nF)\ln Q$$

where E is the electrical potential at the experimental conditions, $E°$ is the standard potential from a table, R is the ideal gas constant equal to (8.31 J/K·mol), T is temperature in kelvins, n is the number of electrons being transferred, F is Faraday constant equal to 9.65×10^4 J/V·mol. $Q = C^cD^d/A^aB^b$, where the capital letters represent the actual concentration of each of the reaction components, and the lowercase exponents are the coefficients of the balanced equation

$$aA + bB \rightarrow cC + dD$$

Utilizing the Nernst equation allows the electrical potential to be calculated when given any concentration of the reactants and the products, or temperature.

See also ELECTROCHEMISTRY; MEMBRANE POTENTIAL.

FURTHER READING
Zumdahl, Steven S., and Donald J. DeCoste. *Basic Chemistry*, 6th ed. Boston: Houghton Mifflin, 2008.

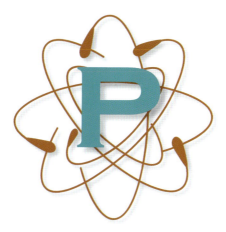

particle physics   This is the field of physics that studies the properties of and forces acting among the elementary particles. Particle physics is also called elementary particle physics, high-energy physics, or physics of particles and fields. The field is divided into three subspecialties: experimental particle physics, particle phenomenology, and theoretical particle physics. This trifold division is unusual, since most other fields of physics are divided only into experimental and theoretical subspecialties and some not even into those. The particle-physics experimentalists perform experiments and collect data. Their tools are mainly particle accelerators, detectors, and powerful computers. The experiments involve bringing about collisions among particles and detecting and analyzing the results of the collisions. The collisions might be between moving particles and particles at rest or between moving particles and particles moving in the opposite direction. The moving particles might be supplied by an accelerator—either directly or as particles produced when accelerated particles collide with stationary ones—or by nature, in the form of particles reaching Earth from beyond, called cosmic rays. The particles emanating from the collisions are detected by devices whose signals are analyzed by computer to determine the identities and properties of the collision products.

The subfield of particle phenomenology is the study and interpretation of experimental data. Phenomenologists plot the results of the experiments in various ways and attempt to find order among the data and to interpret them in light of theoretical ideas and models. In general, it is the results from the phenomenologists that are the grist for the mill of the theoreticians. The basic goal of theoretical particle physics is to understand, in fundamental terms, the properties and forces of the elementary particles, that is, to devise theories of the elementary particles. Currently accepted theories are constantly tested against experimental data. Deviations from expected results are studied as possible guides to modifications of theories or to new theories altogether.

An intimate relation appears to exist between particle physics and cosmology. The types and properties of particles and particle forces that exist at present presumably came about through cosmological processes. On the other hand, certain stages in the evolution of the universe, such as those immediately following the big bang, according to current thinking, must surely have been affected by the properties of the elementary particles and their forces.

Also called fundamental particles, the *elementary particles* include, strictly speaking, only the ultimate constituents of matter and the particles that mediate the fundamental forces among them. Somewhat more loosely, the term also includes the nucleons, which constitute the nuclei of atoms, the particles that mediate the forces holding the nucleus together, and other particles, similar to all those, although they all are composed of more fundamental components. For the looser meaning, the term *subatomic particles* is perhaps more appropriate.

PROPERTIES OF PARTICLES

Physicists characterize the elementary particles by a number of properties, the most important and better known of which are their mass, spin, and electric charge. The electric charges of the elementary particles are all positive or negative integer multiples (0, ±1, ±2, . . .) of the magnitude of the electron's charge, $e = 1.602176463 \times 10^{-19}$ coulomb (C), or, only for quarks, $\pm^1/_3 e$ or $\pm^2/_3 e$. Very often, physicists express

One of the superconducting magnets of the Large Hadron Collider at the European Organization for Nuclear Research (CERN) is being transported to its position in the assembly of the immense device. This accelerator, when completely set up, will join the search for new effects in particle physics. *(Image by Maximilien Brice/ Copyright CERN)*

the value of a particle's electric charge simply as, say, +1 or –2/3, with the *e* factor implied. Spin is an intrinsic angular momentum of a particle, as if the particle were rotating. The values of spin that particles possess are positive and limited to integer multiples (0, 1, 2, . . .) or half-integer multiples (1/2, 3/2, . . .) of the fundamental unit of angular momentum $h/(2\pi)$, where h denotes the Planck constant, whose value is $6.62606876 \times 10^{-34}$ joule-second (J·s). Spin is normally specified simply by stating the multiplier of $h/(2\pi)$, such as ½ for the electron and 1 for the photon. Masses of elementary particles are commonly designated in the unit mega-electron-volt divided by the square of the speed of light in vacuum, denoted MeV/c^2, where c represents the speed of light in vacuum. This is the mass equivalent, according to special relativity, of the energy of 1 million electron-volts and equals 1.7827×10^{-30} kilogram (kg).

Physicists characterize elementary particles by additional kinds of charge, some of which are presented later in this entry. Present understanding of the elementary particles has them divided into the two broad categories of matter particles and force

particles. The former are all fermions, while the latter are all bosons.

Named for the Italian-American physicist Enrico Fermi, a fermion is any particle—whether an elementary particle or a composite particle, such as a nucleus or an atom—that has a half-integer value of spin, that is, 1/2, 3/2, ... All of the elementary particles that form matter are fermions. Among the most common are the protons and neutrons, which are spin-½ particles and are the building blocks of atomic nuclei. Together with nuclei, electrons, which are spin-½ particles too, form atoms. Protons and neutrons consist of quarks, which are spin-½ particles as well. Collections of identical fermions are governed by the Pauli exclusion principle—which states that no two fermions can exist in exactly the same quantum state—and by Fermi-Dirac statistics—which describes the statistical properties of collections of identical fermions.

A boson is any particle—whether an elementary particle or a composite particle, such as a nucleus or an atom—that has an integer value of spin: that is, 0, 1, 2, ... This type of particle is named for the Indian physicist Satyendra Nath Bose. All the particles that mediate the various forces among the elementary particles are bosons. In particular, they are the gluons, which mediate the strong force among quarks; the intermediate vector bosons W^+, W^-, and Z^0, which mediate the weak force; the photon, which mediates the electromagnetic force; and the graviton, which mediates gravity, or the gravitational force. Except for the graviton, these are all spin-1 bosons and are called vector bosons. The graviton is a boson with spin 2. At the level of atomic nuclei, the nucleons constituting the nuclei are held together through the mediation of pions, which are spin-0 bosons. (More fundamentally, this force derives from the strong force among quarks, mediated by gluons, since nucleons and pions consist of quarks.) Collections of identical bosons are governed by the rules of Bose-Einstein statistics. At sufficiently low temperatures, very close to 0 kelvin (K), a well-isolated collection of identical bosons can become so mutually correlated that they lose their individual identities and form what amounts to a single entity. That state is called a Bose-Einstein condensate and has been created and studied in laboratories.

FUNDAMENTAL PARTICLES

Restricting the discussion to the particles that do not appear to be composed of constituents (i.e., to what seem—at least for the present—to be truly fundamental particles), the matter particles are subdivided into quarks and leptons. The leptons consist of the electron, muon, tau, and their corresponding neutrinos. Both the quarks and the leptons are further subdivided into generations. The first generation consists of the up quark (denoted u), the down quark (d), the electron (e^-), and the electron type neutrino (ν_e). The second generation is made up of the strange quark (s), the charm quark (c), the muon (μ), and the muon type neutrino (ν_μ). The third generation of matter particles consists of the top quark (t), the bottom quark (b), the tau (τ), and the tau type neutrino (ν_τ). Lumping generations together, one has six kinds—called flavors—of quark (up, down, strange, charm, top, bottom) and six leptons (electron, electron type neutrino, muon, muon type neutrino, tau, tau type neutrino). All of these possess spin ½, and an antiparticle exists for each. The antiparticles are denoted by adding a bar on top of the symbol of the corresponding particle, such as $\bar{u}$, $\bar{\mu}$, $\bar{\nu}_e$. An exception to this convention is that the electron's antiparticle, the positron, is often denoted e^+. In addition, the muon is sometimes denoted μ^- and the antimuon μ^+. Each antiparticle type possesses the same mass, spin, and magnitude of electric charge as the corresponding particle type, but its electric charge has the opposite sign. Refer to the tables "Properties of Quarks" and "Properties of Leptons."

The quarks all are stable, possess mass, and are electrically charged. Their electric charges are fractions of the magnitude of the electron's charge: the up, charm, and top quarks carry charge +2/3, while the charge of the down, strange, and bottom quarks is –1/3. In addition, all quarks carry color charge, which has three values, denoted red, blue, and green. (The use of color names is a matter of physicists' whimsy and has nothing to do with real colors.) Quarks and antiquarks in various combinations make up the subatomic particles called hadrons (discussed later). It appears that—except possibly under special, extreme conditions—quarks cannot exist independently outside hadrons.

PROPERTIES OF QUARKS

Name	Symbol	Spin	Electric Charge	Mass (MeV/c^2)
First generation				
Up	u	1/2	+2/3	2
Down	d	1/2	-1/3	5
Second generation				
Strange	s	1/2	-1/3	95
Charmed	c	1/2	+2/3	1.3×10^3
Third generation				
Bottom	b	1/2	-1/3	4.2×10^3
Top	t	1/2	+2/3	190×10^3

PROPERTIES OF LEPTONS

Name	Symbol	Spin	Electric Charge	Mass (Meν/c^2)	Lifetime (s)	Principal Decay Modes
First generation						
Electron	e^-	½	–1	0. 511	Stable	
Electron neutrino	ν_e	½	0	$< 2 \times 10^{-6}$	Stable	
Second generation						
Muon	μ	½	–1	106	2.2×10^{-6}	$e^- \bar{\nu}_e \nu_\mu$
Muon neutrino	ν_μ	½	0	< 0.19	Stable	
Third generation						
Tau	τ	½	–1	1.78×10^3	2.9×10^{-13}	$\mu^- \bar{\nu}_\mu \nu_\tau, e^- \bar{\nu}_e \nu_\tau$
Tau neutrino	ν_τ	½	0	<18.2	Stable	

The leptons all have mass, too, where the masses of the neutrinos are considerably smaller than those of the electron, muon, and tau. The latter three carry one negative unit of electric charge each, while the neutrinos are electrically neutral. The electron is a stable particle, while the muon and the tau both decay. The neutrinos do not decay, but they do undergo the process of mixing, whereby, as they travel, they continuously convert into varying quantum mixtures, or superpositions, of all three neutrino types.

The fundamental force particles are the particles that serve to intermediate the four fundamental forces. They perform their mission by being exchanged by the interacting particles. The photon (γ) is the intermediary of the electromagnetic force. It is massless, is electrically neutral, and has spin 1. It couples to (i.e., is emitted and absorbed by) particles possessing electric charge and/or magnetic dipole moment (which even some electrically neutral particles possess). The carriers of the strong force, or color force, are the eight gluons (g). Like the photon, they are massless, are electrically neutral, and possess spin 1. They carry various combinations of color charge. They couple only to particles carrying color charge (i.e., only to quarks and to themselves).

The weak force is mediated by the three intermediate vector bosons, which are named and denoted W^+, W^-, and Z^0. They all have mass and possess spin 1. The particles W^+ and W^- carry electric charge +1 and -1, respectively, while Z^0 is electrically neutral. They couple to all matter particles. Although it has not (yet) been experimentally detected, the graviton is the putative intermediary of gravitation. It should be massless, be electrically neutral, and possess spin 2. The graviton should couple to all mass and energy. For a summary of the fundamental forces and their mediating particles, refer to the table "The Fundamental Forces."

HADRONS

Consider now the hadrons, which are subatomic particles that consist of quarks and antiquarks in various combinations. The hadrons form a family of particles that comprises the baryons, which are matter particles, and the mesons, which are force particles. The baryons all consist of three quarks each. Among the baryons, the most common are the proton (p) and the neutron (n), which are the building blocks of atomic nuclei. The proton consists of two up quarks and one down quark (denoted uud), while the neutron is made of one up quark and two downs (udd). Both particles have mass and possess spin ½. The proton carries electric charge +1, and the neutron is electrically neutral. The baryons include additional particles, such as those named and denoted Λ^0 (called lambda-zero), Ω^- (omega-minus), and various versions of Δ (delta), Σ (sigma), and Ξ (xi). The Λ^0, for example, consists of an up, a down, and a strange quark (uds), while the Ω^- is made of three strange quarks (sss).

The mesons are made of quark-antiquark pairs. Among the mesons, the pions (π^+, π^-, π^0) are most noted for their role in mediating the strong force among protons and neutrons that holds nuclei together. The π^+ consists of an up quark and an anti-down quark ($u\bar{d}$), the π^- is made of an antiup and a down quark ($\bar{u}d$), while the π^0 comprises a quantum

THE FUNDAMENTAL FORCES

Force	Relative Strength	Typical Lifetime for Particle Decay via This Force (s)	Range	Mediating Particle			
				Name	Mass (GeV/c^2)	Spin	Electric Charge
Strong	1	$\leq 10^{-20}$	Short ($\approx 10^{-15}$ m)	Gluon (g)	0	1	0
Electromagnetic	1/137	$\approx 10^{-16}$	Long	Photon (γ)	0	1	0
Weak	10^{-9}	$\geq 10^{-10}$	Short ($\approx 10^{-18}$ m)	$W^\pm$, Z^0 Bosons	80.4, 91.2	1	± 1, 0
Gravitational	10^{-38}	?	Long	Graviton	0	2	0

mixture of up-antiup (u$\bar{u}$) and down-antidown (d$\bar{d}$). The pions π^+, π^-, π^0 possess mass and have spin 0. Their electric charges are +1, −1, and 0, respectively. There are additional kinds of meson, as well, two of which are named and denoted η (eta) and η′ (eta-prime), and various forms of the kaon (K). The η′, for instance, is composed of a strange-antistrange pair (s$\bar{s}$).

Among the hadrons, only the proton seems to be stable (although its long-term stability is put into question by certain elementary-particle theories). The neutron is basically unstable and decays via beta decay. However, in nuclei it can be stable, allowing for the existence of stable nuclei.

To every species of hadron there corresponds an antiparticle, denoted by adding a bar above the particle symbol. Its quark composition matches that of the corresponding particle, with every quark replaced by its antiquark, and vice versa. The antiproton ($\bar{p}$), for example, consists of two antiups and an antidown ($\bar{u}\bar{u}\bar{d}$), the π^0 is its own antiparticle, and the π^+ and π^- are each other's antiparticles. Refer to the table "Some Hadrons and Their Properties."

HISTORY

The history of human understanding of the elementary particles commenced with the 1897 discovery of the electron by the English physicist Sir J. J. Thomson, for which he was awarded the 1906 Nobel Prize in physics. Soon after, physicists recognized the hydrogen nucleus as the proton. In 1932, the English physicist James Chadwick discovered the neutron, whose existence had been predicted earlier, in order to understand the composition of atomic nuclei. He received the 1935 Nobel Prize in physics for his work. One of the principal predictors of the proton was the New Zealand–born English physicist Sir Ernest Rutherford, who was awarded the 1908 Nobel Prize in chemistry.

Next to be discovered was the positron, the electron's antiparticle. In 1928, the English physicist Paul Dirac had proposed an equation to describe the electron, a negatively charged particle. This equation appeared to predict the existence of a particle that was identical to the electron, but positively charged. At first, it was thought that the proton might be the predicted particle, since it was the only positively charged particle known then and it did possess the same spin as the electron. However, it was almost 2,000 times too massive. The American physicist Carl Anderson discovered the electron's true antiparticle, the positron, in 1932, among the particles produced by energetic particles from space, called cosmic rays, as they pass through Earth's atmosphere. For this discovery, Anderson shared the 1936 Nobel Prize in physics (with Victor Hess, who discovered cosmic rays). Dirac shared the 1933 Nobel Prize in physics for his theoretical work predicting the positron. (His partner for the prize was Erwin Schrödinger, who proposed the Schrödinger equation.)

In the early 20th century, physicists developed the theory of quantum electrodynamics, according to which the electromagnetic force resulted from the exchange of photons. Sin-Itiro Tomonaga (Japanese) and Julian Schwinger and Richard Feynman (both American) received the 1965 Nobel Prize in physics for their work in quantum electrodynamics. By the early 20th century, it was clear that some attractive force must exist among the protons and neutrons forming nuclei. Otherwise, the mutual electric repulsion among the positively charged protons would prevent the formation of nuclei. This nuclear force clearly possessed a short range, of the order of the size of a nucleus, which was very different from the long-range character of the electromagnetic force. The latter was associated with the masslessness of the photon. By analogy with quantum electrodynamics and taking into consideration the range of the nuclear

SOME HADRONS AND THEIR PROPERTIES

Name	Symbol	Quark Composition	Mass (MeV/c^2)	Spin	Electric Charge	Lifetime (s)	Principal Decay Modes
Mesons							
Pion	π^0	$u\bar{u}, d\bar{d}$	135.0	0	0	8.4×10^{-17}	2γ
	π^+	$u\bar{d}$	139.6	0	+1	2.60×10^{-8}	$\mu^+ \nu_\mu$
	π^-	$\bar{u}d$	139.6	0	−1	2.60×10^{-8}	$\mu^- \bar{\nu}_\mu$
Kaon	K^+	$u\bar{s}$	493.7	0	+1	1.24×10^{-8}	$\mu^+ \nu_\mu$
	K^-	$\bar{u}s$	493.7	0	−1	1.24×10^{-8}	$\mu^- \bar{\nu}_\mu$
	K^0_S	$d\bar{s}, \bar{d}s$	497.9	0	0	8.9×10^{-11}	$\pi^+ \pi^-, 2\pi^0$
	K^0_L	$d\bar{s}, \bar{d}s$	497.9	0	0	5.2×10^{-8}	$\pi^\pm e^\mp \bar{\nu}_e, 3\pi^0, \pi^\pm \mu^\mp \bar{\nu}_\mu$
Eta	η	$u\bar{u}, d\bar{d}$	548.8	0	0	$< 10^{-18}$	$2\gamma, 2\pi$
	η'	$s\bar{s}$	958	0	0	2.2×10^{-21}	$\eta\pi^+\pi^-$
Baryons							
Proton	p	uud	938.3	$\frac{1}{2}$	+1	Stable	
Neutron	n	udd	939.6	$\frac{1}{2}$	0	887	$pe^- \bar{\nu}_e$
Lambda	Λ^0	uds	1,116	$\frac{1}{2}$	0	2.6×10^{-10}	$p\pi^-, n\pi^0$
Sigma	Σ^+	uus	1,189	$\frac{1}{2}$	+1	8.0×10^{-11}	$p\pi^0, n\pi^+$
	Σ^0	uds	1,193	$\frac{1}{2}$	0	6×10^{-20}	$\Lambda^0\gamma$
	Σ^-	dds	1,197	$\frac{1}{2}$	−1	1.5×10^{-10}	$n\pi^-$
Xi	Ξ^0	uss	1,315	$\frac{1}{2}$	0	2.9×10^{-10}	$\Lambda^0 \pi^0$
	Ξ^-	dss	1,321	$\frac{1}{2}$	−1	1.6×10^{-10}	$\Lambda^0\pi^-$
Delta	Δ^{++}	uuu	1.232	$\frac{3}{2}$	+2	$\approx 10^{-23}$	$p\pi^+$
Omega	Ω^-	sss	1,672	$\frac{3}{2}$	−1	8.2×10^{-11}	$\Lambda^0 K^-, \Xi^0\pi^-, \Xi^-\pi^0$

force, the Japanese physicist Hideki Yukawa proposed in 1935 that the nuclear force results from the exchange of a mediating particle, whose mass should be around 250 times the mass of the electron. The following year such a particle, the muon, was discovered, with a mass of about 207 times the electron's mass. It soon became clear that the muon was not the particle that Yukawa predicted, because it interacted with nuclei only weakly. In 1947, Yukawa's particle was discovered in the form of the pion (also called pi meson), whose mass is around 270 times that of the electron. Yukawa received the 1949 Nobel Prize in physics for his work on the nuclear force.

In the mid-20th century, experiments using increasingly powerful accelerators revealed an embarrassing wealth of baryons (the family that includes the proton and neutron), mesons (similar to the pions), and leptons (including the electron and muon), numbering a hundred or so and exasperatedly referred to at the time as the particle zoo. In 1961, the American physicist Murray Gell-Mann and, independently, the Israeli physicist Yuval Ne'eman proposed a successful classification scheme for the hadrons (baryons and mesons) called the eightfold way. Gell-Mann received the 1969 Nobel Prize in physics for this classification. The particle picture remained complex, however, and was begging for simplification. In the late 20th century, the situation with the hadrons became greatly simplified by the introduction of the idea that quarks were the fundamental particles from which hadrons were formed. The strong force among quarks was mediated by gluons, and the nuclear force was understood as an aspect of the strong force. The weak force, responsible for such phenomena as the decay of the neutron, became understood as a separate force mediated by exchange of intermediate vector bosons (W and Z). The lepton picture became complete (as it seems today) with the discovery of the tau and the three kinds of neutrino. And physicists recognized the analogy between the quarks and leptons.

Additional developments during the late 20th century followed. Physicists came to understand the fundamental forces among the elementary particles in terms of gauge theories, which are quantum theories of fields that possess a space-time-dependent symmetry called gauge symmetry. The electromagnetic and weak forces became understood as aspects of a more general force, called the electroweak force. Physicists developed quantum chromodynamics, the gauge theory of the strong force. This picture of the particles and forces described earlier, with the forces understood in terms of gauge theories, is called the standard model. It indeed serves as the present standard against which new ideas are compared. Although physicists generally assume that a further unification of the electroweak and strong forces should be accomplishable, that has not yet happened. Such a unification is referred to as a grand unified theory (GUT). If it is ever accomplished, an even farther-reaching unification of the GUT force with gravitation is envisioned and is dubbed a theory of everything (TOE). Some physicists developed and are working on string theory as an attempt to understand what underlies the fundamental forces in terms of even more fundamental entities, extremely tiny vibrating objects, requiring more than three spatial dimensions for their existence. This approach has yet to prove itself useful.

Some of the physicists who contributed to the understanding of the elementary particles and their forces since the mid 20th-century, as reflected by the Nobel Prizes in physics that were awarded, are the following:

- The Chinese-Americans Chen Ning Yang and Tsung-Dao Lee investigated the weak force and especially its violation of the reflection (left-right) symmetry that had been assumed to hold (1957 prize).
- The Americans Emilio Segrè and Owen Chamberlain discovered the antiproton (1959 prize).
- The American Robert Hofstadter studied the internal structure of nucleons (protons and neutrons) by scattering electrons from atomic nuclei (shared the 1961 prize).
- The Americans Burton Richter and Samuel Ting fruitfully investigated hadrons (1976 prize).
- The American Sheldon Glashow, the Pakistani Abdus Salam, and the American Steven Weinberg contributed to the unification of the weak and electromagnetic forces to the electroweak force (1979 prize).
- The Americans James Cronin and Val Fitch discovered that the decay of the neutral kaons violated a hitherto assumed symmetry involving reflection (left-right) and particle-antiparticle exchange (1980 prize).
- The American Carlo Rubbia and the Dutch Simon van der Meer discovered the mediating particles of the weak force, the W and Z (1984 prize).
- The Americans Leon Lederman, Melvin Schwartz, and Jack Steinberger investigated neutrinos and discovered the muon neutrino (1988 prize).
- The Americans Jerome Friedman and Henry Kendall and the Canadian Richard Taylor furthered the investigation into the internal structure of the proton and neutron by scattering electrons from them, a process that served to support the quark model (1990 prize).
- The Americans Martin Perl and Frederick Reines experimentally investigated the leptons and discovered the tau (Perl) and the muon neutrino (Reines) (1995 prize).
- The Dutch Gerardus 't Hooft and Martinus Veltman carried out theoretical investigations of the electroweak force (1999 prize).
- The Americans David Gross, David Politzer, and Frank Wilczek investigated the strong force theoretically and showed why single quarks do not appear under ordinary conditions (2004 prize).

See also ACCELERATORS; BIG BANG THEORY; BOSE-EINSTEIN STATISTICS; COSMOLOGY; ELECTRICITY; ELECTROMAGNETISM; ENERGY AND WORK; FERMI-DIRAC STATISTICS; FERMI, ENRICO; FEYNMAN, RICHARD; GELL-MANN, MURRAY; INVARIANCE PRINCIPLES; MASS; MATTER AND ANTIMATTER; MOMENTUM AND COLLISIONS; NUCLEAR PHYSICS; PAULI, WOLFGANG; PHYSICS AND PHYSICISTS; QUANTUM MECHANICS; RADIOACTIVITY; ROTATIONAL MOTION; RUTHERFORD, SIR ERNEST; SCHRÖDINGER, ERWIN; SPECIAL RELATIVITY; SYMMETRY; THEORY OF EVERYTHING; THOMSON, SIR J. J.

FURTHER READING

Barnett, R. Michael, Henry Muehry, and Helen R. Quinn. *The Charm of Strange Quarks: Mysteries and Revolutions of Particle Physics.* New York: Springer-Verlag, 2000.

Kane, Gordon L. *The Particle Garden: Our Universe as Understood by Particle Physicists.* Boston: Addison Wesley, 1995.

Ne'eman, Yuval, and Yoram Kirsh. *The Particle Hunters,* 2nd ed. Cambridge: Cambridge University Press, 1996.

Schwarz, Cindy, and Sheldon Glashow. *A Tour of the Subatomic Zoo: A Guide to Particle Physics,* 2nd ed. New York: Springer-Verlag, 1997.

Pasteur, Louis (1822–1895) French *Chemist*

Louis Pasteur was born in Dôle, France, on December 27, 1822, to hardworking parents named Jeanne and Jean-Joseph Pasteur. His father was a tanner by trade and his family lived a modest life. As his parents valued learning, he was encouraged to attend school and pursue higher education, although he was not considered a model student, often preferring fishing to studies. Pasteur attended school in the town in which he lived (Arbois) and was encouraged by the headmaster to take the entrance test for the École Normale Supérieure, a competitive arts and science college. Pasteur was sent to Paris to study for the entrance test but returned home shortly afterward as a result of homesickness. Pasteur continued his local studies and received his bachelor's degree from the Royal College of Besançon in 1840. He later received acceptance into the École Normale Supérieure in 1843 for his doctoral work. Pasteur received his doctorate in 1847 and took a position at the University of Strasbourg, where he began his work on the study of fermentation. In 1854, he became a chemistry professor at the University of Lille. In 1857, he accepted a professorship at his alma mater, École Normale Supérieure. He later was given a position as dean when he returned to the University of Lille in 1863. In 1888, Pasteur founded the Pasteur Institute in Paris and personally ran and worked at the institute until his death. Pasteur is responsible for the development of multiple new fields of science, as well as the prevention of many diseases.

Louis Pasteur married Marie Laurent in 1849. They had five children, three of whom died while very young. Many believe that these tragedies led to Pasteur's steadfast mission of curing diseases later in his career. One of his most valuable discoveries related to actually preventing diseases rather than later to curing them, leading to the development of vaccinations. His work on microorganisms and his commitment to solving the problems that they cause were consistent themes throughout his life.

Louis Pasteur was a 19th-century French chemist who contributed greatly to many areas of chemistry. He helped refute the concept of spontaneous generation, contributed to the wine industry with his work on fermentation, and developed vaccinations for diseases such as rabies. *(National Library of Medicine)*

CONTRIBUTIONS TO SCIENCE

Pasteur's initial work at École Normale Supérieure was on the optical activity of crystals. He noticed that synthetically produced crystal samples of tartaric acid, the main acid in wine, were not optically active, meaning they were not able to rotate plane-polarized light. White light travels in all directions, but plane-polarized light has passed through a polarizing filter and only the light that vibrates in the same direction passes through the polarizing filter. The filter blocks passage of light traveling in other directions. Placing a compound with an asymmetric carbon (having four unique substituents attached) in the path of plane-polarized light rotates the direction of the pathway of the light. When Pasteur closely studied this phenomenon and isolated the individual crystal types in tartaric acid mixtures, he found that the tartaric acid crystals contained two oppositely rotating crystals; this was the first example of a racemic mixture, one that contains both optically active forms of a com-

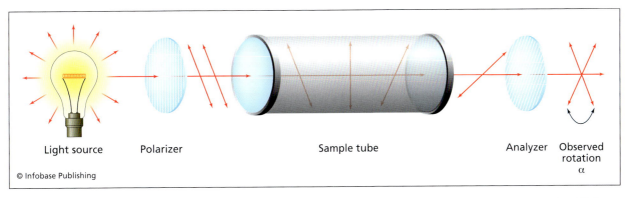

Light source   Polarizer   Sample tube   Analyzer   Observed
                                                                                                rotation
                                                                                                    α

© Infobase Publishing

A compound is considered to be optically active when it rotates the polarization plane of plane-polarized light. Ordinary light vibrates in all directions. When passed through a polarizer, the light vibrates in a single plane. Passing the polarized light through a sample of an optically active substance causes the plane of polarization to be rotated, something that can be measured by an analyzer. The amount of rotation and the direction of rotation reveal information about the structure of the sample.

pound. These different structural arrangements with the same molecular formula are known as stereoisomers. Since the tartaric acid crystals rotated light in both directions, they cancelled one another out and created an optically inactive substance.

When a synthetic tartaric acid (which normally is not optically active) sample was inoculated with microorganisms, Pasteur found that the optical activity of the sample increased. He hypothesized that the change in the optical activity of the sample was due to the microorganism's preferentially utilizing only one form of the acid. As the one form was depleted from the sample, it left behind the other form of the crystal, which could then rotate the plane-polarized light without interference from the other stereoisomer. Pasteur concluded that living organisms can only utilize one type of crystal structure. Work on stereoisomers by Pasteur set the stage for the development of a field of chemistry called stereochemistry, which deals with the three-dimensional structure of molecules.

The discovery in optical activity allowed Pasteur to expand his work on the synthetic tartaric acid experiments to a living system. Pasteur investigated the optical activity of tartaric acid found on grapes in the vineyards of France. At the time of his study, the wine industry was suffering serious losses due to a mysterious illness in their grape vineyards. Pasteur determined the optical activity of the tartaric acid in the grape samples and realized that it was becoming increasingly optically active, indicating that the grape blight was being caused by a microorganism (a microscopic organism such as bacteria, yeast, or virus). This was the beginning of what became known as the germ theory, which states that diseases are caused by microorganisms. The germ theory has been called the single most important contribution to medical

science. Developments in germ theory led to the birth of the branch of life science known as microbiology. Pasteur's work continued to revolve around microorganisms and the control of their growth for years.

Other work of Louis Pasteur was pivotal in disproving a commonly accepted phenomenon known as spontaneous generation. Believers in spontaneous generation suggested the reason that meat rotted, for instance, was that the meat gave rise to maggots. As supporting evidence, they noted that everytime that meat was left out, maggots appeared on it, demonstrating spontaneous generation. Work by Francesco Redi and Louis Pasteur helped disprove this long-held concept. Redi designed an experiment with sealed jars containing meat. When the meat was placed in a sealed jar, no maggots appeared; when the meat was exposed to the outside air, maggots formed. This suggested that the meat needed to be in contact with the air in order to have maggots. Proponents of spontaneous generation stated that the reason nothing grew in the closed container was that sealing the container prevented air from reaching the maggots, and that air contained a "life force" that was required for growth. This is where Pasteur became involved.

Scientists knew that if broth was left exposed to air, bacteria would quickly fill the flask. Pasteur designed an experiment using swan-necked flasks that enabled air to travel through them but prevented microorganisms or dust from entering. The air and the so-called life force were still able to enter the flask. When Pasteur placed regular broth inside these flasks, bacteria grew, as expected. Pasteur contended that the broth already contained the microorganisms when it was added to the flask. To support this idea, he then placed boiled broth, which would contain no live microorganisms, in the swan-necked flask and saw that no bacterial growth occurred in the broth. Further evidence that an external factor caused the broth to spoil was collected when Pasteur broke the gooseneck top off the flask, or tilted the flask so the broth could have contact with microorganisms in the neck of the flask, in order to allow outside microorganisms into the flask. The result was bacterial growth in the broken or tilted flasks. Pasteur's original flasks still exist and the broth remains uncontaminated.

CONTRIBUTIONS TO THE WINE INDUSTRY

Pasteur next turned his attention to problems affecting the wine industry. Pasteur's contributions to the wine industry are significant in that they led to a better understanding of the metabolism of bacteria and yeast. Brewers and wine producers did not fully understand what caused some batches of beer or wine to spoil or sour when others were fine. Fermentation is a process responsible for converting sugar into alcohol and carbon dioxide gas. At the time, it was thought that this was simply a chemical process, though people thought that yeast played some sort of role. Pasteur applied his knowledge of microorganisms to his work as he examined samples of the sour products. He observed that in addition to the yeast, a large amount of bacteria were present in the soured samples. When the bacteria overran the yeast in the culture, the product spoiled. Pasteur suspected that fermentation was a biological process: that the yeast were living beings, and that they converted the sugar into alcohol by their normal metabolic processes. From these studies, Pasteur developed his famous germ theory of fermentation, which stated that microorganisms were the cause of fermentation and that specific types of microorganisms were the cause of specific types of fermentation.

Pasteur did not stop there; he diagnosed the problem (too much bacterial growth), as well as supplied a solution. He recommended that if the brewers would heat their samples to 131–140°F (55–60°C), no harm would come to the beverage, but the bacteria would be killed and spoiling could be prevented. This process became known as pasteurization, reflecting his contribution, and it is utilized in beer, milk, and food production to this day.

OTHER CONTRIBUTIONS OF PASTEUR

Louis Pasteur was responsible for restoring the French silk industry using his knowledge of microorganisms. Damage to the larvae of silkworms was negatively affecting silk production in France at the time, and this negatively impacted the economy. Pasteur recognized that the reason the silkworms were not effectively producing silk was that they were being killed by a microorganism known as *pébrine*.

Pasteur once again not only diagnosed the problem but also worked with silk producers to maintain healthy environments for their silkworms in order to prevent reinfestation. Pasteur's work clearly defined the link between microorganisms and diseases. This led the way for dramatic improvements in medical science.

The link caught the attention of an Edinburgh surgeon, Lord Joseph Lister, who was concerned with the number of patients dying in hospitals after surgical procedures. Lister had observed that the cleanliness of the operating room, equipment, and surgeon affected the outcome of the patient and the likelihood that the patient would contract an infection. Pasteur's work and the implications that microorganisms caused diseases led to dramatic overhauls of treatment. Knowing how bacteria and other microorganisms could be destroyed led to improved medical treatment and a reduction in the number of surgery related deaths.

PASTEUR AND VACCINATIONS

Pasteur was asked to help find a cure for anthrax, a deadly disease in cattle that is now known to be caused by the bacterium *Bacillus anthracis*. While observing the cows, Pasteur recognized that not all the cows in a herd were affected to the same extent by the disease. He hypothesized that something made certain individuals better at fighting off the disease than the cows that were killed. As Pasteur investigated the situation, he noticed that the cows that had survived an anthrax outbreak were those that had previously contracted a mild form of the disease. Pasteur investigated whether the previous infection enabled them in some way to fight off the infection. Pasteur tested a small dose of weak anthrax and injected it into healthy cows. He then exposed the cows to full-fledged anthrax and found that they did not contract the disease. He had essentially made the cows resistant to the disease by giving them a small dose of the disease-causing microorganism prior to exposure. These results led the way for the future of vaccination and disease prevention for centuries to come.

Another disease that Pasteur solved was rabies, a viral infection that attacks the central nervous system. Left untreated, rabies is a deadly disease to both animals and humans. By removing samples from the spinal column of infected dogs, Pasteur collected full-strength virus, upon which he then experimented to create weakened strains. These weakened (attenuated) forms of the rabies virus were then injected into healthy animals to prevent rabies. Common practice today includes rabies vaccinations for all domesticated animals. The first human was treated with a rabies vaccination in 1885 and was cured of the disease after a 10-day course of the vaccine. Such a simple shot can cure such a deadly disease because of the persistence and vision of one man, Louis Pasteur. The principles of vaccination developed by Louis Pasteur led to the production of vaccinations for multiple deadly, contagious, and debilitating diseases such as smallpox and polio.

See also BIOCHEMISTRY; CHEMISTRY IN WINE-MAKING.

FURTHER READING

Debre, Patrice. *Louis Pasteur.* Baltimore: Johns Hopkins University Press, 2000.

Pasteur, Louis, and Joseph Lister. *Germ Theory and Its Applications to Medicine and On the Antiseptic Principle of the Practice of Surgery* (Great Minds Series). Amherst, N.Y.: Prometheus Books, 1996.

Pauli, Wolfgang (1900–1958) Austrian/Swiss/American *Physicist*

Wolfgang Pauli was an intellectual genius who has been compared to Albert Einstein—some claimed that he even out-geniused Einstein. They both possessed a phenomenal gut intuition for the physical world, being able quickly to assess whether an idea or theory fit into "the scheme of things" or not. But whereas Einstein tended toward formal courtesy, Pauli was a sarcastic, outspoken critic with grand disdain for authority. He would tell someone to his face that his idea was "all wrong," or, in the worst case, "not even wrong." Pauli's contributions to physics were great, although he tended to publish little and preferred to develop and spread ideas through lengthy communications with colleagues. For example, his exclusion principle, for which he was awarded the Nobel Prize, was vital to the correct understanding of atomic structure, and he proposed the existence of the neutrino from the properties of radioactive beta decay.

To his delight, Pauli was the subject of a legend known as the "Pauli effect." He had the reputation that when he merely approached a laboratory, glass apparatus would shatter, vacuum systems would spring leaks, accidents would happen, and things would generally go very wrong. At the instant of his train's arrival at the Göttingen railroad station, an explosion reportedly occurred in a laboratory in the same city. The Pauli effect caused no harm to Pauli himself, however. This was demonstrated by the attempt of students to stage the Pauli effect with a complicated mechanism that would cause a chandelier to crash when Pauli arrived at a reception. When he did arrive, however, the device jammed and the chandelier never fell.

Wolfgang Pauli contributed to the foundations of quantum field theory. He discovered the exclusion principle, named for him, for which he was awarded the 1945 Nobel Prize in physics. *(AP Images)*

PAULI'S LIFE

Wolfgang Ernst Pauli was born in Vienna, Austria, on April 25, 1900, to Wolfgang Joseph Pauli and Berta Camilla Pauli, whose maiden name was Schütz. (His middle name, Ernst, was in honor of his godfather, the famous physicist Ernst Mach.) The young Wolfgang attended gymnasium (equivalent to high school in the United States) in Vienna and proved to be a prodigy. While still in high school, he started a serious study of physics and mathematics. Ordinary schoolwork bored him, and instead of paying attention in class, he studied such material as Albert Einstein's papers on relativity. That did not prevent him from graduating with distinction in 1918, at the age of 18. Two months after his graduation from gymnasium, Pauli published his first paper, which was on the general theory of relativity.

Upon graduation, Pauli enrolled in the Ludwig Maximilian University of Munich, Germany, where he studied theoretical physics under the well-known German physicist Arnold Sommerfeld. The latter clearly held his student in high esteem, since he asked Pauli, then only in his second year at the university, to write a review article on Albert Einstein's theories of relativity for the *Encyklopädie der mathematischen Wissenschaften* (Encyclopedia of Mathematical Sciences). Pauli wrote the article over about a year. It ran to 237 pages and received Einstein's enthusiastic and glowing praise. Pauli, in turn, highly respected his teacher Sommerfeld, probably one of the very few people he did respect. Pauli graduated from the university in 1921, after only three years of study and at the age of 23, having earned a doctorate for a dissertation on the quantum theory of the ionized hydrogen molecule.

Immediately after graduation at Munich, Pauli went to the University of Göttingen, in Germany, to work as assistant to Max Born for a year, and afterward to Copenhagen, Denmark, for a year with Niels Bohr. Both of these world-famous physicists were pioneers in the field of quantum mechanics. It was then 1923, and at the age of 23, Pauli took a position as lecturer at the University of Hamburg, in Germany, for five years, until 1928. During this period, in 1927, Pauli's mother committed suicide. That was particularly hard for him, as he had been very close to her. The next year Pauli's father remarried, but the son was apparently not very happy with his father's new wife, referring to her as his "evil stepmother." During that year, 1928, at the age of 28, Pauli was appointed to a professorship in physics at the Eidgenössische Technische Hochschule (ETH-Federal Institute of Technology) in Zurich, Switzerland. At age 29, Pauli married Käthe Margarethe Deppner in 1929. But the marriage did not work out, and the couple divorced in 1930, after less than a year together and with no children.

Soon after his divorce, in early 1931, Pauli suffered a severe breakdown. He consulted the psychiatrist Carl Jung, who analyzed hundreds of Pauli's dreams. He entered into a correspondence with Jung, in which they attempted to clarify Jung's thoughts on such topics as mind, collective consciousness, and synchronicity, the supposed meaningful simultaneous occurrence of events that do not bear any causal relation to each other. Pauli's recovery allowed him to take a visiting position in 1931 at the University of Michigan. In 1934, at the age of 34, he married Franciska Bertram. This time, the marriage lasted, for the rest of Pauli's life. They had no children.

While still at the ETH, Pauli took a visiting position at the Institute for Advanced Study in Princeton, New Jersey, in 1935. In 1938, Germany annexed Austria, making Pauli, then in Zurich, a German citizen. That caused him difficulty when the Second World War broke out in 1939. But an offer of a position at the Institute of Advanced Study in Princeton allowed Pauli and his wife to move to the United States. While at Princeton, he served as visiting pro-

fessor at the University of Michigan and Purdue University, in 1941 and 1942, respectively. In 1945, Pauli was awarded the Nobel Prize in physics for "the discovery of the Exclusion Principle, also called the Pauli Principle." Einstein nominated him for the prize. Pauli became a naturalized U.S. citizen. After the war, in 1946, he and Franciska returned to Zurich, where he remained until his death of pancreatic cancer on December 15, 1958.

PAULI'S WORK
After Pauli's masterful review of Einstein's theories of relativity, described earlier, his first major contribution to physics was in connection with atomic structure. In order to understand atomic spectra as reflecting the energy states of atoms, Pauli proposed, in 1924, that each electron that forms part of an atom possesses a fourth quantum number, in addition to the three that were already known. This quantity has two values, taken as $+\frac{1}{2}$ and $-\frac{1}{2}$, which represent the two spin states of an electron, up and down with respect to the plane of its orbit. Pauli also postulated his exclusion principle, which states that no two electrons in an atom can exist in the same state. In other words, no two electrons in an atom can possess identical values for their sets of quantum numbers, including his newly proposed spin quantum number. If an electron occupied the lowest-energy state available to electrons in an atom, for instance, no other electron could share this state. In the atom's lowest-energy configuration, its electrons would fill the available states consecutively, starting from the state of lowest energy. Pauli's ideas brilliantly shed light on and gave order to atomic structure.

Later, it became clear that Pauli's exclusion principle applies to electrons that are assembled together in any situation, not only in atoms. The electrons in a solid, a metal, for example, cannot all occupy the state of lowest electron energy that the structure of the solid allows. Here, too, in the lowest-energy configuration, meaning at the temperature of absolute zero, the electrons fill the allowed states consecutively from lowest energy up. Without this behavior of electrons, all solids would be electric insulators, and there would be no conductors or semiconductors. That would happen, because all the electrons would then share the lowest energy states, in which they are not free to move about within the solid and conduct electricity. A large amount of energy would be required to "raise" a significant number of electrons to states of sufficiently high energy for those electrons to move freely. It also became clear that Pauli's principle governs the behavior of collections of identical fermions of any kind. (A fermion is any particle whose spin value is half an odd integer, e.g.,

1/2, 3/2, 5/2, ...) No two neutrons in a nucleus can exist in the same state, for example—and the same applies to the protons in a nucleus, thereby affecting the properties of nuclei.

Another major contribution of Pauli was in the understanding of beta decay, a type of radioactivity in which an unstable nucleus emits an electron (historically referred to as a beta particle) and increases its atomic number by 1 with no change in its mass number. Scientists knew that a neutron in the nucleus spontaneously decays into a proton, which remains in the nucleus, and an electron, which is emitted. Thus the atomic number rises by 1—an additional proton appears in the nucleus—while the mass number does not change—the total number of nucleons (protons and neutrons) remains the same. The mystery was that the electrons that were emitted by decaying nuclei of the same kind, say, carbon 14, left the nucleus with a range of kinetic energies up to some maximum. However, in a two-body decay, such as a neutron decaying into a proton and an electron, the laws of conservation of energy and momentum require that the kinetic energy of the emitted electrons will always have the same value. Why, then, were they emitted with a range of energies? Pauli proposed in 1930 that the decay is, in fact, a three-body decay: that the neutron decays into a proton, an electron, and a third particle, which is neutral and possesses a very small mass. In 1934, Enrico Fermi incorporated Pauli's idea in his theory of the weak force, calling the new particle a neutrino. The neutrino was discovered experimentally in 1956, while Pauli was still alive and able to appreciate the discovery. In more detail, the particle emitted together with an electron in beta decay is the antiparticle of a type of neutrino called an electron neutrino. The other types of neutrinos, which were discovered later, are the muon neutrino and tau neutrino.

In addition to those major contributions, Pauli played a large role in understanding the foundations of quantum theory in general and of quantum field theory. His work in those fields is of a very technical nature. One accomplishment is his theoretical proof that identical fermions obey Fermi-Dirac statistics while identical bosons (bosons are particles possessing an integer value of spin, e.g., 0, 1, 2, . . .) obey Bose-Einstein statistics.

PAULI'S LEGACY

Wolfgang Pauli is considered one of the great physicists of the 20th century. His exclusion principle, for which he was awarded the Nobel Prize in physics, plays a pivotal role not only in understanding atomic structure, but also in explaining diverse phenomena, such as semiconductors, nuclear properties, neutron stars, and white dwarfs. Pauli's theoretical investi-

gation of beta decay led to its understanding in the context of the weak force and to the neutrino particle types. His other contributions to theoretical physics, although highly technical, have led to further investigations that have proved fruitful and are greatly valued by theoretical physicists.

Many theoretical physicists should refrain from entering laboratories, where they would pose a danger both to themselves and to the lab. The Pauli effect was an extreme manifestation of that. Pauli's presence in the vicinity of a laboratory—he did not have to actually enter the lab—would, according to legend, cause weird, unexplained, and sometimes catastrophic happenings. But he himself seemed to be immune to the effect and was never harmed by it.

See also ATOMIC STRUCTURE; BOSE-EINSTEIN STATISTICS; CONSERVATION LAWS; EINSTEIN, ALBERT; ELECTRICITY; ENERGY AND WORK; FERMI-DIRAC STATISTICS; FERMI, ENRICO; GENERAL RELATIVITY; MOMENTUM AND COLLISIONS; NUCLEAR PHYSICS; PARTICLE PHYSICS; QUANTUM MECHANICS; RADIOACTIVITY; SPECIAL RELATIVITY.

FURTHER READING

Cropper, William H. *Great Physicists: The Life and Times of Leading Physicists from Galileo to Hawking.* New York: Oxford University Press, 2001.

Enz, Charles P. *No Time to Be Brief: A Scientific Biography of Wolfgang Pauli.* Oxford: Oxford University Press, 2002.

Heathcote, Niels, and Hugh de Vaudrey. *Nobel Prize Winners in Physics 1901–1950.* Freeport, N.Y.: Books for Libraries Press, 1953.

The Nobel Foundation. "The Nobel Prize in Physics 1945." Available online. URL: http://nobelprize.org/physics/laureates/1945/. Accessed February 20, 2008.

Pauli, Wolfgang. *Theory of Relativity.* Mineola, N.Y.: Dover, 1981.

Pauli, Wolfgang, and Carl G. Jung. C. A. Meier, ed. *Atom and Archetype: The Pauli/Jung Letters, 1932–1958.* Princeton, N.J.: Princeton University Press, 2001.

Pauling, Linus (1901–1994) American *Quantum Chemist, Biochemist* A charismatic personality exhibiting intellectual genius in multiple disciplines, Linus Pauling was classified by most as a quantum chemist and converted biochemist. Pauling saw himself as also a crystallographer, a molecular biologist, medical researcher, and pacifist. Pauling pioneered the application of quantum mechanics to chemistry, receiving the 1954 Nobel Prize in chemistry for his work describing the nature of chemical bonds. Regarded by many as one of molecular biology's founders as well, Pauling made important contributions to the field by using X-ray crystallography to determine protein structures. Although proven later to be incorrect, he proposed the first published model for the structure of deoxyribonucleic acid (DNA) in the early 1950s, prior to elucidation of the double helix by James Watson and Francis Crick. He possessed a level of expertise in multiple topics—inorganic chemistry, organic chemistry, metallurgy, immunology, psychology, radioactive decay, and consequences of fallout. Later in life, as a consequence of his wife's strong pacifistic tendencies, Pauling actively campaigned against aboveground nuclear testing, leading to a treaty between the United States and the USSR in 1963 banning those tests. Because of his efforts, Pauling received the Nobel Peace Prize in 1962.

Born on February 28, 1901, to Herman Henry William Pauling and Lucy Isabelle Darling, Linus Carl Pauling was the oldest of three children growing up in the Portland, Oregon, area. His father died of a perforated ulcer when Pauling was only nine, and his mother supported the family by running a boarding house on the outskirts of Portland. While attending public elementary schools, Pauling found solace in books and a growing interest in chemistry. A grammar school friend, Lloyd Jeffress, had a small chemistry laboratory in his bedroom that mesmerized Pauling, and he later said that the days spent in Jeffress's room inspired him to become a chemical engineer. In high school, a teacher recognized his potential and encouraged Pauling to pursue a career in chemistry. Before dropping out of high school at the age of 16, Pauling loved to perform chemistry experiments in the basement of the boarding house, borrowing much of the equipment and materials from an abandoned steel company where his grandfather served as night watchman. With no high school diploma, because he failed to take several American history courses, Pauling entered Oregon Agricultural College (OAC), now Oregon State College, in 1917. Pauling worked full-time to support himself and to pay tuition while carrying a full load of courses. After his second year at OAC, Pauling intended to take a job in Portland to help ease his family's financial burdens, but he accepted a position teaching quantitative analysis that permitted him to continue his studies and to help his mother. During the last two years at OAC, Pauling became aware of the work of Gilbert N. Lewis and Irving Langmuir on electronic structure of atoms and the bonding of atoms to form molecules. Pauling envisioned determining how the physical and chemical properties of substances related to the structure of atoms from which they are composed, an endeavor that would combine his love of quantum mechanics and chemistry.

After his graduation from OAC in 1922 with a bachelor of science degree in chemical engineering, Pauling enrolled at the California Institute of Technology (Caltech) to pursue his doctorate under the direction of Roscoe G. Dickinson. For his dissertation, Pauling performed X-ray diffraction studies of crystals and published seven papers on the crystal structure of minerals during graduate school. After he obtained his Ph.D. (graduating summa cum laude) in physical chemistry with minors in mathematics and physics in 1925, Pauling received a Guggenheim Fellowship to travel to Europe and develop his breadth of knowledge in quantum mechanics by studying with experts in the field—the German physicist Arnold Sommerfeld in Munich, the Danish physicist Niels Bohr in Copenhagen, and the Austrian physicist Erwin Schrödinger in Zurich. He participated in the first quantum mechanical analyses of bonding in hydrogen molecules by Walter Heitler and Fritz London.

When he returned to the United States, Pauling accepted a position at Caltech as assistant professor of theoretical chemistry, becoming one of the first scientists in the field of quantum chemistry and a pioneer in its application to molecular structures. In his first five years at Caltech, Pauling published 50 papers reporting X-ray diffraction studies and quantum mechanical analyses of atoms and molecules. He also developed five rules, known as Pauling's rules, that defined common atomic arrangements in crystals with ionic bonds, where negatively charged electrons moved from one atom to the other while orbiting positively charged nuclei. In 1929, Caltech promoted him to associate professor, and the university promoted him to full professor the following year. The American Chemical Society awarded Pauling the Langmuir Prize, an honor given to the most significant body of work in pure science by a person 30 years old or younger in 1931.

Pauling spent the rest of the 1930s trying to decipher the nature of chemical bonds. In 1932, Pauling published a paper that he later believed was his most significant work, outlining the concept of hybridization of atomic orbitals, or orbital hybridization. Physical scientists discuss electrons in atoms in terms of single orbitals, such as s, p, d, or f orbitals. When describing bonds between atoms, Pauling proposed that properties from multiple orbitals be combined to form hybrids of two individual orbitals. For example, when studying carbon atoms, one $2s$ and three $2p$ orbitals in a carbon atom combine to form four equivalent orbitals, called sp^3 hybrid orbitals, which would describe carbon compounds such as methane (CH_4). Another scenario might involve the $2s$ orbital combining with only two of the $2p$ orbitals to make three equivalent orbitals called sp^2 hybrid orbitals with one $2p$ orbital remaining unhybridized, which would describe unsaturated carbon compounds such as ethylene (C_2H_4). Countless hybridization schemes exist for other molecules as well.

Pauling ascertained that electrons transferred from one atom to another in ionic bonds and that the atoms shared electrons in covalent bonds. He also determined that most interactions between atoms fell somewhere between these two extremes. Trying to quantify the chance that a pair of atoms would form either an ionic or a covalent bond, Pauling developed a second concept, called electronegativity, that could be used to predict the nature of bonds between atoms in molecules. Taking into account various properties of molecules (such as the energy required to break bonds and the dipole moments of molecules), he established a scale and gave most elements a numerical value for comparative analyses. These analyses suggested that the type of bond that two atoms formed was dependent on their individual magnetic properties. A smaller difference in electronegativity between two atoms results in a greater chance that the bond that forms will be covalent.

Pauling also performed structural analyses of aromatic hydrocarbon structures, particularly benzene (C_6H_6). When the German chemist Friedrich Kekulé studied aromatic compounds, he concluded that two populations of the molecule exist with alternating single and double bonds—where a double bond exists in one structure, a single bond exists in the second structure. Using quantum mechanics, Pauling demonstrated that intermediate structures existed with aspects of both individual structures; in other words, an aromatic structure existed that was a superposition of two structures rather than a rapid interconversion between two individual molecules. Pauling named this phenomenon, in which a hybrid bond (a bond that is an intermediate between a single and double bond) exists between two atoms in a bond network, resonance. Because it combines more than one electron structure to achieve an intermediate result, resonance resembles orbital hybridization, and it contributes to the structural geometry and ultimate stability of a molecule.

With his conclusions about the nature of chemical bonds in mind, Pauling tried his hand at writing a textbook that embodied a modern theoretical approach to chemistry in words that a college freshman would understand. The textbook, *The Nature of the Chemical Bond* (1939), based on his Nobel Prize–winning research (he received the 1954 Nobel Prize in chemistry for his work regarding the chemical nature of bonds) and its applications, became one of the most influential chemistry textbooks ever published, being translated into 13 different languages. In the first 30 years after its publication, scientists cited it more than 16,000 times. The combination of this text and Pauling's engaging lecture style, bordering on carnival-like antics with occasional pyrotechnical displays during lab demonstrations, made his freshman chemistry course at Caltech a popular one during this time.

Pauling delved into a new area of interest, the study of biological molecules, in the mid-1930s; after interactions with the Caltech biologists such as Thomas Hunt Morgan and Alfred Strurtevant. First, Pauling showed that hemoglobin (an oxygen-carrying protein) undergoes conformation changes when it gains or loses an oxygen atom. To determine the specific structural changes that occur, he turned back to X-ray diffraction studies to elucidate the answer. Pauling immediately discovered that proteins are far less amenable to X-ray studies than inorganic crystalline minerals, such as those he used for his doctoral studies. More than a decade passed before a breakthrough came; Pauling solved the conundrum by eludicidating the conformation of two structural motifs in proteins, the alpha helix and the beta sheet. Pauling knew the structures of amino acids and the planarity of the peptide bond that links individual amino acid residues together. While serving as guest lecturer at Oxford University in 1948, Pauling worked with a sheet of paper as a model peptide chain, which he folded to represent what he knew about its intrinsic properties from chemical bonding considerations and X-ray diffraction data. Thinking out of the box, he assumed that it was possible for a chain to coil around itself in such a way that a nonintegral number of residues made up each turn. The unorthodox assumption paid off, and he correctly proposed the structure of the alpha helix and a second configuration common in proteins, the beta sheet. Later X-ray diffraction studies confirmed that his models were accurate, as well as showing the prevalence of these two structures in globular and fibrous proteins.

Pauling attempted to solve the structure of another important biological molecule, deoxyribonucleic acid (DNA). Uncharacteristically, the model that he proposed had several glaring flaws. The building blocks of DNA (nucleotides) consist of a phosphate group, a sugar, and nitrogenous bases aromatic in nature. Pauling proposed that DNA consisted of a triple-stranded helix with hydrophilic (water-loving), sugar-phosphate chains running down the center of the molecule and hydrophobic (water-hating) nitrogenous bases protruding outward toward the aqueous environment. The structure proposed by Pauling would be thermodynamically unstable, an undesirable characteristic for a molecule that must faithfully transmit genetic information from one generation to the next. By placing the phosphate groups into the middle of the molecule, he also neutralized the phosphate group, making DNA lose its acidic nature. Across the Atlantic Ocean, James Watson and Francis Crick had access to X-ray diffraction studies by Rosalind Franklin and Maurice Wilkins that clearly indicated that DNA was a helix. In 1953, using model building as Pauling had for the alpha helix, Watson and Crick correctly proposed the structure, a two-stranded helix with the sugar phosphate backbones facing outward and the nitrogenous bases inward, making complementary base pairs.

Four years before the structure of DNA was determined (1949), Linus Pauling, Harvey Hano, S. J. Singer, and Ibert Wells published the first evidence that a human disease resulted from a change in a specific cellular protein. Using electrophoresis, Pauling and his colleagues demonstrated that individuals with sickle cell disease have a different form of hemoglobin protein in their red blood cells than healthy individuals. These proteins distort the shape of the red blood cells into a sickle, or crescent moon, shape, rather than a doughnut-shaped disk, as seen in unaffected cells. The sickle cells stick together to form long rods that stick together and interfere with blood flow. Pauling and his colleagues also used electrophoresis to show that individuals who carry the sickle trait have both normal and abnormal hemoglobin proteins, as one would predict if the trait follows the rules of Mendelian genetics. With the results from this study, Pauling ushered in another new field, molecular genetics, when he demonstrated the correlation between a specific protein and a disease and the correlation between the protein composition of an individual and classical inheritance patterns first described by Mendel in the late 1800s.

As he grew older, Pauling became an outspoken pacifist, and many members of the scientific community thought that his work suffered as a result. An avid anti-Nazi at the beginning of World War II, he designed explosives and rocket propellents to be used in the war effort. His wife, Ava Helen, whom he married in 1923, was a fervent pacifist. The theoretical physicist and director of the Manhattan Project (a war initiative whose goal was to design an atomic bomb) Robert Oppenheimer asked Pauling to be in charge of the chemistry division of the project. Pauling declined, telling Oppenheimer that he was a pacifist. After the war, this pacifism became evident in his words and his actions. In 1946, he joined the Emergency Committee of Atomic Scientists, chaired by Albert Einstein, whose mission was to warn the public of the dangers associated with the development and use of atomic weapons. His political activism prompted the U.S. Department of State to deny Pauling a passport to England in 1952, a trip that might have given Pauling the opportunity to see

X-ray diffraction data that would have helped him solve the structure of DNA. The State Department restored the passport in 1954, in time for Pauling to travel to Stockholm to accept the Nobel Prize in chemistry. In 1958, Pauling and his wife presented the United Nations with a petition signed by more than 11,000 scientists calling for an end to aboveground nuclear weapon testing. Public pressure led to a moratorium on the testing followed by the Partial Test Ban Treaty signed in 1963 by John F. Kennedy and Nikita Khrushchev. On the day that the treaty went into effect, the Nobel Prize Committee awarded Pauling his second Nobel, the Nobel Peace Prize, for his ceaseless campaign against all warfare, especially nuclear warfare, as a means to solve international conflict. Interestingly, the Caltech chemistry department, wary of his political views and activism, never formally congratulated him for the second Nobel.

Because he felt misunderstood and slighted, Linus Pauling left Caltech after four decades, going first to a think tank in Santa Barbara, California, then ultimately to Stanford University, where he stayed until his retirement as professor emeritus of chemistry in 1974. In these later years, Pauling's research waned. He spent a fair amount of time studying the benefits of vitamin C to a person's health. In 1973, with the help of Arthur B. Robinson, he founded a nonprofit biomedical research organization that bears his name—the Linus Pauling Institute of Science and Medicine. Pauling established the institute to conduct research and education in alternative medicine, following his belief that nutrition could prevent or ameliorate many diseases, decrease suffering, and slow the aging process. In 1996, the institute moved from California to Oregon State University, where current research focuses on heart disease, cancer, aging, and neurodegenerative diseases.

A multifaceted genius with a zest for life, Linus Pauling became one of the most visible, vocal, and accessible American scientists in modern history. Remembered often by his trademark black beret over a shock of white hair, he possessed a pair of vivid blue eyes that conveyed his interest in challenging topics to those around him. He never tired of explaining difficult concepts to people genuinely interested in understanding. He was a true 20th-century Renaissance man, well versed in and adept at a multitude of disciplines.

Outside his professional life, he made time for a family, too. During his senior year at OAC (1922), he taught a "chemistry for home economics majors" course, where he met Ava Helen Miller of Beaver Creek, Oregon. Pauling married Ava Helen the following year, and they had four children (Linus Carl Jr., Peter Jeffress, Linda Helen, and Edward Crellin), 15 grandchildren, and 19 great-grandchildren. After the death of his wife in 1981, Pauling spent a great deal of his time at his ranch near Big Sur, California. Diagnosed with prostate cancer in 1991, Pauling battled the disease for three years. The world lost one of its greatest scientists and humanitarians on August 19, 1994, when Linus Pauling passed away at the age of 93, at his ranch near the coast.

See also ATOMIC STRUCTURE; BIOCHEMISTRY; BONDING THEORIES; COVALENT COMPOUNDS; CRICK, FRANCIS; EINSTEIN, ALBERT; ELECTRON CONFIGURATIONS; FRANKLIN, ROSALIND; IONIC COMPOUNDS; NUCLEIC ACIDS; ORGANIC CHEMISTRY; PROTEINS; QUANTUM MECHANICS; WATSON, JAMES; WILKINS, MAURICE; X-RAY CRYSTALLOGRAPHY.

FURTHER READING

Pasachoff, Naomi E. *Linus Pauling: Advancing Science, Advocating Peace (Outstanding Science Trade Books for Students K-12)*. Berkeley Heights, N.J.: Enslow, 2004.

Pauling, Linus, and Barbara Marinacci. *Linus Pauling in His Own Words: Selections from His Writings, Speeches, and Interviews*. N.Y.: Touchstone, 1995.

periodic table of the elements The periodic table of the elements, or periodic table, is a diagrammatic representation of the known elements listed according to atomic number (the number of protons) of the element. (*See* APPENDIX: PERIODIC TABLE OF THE ELEMENTS). This arrangement of the elements is based on the periodic law, which states that the physical and chemical properties of the elements are periodic functions of their atomic number. The position of each element on the table also indicates the electron configuration of the element. For example, the element carbon, with an atomic number of 6 and an atomic mass of 12.011, has an electron configuration of $1s^2 2s^2 2p^2$.

The periodic table is a powerful tool for understanding chemistry. Within a single family, the elements display similar chemical characteristics. Rather than memorizing information about each element, a chemist has only to learn about a single element in the family. If sodium and chlorine combine to form table salt, NaCl, other members of the sodium family will also combine with chlorine to form salts. Because positioning of the elements on the periodic table visually represents the periodic law, a simple glance allows chemists to predict how one element will react with another.

The 60 or so elements familiar to 19th-century scientists exhibited a wide range of chemical and physical properties. The periodic table resulted from the work of many chemists, who struggled to find patterns and determine relationships among the elements to establish order in the then-chaotic field of chemistry. The British chemist John Newlands was one of the first to arrange the elements into the form of a table. He based his table on the law of octaves, after recognizing that the properties of the elements repeated every eighth element. Building upon the work of Newlands and others, the Russian chemist Dmitry Mendeleyev assembled all the pieces of information together to develop the first periodic table that resembles the familiar modern version. Mendeleyev first arranged the 63 known elements according to atomic mass. He also took into account the law of octaves by arranging elements into horizontal rows called periods in his table. Elements positioned in the same vertical column, or group, had similar properties. The table that Mendeleyev published in 1871 did not differ significantly from Newlands's, except that Mendeleyev left gaps and made bold predictions for the undiscovered elements. Because of his tremendous intuition in recognizing a periodic dependence of the elements based on their atomic weights, Mendeleyev is now known as the father of the periodic table.

The British physicist Henry Moseley x-rayed crystals of various elements and found out that the diffracted wavelengths correlated in a systematic manner with the atomic numbers of the elements, demonstrating that atomic numbers represented more than simply increasing mass. Moseley recognized that the number of protons and electrons gave an element its properties.

As Moseley discovered, the arrangement of the elements into specific groups and periods directly relates to their electron configuration. This arrangement of electrons into specific atomic orbitals gives elements within a group the similar physical and chemical properties seen by early scientists. The periods on the periodic table are numbered 1–7 from top to bottom. Elements in a period have their valence electrons in the same energy level. The groups are numbered in an old-style format, where the representative elements are given names IA–VIIIA, and the transition metals are named IB–VIIIB. The old-style format has been replaced by the IUPAC numbering system, where the groups are labeled 1–18 consecutively. The elements within a group have similar physical and chemical properties, but with some slight differences moving down the rows represented in the group based on the number of energy levels the element contains.

The periodic table is conveniently divided into two distinct regions, metals and nonmetals, separated by the stair-step line. The metals, found beneath the line, include all members of groups 1–12 as well as some elements of groups 13–15. Most of the known elements are metals and share the characteristics of being malleable and ductile and being good conductors of electricity and heat. Except mercury, all metals are solid at room temperature.

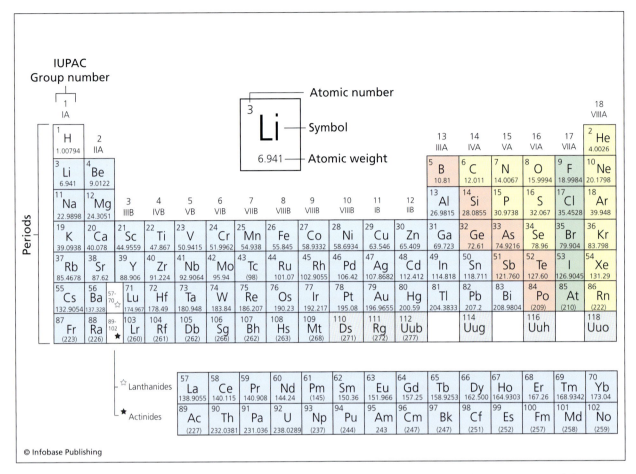

The periodic table is arranged by increasing atomic number. The columns of the table are groups and the rows of the table are periods.

The second region of the periodic table contains the nonmetals, which are found to the right of the stair-step line in the upper-right corner of the periodic table. This group is made up of the halogens, noble gases, and hydrogen, carbon, nitrogen, oxygen, phosphorus, sulfur, and selenium. Nonmetals are considered insulators. They form ionic bonds with metals as a result of their high electronegativities, and they form covalent bonds with other nonmetals. Many nonmetals also exist in nature as diatomic molecules: hydrogen, oxygen, bromine, fluorine, iodine, nitrogen, and chlorine. Different nonmetals exist as gases, liquids, or solids at room temperature.

Elements known as metalloids, or semiconductors, lie along the stair-step line separating metals and nonmetals. As their name implies, these elements conduct electricity better than nonmetals but not as well as metals. Other properties also lie somewhere between the properties of the metals and nonmetals. The metalloids include boron, silicon, germanium, arsenic, antimony, tellurium, and polonium.

The elements of the periodic table can also be divided into the *s*-block, *p*-block, *d*-block, and *f*-block elements. This refers to the sublevel in which the elements have their valence electrons. Because they illustrate the entire range of chemical properties, the *s*-block and *p*-block elements are known as representative elements. For all of the representative elements, the group number in the old-style numbering system equals the number of electrons in the outer energy level. The period number of the representative elements indicates the energy level of the valence electrons of that element.

The first group of representative elements is the alkali metals (Group IA, IUPAC Group 1). This group consists of the elements lithium, sodium, potassium, rubidium, cesium, and francium. The elemental forms of the alkali metals are highly reactive. The alkali metals have a low density and can explode if they react with water. Each element in this group has one valence electron, which is located in an *s* sublevel. The alkali metals form +1 cations by losing their s^1 electrons and readily form ionic compounds (salts) with nonmetals. Salts of alkali metals are soluble in water.

The alkaline earth metals (Group IIA, IUPAC Group 2) include the elements beryllium, magnesium, calcium, strontium, barium, and radium. They are gray-white soft solids and, as do the alkali metals, have a low density. The alkaline earth metals are less soluble in water than the alkali metals. These elements are always found combined with other elements in nature because they are considered very reactive, although they are less reactive than the alkali metals. Each element in this group has two valence electrons, located in an s sublevel, and therefore form +2 cations by losing their s^2 electrons.

The next group of representative elements, often called the boron group (Group IIIA, IUPAC Group 13), consists of boron, aluminum, gallium, indium, and thallium. The elements in this group exhibit a range of properties. Boron is considered a metalloid, while the other members of this group are considered poor metals. Boron is black, lustrous, and brittle. Aluminum, the most abundant element in the Earth's crust, is strong and, like all metals, is ductile, and, therefore, very useful in a variety of manufacturing processes. Gallium is a semiconductor and has a melting point of 86°F (30°C) and a boiling point of 3,999°F (2,204°C). All of the elements in this group have three valence electrons located in the s and p sublevels. They usually form +3 cations by losing their s^2 and p^1 electrons. The carbon group (Group IVA, IUPAC Group 14) contains carbon, silicon, germanium, tin, and lead. Each of these elements has four electrons in its outer energy level in s and p sublevels. The elements in this group generally do not form ions, although they can. They are more likely to form covalent compounds. Carbon can form either cations or anions with -4, +2, or +4 oxidation states by gaining or losing its valence electrons. The element carbon is fundamental to all living organisms, as it forms the framework of organic molecules. Other forms of carbon are graphite and diamonds. Silicon, which is very abundant in the Earth's crust, acts as a semiconductor, as does germanium; both of these elements have played a pivotal role in the development of the computer chip. The final metals in this group, tin and lead, are relatively nonreactive.

The elements in the nitrogen group (Group VA, IUPAC Group 15) all have five electrons in their outer shell. This group includes nitrogen, phosphorus, arsenic, antimony, and bismuth. The nitrogen group contains nonmetals (nitrogen and phosphorus), metalloids (arsenic and antimony), and metals (bismuth). The elements in this group have five valence electrons in both s and p sublevels.

The chalcogen group (Group VIA, IUPAC Group 16) is also called the oxygen family after the most abundant element on Earth. The group includes oxygen, sulfur, selenium, tellurium, and polonium. All have six valence electrons located in the s and p

sublevels. The elements in this group typically form anions with a -2 charge by gaining two electrons from a metal. Oxygen and sulfur are nonmetals, whereas selenium, tellurium, and polonium are metalloids.

The halogens (Group VIIA, IUPAC Group 17) participate in the formation of many common compounds. The elements in the halogen group are fluorine, chlorine, bromine, iodine, and astatine. Each has seven valence electrons with two in the s sublevel and five in the p sublevel. The elements in this group will become anions with -1 charge when they ionize by gaining an electron from metals. This allows them to form ionic compounds with metals easily. Fluorine and chlorine are gases, bromine is a liquid, iodine is a solid, and astatine is a rare radioactive solid.

The final group of representative elements (Group VIIIA, IUPAC Group 18) is the noble gases. This group contains helium, neon, argon, krypton, xenon, and radon. The noble gases have a filled outer energy level with two electrons in the s sublevel and six in the p sublevel. Since they have a full outer shell, these elements do not gain or lose electrons and are therefore considered nonreactive.

The d-block elements (Group IIIB–IIB, IUPAC Group 3–12), the transition metals, are malleable and ductile and are good conductors of electricity. Many elements in this block have multiple oxidation states. The outer energy level of transition metals often has a full s sublevel, and some electrons exist in the d sublevel of the second highest energy level as well. The transition metals contain a wide variety of elements including copper, gold, iron, silver, tungsten, zinc, mercury, nickel, and cobalt.

The f-block elements are called inner transition metals and are divided into two periods called the lanthanides and actinides. The lanthanides were once called the rare earth metals despite the fact that they are not very rare. The lanthanides have high melting points and silver color. The lanthanides include elements with atomic numbers of 57–71: lanthanum, cerium, praseodymium, neodymium, promethium, samarium, europium, gadolinium, terbium, dysprosium, holmium, erbium, thulium, ytterbium, and lutetium. All of the elements in this period have a full $6s$ sublevel, and most are adding electrons to their $4f$ energy sublevel. They are usually depicted as a block of boxes below the central portion of the periodic table.

The actinides include elements with atomic numbers 89–103: actinium, thorium, protactinium, uranium, neptunium, plutonium, americium, curium, berkelium, californium, einsteinium, fermium, mendelevium, novelium, and lawrencium. The actinides are all radioactive and have electrons in their $5f$ electron shell. They have a wider range of physical properties than the lanthanides. As the lanthanides are, the actinides are depicted as a block of boxes below the central portion of the periodic table.

Hydrogen, the most common element in the universe, behaves as no other element in the periodic table does because it has only one proton and one electron. While it is listed as a group 1 element, hydrogen is not an alkali metal. Hydrogen can react with many elements, including oxygen. Hydrogen can become a cation with a +1 charge (essentially, a proton), or it can become an anion known as hydride with a -1 charge. Hydrogen gas and oxygen gas react violently to form water. Hydrogen is also found in all hydrocarbons and biomolecules, such as fats, proteins, lipids, and carbohydrates, that are essential to the chemistry of life. For these reasons, hydrogen exists in a class on the periodic table all by itself.

See also Appendix IV; ATOMIC STRUCTURE; DALTON, JOHN; ELECTRON CONFIGURATIONS; MENDELEYEV, DMITRY; PERIODIC TRENDS.

FURTHER READING

Wilbraham, Antony B., Dennis D. Staley, Michael S. Matta, and Edward L. Waterman. *Chemistry.* New York: Prentice Hall, 2005.

periodic trends The periodic table of the elements is a visual representation of the atomic structure of the known chemical elements. The table is arranged in order of increasing atomic number, with elements having the same number of valence electrons found in the same column or group, and elements with valence electrons in the same energy level found in the same row or period. Periodic trends become evident when the chemical and physical properties of the elements vary according to their position on the periodic table. Types of periodic trends that can be determined by an element's position on the periodic table include atomic mass, atomic radii, ionic radii, density, ionization energy, electronegativity, and electron affinity. Each of these properties contributes to the chemical reactivity of the element.

The arrangements of the elements on the earliest versions of the periodic table depended on increasing atomic mass, and then in 1914 a British chemist named Henry Moseley rearranged the table and based his arrangement on atomic number. The current model of the table shows that most elements follow the trend of increasing atomic number across the periods from left to right and down the groups from top to bottom. Exceptions include cobalt and nickel, tellerium and iodine, and several of the newly discovered elements. In general, the periodic trend is that the atomic mass of the element increases with the atomic number.

Many of the periodic trends of elements result from the shielding effect of the electrons. Every nucleus exerts a positive attraction, called the effective nuclear charge, for the electrons of the atom. The existence of electrons positioned between the nucleus and the energy level of the electron in question weakens the force of this attraction. The electrons located in the inner energy levels, between the electrons in the outer shells and the nucleus, block the force of the protons' attraction; simply put, they get in the way. This phenomenon is known as electron shielding, and it reduces the hold of the protons on the outer electrons. The shielding effect increases when more energy levels are present in the atom. The repulsive forces between electrons in the same energy level have a similar shielding effect, although not as strong as that of the electrons in lower energy levels.

Atomic radius is a measure of the size of the atom, and for metals it is one-half the distance between two nuclei in a diatomic molecule. The atomic radius indicates the distance of the valence electrons from the nucleus. The hold on electrons located farther away from the nucleus is weaker, so the distance from the nucleus to the outside of the atom or the outer energy level plays an important role in determining chemical characteristics of an element. The periodic trend of atomic radii increases in a group from top to bottom. As one moves down the group, the elements at each step have one more energy level than the elements in the previous (higher) row, and, therefore, the atomic radius increases. One might expect the atomic radius of the elements to increase across a period as well, but that is not the case—the atomic radius decreases across a period. Increasing the number of electrons in an outer energy level (as one moves from left to right across a period) does not increase the overall size of the atom. The elements across the period have more protons and electrons, and the addition of more protons gives the positively charged nucleus a stronger attractive force for the electrons in the outer energy level with the same relative amount of shielding from lower-energy-level electrons. Those electrons are held more closely than the electrons of an element in an equivalent energy level and with fewer protons, explaining why the atomic radii decrease from left to right within a period.

The ionic radius of an element is the radius of a cation or an anion. When an element gains an electron and becomes an anion, the radius of the anion is larger than that of the original atom. This is because the atom and ion have the same number of protons and, therefore, the same effective nuclear charge, but the ion now has one more electron that will increase the repulsion between the electrons and make the outer shell larger. When an element loses an electron and becomes a cation, the radius of the ion is smaller

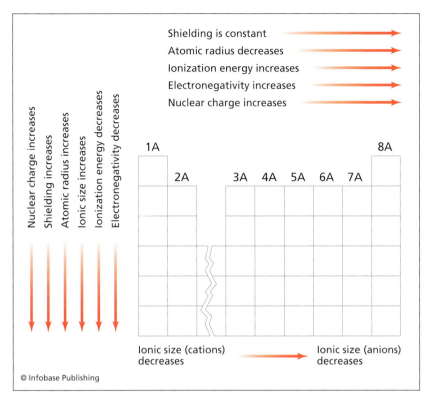

Shielding is constant
Atomic radius decreases
Ionization energy increases
Electronegativity increases
Nuclear charge increases

Nuclear charge increases
Shielding increases
Atomic radius increases
Ionic size increases
Ionization energy decreases
Electronegativity decreases

1A   2A   3A 4A 5A 6A 7A   8A

Ionic size (cations) decreases   Ionic size (anions) decreases

© Infobase Publishing

The properties of elements vary with their position on the periodic table.

ionization energy must be measured in the gaseous state of the atom to avoid any intermolecular forces between neighboring atoms. Ionization energy is measured in units of kilojoules per mole. The tighter the hold a nucleus has on a valence electron, the harder it is to remove that electron from the element, and therefore the ionization energy will be greater. The ionization energy varies across the periods, increasing from left to right. If the elements are in the same period, then one can predict the ionization energy of the elements relative to one another. Elements on the right side of the periodic table have the highest ionization energy, meaning energy must be supplied in order to remove an electron. These elements have a higher effective nuclear charge and a smaller atomic radius than atoms to their left, and, therefore, they hold their electrons more tightly than atoms with the same number of energy levels and fewer protons. Because the atomic radii are smaller, the elements on the right of the table require more energy input to remove one of their electrons. Fluorine, with nine protons, being shielded by only two inner shell electrons, holds onto its outer electrons more tightly than lithium, with only three protons being shielded by its two inner shell electrons. The noble gases have the highest ionization energies of all of the elements on the periodic table, and helium has the highest first ionization energy. The high ionization energy of nonmetals explains why they become anions when they ionize. The ionization energy decreases down the groups of the periodic table. The farther down the table, the more energy levels shielding the nucleus from the valence electrons, and the harder it is for the protons to hold onto the valence electrons. The very low ionization energies of the alkali metals allow them to donate an electron readily. This periodic property explains why metals form cations when they ionize. If an element is able to lose more than one electron, each electron lost has a characteristic ionization energy. The amount of energy required to remove the first electron is known as the first ionization energy, the energy required to remove the second electron is known as the second ionization energy, and so on. Once an atom loses its first electron, the amount of shielding due to the repulsion of the electrons is reduced, so the effective nuclear force

than that of the original atom because the cation contains one less electron but has the same effective nuclear charge. This causes the electrons of the cation to be pulled more tightly toward the nucleus. While the ionic radius varies across the periods of the periodic table, both anions and cations increase their ionic radius as one progresses down a group. The ionic radius increases with the increased number of energy levels added to the atom, just as atomic radius does. When the ionic radius is compared for different elements across a period, it is necessary to compare elements that contain the same number of electrons. When a cation and an anion contain the same number of electrons, the cation will always have the smaller ionic radius because it has the higher number of protons, thus a higher effective nuclear charge.

The density of an element is defined as the mass of an atom of the element divided by the volume the atom occupies. For some elements, the density can be determined by their location on the periodic table. The density of the nonmetals decreases significantly and the density of the gases and noble gases drops off the charts, with the result that the least dense elements are located on the right side of the periodic table.

The ionization energy of an atom is the amount of energy required to remove an electron from the neutral atom in its gaseous state and to form an ion. The

of the protons increases and the remaining electrons are held more tightly. This means that the second ionization energy is higher than the first, and the third ionization energy is higher than the second.

Electronegativity is the measure of the attraction of an atom for the electrons shared in a covalent bond. If an atom has a very high electronegativity, it is highly attracted to shared electrons and is likely to "hog" the electrons when forming a bond. The difference in electronegativity between the two elements involved in the bond determines whether they form ionic or covalent compounds. In 1932, the American chemist Linus Pauling developed a measurement system for electronegativities that gives the most electronegative element on the periodic table, fluorine, a value of 4, and lithium, which is located to the far left in the same period on the table, is given a Pauling value of 1. The least electronegative element on the periodic table has a Pauling value of 0.7. If the electronegativity difference between the two atoms involved in a bond is high enough, the more electronegative element can "steal" the electron from the less electronegative element, and an ionic bond will form. An electronegativity difference that is greater than or equal to 1.7 means an ionic bond will form. If the difference is between 1.4 and 0.4, a polar covalent bond will form. An electronegativity difference of less than 0.2 will create a true covalent bond where the bonding electrons are equally shared between the two elements. The electronegativity of an atom is dependent on the effective nuclear charge, related to its number of protons. If the element is able to hold on tightly to its own electrons, that same nuclear charge will be able to hold additional electrons gained by forming a covalent bond closely to that atom.

Electronegativity increases from left to right across the periodic table. The elements in a period essentially have the same shielding but with an increased nuclear charge as one moves across the table. Electronegativity values decrease as one proceeds down a group on the periodic table. The shielding caused by the increased number of energy levels of electrons causes the elements farther down the group not to have enough of a nuclear charge to attract the bonded electrons. The decrease down the group of electronegativity helps explain the difference in elemental properties on groups of the periodic table. Some groups, or families, contain nonmetal, metalloid, and metal elements. This is due to the fact that as more energy levels are added, the atoms lose their ability to attract bonded electrons.

Electron affinity is the attraction that an element has for additional electrons, a periodic property closely related to ionization energy. While the ionization energy is the amount of energy required

to remove an electron from a neutral atom, electron affinity indicates the likelihood of an atom to take on another electron. In order to measure electron affinity, one measures the amount of energy required to remove an electron from the ionized gaseous state of the atom. The difference between this and measuring the first ionization energy is that here electrons are being removed from an ion rather than a neutral atom. A large, positive electron affinity means that a lot of energy is required to remove the electron from the ionized state, so that atom must have a high affinity for the extra electron. The harder it is to remove the extra electron from the ion, the higher the atom's affinity for that electron. The variation of electron affinity down the groups is relatively small. All of the elements in a group have the same valence electron configuration, and so they exhibit a similar tendency to accept more electrons. The halogens have the highest electron affinities of all of the groups on the periodic table because they have seven valence electrons and the acceptance of one more electron gives them the same stable electron configuration as that of the next noble gas. The electron affinities of the noble gases are near 0. Because the noble gases already have a full octet of valence electrons, they have no inclination to take on another electron. All metals including the alkali metals (Group IA) and alkaline earth metals (Group IIA) have electron affinities lower than those of the nonmetals.

See also ATOMIC STRUCTURE; COVALENT COMPOUNDS; ELECTRON CONFIGURATIONS; IONIC COMPOUNDS; MENDELEYEV, DMITRY; PAULING, LINUS; PERIODIC TABLE OF THE ELEMENTS.

FURTHER READING
Chang, Raymond. *General Chemistry: The Essential Concepts,* 4th ed. New York: McGraw Hill, 2006.
Silberberg, Martin. *Chemistry: The Molecular Nature of Matter and Change,* 4th ed. New York: McGraw Hill, 2006.

Perutz, Max (1914–2002) Austrian-British *Biochemist* With a lingering Austrian accent, Max Perutz unpretentiously led one of the most successful biochemistry laboratories in Great Britain for 30 years after World War II. Marrying the techniques of physical science to the problems of biology, he became a world-renowned leader in the new field of molecular biology. His lifelong interest in the protein hemoglobin resulted in its structural determination using X-ray crystallography, earning him the 1962 Nobel Prize in chemistry, along with his colleague John Kendrew. Perutz's research interests consistently found him at the frontiers of scientific development in spectroscopy, protein chemistry, three-dimensional imaging of macromolecules, and molecular genetics. His success could be attributed to several outstanding qualities, not all intellectual in nature, that he possessed. Perutz had an uncanny knack of gentle persuasion. Before the structures of deoxyribonucleic acid (DNA) and hemoglobin were determined, Perutz convinced the physicist Sir Lawrence Bragg (the director of the Cavendish Laboratory at Cambridge University and a member of the Medical Research Council) that a unit for molecular biology, albeit staffed by only him and Kendrew, was needed, and the unit came into existence in 1947. Despite his strong accent, Perutz communicated ideas, whether his own or a member of his staff's, with clarity and simplicity. Said by many to have a "golden pen," he had a mastery of the written word later in life that made reading his writings a pleasure as they were elegant, compelling, and stimulating. Most important, his style of leadership left his staff and colleagues with the freedom to decide their own way forward, leading to countless discoveries for the laboratory. Perutz believed that he needed only to give his colleagues a vision of long-term goals to be accomplished in their topics of interest, leaving the details up to them. Perutz also developed an uncanny insight into the potential of young researchers seeking to work with him. At different times, he acted as supervisor to Francis Crick, James Watson, Aaron Klug, John Walker, and Fred Sanger (all great scientists in their own right). After his retirement as chairman of the molecular biology laboratory that he founded, he still worked most days at his bench, increasing his knowledge and understanding of hemoglobin until his death.

The son of Hugo Perutz and Dely Goldschmidt, Max Ferdinand Perutz was born May 19, 1914, in Vienna, Austria. Both his parents were members of affluent families of textile manufacturers who had made their fortunes in the 19th century when spinning and weaving became mechanized. The young Max attended Theresianum, a grammar school whose origins could be traced back to an officer's academy formed during the days of Empress Maria Theresia. Thinking Max would enter the family business after completing his schooling, his parents suggested that he consider law as a profession. A schoolteacher sparked Max's interest in chemistry during grammar school, and Max persuaded his parents to let him pursue a career in science instead. Max enrolled at the University of Vienna in 1932. After trudging through five semesters of inorganic chemistry courses, Max Perutz found it sheer delight to take courses in organic chemistry. One of these courses, a biochemistry course taught by F. von Wessely, introduced Perutz to fascinating research in chemistry and physics being performed at Cambridge University in

Max Perutz is an Austrian-born biochemist who worked in Great Britain for the majority of his career. He received the Nobel Prize in chemistry for his X-ray crystallographic determination of the structure of the hemoglobin molecule. Max Perutz is shown here in his lab at Cambridge, November 1, 1962. *(AP Images)*

Great Britain. Wanting to be part of that intellectual community, Perutz enrolled at Cambridge University in September 1936 to pursue a doctoral degree at the Cavendish Laboratory under the supervision of the physicist John Desmond Bernal. A pioneer in the field of X-ray crystallography, Bernal encouraged Perutz to use X-ray diffraction studies to determine the structure of two proteins—chymotrypsin and hemoglobin. After G. S. Adair obtained pure crystals of horse hemoglobin, Bernal taught Perutz how to take X-ray photographs of crystals and how to interpret the resulting data. Early in 1938, Bernal and Perutz published a paper containing promising X-ray diffraction data from their studies of chymotrypsin and hemoglobin; particularly promising were the data from the hemoglobin crystals. Perutz, taking the advice of Bernal, decided to focus on determining the structure of hemoglobin alone. Hitler's annexation of Austria and Czechoslovakia threatened Perutz's remaining graduate studies because the German leader seized the Jewish family's business, leaving his parents penniless wartime refugees. A Rockefeller Foundation fellowship appointing Perutz as a research assistant to Sir Lawrence Bragg, the Cavendish Laboratory director, saved his career. Perutz overcame the remaining obstacle of studying a biological molecule using X-ray crystallography when he persuaded David Keilin, a professor of biology and parasitology as well as the director of the Molteno Institute at Cambridge, to permit Perutz and his colleagues access to the biochemical laboratory facilities that the Cavendish Laboratory (a physics laboratory) lacked. From 1938 until the early 1950s, Perutz divided his time between the two laboratories. He performed his protein chemistry experiments at

Keilin's Molteno Institute, and he performed his X-ray diffraction studies at the Cavendish Laboratory. To overcome the physical distance between the two laboratories on Cambridge's campus, Perutz found his bicycle extremely useful.

After earning his Ph.D. in 1940, Perutz chose to remain at Cambridge University, where he continued his studies of hemoglobin, a familiar protein molecule with an extraordinary range of properties relevant to the physical as well as the biological sciences. Envisioning a new branch of science that solved biological problems with physical science techniques, Perutz convinced the secretary of the Medical Research Council (MRC), Sir Edward Mellanby, to establish a new unit, the MRC Unit for Molecular Biology in 1947. At its founding, the unit consisted of two individuals: Max Perutz, serving as its head, and his colleague, John Kendrew. Both men studied hemoproteins; Perutz focused on hemoglobin, and Kendrew was interested in myoglobin, a smaller, heme protein found in muscle. By the early 1950s, Perutz had expanded the staff of the MRC molecular biology unit to include an extraordinary group of scientists. Acting as Perutz's senior colleague, Kendrew was a chemist trained in X-ray crystallography. With a strong driven personality, Kendrew had exceptional organizational skills that allowed him to surpass Perutz at one point in structural analyses of myoglobin. Perutz served as the adviser to Francis Crick, a physicist working on a Ph.D. in the biological sciences, who exhibited a dazzling intellect constantly darting from one problem to another. During this period, Perutz also accepted a postdoctoral researcher from Chicago, Illinois, a 23-year-old whiz kid named Jim (James) Watson. Only 10 years later, all four of these men (Perutz, Kendrew, Crick, and Watson) received a Nobel Prize for work performed in the MRC unit. Kendrew and Perutz shared the 1962 Nobel Prize in chemistry for their structural studies of hemoglobin and myoglobin; Crick and Watson (along with Maurice Wilkins, who belonged to another laboratory) shared the 1962 Nobel Prize in physiology or medicine for determining the structure of deoxyribonucleic acid (DNA) and suggesting a feasible mechanism for its replication.

In 1953, while Watson and Crick worked out the molecular structure of DNA and a mechanism for its replication, Perutz and Kendrew hit a roadblock in their crystallographic studies. The X-ray diffraction methods that worked so well with small molecules, such as penicillin, yielded little, if any, useful data on larger protein molecules. Specifically, X-ray data were hard to justify because multiple phases in the diffraction patterns exist in larger molecules that are missing in smaller, less complex molecules. As Watson and Crick basked in their success, Perutz thought of a way to circumvent the phase problem and to get the diffraction data necessary for determining hemoglobin's structure. If he could attach a heavy atom (mercury) to a specific site in the hemoglobin without altering its overall structure, he would crystallize this modified hemoglobin and perform X-ray diffraction studies on it. Then, he would compare hemoglobin and the modified mercury-hemoglobin by subtracting out differences in the diffraction patterns (ridding the data of multiple phases) to determine the structure of the molecules. Using Perutz's solution to the phase problem in larger molecules, Kendrew solved the simpler myoglobin structure in 1957, and Perutz solved the structure of hemoglobin in 1959. Perutz's studies showed that hemoglobin had a tetrameric structure that consisted of four separate polypeptides with heme groups near the surface of each peptide chain. Significantly, he discovered that the individual polypeptide units took on different conformations in the presence and absence of oxygen. This finding demonstrated what is now known to be a common principle in enzymes, the allosteric effect, in which a protein undergoes a structural change to switch a molecular process on or off. His studies also ascertained that the molecular mechanism of oxygen transport required all four polypeptides in hemoglobin to be fully oxygenated or fully reduced, making it an ideal oxygen transporter with this all-or-none phenomenon.

Perutz had the pleasant problem of running out of space to house the flourishing research taking place in the MRC unit for molecular biology. The unit started out in a small hut near a Cambridge University parking lot. By the late 1950s, the MRC unit occupied every empty room, shed, and even Sir Ernest Rutherford's stable to continue its studies. Long before the nomination for the Nobel, Perutz convinced Sir Harold Himsworth that a newly constructed laboratory was in order for the unit. Heeding his request, Himsworth approved the construction of a laboratory of molecular biology on the edge of Cambridge near Addenbrooke's Hospital; it was completed in 1962. The laboratory, with Perutz serving as its chairman, continued to be prolific and successful in all its research endeavors. Under Perutz's guiding hand, Hugh Huxley used electron microscopy to study the mechanism of muscle contraction, Cesar Milstein and George Köhler created monoclonal antibodies, Aaron Klug studied the organization of nucleic acids into large protein–DNA complexes called chromatin, John Walker characterized adenosine triphosphate (ATP) synthase, and Fred Sanger invented a new way to determine the sequence of nucleotides in a DNA molecule (leading to a Nobel Prize). In 1979, Max Perutz decided to retire as chairman of the laboratory, although he still intended to work on hemoglobin in the unit. At his retirement, the MRC had an inflexible rule that did not allow directors to work in the laboratory from which they retired. Because Perutz never assumed the directorship (he was always chairman instead), he continued working in the laboratory that he founded to the delight of his staff and to the consternation of the MRC.

In retirement, Perutz became a prolific writer over a broad range of topics, including a collection of essays, *I Wish I'd Made You Angry Earlier: Essays on Science, Scientists, and Humanity* (2002), discussing the pursuit of scientific knowledge by describing the passionate work of the scientists involved in famous discoveries. Above all, he pursued the endless ramifications of his lifelong research of hemoglobin and the many human diseases linked to it. A deeply humane man with exceptional powers of analysis, planning, and leadership, he was dearly loved by his colleagues, friends, and family. While his achievements were many, Perutz was a humble man by nature. To his death, he refused any honor that would give him a title; in fact, everyone around him invariably addressed him as simply "Max." Perutz lived a quiet, unostentatious life with his wife, Gisela (whom he married in 1942); son, Robin; and daughter, Vivien, walking from home to the laboratory daily until a few months prior to his death. His mind remained razor-sharp even in his later years. He presented many thrilling lectures, and he made important contributions to the understanding of Huntington's disease, based on ideas of crystal nucleation. A robust and confident mountaineer earlier in his life, Perutz had studied glacier flow so that he could work in the Alps and get in some occasional skiing. His love for the mountains, hiking, and the outdoors stayed within him until the end. Survived by his wife and children, Max Perutz died February 6, 2002, at the age of 88.

See also BIOCHEMISTRY; CRICK, FRANCIS; ENZYMES; NUCLEIC ACIDS; PAULING, LINUS; PROTEINS; SPECTROPHOTOMETRY; WATSON, JAMES; X-RAY CRYSTALLOGRAPHY.

FURTHER READING

Perutz, Max. *I Wish I'd Made You Angry Earlier: Essays on Science, Scientists, and Humanity.* New York: Cold Spring Harbor Laboratory Press, 2002.
———. *Science Is Not a Quiet Life: Unraveling the Mechanism of Haemoglobin.* (Series in 20th Century Biology). Hackensack, N.J.: World Scientific, 1998.

pharmaceutical drug development Pharmaceutical drug development is the process by which new pharmaceutical medicines are designed or dis-

covered and experiments are performed to determine their tolerability, safety, and efficacy for a specified patient population. The development consists of two main stages. The first stage is preclinical development, during which chemists develop a compound for study using Good Laboratory Practices (GLP). Preclinical development often includes the use of laboratory animals whose body systems are similar to humans', although reaction to the use of laboratory animals has limited many preclinical studies to the use of rodents and dogs. The goal of preclinical development is to determine tolerable initial doses for humans and study the pharmacological effects of the substance on the organs and body systems. Researchers examine the chemical structure of the compound and propose hypotheses about its safety and efficacy. After completion of the preclinical stage of development, an investigational new drug (IND) application is filed with the Food and Drug Administration (FDA) requesting permission to proceed to clinical drug development on humans in the United States.

Clinical drug development differs from preclinical development because it uses human volunteers as subjects. Clinical development has four main phases, which often overlap. Phase I involves healthy adult volunteers as the subjects, and the goal is, first, to determine the tolerability of the substance and then determine an initial safety profile of the medicine. Often the subjects are administered the drug in an inpatient clinic, where they can be continually examined on site for 24 to 48 hours after drug administration. Repeated sampling of the blood at regular intervals allows for the determination of the amount of substances present in healthy individuals. A profile can be determined to examine the rate of absorption of the new substance, and mathematical models are used to describe this rate. Repeated urine samples can

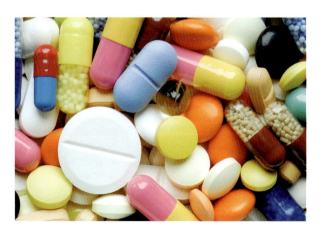

In the United States, the Food and Drug Administration (FDA) regulates the production and sale of prescription medications. *(ajt, 2008, used under license from Shutterstock, Inc.)*

help an investigator to determine how quickly and how much of the drug is excreted from the system. Phase I studies often involve a dose escalation from the initial tolerable dosage to find a maximal tolerable dosage (MTD) that does not overly compromise the safety profile of the drug. This escalation may involve individual escalations or continued enrolling of groups of patients at repeatedly higher dosages until a compromise of the safety profile occurs.

Phase II of clinical drug development assesses the overall safety of the medicine and provides indications of the drug's effectiveness, or efficacy. This phase is conducted with patients who are part of the drug's intended patient population. Common safety measures include examining the blood chemistry, urinalysis, liver function tests, and, most important, reported adverse events. The severity and relatedness of the adverse events to the medication are measured. Interactions between concomitant medications (medications taken at the same time as the trial drug) used

by the patient population and the new treatment medication are examined. Dosage is reevaluated for this intended population and possible side effects are determined. This phase is often separated into A and B segments, where phase IIA is more concerned with the safety parameters, and phase IIB is focused on effectiveness. Phase II trials are typically larger than the phase I trials in order to get a better understanding primarily of the medicine's safety but also of the effectiveness of the medicine in treating the illness or disorder.

Phase III trials are even larger studies, which are often coordinated at multiple centers and are designed primarily to study the large-scale efficacy of the substance and confirm the safety risks. The sample size is predetermined and based on prior studies to maximize the potential for achieving definitive efficacy results. Examples of primary efficacy end points include mortality in cancer studies or diastolic and systolic blood pressure measurements in hyper-

PHARMACOGENETICS: THE FUTURE OF DRUG DESIGN

by Lisa Quinn Gothard

The pharmaceutical industry is a multibillion-dollar industry that affects all people at some point in their lives. The methods used to develop an effective and safe supply of medications but keep those medications affordable are important for everyone's health. New advances in pharmacogenetics, the personalizing of medications to match individual genetic profiles, have streamlined the process of drug design, leading to higher success rates for individual patients.

Pharmaceutical drug development is the process by which new pharmaceutical medicines are designed or discovered and experiments are performed to determine their tolerability, safety, and efficacy for a specified patient population. In the United States, drug development is controlled by the Food and Drug Administration (FDA), and it is an expensive, time-consuming process. Before medications can be tested on humans, preclinical development in laboratory animals is required to determine initial doses for

humans and the effects of the medication on body systems. In addition to carrying out animal studies, researchers examine the chemical structure of the compound, propose hypotheses about its safety and efficacy, and file an investigational new drug (IND) application. If approved, the IND allows the researchers to begin human testing. Phase I clinical trials involve healthy adult volunteers as the subjects, and the goals include determining how the drug is tolerated and initial safety standards of the medicine. Phase II of clinical drug development assesses the overall safety of the medicine and provides indications of the drug's effectiveness, or efficacy. This phase is conducted using patients who are part of the drug's intended patient population. Phase III trials are even larger studies that are often coordinated at multiple centers and are designed primarily to study the large-scale efficacy of the substance and confirm the safety risks.

The present method of drug development has been predominantly led by the shotgun approach, in which scientists test

tens of thousands of potential chemicals in relation to the infected cells or target cells and determine which are effective. This is incredibly time-consuming and is relatively ineffective. This method of drug design requires on average a cost of $802 million for one medication and takes approximately 10–15 years.

Drug design is a complicated process whose goal is to ensure a safe drug supply in the United States. The completion of the sequencing of the human genome by the Human Genome Project in 2003 has allowed the complete rethinking of the chemical design of pharmaceuticals. Rather than depending on the shotgun approach to isolate effective chemicals as potential drugs that are then tested on all people, pharmacogenetics determines the genetic makeup of the individual needing treatment and determines which medication would most effectively work for him or her. The specific choice of a medication based on an individual's genetic makeup allows the correct medication to be prescribed and has the potential to prevent side effects.

tension studies. Previous larger phase IIB trials give some insight into the larger-scale potential efficacy of the new medication and are extremely useful in designing the phase III trials. The purpose of phase III trials is to show definite efficacy results without an increase in safety concerns.

The FDA requires the approval of a new drug application (NDA) before the medicine may be sold and shipped across state boundaries in the United States. This application incorporates all the preclinical and clinical data and a statistical summary of the safety and efficacy provided in integrated summaries of safety (ISS) and efficacy (ISE). The efficacy of the drug is weighed against any contraindications during the approval process. More studies may be required by the regulatory body until approval is finally granted, or the drug approval may be denied. If it is approved, postlaunch surveillance studies constitute phase IV of the drug development process, during which the much broader

effects of the medicine are assessed after it has been made available to the general patient population. The longer-term effects of the drug can be determined through these studies, which may or may not be mandated by the regulatory body in charge. Other indications can be examined to provide larger markets for the drug. Since the drug has already established a clear safety profile with regulatory bodies, these studies can focus on the relative effectiveness of the drug for different treatment areas. Sometimes the focus is on the cost of the drug relative to accepted medicines already on the market, on increased quality of life measures, or on other tertiary efficacy objectives.

Clinical trials are performed under highly regulated conditions in order to protect the general population. The Declaration of Helsinki requires that patients provide informed consent to agree to participate in the trial. Additionally, clinical trials must follow a predetermined protocol, and any deviance from

---

Patients can be screened for their genetic makeup prior to volunteering for a clinical trial, and this allows one to know how a particular drug will take effect. Those who are not expected to respond to the medication because of their genetic makeup are excluded from the drug trials, reducing the amount of testing required. Genetic screening to determine how one will react to medication allows one to understand what dosage is appropriate for an individual as well as whether the particular drug can be used safely by that patient. Pharmacogenetics has decreased the amount of testing required for a particular drug and thus lowered the cost and lessened the time involved in designing new drugs.

Scientific study of the effect of differences in genetic makeup on individuals' reactions to a particular medication has given insight to why throughout the history of medical science some patients respond appropriately to medications, others require a higher dosage, and still others have dramatically serious side effects. When this is due to a genetic difference between the patients, knowing whether a patient will have an adverse reaction to a medication prior to the deliv-

ery of that drug protects the patient and allows him or her to be given an alternative medication sooner.

Scientists are working on developing screening tools in order for a patient to be tested prior to administering a drug. One example is a common chemotherapy drug that is routinely used for children who have leukemia. The unexplained problem with this medication was that it was effective in some individuals, ineffective in others, and toxic in some. Without an explanation, the physicians treating the patient would not have a clear course of action to treat the leukemia patient.

In 1982, the American physician Richard M. Weinshilboum discovered the gene that was responsible for this differing response from patients. The thiopurine methyltransferase (*TPMT*) gene is responsible for the production of an enzyme that breaks down the chemotherapy drug in order for the patient's body to remove it after treatment. The patients for whom the chemotherapy drug was fatal did not have functional copies of the *TMPT* gene and, therefore, could not metabolize the chemotherapy drug, inability that led to the toxic effects. Those for

whom the drug was ineffective had one functional copy of the gene. The patients with two functional copies of the gene benefited from and were able to break down the medication. A simple screening for the patients prior to chemotherapy will allow their physicians to know which versions of the genes they carry and whether they can tolerate the chemotherapy. This is one of the first examples of pharmacogenetics in action and demonstrates the potential for the reduction of side effects and appropriate administration of medications to patients when a genetic link to drug effectiveness can be found.

FURTHER READING

Learn.Genetics, Genetics Science Learning Center, The University of Utah. "Personalized Medicine: Drugs Designed Just for You." Available online. URL: http://learn.genetics.utah.edu/units/ pharma/index.cfm. Accessed August 20, 2008.

National Institute of General Medical Sciences, National Institutes of Health. Pharmocogenetics Research Network home page. Available online. URL: http://www. nigms.nih.gov/Initiatives/PGRN/. Last updated April 4, 2008.

that protocol should be documented and reported. The experimental design must be clearly explained in the protocol and all the safety and efficacy parameters clearly stated. A well-designed experiment should incorporate the use of a control group as a means of comparison to the treatment group. If no accepted treatment is available, then the control group commonly receives a placebo, an inert substance, often a sugar pill. Studies have shown that some patients respond positively after receiving placebo even when the response is an objective measure such as blood pressure. More subjectively measured outcomes can be strongly affected by taking the placebo. This positive response that patients sometimes display after receiving placebo is called the placebo effect and is the main reason that a comparison group is needed to demonstrate the added benefits of the medicine over receiving the placebo.

When an approved and accepted treatment already exists, an active control comparison is commonly used. An active control is a group that receives the accepted treatment and not a placebo or other inert substance. An experimental design that randomly allocates patients to treatment or control groups is preferable to prevent bias in the selection process. Sometimes patients are grouped in blocks by similar traits and then randomly assigned from within the blocks to different groups. This block design allows for more uniform treatment and control groups as similar characteristics are evenly spread out over the two groups. The patient should not know to which group, treatment, or control he or she is assigned; nor should the evaluator of the patient's results know. This prevents the patient's biases about the treatment and control from affecting his or her perceived response to the treatment. Blinding the evaluators prevents their own biases from affecting their evaluations of the response to the treatment or control. This ideal situation where both patient and evaluator are unaware of their group assignment is called double-blinding but is not always possible in practice. The gold standard for clinical trials is a randomized, comparative, double-blinded experimental design that incorporates all of the aforementioned properties. Blocking of the subjects by similarities is included in this gold standard experimental design. Additional regulations require an independent review board to ensure that the clinical trial follows the protocol and is responsible for discontinuing the trial if the patients in the trial are placed at undue risk of harm. Severe adverse events must be reported to medical experts, who help determine whether the safety of the patients involved in the trial is compromised.

The FDA has a fast-tracking procedure whereby promising medicines in specific treatment areas are moved more rapidly through the approval process. This way a breakthrough medicine can be provided sooner to patients. Regulatory guidelines often focus on the "do no harm" principle toward patients involved in clinical trials and when clinical trials should be halted prematurely because of increased safety risks. Fast-tracking potentially beneficial medications can be thought of as harming those patients who are not able to receive the benefit of these new drugs and is therefore a corollary to the do no harm principle.

The International Conference on Harmonization of Technical Requirements for Registration of Pharmaceuticals for Human Use (ICH) and the World Health Organization (WHO) have provided guidelines for conducting clinical research and Good Clinical Practices (GCP). The purpose of these guidelines is to standardize the way clinical research is conducted in Europe, Japan, and the United States. This prevents the unnecessary duplication of trials while setting common guidelines that speed the approval process of a new drug across multiple regulatory bodies. Manufacturing practices are established for production of the new medicine that avoid impurities and ensure the proper amounts of active substances are present in the medication. The Good Manufacturing Practices (GMP) ensures that the drug that was approved is chemically the same as the drug that is manufactured and sold by pharmacies. Efficacy guidelines include dose-response evaluations so that the relative safety costs and efficacy benefits can be more readily determined. They allow safe dosages to be determined across a variety of patient subsets.

The analyses of efficacy are often grouped by different patient characteristics such as age, race, and gender in order to see how consistently the medication acts across these different subgroups. Safety analyses are often subgrouped by similar characteristics to look for uniform safety profiles. Additional guidelines require that the database used in the statistical tabulations accurately reflects the data collected in the individual case report forms. Case report forms are the forms on which all the data are written when collected. Double-key data entry is performed when two people independently enter the data into separate databases. The databases are then merged and any discrepancies are noted and resolved by a third party. All the information from clinical trials that is electronically entered into databases is audited, and the database must be shown to be in concordance with the data recorded on paper. Any fabrication of data is considered a federal offense and is prosecutable under federal law.

Safety guidelines include the study of the carcinogenic nature of the compound. A common medical dictionary, MedDRA, has been developed to code clinical trial safety data universally. All concomitant medications taken by patients must be coded universally so that interactions between the explored medication and accepted medications can be determined. All adverse events are coded to determine whether particular types of adverse events are prevalent with the medication. Determination of the severity and relatedness of adverse events to the exploratory medication are also required in order to assess causality between severe adverse events and the medication.

In the 19th century, states were responsible for regulating domestically produced food and drugs. Dishonest individuals started bilking people by promises of effective and safe medicines. The term *snake oil salesman* was used to describe these people, who made boastful claims about their wares, only to leave the towns they visited unfulfilled and with lighter wallets. The Bureau of Chemistry, a governmental agency, started researching fraudulent claims as early as 1867. Harvey Washington Wiley, chief chemist in the Bureau of Chemistry, was responsible for uniting state and other officials behind passage of a federal law prohibiting the misbranding of food and drugs. In 1906, President Theodore Roosevelt signed the Food and Drugs Act, also known as the Wiley Act, regulating medical efficacy claims and the ingredients of foodstuffs. This effectively ushered in the modern era of the FDA. In 1938, President Franklin D. Roosevelt signed the Food, Drug, and Cosmetic Act in response to an unsafe medication that killed more than 100 people in 1937. This law required approval of all drugs before allowing the drugs to be marketed and sold inside the United States and required that drugs have labeling materials that gave directions for safe usage. This legislation put in place the regulatory approval process that has since been standardized through accepted practice and input from the ICH and WHO organizations. The Food, Drug, and Cosmetic Act required a formal approval process including the application for a new drug approval and investigation new drug application that is used today by the FDA.

See also SCIENTIFIC METHOD; TOXICOLOGY.

FURTHER READING

"The International Conference on Harmonisation of Technical Requirements for Registration of Pharmaceuticals for Human Use." Available online. URL: http://www.ich.org. Accessed July 25, 2008.

U.S. Food and Drug Administration home page. Available online. URL: http://www.fda.gov. Accessed July 25, 2008.

photoelectric effect   The emission of electrons from metals as a result of illumination by light or other electromagnetic radiation is the photoelectric effect. The 20th-century German-Swiss-American physicist Albert Einstein explained this effect as a quantum phenomenon in which the metal absorbs one or more photons, with practically all of their energy taken up by an electron that subsequently exits the metal. At sufficiently low light intensities, the kinetic energy of an emitted electron is from a single photon, and the effect is designated the single-photon photoelectric effect. (Lasers can produce the multiple-photon photoelectric effect.) In the single-photon case, even for light of a definite frequency (i.e., monochromatic radiation), each of whose photons possesses the same amount of energy, the emitted electrons carry a range of kinetic energies up to some maximal value.

For all metals the relation between maximal electron kinetic energy and light frequency is found to have the form

$$\text{Maximal kinetic energy} = hf - \phi$$

where energy is in joules (J), frequency f is in hertz (Hz), and h represents the Planck constant, whose value is $6.62606876 \times 10^{-34}$ joule-second (J·s), rounded to 6.63×10^{-34} J·s. The quantity ϕ, in joules (J), is called the work function of the metal and differs from metal to metal. This equation is explained by the model of a photon, whose energy is hf, entering the metal and being absorbed by an electron. The photon's energy is practically fully converted to the electron's kinetic energy. If that electron subsequently leaves the metal, it will suffer energy losses along the way to its emergence. Electrons near the surface suffer the least loss and emerge with the greatest amount of kinetic energy. The minimal amount of energy needed to liberate an electron from the metal is the work function of the metal, which varies from metal to metal. The electrons that are emitted with the most kinetic energy are those that lost the least of their initial kinetic energy hf on the way out. That least lost kinetic energy is the work function ϕ. So the equation is explained.

Because of the work function, every metal possesses a threshold frequency below which incident light produces no single-photon photoelectric effect. A single photon of such radiation simply does not possess sufficient energy to liberate an electron. The threshold frequency f_0 is given by

$$f_0 = \frac{\phi}{h}$$

and the frequency dependence of the maximal electron kinetic energy can be expressed as

$$\text{Maximal kinetic energy} = h(f - f_0)$$

The photoelectric effect stands in stark contrast to what might be expected to occur according to classical electromagnetism. The classical picture has the electromagnetic wave causing the electron to oscillate and gradually absorb energy, until it possesses sufficient energy to leave the metal. Then it is expected that

- The maximal kinetic energy of the emitted electrons will depend on the intensity of the light and should increase with increasing intensity.
- There will be no simple dependence of the effect on the frequency of the light.
- At low light intensity, there will be a time delay from the start of the illumination to the onset of electron emission.

In reality, however, the situation is very different.

- The maximal electron kinetic energy is independent of light intensity. Rather, it depends on the frequency of the incident light and only on the frequency (for any particular metal), according to the preceding equation. What the intensity does affect is the number of electrons emitted per unit time.
- The dependence of the effect on frequency is indeed a simple one, a linear relation between maximal kinetic energy of the emitted electrons and light frequency.
- Even at extremely low light intensities, there is no delay between the start of illumination and onset of electron emission.

Einstein's model explains what classical electromagnetic theory could not:

- Greater light intensity means a larger number of photons entering the metal per unit time, which results in a larger number of electrons emitted per unit time. Intensity does not affect the energies of the individual photons, which depend only on the frequency of the light, and it is the energy of an individual photon that gives an individual electron its initial kinetic energy, as explained previously.
- The linear relation follows straightforwardly from simple energy considerations.
- The acquisition of energy by an electron is not a gradual process that requires a time delay; rather, it is an instantaneous process of photon absorption by an electron.

Einstein's explanation of the photoelectric effect as the total absorption of the energy of a single photon by a single electron helped convince physicists that the photon picture of light was valid. In 1921, Einstein was awarded the Nobel Prize in physics, not for his theories of relativity, for which he is most famous, but "for his services to theoretical physics, and especially for his discovery of the law of the photoelectric effect."

The best-known practical application of the photoelectric effect is the photocell, or photoelectric cell, also called the "electric eye." This is a device that, as long as light shines on it, conducts an electric current due to the electrons emitted by the photoelectric effect. When no light is incident on it, it does not conduct. A common use of this device is to detect the presence of an object that breaks a continuous beam of light shining on the cell. That might serve to open a door when a person approaches or to warn of a burglar. By now, however, photocells have been supplanted by smaller, cheaper, and more efficient semiconductor devices called photodiodes and phototransistors. These operate by a variation of the photoelectric effect in which photons cause nonconducting electrons to become conducting within the material, without being emitted.

Additional methods of causing electron emission from metals include field emission and thermionic emission. Field emission occurs when a strong external electric field pulls electrons off the surface of a metal. In thermionic emission, the metal is heated, thereby endowing its electrons with sufficient kinetic energy to leave the metal.

See also CLASSICAL PHYSICS; EINSTEIN, ALBERT; ELECTRICITY; ELECTROMAGNETIC WAVES; ELECTROMAGNETISM; ELECTRON EMISSION; ENERGY AND WORK; FERMI-DIRAC STATISTICS; HEAT AND THERMODYNAMICS; PLANCK, MAX; QUANTUM MECHANICS.

FURTHER READING

Brush, Stephen G. "How Ideas Became Knowledge: The Light-Quantum Hypothesis 1905–1935." *Historical Studies in the Physical and Biological Sciences* 37, no. 2 (March 2007): 205–246.

The Nobel Foundation. "The Nobel Prize in Physics 1921." Available online. URL: http://nobelprize.org/physics/laureates/1921/. Accessed July 27, 2008.

Serway, Raymond A., Clement J. Moses, and Curt A. Moyer. *Modern Physics*, 3rd ed. Belmont, Calif.: Thomson Brooks/Cole, 2004.

photosynthesis Every living organism requires energy in a suitable form to survive, grow, and reproduce. Heterotrophic organisms, including humans and all other animals, fungi, and other microorganisms, must obtain their energy in the form of organic molecules, such as carbohydrates or proteins. After ingesting these caloric substances as food and digesting them into smaller components, the cells of the organism oxidize the organic molecules to release the energy stored in the chemical bonds in order to synthesize adenosine triphosphate (ATP). Cells use the ATP as their energy currency, burning it when they require energy for cellular processes such as forming new chemical linkages or transporting molecules against a gradient.

Autotrophic organisms do not depend on an outside supply of organic material in order to draw energy into the cell for ATP synthesis. They are capable of synthesizing their own organic molecules using other sources of energy. Photoautotrophs are autotrophic organisms that possess specialized molecules that absorb energy from sunlight and convert it into chemical energy that is used to synthesize reduced carbon compounds such as simple carbohydrates $(CH_2O)_n$ from inorganic carbon sources, such as carbon dioxide (CO_2) and water (H_2O). This process, known as photosynthesis, also generates molecular oxygen (O_2) and can be summarized by the following simplified equation:

$$H_2O + CO_2 \xrightarrow{\text{light}} (CH_2O) + O_2$$

Carried out by plants, algae, and some bacteria, photosynthesis is one of the most important biochemical processes that occur on Earth. Photosynthesis produces an estimated 160–170 billion metric tons (176–187 billion tons) of carbohydrates globally each year and maintains the level of oxygen in the atmosphere at approximately 21 percent. Without photosynthesis, most organisms would die of lack of food and insufficient oxygen in the atmosphere. Only chemosynthetic bacteria that obtain their energy from reduced inorganic chemical compounds could survive in its absence. Creation of fossil fuels such as coal, oil, and gas also depends on photosynthesis, which evolved approximately 3 billion years ago. Past life forms that were photosynthetic or that fed on photosynthetic organisms died and became part of the Earth's crust. Over millions of years, geological and chemical processes slowly turned the organic material into fossil fuels that supply energy for daily activities.

The process of photosynthesis consists of two major stages. The first stage is referred to as the light-dependent reactions, because during this stage, electrons absorb radiant energy from the Sun. During the second stage, the light-independent reactions (sometimes loosely referred to as the dark reactions), the cell uses the high-energy electrons to form covalent bonds between carbon atoms, building simple carbohydrates $(CH_2O)_n$. Water serves as the original source of low-energy electrons, and O_2, a compound crucial for aerobic life forms, is a waste product.

LIGHT-DEPENDENT REACTIONS

In eukaryotic cells such as those in plants and algae, photosynthesis occurs in chloroplasts, organelles whose structure is uniquely adapted to perform this important function. A double membrane envelops the chloroplast, with the outer membrane containing porins that allow for the passage of certain molecules and the inner membrane being relatively impermeable. The stroma, or the space bounded by the double membrane, contains thylakoids, flattened sacs organized into structures called grana that resemble stacks of pancakes. The membrane system of the thylakoids contains pigments, colored molecules that absorb light of specific wavelengths. The principal light-absorbing pigments in green plants and algae are chlorophylls. Consisting of four pyrrole rings arranged into a larger ring called a porphyrin ring, chlorophyll contains many alternating single and double bonds, a characteristic of polyenes. The nitrogen atoms of the pyrrole rings all interact with a magnesium atom, and a hydrophobic, 20-carbon-long alcohol called phytol extends from an acidic side chain of one pyrrole ring. Two forms, chlorophyll *a* and chlorophyll *b*, differ by the presence of either a methyl group attached to another pyrrole ring (in chlorophyll *a*) or a formyl group (in chlorophyll *b*). The unique structure of chlorophyll allows it to absorb light in the visible range of the electromagnetic spectrum, most intensely in the ranges of 400 to 500 nanometers (nm) and 600 to 700 nm, the violet-blue and red regions of the spectrum, respectively. Because chlorophyll only weakly absorbs light with a wavelength between 500 and 600 nm, the green region of the spectrum, this color is reflected; thus, most plants and algae appear green. Other pigments absorb and reflect different wavelengths of light, making some organisms or parts of organisms that contain them appear yellow, orange, or red.

Hundreds of chlorophyll molecules work together with other pigment molecules, small organic molecules, and proteins in photosynthetic units called photosystems, embedded in the thylakoid membranes. When light hits a chlorophyll molecule, the porphyrin ring absorbs a photon, a packet of radiant energy, and enters an excited state. The excited pigment transfers the energy to a series of other pigments and eventually to the reaction center of the photosystem. A protein complex, two chlorophyll *a* molecules, and a primary electron acceptor make up the reaction center. Within picoseconds (10^{-12} second), the excitation energy reaches the reaction center. The energy levels of the two chlorophyll molecules located there are lower than those of the other

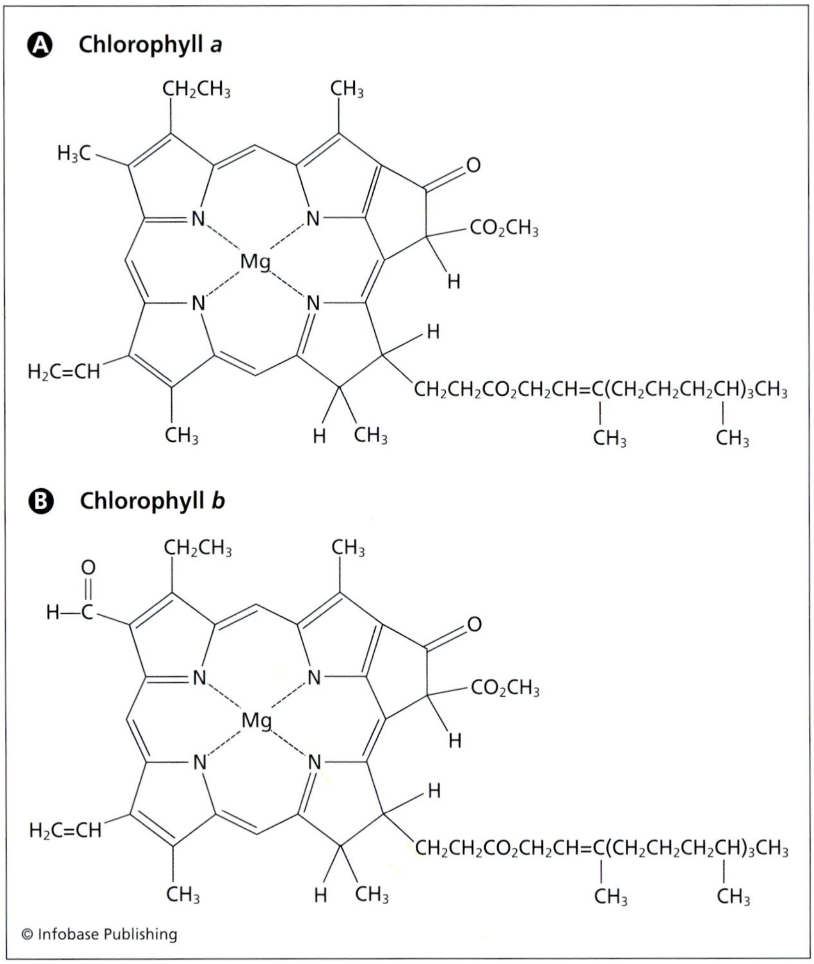

A Chlorophyll *a*

B Chlorophyll *b*

© Infobase Publishing

Because chlorophyll molecules contain networks of alternating single and double bonds, they strongly absorb visible light. This figure shows the molecular structure of chlorophyll *a* and of chlorophyll *b*.

chlorophyll molecules in the photosystem, and, thus, they are able to trap the energy associated with the photons of light, boosting one of their electrons to a higher energy level. Nearby, a primary electron acceptor molecule is poised to accept the excited electrons from the two chlorophyll molecules, which then return to their ground (resting) state.

The thylakoid membranes have two types of photosystems, photosystem I (PSI) and photosystem II (PSII), which cooperate to drive the light-dependent reactions. The roman numerals indicate the order in which they were discovered, not their order of function. PSII operates first, and the chlorophyll *a* in its reaction center is called P680 because it best absorbs light at a wavelength of 680 nm. The chlorophyll *a* in PSI best absorbs light at 700 nm and is called P700. The structures of the two pigment molecules are identical, but their positioning and association with nearby molecules affect the distribution of their electrons and give them slightly different properties. PSII produces a strong oxidant and

leads to the formation of O_2. PSI produces a strong reductant that leads to the formation of reduced nicotinamide adenine dinucleotide phosphate (NADPH).

Once electrons become excited, they can follow a noncyclic or a less common cyclic pathway. Noncyclic electron flow utilizes both photosystems and begins with the immediate transfer of an excited electron from P680 to pheophytin, which, as does chlorophyll, has a porphyrin ring but no magnesium atom in its center. The cation radical P680$^+$ is the strongest known biological oxidizing agent, and the hole left behind by the lost electron must be filled. P680$^+$ pulls the electrons from a molecule of water, splitting it into two electrons (e$^-$), two hydrogen ions (H$^+$), and one oxygen atom (O). The two electrons replace the electrons lost by the two chlorophyll molecules in the reaction center, and the oxygen atom immediately combines with another oxygen atom from another split water molecule, generating O_2, a by-product that can diffuse out of the thylakoids, out of the chloroplasts, and out into the atmosphere.

Meanwhile, the original electrons travel down an electron transport chain, passed from one electron carrier to another at a lower energy level with a higher reduction potential. Pheophytin donates the electrons to plastoquinone (PQ) embedded in the thylakoid membrane, which then feeds them to the cytochrome *bf* complex. This multisubunit complex catalyzes the transfer of the electrons from plastoquinone to a lower energy state in plastocyanin (PC). The energy released during this step is used to pump H$^+$ from the stroma to the lumen of the thylakoid, creating an electrochemical gradient.

Plastocyanin carries the now-low-energy electrons to PSI, where they replace electrons lost when chlorophyll molecules (P700) in the PSI reaction center capture photons of light and boost electrons to a higher energy level. Replacing the electrons prepares P700 to be excited once again. As in PSII, the excited electrons from PSI also make their way down an electron transport chain. The cell does not use energy carried by the electrons in PSI to transport protons across the thylakoid membrane but instead uses it to generate NADPH. A series of oxidation-reduction reactions transfers electrons to ferredoxin (Fd), a strong reductant. The enzyme ferredoxin-$NADP^+$ reductase catalyzes the reduction of $NADP^+$ to NADPH, thus accomplishing one goal of the light-dependent reactions.

The goals of the light-dependent reactions are the generation of the reduced form of NADPH and the synthesis of ATP. Energy captured by PSII splits water molecules, releasing O_2 as a by-product and sending electrons down a series of transporters to reduce NADPH in the stroma. The fate of the protons contributes to the creation of a proton-motive force. In addition to the protons created by splitting water molecules, membrane proteins have pumped additional protons across the thylakoid membrane into the lumen, creating a pH gradient. While a membrane potential, or electrical gradient, is also created, the pH gradient contributes most of the force in chloroplasts because thylakoid membranes are relatively permeable to other ions that help maintain electrical neutrality. In a process called chemiosmosis, this gradient powers the transport of protons back across the membrane into the stroma through channels that have an associated ATP synthase activity. As the H^+ diffuse back across the membrane, ATP synthase attaches an inorganic phosphate to a molecule of adenosine diphosphate (ADP) to generate ATP by photophosphorylation and releases it into the stroma.

Noncyclic electron flow results in equal molar quantities of NADPH and ATP, but the light-independent reactions utilize more ATP than NADPH. Cyclic electron flow compensates for this extra ATP demand in addition to supplying ATP for other metabolic processes. By rerouting the electron traffic from PSI back to the cytochrome *bf* complex located in between PSII and PSI, $NADP^+$ reductase never gains access to the electrons. Instead, ferredoxin delivers them back to plastoquinone, which carries them to the cytochrome *bf* complex, where they are transferred to a lower energy state. So, the transport of the electrons through this cyclic pathway still allows for a gradient to be created, driving the production of ATP by chemiosmosis.

LIGHT-INDEPENDENT REACTIONS

During the light-dependent reactions, the chloroplast harvests energy from sunlight, electron transport moves low-energy electrons from water to a higher potential energy state in NADPH, chemiosmosis synthesizes ATP, and O_2 is produced as a waste product. The light-independent reactions, sometimes called the dark reactions because they do not require sunlight, occur in the stroma of the chloroplasts of eukaryotic cells. The production of NADPH and the ATP occurs in the stroma during the light-dependent reactions. The overall goal of the light-independent reactions is to use the chemical energy created by the light-dependent reactions to synthesize carbohydrates from CO_2. The incorporation of carbon from an inorganic source (CO_2) into an organic molecule (e.g., sugar) is called carbon fixation, and it requires a lot of energy.

The Calvin cycle, named after its discoverer, the American chemist Melvin Calvin, is a cyclical biochemical pathway responsible for the assimilation of inorganic carbon into organic molecules. Also called the reductive pentose phosphate cycle, the Calvin cycle occurs in three stages: carbon fixation, reduction, and regeneration of ribulose-1,5-bisphosphate. The enzyme ribulose-1,5-bisphosphate carboxylase/oxygenase (shortened to *rubisco*) is the most abundant protein in chloroplasts and is embedded into the stromal side of the thylakoid membranes. The addition of a CO_2, distinct from the CO_2 molecule that acts as a substrate, to a specific lysine residue of the rubisco activates the enzyme. The activated rubisco catalyzes the first step of the Calvin cycle, the condensation of CO_2 with the five-carbon sugar ribulose-1,5-bisphosphate, to form an unstable six-carbon intermediate that rapidly hydrolyzes, or breaks apart by the addition of a water molecule, into two three-carbon molecules of 3-phosphoglycerate. This step is the carbon fixation stage, because carbon from inorganic CO_2 is assimilated into an organic molecule.

In the reduction stage of the Calvin cycle, first an additional phosphate group is transferred from one of the ATPs generated during the light-dependent reactions to 3-phosphoglycerate to make 1,3-bisphosphoglycerate. Then, NADPH donates the two electrons it has been carrying since the light-independent reactions to reduce 1,3-bisphosphoglycerate to glyceraldehyde-3-phosphate. For every one molecule of glyceraldehyde-3-phosphate that exits the cycle to be used by the plant cell, five molecules must continue cycling to regenerate the carbohydrate intermediates that were already present. This is because one CO_2 is incorporated into an organic molecule with each turn of the cycle, it takes three CO_2 to make one new carbohydrate, but each one of them is added to a five-carbon molecule, so 15 original carbons have already been incorporated into organic molecules at the beginning of the cycle. So, of 18 total carbons in six three-carbon carbohydrates, there is only a net production of one three-carbon sugar for every six molecules of glyceraldehyde-3-phosphate.

In summary, the fixation of three successive CO_2 molecules results in the formation of six three-carbon molecules, the ultimate reduction of which leads to the creation of one three-carbon molecule of glyceraldehyde-3-phosphate. Five of these three-carbon molecules continue cycling and participate in the regeneration of ribulose 1,5-bisphosphate, which is the purpose of the rest of the cycle. This is accomplished by rearranging the carbon skeletons of the five cycled glyceraldehyde-3-phosphate molecules and the hydrolysis of more ATP.

In the end, in addition to three molecules of CO_2, six molecules of NADPH and nine molecules of ATP are consumed for every molecule of glyceraldehyde-3-phosphate produced. Synthesis of a six-carbon sugar (a hexose), such as glucose, requires two molecules of glyceraldehyde-3-phosphate; therefore, in order to create one molecule of glucose, the net reaction is as follows:

$$6CO_2 + 18ATP + 12NADPH + 12H_2O \rightarrow C_6H_{12}O_6 + 18ADP + 18Pi + 12NADP^+ + 6H^+$$

Once the carbon is fixed, or incorporated into the organic compound glyceraldehyde-3-phosphate, other enzymes convert it into glucose and other organic compounds. The process of transferring energy from sunlight into chemical energy stored in

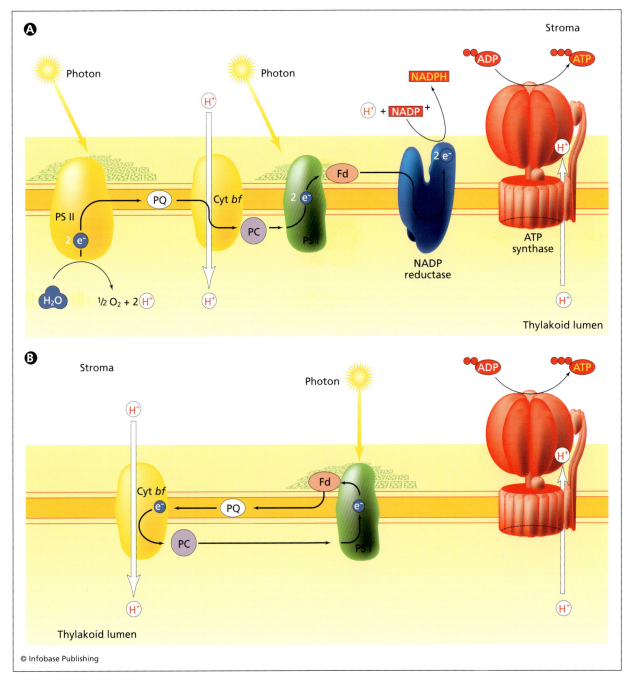

During the light-dependent reactions of photosynthesis, high-energy electrons create energy-rich molecules. In noncyclic electron flow, (a), both NADPH and adenosine triphosphate (ATP) are made, whereas in cyclic electron flow (b) only ATP is synthesized.

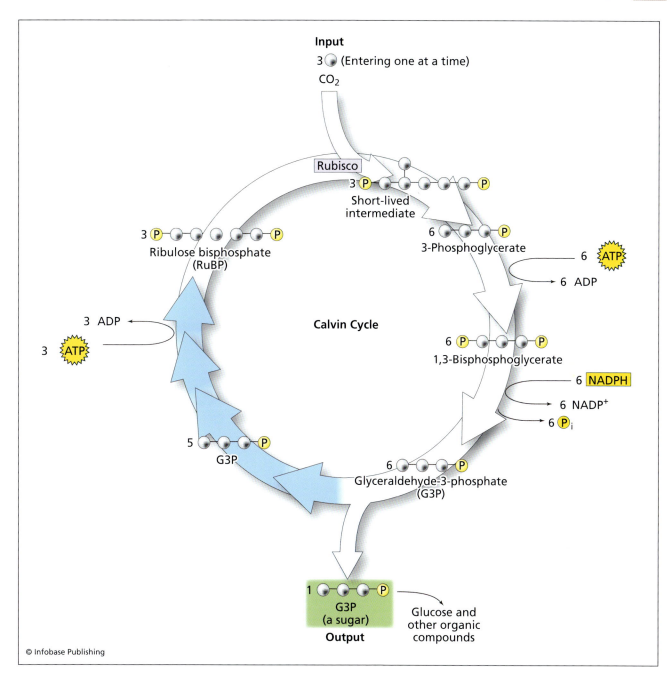

Input

3 ⬤ (Entering one at a time)

CO_2

Rubisco

3 (P)⬤⬤⬤⬤(P)

Short-lived
intermediate

3 (P)⬤⬤⬤⬤(P)

Ribulose bisphosphate
(RuBP)

3 ADP

3 ATP

Calvin Cycle

6 ⬤⬤⬤(P)

3-Phosphoglycerate

6 ATP

6 ADP

6 (P)⬤⬤⬤(P)

1,3-Bisphosphoglycerate

6 NADPH

6 $NADP^+$

6 (P)$_i$

6 ⬤⬤⬤(P)

Glyceraldehyde-3-phosphate
(G3P)

5 ⬤⬤⬤(P)

G3P

1 ⬤⬤⬤(P)

G3P
(a sugar)

Output

Glucose and
other organic
compounds

© Infobase Publishing

The Calvin cycle uses high-energy molecules generated during the light-dependent reactions of photosynthesis to incorporate carbon from CO_2 into organic molecules.

organic compounds is now complete. The cell can use the organic compounds as fuel that can be burned as needed to generate ATP or as structural components to build new cell parts during growth.

See also CARBOHYDRATES; ELECTROMAGNETIC WAVES; ELECTRON TRANSPORT SYSTEM; ENZYMES; OXIDATION-REDUCTION REACTIONS.

FURTHER READING

Berg, Jeremy M., John L. Tymoczko, and Lubert Stryer. *Biochemistry*, 6th ed. New York: W. H. Freeman, 2006.

pH/pOH In 1887, the Swedish chemist Svante Arrhenius developed simple working definitions for acids and bases. The Arrhenius definition of an acid is a substance that loses a hydrogen ion (H^+) when dissolved in aqueous solution. The definition of a base according to Arrhenius's system is a substance that loses a hydroxide ion (OH^-) when dissolved in aqueous solution. The production of these ions depends on the extent to which H^+ and OH^- can be lost from a substance. The hydrogen ions associated with an acid are known as ionizable hydrogen atoms because they are easily lost (ionized) when water dissolves the acid. Other hydrogen atoms are not easily lost; for example, a molecule of sucrose ($C_{12}H_{22}O_{11}$) has 22 covalently attached hydrogen atoms, but sucrose's hydrogens do not dissociate in water and are, therefore, not ionizable, or acidic, hydrogens. When an acid contains one ionizable hydrogen, it is known as a monoprotic acid; when it contains two, it is a diprotic acid; and if it has three ionizable hydrogens, it is a triprotic acid. Hydrochloric acid (HCl) is a monoprotic acid. Sulfuric acid (H_2SO_4) is a diprotic acid, and phosphoric acid (H_3PO_4) is a triprotic acid.

The extent of dissociation of acids and bases can be quantified by measuring the concentration in molarity of hydrogen and hydroxide ions in solution. Water will dissociate to a small extent and lose a proton according to the following formula:

$$H_2O \leftrightharpoons H^+ + OH^-$$

The concentrations of hydrogen ions, $[H^+]$, and hydroxide ions, $[OH^-]$, present in the solution are related by the following formula:

$$[H^+] \times [OH^-] = 1 \times 10^{-14} \ M^2$$

This value is known as the ion-product constant of water (K_w). As the hydrogen ion concentration increases, the hydroxide ion concentration decreases. Conversely, when the hydrogen ion concentration decreases, the hydroxide ion concentration increases.

Molarity values showing hydrogen ion concentrations are useful, but they also complicate mathematical calculations. In 1909, Soren Sorenson developed a system to compare these ion concentrations in solution when he devised the pH scale, based on the natural dissociation of water molecules. The pH scale is a logarithmic scale based on the $[H^+]$ present in an acidic solution. The pH scale ranges in values from 0 to 14, and these whole numbers are much simpler to interpret than the scientific notation values of the initial concentrations. The formula to calculate pH is

$$pH = -\log[H^+]$$

where $[H^+]$ stands for the concentration of hydrogen ions in moles/liter (M).

The logarithmic scale (log) is the inverse of an exponential base 10 scale. In other words, the log of 10^3 is 3. The logarithm illustrates to what power the base of 10 is raised. As defined by the logarithmic scale, each 1-point difference in pH value represents a 10-fold difference in $[H^+]$. For instance, the difference of 1 between pH values 4 and 5 actually represents a 10-fold difference in $[H^+]$. A solution with a pH value of 4 has an $[H^+]$ of 1×10^{-4} M, and a pH value of 5 has a $[H^+]$ of 1×10^{-5} M. These numbers are logarithms that represent actual values of 0.0001 and 0.00001 and therefore have a 10-fold difference in $[H^+]$.

When the values of $[H^+]$ and $[OH^-]$ equal each other, the solution is neutral, and the $[H^+]$ and $[OH]$ concentrations equal 1×10^{-7} M. By plugging in the neutral value for $[H^+]$, it is shown that the pH value of a neutral solution is 7.

$$pH = -\log(1 \times 10^{-7})$$

$$pH = -(-7)$$

$$pH = 7$$

A solution with the maximal $[H^+]$ concentration (1×10^0 M) would have a pH value of 0, as shown in the following calculation.

$$pH = -\log(1 \times 10^0)$$

$$pH = -(-0)$$

$$pH = 0$$

All solutions with a pH value below 7 have an $[H^+]$ greater than $[OH^-]$ and are considered acidic. A pH value of 0 is the highest $[H^+]$ concentration possible and, therefore, the most acidic type of solution.

An $[H^+]$ that is as low as possible (1×10^{-14} M) has a pH value of 14. The pH calculation is shown.

$$pH = -\log (1 \times 10^{-14} \ M)$$

$$pH = -(-14)$$

$$pH = 14$$

pH values greater than 7 have an $[H^+]$ less than $[OH^-]$ and are considered basic.

Sometimes experimentally measuring the concentration of OH^- is more convenient than using $[H^+]$. The $[OH^-]$ can be used to calculate another indicator of the acidity/basicity of a solution, the pOH. The formula for pOH is

$$pOH = -\log[OH^-]$$

Similar calculations to the pH scale will use the $[OH^-]$ to give the pOH value. A solution with an $[OH^-]$ value of 1×10^{-8} will have a pOH value of 8. The relationship between the pOH and the pH values is given by the following formula:

$$pH + pOH = 14$$

More basic / More acidic	pH	$[H^+]$	$[OH^-]$	pOH
	14	1×10^{-14}	1×10^{-0}	0
NaOH, 0.1*M*; Household bleach	13	1×10^{-13}	1×10^{-1}	1
Household ammonia	12	1×10^{-12}	1×10^{-2}	2
	11	1×10^{-11}	1×10^{-3}	3
Lime water; Milk of magnesia	10	1×10^{-10}	1×10^{-4}	4
Borax	9	1×10^{-9}	1×10^{-5}	5
Baking soda; Egg white, seawater	8	1×10^{-8}	1×10^{-6}	6
Human blood, tears; Milk	7	1×10^{-7}	1×10^{-7}	7
Saliva; Rain	6	1×10^{-6}	1×10^{-8}	8
Black coffee; Banana	5	1×10^{-5}	1×10^{-9}	9
Tomatoes	4	1×10^{-4}	1×10^{-10}	10
Wine; Cola, vinegar	3	1×10^{-3}	1×10^{-11}	11
Lemon juice	2	1×10^{-2}	1×10^{-12}	12
Gastric juice	1	1×10^{-1}	1×10^{-13}	13
	0	1×10^{0}	1×10^{-14}	14

© Infobase Publishing

The pH value of solutions varies with $[H^+]$. pH values above 7 are basic and pH values below 7 are acidic.

So, this solution would have a pH value of 6 and would be considered acidic. Given one of the four possible values to measure solution acidity, either pH, pOH, $[H^+]$, or $[OH^-]$, one can calculate each of the other values as shown in the square.

The pH of a solution can be measured by a variety of methods, including liquid chemical indicators, pH indicator paper, and pH meters. Chemical indicators are molecules that change color, depending on the pH of the solution in which they are placed. Indicators are organic molecules that can change colors on the basis of their ionized state. Examples of pH indicators are phenolphthalein, bromothymol blue, bromophenol blue, methyl red, methyl orange, cresol red, and many others. Chemical indicators can also be produced from everyday natural sources. For example, red cabbage, red beets, strawberries, rose petals, carrots, and blueberries can be turned into pH indicators. The hydrangea is a popular plant that blooms with different color flowers depending on the pH of the soil in which it is grown.

Liquid chemical indicators can be added to a solution to reveal when the pH of the solution changes. These are primarily used in titration experiments, as described later. The easiest way to use an indicator is with pH indicator paper. pH paper (litmus paper is one example) is paper that has been saturated with chemical indicators so the color will change when exposed to solutions of different pH. Different pH papers are optimized for detecting specific ranges of pH. Litmus paper simply turns red in the presence of acid and blue in the presence of base. Universal indicators and pH papers with broad-range pH levels contain a mixture of many different indicators in order to give a color change across a broad range of pH levels.

Indicators offer a quick, easy, and inexpensive glimpse into the pH of the solution, but the most accurate method of reading pH is using a pH meter. The electric potential varies with the pH of the solution. A pH meter contains two electrodes that can measure the change in electric potential in a solution in millivolts. An internal computer calculates the pH value associated with the voltage, and the meter displays this value on either a digital or an analog display.

The reactions of acids with bases are called neutralization reactions. When the number of moles of hydrogen ions donated by the acid equals the number of moles of hydroxide ions donated by the base, then the two different ions balance each other out and the solution remains neutral. The reaction between a strong acid and a strong base is shown:

$$NaOH + HCl \rightarrow NaCl + H_2O$$

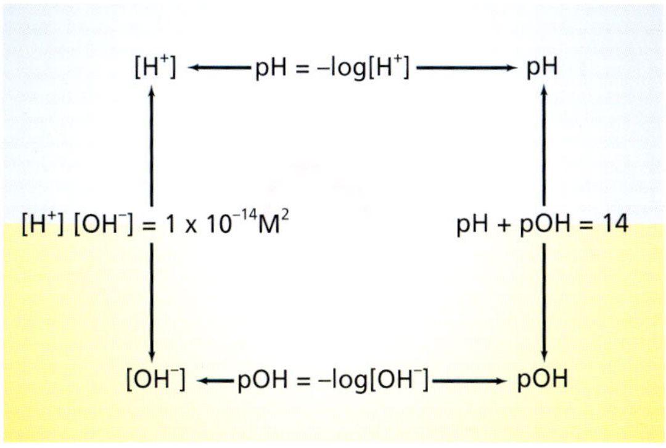

© Infobase Publishing

Given pH, pOH, $[H^+]$, or $[OH^-]$, each of the other three values can be calculated.

The use of pH paper simplifies the determination of the pH of a solution. The paper contains chemical indicators that change colors depending on the solution's pH. The paper can then be compared to a color reference chart to determine the pH. *(Christina Richards, 2008, used under license from Shutterstock, Inc.)*

The concentration of an acidic or basic solution can be calculated using a known concentration of an acid and base and then measuring when neutralization occurs. This process is known as a titration. A hydrochloric acid (HCl) solution of unknown concentration can be titrated by first adding an indicator to the acidic solution. The indicator must have a color change in the pH range of interest (pH = 7), allowing for quick identification of the neutraliza-

tion point. A known concentration of base, such as NaOH, can be added using a buret so the exact volume of the base added can be measured. Once the number of moles of OH^- is equal to the number of moles of H^+ present in the original unknown solution, known as the equivalence point, the pH of the solution will be neutral, or 7, and the indicator will change the color of the solution. Calculation of the number of moles of NaOH required to neutralize the volume of HCl gives the concentration of HCl in the original solution. The following example demonstrates how to calculate the concentration of an acid after a titration reaction.

During a titration reaction, 35 mL of 1.0 M NaOH were added to 50 mL of a HCl solution before the solution turned pink, indicating neutralization had occurred.

1. Calculate the number of moles of NaOH that were required to neutralize the acid.
$$(35 \text{ mL}) \times [(1 \text{ mole})/(1{,}000 \text{ mL})] = 0.035 \text{ moles of NaOH}$$

2. Calculate the ratio of H^+ and OH^-.
Every HCl gives one H^+, and every NaOH gives one OH^-, so the ratio is 1:1.

3. 0.035 mole of HCL must have been in the original solution.

4. Calculate the molarity of the HCl solution by dividing the number of moles by the number of liters.
$$(0.035 \text{ moles})/(0.050 \text{ L}) = 0.7 \text{ M HCl}$$

HCl is a monoprotic acid, and when it is neutralized using NaOH, which gives off one OH^- for every mole of base, the ratio is always 1:1. Titrating a solution of a diprotic or triprotic acid would require incorporating the number of moles of hydrogen or

NATURAL INDICATORS AND CHEMICAL INDICATORS FOR PH

Indicator Name	pH Range	Color below Range	Color above Range
Phenolphthalein	8.2–10.0	Clear	Pink
Bromothymol blue	6.0–7.6	Yellow	Blue
Bromophenol blue	3.0–4.6	Yellow	Violet
Methyl red	4.2–6.3	Red	Yellow
Methyl orange	3.1–4.4	Red	Yellow
Litmus	4.5–8.3	Red	Blue
Congo red	3.0–5.2	Blue	Red

hydroxide ions in the calculation. H_2SO_4 is a diprotic acid. If it were substituted into the previous problem, the ratio of $[H^+]$ to $[OH^-]$ would be 2:1, and the number of moles of H_2SO_4 in the original solution would have been

(0.035 mole)/2 = 0.0175 mole (0.018 with the correct number of significant figures)

The concentration of the original solution would be

(0.018 mole)/(0.05 L) = 0.36 M H_2SO_4

pH balance is critical to the functioning of many living systems. The pH of human blood must be maintained within a very narrow range of values or risk damaging the cells and tissues. Enzymes and other cellular proteins will break down if the pH changes dramatically. The normal range of pH values in human blood is between 7.35 and 7.45. If the pH drops to 6.8 or rises to 7.8, death can result. Diabetes mellitus can alter the pH of the blood. Low levels of insulin in the blood require cells to metabolize fat for energy. Excessive fat metabolism creates acidic by-products that lower the pH of the blood, causing a condition known as ketoacidosis. Under normal conditions, the human body uses buffer systems to maintain a constant level of blood pH.

See also ACIDS AND BASES; CONCENTRATION.

FURTHER READING
Brown, Theodore, H. LeMay, and B. Bursten. *Chemistry: The Central Science,* 10th ed. Upper Saddle River, N. J.: Prentice Hall, 2006.
Crowe, Jonathon, Tony Bradshaw, and Paul Monk. *Chemistry for the Biosciences.* Oxford: Oxford University Press, 2006.

physical chemistry   Physical chemistry is a branch of chemistry concerned with the properties and reactions of chemistry based on the physical laws governing the motion and energy of the atoms and molecules involved. The field of physical chemistry is highly mathematical and also involves physics. Chemists who explore this field utilize many fundamental principles of chemistry and physics to determine the answers to questions such as how proteins will fold, how medications will function in the body, or how a product can best be manufactured. Physical chemists do work similar to engineers, in that they use their knowledge of physical science to find solutions to problems. Modeling is important to physical chemists, as is the use of computers.

EDUCATION AND TRAINING
Education and training in the field of physical chemistry require a strong background in general chemistry, physics, computer science, and mathematics. Extensive laboratory training is critical, and a physical chemistry education also involves the study of computer techniques. A typical physical chemistry lab contains multiple types of high-tech equipment, including mass spectrometers, nuclear magnetic resonance spectrometers, and multiple computers for data analysis. The type of degree required for employment as an analytical chemist depends upon the type of job desired. Degrees in physical chemistry range from four-year bachelor of science degrees (in chemistry), to research-based master of science degrees and doctoral degrees. In general, laboratory technician positions require a bachelor's or master's degree. Supervisory roles require either a master's degree or a doctorate (Ph. D.). Full-time academic positions generally require a Ph.D., and many require additional postdoctoral work in the field.

Some consider physical chemists the most diverse of all chemists. Physical science knowledge can be applied to all areas of chemistry, including biochemistry, environmental chemistry, analytical chemistry, and organic chemistry. Understanding the atomic and molecular interactions that cause chemical reactions to occur and that confer specific physical properties on a substance is necessary in every field of chemistry. Students of physical chemistry are encouraged to have a diverse background, as their physical chemistry knowledge can be applied to a variety of fields. Physical chemists hold positions in every area of industry, academic institutions, and governmental research institutions.

According to the American Chemical Society, physical chemists and physical chemistry students are trained in fundamental physical principles, including thermodynamics, kinetic theory of gases, quantum mechanics, bonding theories based on electron interactions, and chemical kinetics.

THERMODYNAMICS
The driving principle in physical chemistry models involves thermodynamics, the study of energy transformations that occur in physical and chemical systems. The field of thermodynamics is governed by a set of laws known as the laws of thermodynamics. The first law of thermodynamics concerns the amount of energy in a system. The change in the energy of a system is equal to the amount of heat added to the system minus the work done by the system. This principle is based on the law of conservation of energy. The total amount of energy in

the system must be accounted for; it can neither be created nor be destroyed. The second law of thermodynamics describes the entropy (measure of disorder) of a system as always increasing. The nature of the physical world is that the amount of disorder is increasing. Imagine one's bedroom. A large amount of energy is required to clean up the room and put things in order. On the other hand, the room seems to become disordered within no time.

KINETIC THEORY OF GASES

Gases are the most energetic of all the states of matter, with the exception of plasma. The study of gases can include the overall motion of a large number of gas molecules or atoms. Factors that affect a gas include pressure, volume, temperature of the gas, and the number of moles of gas being studied. The relationship of these variables for a sample of a gas is given by the relationship known as the ideal gas law:

$$PV = nRT$$

where P stands for the pressure of the gas in kilopascals, V represents the volume in liters, n represents the number of moles of the sample, T is the absolute temperature of the gas (in kelvins), and R is the ideal gas constant (8.31 L × kPa/K × mol). When physical chemists study gases, understanding the motion of the individual atoms or molecules is necessary. The kinetic theory of gases involves the motion and interactions of the particles of gas. The principles of the kinetic molecular theory include the following:

● All gases are made up of atoms or molecules that are small relative to the large distances between them.
● These atoms and molecules of a gas are always in constant random motion.
● All collisions among the atoms or molecules in a gas are perfectly elastic, meaning that no energy is lost during the collision.

The temperature of a gas is the measure of the average kinetic energy of the particles in a gas sample. As the kinetic energy of the particles of a gas (and thus its temperature) increases, the number of collisions that occur in that sample increases. The pressure exerted by the gas on the container is determined by the number of collisions the gas has with the container. When the temperature of a gas sample increases, the number of collisions increases, causing the pressure on the container to increase.

QUANTUM MECHANICS

Physical chemists utilize the quantum mechanical model of the atom to help explain subatomic structure as well as predict how atoms will interact with one another. As the model of the atom changed through the early 1900s, the planetary model of the atom put forth by the Danish physicist Niels Bohr was insufficient to explain the true nature of the atom. Quantum mechanical theory was developed to help explain an appropriate atomic model. Through the work of the German physicist Werner Heisenberg and the Austrian physicist Erwin Schrödinger, the quantum mechanical models of the atom take into consideration the motion of the subatomic particles, namely, electrons, in two ways. One can consider the motion of these electrons as particles or as a wavelike motion; this concept became known as the wave-particle duality of nature.

BONDING THEORIES

Bonding theories explain how atoms interact when they are bonded together. The types of atoms involved in chemical bonds determine the nature, strength, and geometry of the bond formed. Several models exist to explain how atoms bond including the Lewis theory, the valence shell electron pair repulsion (VSEPR) theory, valence bond theory, and molecular orbital theory. Understanding bonding theories helps physical chemists understand how the atoms interact with one another and allows them to predict how different types of atoms will react.

Lewis electron dot structures depict the valence electrons (those in the outermost energy level) of an atom. They can be used to draw simple representations of the interactions of the atoms in a covalent bond by showing how the valence electrons of each atom interact with those of other atoms.

The valence shell electron repulsion theory (VSEPR) reveals three-dimensional geometrical approximations based on electron pair repulsions. Predicting the bond angles and bond lengths helps one determine the spatial relationships among the atoms in a molecule. Using this model, one can accurately predict the three-dimensional shape of the molecule.

Valence bond theory is based on the overlap of electron orbitals, the allowed energy states of electrons in the quantum mechanical model. Chemists developed this type of model to explain further the differences in bond length and bond energy. Molecular orbital theory is one of the most sophisticated bonding theories, as it takes the three-dimensional structure of the atom into account. The main principle of this theory involves the formation of molecular orbitals from atomic orbitals.

REACTION RATES

The speed at which a chemical reaction takes place is known as the reaction rate. The study of reaction rates falls into the realm of chemical kinetics. Physical chemists are interested in understanding how quickly a particular reaction occurs. Rate laws are the physical laws that govern the speed at which reactants are converted into products. They can be used to establish the speed of the reaction and therefore the amount of product that will accumulate after a given period. Factors that affect the rate of a reaction include the temperature of the reaction, the concentration of the reactants, the total surface area of the reactants, and the presence of a catalyst. In general, the higher the temperature of the reaction, the more quickly the reaction will proceed. When the concentration of the reactants increases, the reaction will proceed toward the right, or toward the formation of products. An increased surface area will also increase the rate at which the reactants form products. A catalyst, such as an enzyme, is able to increase the rate of a reaction without itself being consumed or altered in the reaction.

The rate law demonstrates how the speed of a reaction depends on the concentration of the reactants and products. Consider the following reaction:

$$A + B \rightarrow C$$

given a rate law of

$$Rate = k[A][B]$$

where k is a rate constant specific for that reaction, and the concentrations of A and B are for a given time point. This example demonstrates that as the concentration of either A or B is increased, the rate of the reaction increases proportionally. Physical chemists adjust the reaction conditions in order to favor a desired product formation.

TECHNIQUES OF PHYSICAL CHEMISTRY

Physical chemistry depends heavily on high-tech machinery, including computers capable of synthesizing large quantities of data and molecular modeling programs. Physical chemists employ many analytical chemistry techniques such as mass spectroscopy, which separates ionized compounds on the basis of mass/charge (m/v) ratio of the ions, in their study of the interactions between atoms and their chemical properties. Nuclear magnetic resonance (NMR) is another commonly used technique in physical chemistry. NMR reveals the location and arrangements of atoms in a molecule, which are based on their spin in a magnetic field when they are bombarded with radio waves.

See also BONDING THEORIES; MASS SPECTROMETRY; NUCLEAR MAGNETIC RESONANCE (NMR); QUANTUM MECHANICS; RATE LAWS/REACTION RATES.

FURTHER READING
Engel, Thomas, and Philip Reid. *Physical Chemistry*. Upper Saddle River, N.J.: Pearson Education, 2005.

physics and physicists   This entry discusses the nature of physics and the position of physics among the natural sciences. It then continues with a discussion about those who practice physics—the physicists: what they do and the education required of them. Finally, the entry presents and describes the American Physical Society and the International Union of Pure and Applied Physics.

PHYSICS

Physics is the branch of natural science that deals with the most fundamental aspects of nature. Natural science—some prefer simply the term *science*—is the human endeavor of attempting to comprehend rationally—in terms of general principles and laws—the reproducible and predictable aspects of nature. Physics, as a branch of natural science, concentrates on those aspects of nature that are, or at least appear to be, the most fundamental, in the sense that they underlie all the rest. The notion of fundamentality here is a component of the view that some aspects of nature indeed underlie others, that not all aspects of nature are equal in this regard. That idea is well reflected in the hierarchical organization of science and seems to form a valid picture of nature. As an example, chemistry is the study of matter and its transformations at the atomic, molecular, and larger scales. Chemistry is quite autonomous in that it can and does achieve a great deal of understanding without needing to probe into the nature of atoms. So, chemistry forms an independent level in the hierarchy of science. Nevertheless, it is understood by all chemists that everything going on in their field of interest ultimately does derive from the nature and properties of atoms. Physics studies the nature and properties of atoms, among other phenomena. Thus physics is a more fundamental branch of science than is chemistry. In similar vein, ascending the hierarchy ladder, chemistry underlies biochemistry, which underlies biology, which might even be viewed as underlying psychology, although the latter is not commonly considered a natural science.

In its study of the atom, physics finds electrons and nuclei and studies them, too. Nuclei are discovered to be composed of protons and neutrons, which physics investigates, as it does also the other elementary particles. Some elementary particles, including the proton and neutron, are found to be composed of quarks. Consequently, physics studies quarks. In this manner, physics probes nature at smaller and smaller scales, which translate into higher and higher energies. On the other hand, physics looks at larger scales by investigating aggregations of particles, atoms, and molecules. In this connection, here is another example of the relation between physics and chemistry, as a further illustration of their relative positions in the science hierarchy. One interest of chemists is the various states of matter (gas, liquid, solid) and transitions from one to another, taking them as given. Physics studies the formation of states of matter and the basic principles underlying their transitions, as part of its investigation of the general behavior of conglomerations of atoms and molecules.

Proceeding toward larger and larger scales, physics studies the formation, properties, and behavior of astronomical bodies, such as planets, moons, and stars. Ever-greater-size groups of such bodies are investigated: solar systems, galaxies, clusters of galaxies, and superclusters. That leads to everything, to the universe itself, whose study, cosmology, forms a field of physics. And here physics is revealing an amazing circularity in nature. Since the universe comprises everything, its properties and behavior must tie in with the principles and laws of nature at all scales, and in particular at the smallest and most fundamental scale, that of the elementary particles. Present understanding does indicate an intimate meshing of the largest-scale properties and behavior of the universe with the properties and forces of the elementary particles. That interdependence is understood to have developed during the earliest stages in the evolution of the universe following the big bang.

Yet even the universe does not constrain the imagination of physicists. Space and time themselves are subjects of study, and cosmologists are actively investigating cosmological models in which our universe forms but a part. Some examples: a multiverse, in which other universes accompany ours; extra dimensions beyond the four dimensions of space and time; and cyclic evolution of the universe, with recurring big bangs.

Physics is divided into various and diverse fields of specialization, some of which are included in this work.

PHYSICISTS

A physicist is a person who has studied physics in depth and in breadth and has made physics his or her profession. Most often, a physicist is involved in physics research, physics teaching, application of physics, industrial administration, academic administration, consulting, or some combination of those. As

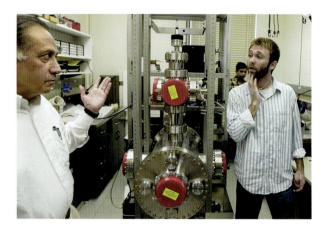

Physicists in a research laboratory discussing their apparatus, which involves mirrors and a high-precision camera for testing them. *(Noah Devereaux/MCT/Landov)*

a rule, a physicist is either an experimental physicist or a theoretical physicist.

Experimental physicists, or experimentalists, probe nature by performing experiments. Every experiment is a question asked of nature. Experimentalists force nature to give answers. They are in direct contact with the natural phenomena. They plan the experiments. Then they design, often manufacture, and construct and assemble the experimental apparatus for performing the experiments. They debug the apparatus and calibrate it. When all is operating smoothly and reliably, they then perform the experiments. That can require very many runs of each experiment. The experimentalists collect data as the experiment progresses, process the data to obtain meaningful results, and publish the results so others can (ideally) confirm them or (unfortunately) prove them wrong. Confirmed results then become grist for the mill of the theoretical physicists, or theoreticians, who try to put the results in broader contexts, relate them to other experimental results, and gain an understanding of them.

The domain of theoretical physics is the realm of ideas, concepts, patterns, relations, generalizations, abstractions, and unifications. Mathematics forms the essential tool for expressing all that in coherent ways and obtaining results that can be compared with experiment. One goal of theoretical physics is to discover patterns and order in natural phenomena and experimental results, which allow the prediction of new phenomena and results. These schemes are called laws. The Stefan-Boltzmann law can serve as an example. This law states that the total power of electromagnetic radiation emitted by a blackbody is proportional to the fourth power of the body's absolute temperature. The law was discovered and formulated in terms of a pattern that was found among the results of many measurements of temperature and radiated power in situations that approximate a blackbody. It allows the prediction of radiated power for all temperatures, such as the temperatures of stars, even if the temperatures are not among those of the set of experimental data that formed the basis for the law. So, the law forms a generalization from a particular set of measured values of radiated power at certain temperatures to the values of radiated power at all temperatures.

Beyond discovering laws, however, a more ambitious and fundamental goal of theoreticians is to devise explanations for laws. Such an explanation is termed a *theory*. A theory provides a fundamental framework from which the validity of a law follows as a logical (i.e., mathematical) consequence. In the case of the Stefan-Boltzmann law, for instance, Max Planck devised a framework for the exchange of energy between matter and the electromagnetic field that led to a formula for the blackbody radiation spectrum. The Stefan-Boltzmann law follows immediately from this formula. Planck's theory contributed to the development of quantum mechanics and the concept of the photon and eventually to quantum electrodynamics (QED), which is one of the most successful theories of physics. QED offers a wide range of theoretical results that are exceedingly precise—they include many significant digits—and accurate—they compare well with experimental measurements. One example of such a result is the value of the magnetic moment of the electron.

In this connection, it should be noted that the meaning of the term *theory* in science, and in physics in particular, is different from the common connotation of hypothesis or speculation, as in "It's only a theory." A theory is an explanation. It might indeed be a speculative theory, as yet not well confirmed, or it might be a solid, accepted one. Albert Einstein's special theory of relativity, for example, is a very well-confirmed theory and universally accepted. It is not Einstein's special hypothesis of relativity or special speculation. It is a well-founded, unifying explanation of a tremendous number of phenomena.

In different fields of physics at different times, the physics might be experiment driven or theory driven. In the former case, the experimentalists of the field are taking the initiative by examining interesting (to them) unexplained phenomena. The theoreticians are, so to speak, running along behind them, trying to make sense of the experimental observations. In theory-driven physics, on the other hand, the theoreticians are in the lead, devising theories of nature that not only explain what is already known, but predict new phenomena. The experimentalists' job is then to check those theories against the real world, to disprove the false ones and to confirm the more successful ones.

In brief caricature: experimentalists tinker with designs and apparatus and use equipment, theoreticians tinker with ideas and mathematics and use paper, while both use computers.

In a typical situation, these theoretical physicists are carrying on a lively discussion about the elementary-particle research in which they are collaborating. *(Peter Ginter/Science Faction/Getty Images)*

The field of particle physics makes use of a specialized intermediate endeavor of phenomenology. Phenomenologists are positioned somewhere between the experimentalists and the theoreticians. They study the experimental results and look for order and regularities. Then the theoreticians take over from there.

The education of a physicist starts with the most general ideas, concepts, laws, and theories that

BEAUTY: ONLY IN THE EYE OF THE BEHOLDER?

by Joe Rosen, Ph.D.

"They performed a beautiful experiment on liquid helium." "His theory of high-temperature superconductivity is a real beauty." "She discovered a beautiful law for the low-pressure behavior of such systems." "Now, that's a beautiful idea!" Such expressions are common among physicists. It would seem there is a lot of beauty in physics.

But what does *beauty* have to do with *physics*? Why should aesthetics enter the picture? After all, physics is a *rational* study of nature. Physics is carried out with reason, logic, and mathematics. As all branches of science do, physics strives for objectivity, and what could be more subjective than beauty, which, as is claimed, lies in the eye of the beholder?

In spite of all that and irrational as it might be, physicists do indeed find beauty in physics. Physics does possess an aesthetic component. Other factors being equal, a physicist will always prefer a beautiful theory to an ugly one. In fact, a physicist might very well prefer a beautiful theory even when an uglier one fits the data better! Many physicists will admit that the pleasure they derive from their profession contains a large aesthetic component, and for some that component dominates.

What, then, is beauty in physics? And in particular, what makes a *theory* beautiful? What is it that arouses in a physicist the feeling of beauty? It is generally agreed that the principal ingredients of beauty for a physicist are simplicity, generality, and unification.

First consider *simplicity,* perhaps the hardest of the three properties to pin down precisely. At least for physics, simplicity might best be correlated with a small number of conceptual ingredients. As an illustration, consider the law, discovered by Galileo Galilei, that the distance d that a uniform sphere rolls from rest down a straight, inclined track during a time interval t is proportional to the square of the time interval,

$$d = bt^2$$

It follows from this that the instantaneous speed v of the rolling sphere is proportional to the elapsed time,

$$v = 2bt$$

This law is considered extremely simple. It is, perhaps, the simplest law that might be imagined for the given situation. Consider various alternatives. The proportionality coefficient b might have different values for different horizontal orientations of the plane. Then there would be a different version of the law for each orientation: a north value of b, a northeast value, an east value, and so on. That is clearly more complicated than a single value of b for all directions. Or the dependence of d on t might not be a power dependence (i.e., might not be expressible as a power of t). It might be, for instance, an exponential dependence (such as e^{bt} or $10^{7.5t}$) or a logarithmic dependence (for instance, $\ln bt$ or $\log \pi t$). Here physicists and mathematicians alike agree that such a dependence would be more complicated than a power law.

If a power law, then perhaps a sum of terms with different powers, such as $bt^2 + ct^{1/2}$? Again, clearly more complicated than a single term. Well, if a single term, why not something like $bt^{2.067}$? Because integer powers are mathematically simpler than others. So, if an integer power, then how about negative integers? That would not work, since negative powers of time give an infinite value of d at $t = 0$ and decreasing d as time progresses, which clearly does not describe what happens. Among the positive-integer powers, the simplest is unity, with d proportional to t. But that describes constant-speed motion, while the rolling ball clearly starts with zero initial speed and accelerates. Thus, the simplest positive-integer power of time that describes this kind of motion is the second power, which is just what is observed.

For another example of simplicity, turn to Albert Einstein's general theory of relativity. This is one of a number of proposed theories of gravitation, the universal force of attraction between all pairs of bodies in the universe, and has been well confirmed as the winner. Although all those theories might appear overwhelmingly complicated, among physicists who deal with such matters, Einstein's theory is generally perceived as also taking the prize with regard to simplicity. Thus, it would stand as the preferred theory, even if the experimental data were ambiguous.

Symmetry contributes to simplicity. In a symmetric situation, there exists equivalence among certain aspects of the situation. In the rolling-sphere example, having the same value for b in all horizontal directions is symmetry, symmetry under all rotations about a vertical axis. Thus, the more symmetric the situation, the simpler it is. That is because more aspects are equivalent, and that is simpler than the aspects' being completely different. Compare the shapes of a sphere and a cow. The former is clearly the simpler by far. As for symmetry, the sphere possesses symmetry under all rotations about any axis through its center, as well as reflection symmetry

underlie physics and with the basic mathematical tools needed to handle them. Then the student learns more advanced approaches to the fundamental theories along with the more advanced mathematics required for them. Some ideas are approached again at an even higher level. Along the way, the student gains laboratory experience, as well. At some point the physics student specializes in a particular field of

---

through any plane passing through its center. All its orientations and reflection images are equivalent. The symmetry of a cow, on the other hand, consists merely of left-right reflection symmetry through the front-back vertical plane down the center. A cow's shape is equivalent solely to its single reflection image.

Now consider *generality* as a beauty-enhancing property. Generality is easier to describe than simplicity: the more general a category, the greater the number of natural phenomena it encompasses. An experiment whose result has broad implications is more general than one that reflects a particular case. A law is more general as it covers a wider range of situations. The law in the rolling-sphere example is very general indeed. It is valid for all spheres, of all materials and all sizes, not merely for ball bearings, marbles, or bowling balls.

For an additional example, consider Johannes Kepler's laws of planetary motion and compare them with Sir Isaac Newton's laws of motion and law of gravitation. The former, as they were originally stated, deal with the solar system. However, they are actually more general than that and are valid also for the moon systems of multimoon planets. Newton's laws, however, deal not only with the solar system and moon systems, but with *all* objects and systems of objects. Thus, the latter are far more general than the former. (Newton's laws form a theory of, i.e., an explanation for, both Kepler's laws and Galileo's rolling-sphere law.)

Unification goes beyond generality. A unifying law or theory must not only encompass a range of phenomena, but also show that they are, in reality, only different aspects of the same phenomenon. Take Kepler's laws, for instance. Before Kepler, the motions of the Sun's various planets appeared to have nothing to do with each other. Kepler unified the motions by showing that they

are all particular cases of the same phenomenon and obey the same laws. Or consider James Clerk Maxwell's equations that form a theory of electromagnetism. This theory unifies electricity and magnetism and shows that they are two aspects of the single phenomenon of electromagnetism.

Those, then, are the three properties that mostly affect physicists' perception of beauty in physics: simplicity, generality, and unification. As mentioned, physicists always prefer and strive for beautiful laws and theories, even when less beautiful ones appear to fit the data better. Moreover, it turns out that correct laws and theories are inevitably beautiful! Whatever it is that causes physicists to perceive beauty as they do somehow attunes them to nature itself. In brief: *nature prefers beauty.* Why nature does so is quite a mystery. Here is an example.

Paul Dirac developed a very beautiful theory of the electron, called the Dirac equation. The theory predicted the existence of another type of elementary particle, one having the same mass, spin, and magnitude of electric charge as the electron, but with opposite sign of electric charge—positive rather than negative. At the time, the only other known types of elementary particle were the proton and the neutron, and neither fitted the specifications of Dirac's predicted particle. Thus the theory was considered false. Nevertheless, Dirac did not abandon his theory, and the eventual discovery of the positron, the electron's antiparticle, proved him right. The moral, according to Dirac, is that "it is more important to have beauty in one's equations than to have them fit experiment."

One who did not follow Dirac's way and lived to regret it was Erwin Schrödinger. Schrödinger devised a beautiful theory to explain atomic phenomena, but when he applied it to the electron in the hydro-

gen atom, the simplest atomic system, he obtained results that were in disagreement with experiment. Then he noticed that a rough approximation to his equation gave results that agreed with experimental observations. So he published his approximate theory, now called the Schrödinger equation, which is a much less beautiful theory than the original. Because of his delay, the original theory was published by others and credited to them. What had happened was this. The original, beautiful theory was not appropriate for the type of elementary particle called fermion, and the electron is a fermion. Rather, the theory was suitable for bosons, a type of elementary particle that had not yet been experimentally discovered. The approximate, uglier theory was insensitive to the difference between fermions and bosons and turned out to be fairly accurate when applied to the hydrogen atom.

And, as a final example, when Einstein proposed his general theory of relativity, there existed very little experimental evidence in its support. Since then, much more evidence has accumulated, and the theory is now very well confirmed. One can imagine Einstein's repeating to himself at the time, "This theory is too beautiful to be wrong!" He would have been absolutely right, of course, as the general theory of relativity is widely considered to be one of the most beautiful theories of physics.

FURTHER READING

Feynman, Richard. *The Character of Physical Law.* Cambridge, Mass.: M.I.T. Press, 1965.

Hatton, John, and Paul B. Plouffe. *Science and Its Ways of Knowing.* Upper Saddle River, N.J.: Prentice Hall, 1997.

Oliver, David. *The Shaggy Steed of Physics: Mathematical Beauty in the Physical World.* 2nd ed. New York: Springer, 2004.

physics and chooses between an experimental path and a theoretical one and then most often remains either an experimentalist or a theoretician for the rest of his or her active life. The culmination of a physicist's full formal education, leading to a Ph.D. degree, is a supervised research project. Nevertheless, many physicists enter the profession with a bachelor's or master's degree. But even a Ph.D. is not the end of education. Those physicists who are taking part in extending the boundaries of knowledge are gaining new insights into the workings of nature. Other active physicists generally keep abreast of the latest developments in their fields of interest.

Beyond the physics itself, a physics education develops in the student a quantitative approach to dealing with matters, proficiency in applying mathematics to problems, analytic thinking, and, in the case of an experimentalist, laboratory skills as well. In this way, a physics education forms an excellent preparation for much more than solely a career in physics. As an often-presented example, holders of physics degrees have taken jobs in the financial world and have done well there.

AMERICAN PHYSICAL SOCIETY

Founded in New York City in 1899 with the mission "to advance and diffuse the knowledge of physics," the American Physical Society (APS) has grown to be the preeminent professional society of physicists in the world. Its members currently number about 46,000, most of whom are in the United States and about a fifth in other countries. The society's headquarters are located in the American Center for Physics building in College Park, Maryland, in suburban Washington, D.C. Along with nine sister societies, the APS is a member society of the American Institute of Physics.

In fulfillment of its mission, the APS publishes some of the world's most widely read peer-reviewed journals and holds more than 20 meetings a year. The journals include the five sections of the *Physical Review,* which cover all fields of physics. It also publishes *Physical Review Letters,* generally thought to be the most prestigious venue to publish important advances in physics research; *Reviews of Modern Physics*; and the all-electronic journal *Special Topics: Accelerators and Beams (STAB).* Except the latter, all the journals appear in both online and print versions, with the online version the official one. The APS maintains an online archive, called PROLA, in which can be found every article published in all the APS journals since their inception.

The APS organizes two general annual meetings, one of which is the largest meeting devoted to physics anywhere in the world, regularly attracting more than 5,000 attendees. The other APS meetings are on more specialized research topics and are typically organized by the individual divisions and topical groups of the APS.

In addition to those core activities, the APS serves the physics community and the general public in a

A scene from a meeting of the American Physical Society. At such meetings, physicists report to each other on their research and discuss it informally, as shown here. *(Ken Cole/APS)*

variety of ways. Physics education is a major concern, and the APS runs a number of education-oriented programs, including a major one aimed at improving the preparation of science teachers. The society is also active in informing the public about physics, as it does through the media and through its Web site. The APS plays an important role in the international physics community, maintaining contacts and reciprocal relations with physics societies throughout the world and monitoring human rights problems affecting physicists and other scientists. The organization bestows about 40 prizes and awards each year for achievements in physics research and for service to the physics community.

The URL of the American Physical Society Web site is http://www.aps.org/. APS's Web site for communicating the importance and excitement of physics to the general public has the URL http://PhysicsCentral.com/. The society's mailing address is One Physics Ellipse, College Park, MD 20740-3844.

INTERNATIONAL UNION OF PURE AND APPLIED PHYSICS

Established in Brussels in 1922, the International Union of Pure and Applied Physics (IUPAP) is an association of national physics communities, whose present membership numbers 48. IUPAP is itself a member of the International Council for Science and collaborates with similar unions from other disciplines. The mission of IUPAP is to assist in the worldwide development of physics, to foster international cooperation in physics, and to help in the application of physics toward solving problems of concern to humanity. IUPAP carries out its mission by sponsoring suitable international meetings and assisting meeting organizing committees; promoting international agreements on symbols, units, nomenclature, and standards; fostering the free movement of physicists; and encouraging research and education, including sponsoring awards.

The work of the union is carried out with the help of 20 subdisciplinary commissions—such as the Commission on Physics Education and the Commission on Computational Physics—and five working groups—including groups such as the International Committee for Future Accelerators and the Working Group on Women in Physics.

The Web site URL for the International Union of Pure and Applied Physics is http://www.iupap.org/. The mailing address of the union's secretariat in the United States is IUPAP, c/o The American Physical Society, One Physics Ellipse, College Park, MD 20740-3844.

See also BIG BANG THEORY; BIOPHYSICS; BLACKBODY; CHEMISTRY AND CHEMISTS; CLASSICAL PHYSICS; COSMOLOGY; EINSTEIN, ALBERT; ELECTRO-MAGNETIC WAVES; HEAT AND THERMODYNAMICS; MAGNETISM; NUCLEAR PHYSICS; PARTICLE PHYSICS; PLANCK, MAX; POWER; QUANTUM MECHANICS; SPECIAL RELATIVITY; STATES OF MATTER; THEORY OF EVERYTHING.

Planck, Max (1858–1947) German *Physicist*

The world observable to human senses normally appears smooth, continuous, and well defined. In contrast, the quantum world, which underlies everything, is discontinuous, random, and uncertain. Because quanta are so tiny, the macroscopic world *appears* to human senses as it does. Quantum physics is the field that describes behavior and activity at and below the atomic level. The German physicist Max Planck proposed the concept of energy quanta at the beginning of the 20th century. His discovery laid the foundation for quantum theory and revolutionized the field of physics.

THERMODYNAMICS

Max Karl Ernst Ludwig Planck was born to Johann Julius Wilhelm von Planck and Emma Patzig on April 23, 1858, in Kiel, in what is now Germany. Johann already had two daughters from his first marriage, and Max was his fourth child from his second marriage. Max began elementary school in Kiel and then continued in Munich at the Königliche Maximilian-Gymnasium in 1867, when his father accepted a position as professor of law at the University of Munich. Max was a talented pianist but reportedly decided against choosing a career in music when he was told by a professional musician that he was not dedicated enough. Max enjoyed playing the piano throughout his life and also enjoyed climbing mountains. He graduated from the Gymnasium in 1874 with very high marks and enrolled at the University of Munich that fall.

At the university, Max began his studies in mathematics but became interested in physics soon afterward. In 1877 and 1878, Max visited the University of Berlin for two semesters. There he was taught by two well-known physicists of the time, Gustav Kirchhoff and Hermann Helmholtz. Max independently studied thermodynamics, the physics of the relationships between heat and other forms of energy. Two main natural laws govern energy transformations of all matter. The first law of thermodynamics states that energy is conserved; energy can be transferred and transformed, but it can neither be created nor be destroyed.

The second law of thermodynamics states that energy transfers or transformations increase entropy. Entropy is a measure of disorder, or a measurement of probability of the state of a system. For example, the second law asserts that heat naturally flows from

Max Planck proposed that energy is exchanged in discrete units, called quanta. That led to the development of quantum physics and earned him the 1918 Nobel Prize in physics. *(Science Source/ Photo Researchers, Inc.)*

a hot object to a cold object. That a process conserves energy does not mean it will occur. Physicists proposed the second law to help predict which processes will occur and which will not. For example, if you put an ice cube into a bowl of hot soup, the heat from the soup will flow to the ice cube. Energy is transferred from the molecules of the hot soup to the molecules of water in the ice cube. The water molecules begin to move around more, and the ice melts. Energy would still be conserved if heat were transferred from the ice cube to the soup, but an ice cube would never remain frozen when placed into a bowl of hot soup because that would violate the second law of thermodynamics. Max was attracted to the generality of these principles and chose to write his doctoral dissertation on the second law of thermodynamics. Planck's first major book was an extension of these studies. Published in 1897, *Vorlesungen über Thermodynamik* (*Treatise on Thermodynamics*) included studies of thermodynamic principles and concepts of osmotic pressure, boiling points, and freezing points.

After earning his Ph.D. from the University of Munich in 1879, Planck remained there as a lecturer

from 1880 to 1885. The pay was not adequate for him to begin his own family, however, and when a job as associate professor of theoretical physics at the University of Kiel was offered to him, he accepted. With sufficient income to support a family, he married his childhood sweetheart, Marie Merck. They eventually had four children together. By fall 1888, Professor Kirchhoff had died, and the University of Berlin invited Planck to succeed him. He was appointed assistant professor and the first director of the new Institute for Theoretical Physics in November 1888 and promoted to full professor in 1892. Planck remained in Berlin until he retired in 1926.

THE ULTRAVIOLET CATASTROPHE

While at Berlin, Planck began analyzing blackbodies, theoretical objects that completely absorb all the electromagnetic radiation that falls on them. They also act as perfect radiators, releasing all the radiation in the form of electromagnetic waves. A blackbody does not appear black, because it radiates electromagnetic waves, possibly in the visible region of the spectrum. For example, as black coals in a barbeque grill heat up, they become red and maybe even orange. As objects become even hotter, they glow yellowish white, as the filament of a lightbulb does. Many wondered why the color changed as the temperature rose. Theorists were having trouble relating the temperature to the spectrum of the emitted radiation.

Planck investigated the relationship between the electromagnetic energy emitted by blackbodies and varying frequencies and temperatures. The radiation properties of a blackbody depended only on its temperature. At lower temperatures, the intensity of the emitted radiation decreased. Less radiation was emitted, and the peak occurred at longer wavelengths. Reds possess longer wavelengths; that is why when an object is initially heated, it becomes red. Then as the body grows hotter, the color changes to orange or yellow and eventually to blue (a shift to shorter wavelengths of the spectrum). At extremely high temperatures, the blue end of the spectrum is most intense. Thus, as the blackbody absorbs more and more heat, the peak of intensity of the radiation it emits shifts across the electromagnetic spectrum to shorter and shorter wavelengths, that is, to higher and higher frequencies.

All of these phenomena were determined experimentally and plotted to form a curve, but classical theory could not explain the shape of the curve. An infinite number of possible frequencies in the high-frequency, small-wavelength range exists. If a blackbody radiated all frequencies of electromagnetic radiation equally—and there seemed no reason for it not to—then most of the energy emitted should be in the high-frequency region of the spectrum. This corresponds with the very short wavelengths such as those for ultraviolet light. However, the empirical curve was not compatible with this prediction. Physicists called this conundrum the ultraviolet catastrophe.

For a decade, physicists had been trying to explain mathematically the radiation of blackbodies. A few scientists had come up with seemingly suitable solutions, but their formulations only worked for restricted ranges of wavelengths. The German physicist Wilhelm Wien proposed a theoretical curve that worked well for high frequencies. An equation proposed by the English physicist Lord Rayleigh and modified by the English mathematician and astronomer James Jeans accounted for the manner of radiation distribution at low frequencies. Planck became dedicated to reconciling these inconsistencies.

THE QUANTUM CONCEPT

Sticking with a field in which he was comfortable, Planck attempted to accomplish this by approaching the issue from a thermodynamic perspective. He was unsuccessful for several years, when finally in desperation he made an assumption that led to a lucky breakthrough. He assumed that the energy of oscillation of electromagnetic waves in the blackbody was not infinitely divisible; rather, it existed in small specific amounts that he called quanta. Planck hypothesized that the size of an energy packet was proportional to its frequency: that is, the energy of a single quantum for an oscillation of a certain frequency equaled the product of that frequency and a constant, which today is referred to as the Planck constant. This formula is

$$E = hf$$

where E represents the energy of the quantum in joules (J), f denotes the frequency of oscillation in hertz (Hz), and h is the Planck constant, whose numerical value is $6.62606876 \times 10^{-34}$ joule-second (J·s). Planck's quantum hypothesis thus suggested that the energy of electromagnetic oscillation at frequency f in the blackbody must be an integer multiple of hf.

As mentioned, in order to perform the necessary calculations successfully, Planck arbitrarily made the assumption that the electromagnetic oscillation energy inside the blackbody occurred in discrete packages, or quanta. At the time, he thought of quanta simply as a mathematical device to complete the calculations and obtain the results that experiments had shown. But the quantum concept proved to be a revolutionary discovery in itself. Not only were the electromagnetic oscillations in the blackbody, or any body, quantized, as Planck proposed, but as Albert

Einstein showed later, a body also absorbed and emitted electromagnetic waves in energy quanta that obeyed the same relation, $E = hf$. Eventually the idea led to the foundation of a new branch of physics, quantum physics, and to its realization as quantum mechanics. The new concept solved the problem of the ultraviolet catastrophe. Blackbodies could easily contain (and emit) low-frequency, high-wavelength reds, since only a small amount of energy would be necessary to form a low-frequency quantum. As temperature increased, higher-energy quanta could be created. However, at very high frequencies, as in the short-wavelength ultraviolet range, it would be difficult to obtain a quantum with enough energy to satisfy $E = hf$ (since as f increases, so does E). Since only whole quanta can exist in the body, there cannot be an energy packet of less than hf. So the abundance of high-frequency emissions predicted by classical physics does not occur.

The concept of quanta was radical because it went against the age-old theme of continuity. Energy was always assumed to exist and be transmitted continuously, as a wave is. Actions that can be observed in the macroscopic world and that are described by classical physics are continuous. For example, planets orbiting the Sun do not skip between positions; they progress along a continuous path. Warmth from an oven does not heat the family room without first gradually warming the kitchen and then traveling to the adjacent room. Despite science fiction fantasies, people cannot be instantaneously beamed from one location to another. The concept of quanta introduced an innovative mechanism of action that has come to describe much of the behavior in the microscopic world, such as that of molecules, atoms, and electrons. Quantum leaps are extremely tiny, as demonstrated by the small value of the Planck constant; thus, these discontinuous jumps are not apparent to observers in the macroscopic world.

Planck initially proposed his solution to the blackbody radiation problem during a physics seminar at the University of Berlin in October 1900, but he did not present a theoretical justification of his radiation law until December. The resulting article was published in the *Annalen der Physik* and is one of the most important physics papers of all time. He later admitted that finding the correct formula had been simply lucky guesswork. A fuller account of his ideas was published in his 1906 book, *Theorie der Wärmestrahlung* (Theory of Heat Radiation).

A decade after Planck's discovery of quantum action, the field of physics had fully accepted and begun building upon the revolutionary idea. The application of his serendipitous discovery led to the resolution of many incongruities between classical theoretical physics and experimental physics. Most notably, Einstein extended Planck's ideas to the wave-particle duality of light, proposing that light was emitted in individual quanta of energy, now called photons, and the Danish physicist Niels Bohr developed the quantum mechanical model of the atom. In 1918, Planck was awarded the Nobel Prize in physics for his discovery of elementary energy quanta. In the 1920s, the new field of quantum mechanics was established. Planck was elected a foreign member of the Royal Society in 1926 and awarded the Copley Medal in 1928.

PERSONAL TRAGEDIES

These scientific achievements and awards could not protect Planck from several personal tragedies during this time. His wife, Marie, had died in 1909. A daughter died during childbirth in 1917. Two years later, her twin sister, who had married her late sister's husband, died in the same manner. Finally, one of his sons died on the battlefield during World War I. Planck did remarry in 1911, his late wife's niece, Marga von Hoessli. Together they had one son.

Most of Planck's achievements after his Nobel-winning science were administrative. Planck did not shy from administrative positions. He felt it was his responsibility as a scientist to promote scientific interests and was instrumental in drawing many talented scientists to Germany. In 1894, he became a member of the Prussian Academy of Sciences, and he served as permanent secretary from 1912 to 1938. He served as editor of *Annalen der Physik,* a world-class physics journal. In 1929, he was awarded what has become the highest distinction bestowed by the German Physical Society, the Max Planck Medal. In 1930, he was appointed president of the Kaiser Wilhelm Society in Berlin. He held this post until 1937, when he was forced to resign after intervening with Adolf Hitler on behalf of his Jewish colleagues. After the war, the society was renamed the Max Planck Society and moved to Göttingen in 1945. Planck served as temporary president during this difficult transitional period until his death in 1947.

Planck did engage in some intellectual pursuits later in his life. He performed research trying to assimilate his quantum concept into Einstein's theory of relativity, for which he became a staunch supporter. He also published works concerning his overall philosophy of science. In 1935, he published a book, *Die Physik im Kampf um die Weltanschauung,* which approached general philosophical, religious, and societal issues through the use of physics. Planck was always attracted to generalizations. He relentlessly sought out constants of nature. He believed in the absolute validity of simple and accurate natural laws. These ideas were conveyed in his 1959 book, *Philosophy of Physics.*

When the Second World War broke out, Planck felt it was his duty to stay in Germany and try to preserve the integrity of scientific research. However, his efforts against the Nazi regime were futile. In 1943, he moved to Rogätz, and, in 1945, some American colleagues took him to Göttingen, where he spent the last two years of his life with his grandniece. Unfortunately, his home had burned down during an air raid, and he lost most of his research manuscripts and books. In 1944, his surviving son from his first marriage was arrested and executed for conspiring to assassinate Adolf Hitler.

PLANCK SCALE

The Planck scale is a set of fundamental units for length, time, mass, energy, temperature, and more, that indicate the respective scales at which physicists expect nature to reveal the quantum character of space, time, and gravitation. On everyday scales, the quantum aspect of nature seems to be played out in the arena of classical space-time and gravitation. Yet, physicists expect that even space, time, and gravitation are not immune to nature's fundamentally quantum character. Their quantum nature does not reveal itself, since physicists are still probing far from the Planck scale.

The units of the Planck scale are determined by the following three fundamental physical constants: the speed of light $c = 2.99792458 \times 10^8$ meters per second (m/s), the Planck constant $h = 6.62606876 \times 10^{-34}$ joule-second (J·s), and the gravitational constant $G = 6.67259 \times 10^{-11}$ N·m^2/kg^2. These constants represent three fundamental aspects of nature, which are, respectively, space-time, nature's quantum character, and gravitation. One obtains the Planck unit of a physical quantity by forming an algebraic combination of the constants that possesses the appropriate dimension.

Planck length. This unit of length is as follows:

$$\text{Planck length} = \sqrt{\frac{Gh}{c^3}} \approx 4.05 \times 10^{-35} \text{ m}$$

For comparison, note that an atomic nucleus is about 10^{-14} m in size. The Planck length is far beyond physicists' present ability to probe and resolve structure. Scientists generally think that if one ever does manage to approach the Planck length in resolution, as one probes at decreasing scale, one will find that space more and more loses its familiar, classical properties: lengths become blurred and directions become undifferentiated. At the Planck scale, the classical picture of space should disappear altogether. A minority view holds that the Planck scale defines a quantum of space (i.e., a minimal amount of space) and thus a lattice structure for space. That would be similar to the appearance of angular momentum only in discrete values, which are integer or half-integer multiples of $h/(2\pi)$.

Planck time. The Planck time is the interval during which a photon, which travels at the speed of light, traverses the distance of one Planck length:

$$\text{Planck time} = \sqrt{\frac{Gh}{c^5}} \approx 1.35 \times 10^{-43} \text{ s}$$

This unit of time is exceedingly shorter than the briefest time interval that is presently resolvable. The general consensus is that should physicists ever be able to resolve time intervals approaching the Planck time, they will find that time more and more loses its classical character: time intervals become undefined and the past-future distinction blurs. At the Planck scale, time should lose its classical character altogether. And at the Planck scale of length and time, space-time should be completely nonclassical, a wildly fluctuating situation often described as "space-time foam." Nevertheless, some hold the view that the Planck time is a quantum of time, the minimal significant interval of time. Then all time intervals would be integer multiples of the Planck time.

Planck mass. This unit of mass is as follows:

$$\text{Planck mass} = \sqrt{\frac{hc}{G}} \approx 5.46 \times 10^{-8} \text{ kg}$$

The significance of the Planck mass is not clear. It is much larger than the masses of the elementary particles, while less than, but relatively not very much less than, everyday masses. The energy equivalent of the Planck mass is the Planck energy.

Planck energy. Its value is as follows:

$$\text{Planck energy} = \sqrt{\frac{hc^5}{G}} \approx 4.91 \times 10^9 \text{ J}$$
$$= 3.06 \times 10^{28} \text{ eV}$$

Planck temperature. The Planck temperature is the temperature at which the average thermal energy of a particle in matter at that temperature equals the Planck energy. Its value is as follows:

$$\text{Planck temperature} = \sqrt{\frac{hc^5}{Gk^2}} \approx 3.56 \times 10^{32} \text{ K}$$

where k denotes the Boltzmann constant, whose value is $1.3806503 \times 10^{-23}$ joule per kelvin (J/K).

HONORS

Max Planck passed away on October 3, 1947, at Göttingen. He was a man respected as much for his personal characteristics of integrity and sense of duty as he was for his scientific accomplishments. In his honor, the Max Planck Society was founded in 1948 in Göttingen as the successor to the Kaiser Wilhelm Society for the Advancement of Science. The society's research institutes perform basic research in the interest of the general public in the natural sciences, life sciences, social sciences, and humanities, particularly in areas that German universities cannot. In addition, the German Physical Society continues to present the prestigious Max Planck Medal annually to an outstanding theoretical physicist.

Today quantum mechanics explains phenomena such as the spectral distribution of electromagnetic radiation and the ways atoms combine into molecules. Its applications have led to technological advancements including bar code readers at the supermarket, lasers, compact disks, and nuclear energy. The natural science of physics has been separated into two eras. Classical physics encompasses everything before 1900, before the development of quantum theory, including phenomena such as heat, light, sound, mechanics, and thermodynamics. Modern physics is the physics that developed in the 20th century and is heavily based on quantum theory. It includes, for example, nuclear physics, relativity, and big bang cosmology. The Planck constant appears in all specifically quantum mechanical formulas, and quantum theory has long been universally accepted, but the concept of quanta is more than a collection of mathematical formulas or the basis of a bunch of modern electronic devices. It explains the foundation of all physical processes.

See also ATOMIC STRUCTURE; BLACKBODY; BOHR, NIELS; DUALITY OF NATURE; EINSTEIN, ALBERT; ELECTROMAGNETIC WAVES; ELECTROMAGNETISM; ENERGY AND WORK; GRAVITY; HEAT AND THERMODYNAMICS; PARTICLE PHYSICS; PRESSURE; QUANTUM MECHANICS; ROTATIONAL MOTION; SPECIAL RELATIVITY.

FURTHER READING

Adler, Robert E. *Science Firsts: From the Creation of Science to the Science of Creation.* New York: John Wiley & Sons, 2002.

Boorse, Henry A., Lloyd Motz, and Jefferson Hane Weaver. *The Atomic Scientists: A Biographical History.* New York: John Wiley & Sons, 1989.

Cropper, William H. *Great Physicists: The Life and Times of Leading Physicists from Galileo to Hawking.* New York: Oxford University Press, 2001.

Gamow, George. *The Great Physicists from Galileo to Einstein.* Mineola, N.Y.: Dover, 1961.

Heathcote, Niels Hugh de Vaudrey. *Nobel Prize Winners in Physics 1901–1950.* Freeport, N.Y.: Books for Libraries Press, 1953.

Heilbron, J. L., *Dilemmas of an Upright Man: Max Planck and the Fortunes of German Science.* New ed. Cambridge, Mass.: Harvard University Press, 2000.

The Nobel Foundation. "The Nobel Prize in Physics 1918." Available online. URL: http://nobelprize.org/physics/laureates/1918/. Accessed July 25, 2008.

Planck, Max. *The Philosophy of Physics.* New York: Norton, 1936.

———. *The Theory of Heat Radiation.* Mineola, N.Y.: Dover, 1959.

———. *Treatise on Thermodynamics.* Mineola, N.Y.: Dover, 1990.

Segrè, Emilio. *From X-Rays to Quarks: Modern Physicists and Their Discoveries.* San Francisco: W.H. Freeman, 1980.

polymers A polymer is a high-molecular-weight substance made up of a chain of smaller monomers in a repeating structure that produces a tough and durable product. Substances such as proteins, nucleic acids, plastics, and rubber are all considered polymers. Two types of reactions, addition reactions and condensation reactions, result in the stepwise synthesis of polymers. Addition reactions use monomers that have one or more double or triple bonds. The double bond is changed to a single bond, and the available electrons form bonds with the next monomer in the chain. Condensation reactions are characterized by the elimination of a small molecule, usually water, during the joining of two monomers. An individual polymer molecule can consist of thousands of repeating units. Polymers are divided into natural polymers and synthetic polymers.

NATURAL POLYMERS

As the name suggests, natural polymers exist naturally and include such compounds as proteins and nucleic acids, and polysaccharides such as starch, glycogen, and cellulose. Proteins, polymers of amino acids, are responsible for most of the cellular activity in a living organism. The most diverse of all the macromolecules, proteins are often considered the workhorses of the cell. There are 20 different amino acids in the human body that make up all of the known proteins. All amino acids have the same basic structure; the difference lies in the side chain (R) of the amino acids. Each amino acid contains a central carbon that is bonded to an amino group, a carboxyl group, a hydrogen atom, and a variable R group.

These amino acids are joined together through a dehydration reaction or condensation reaction (a bond formed with the accompanying removal of water) between the carboxyl group of the first amino acid and the amino group of the next amino acid, creating a peptide bond. Biochemists study the arrangement of amino acids, the folding and function of each of these proteins, and how it can impact the function and well-being of an organism.

Nucleic acids such as deoxyribonucleic acid or ribonucleic acid belong to a class of natural polymers that are responsible for the storage of hereditary information within the cell and the transmittance of that information from the parent cell to any progeny cells it may produce. Nucleotides, the monomers of nucleic acids, each contain a 5-carbon sugar (ribose or deoxyribose), a phosphate group, and a nitrogenous base. Nucleotides are joined together by dehydration synthesis (condensation) reactions between the hydroxyl group located on the 3′ carbon of the sugar and the phosphate bonded to the 5′ carbon of the sugar, forming a phosphodiester bond.

Polysaccharides include such molecules as starch, glycogen, and cellulose. Starch and glycogen are storage forms of glucose in plants and animals, respectively. Glucose is removed from the bloodstream and incorporated by the liver into a polymer of glycogen through a condensation reaction in response to insulin secretion by the pancreas. Starch and cellulose are both glucose polymers in plants. The glucose monomers in starch are bonded together with alpha (α)-linkages, meaning that the monomers are all oriented in the same direction. Cellulose is a plant polymer made of glucose monomers bonded together with beta (β)-linkages, meaning that each monomer is rotated 180 degrees relative to its neighbors. Cellulose is a structural polymer in plants that lends strength and rigidity to cell walls. Starch is found in potatoes, corn, and wheat. The enzyme that cleaves starch is only able to break α-linkages. Humans lack the enzyme necessary to cleave β-linkages and are therefore unable to digest cellulose.

SYNTHETIC POLYMERS

The synthesis of polymers by organic chemists has been a growing field from the mid-1900s to today. Commercial polymers can include plastics or elastomers. Plastics are polymers that can be molded and shaped into commercially useful products. Plastics that can be melted and created into another form are known as thermoplastics. These types of plastics are used in common everyday items such as plastic soda bottles and milk jugs. The plastics are formed into the shape of the bottle or jug. After the item is recycled, it can be melted down to create a completely new form. Thermosetting plastics are chemically altered during their shaping, making them difficult

ethylene can be used in packaging and in plastic soda, water, and juice bottles.

Polyethylene is rated on the basis of its level of crystallinity, the level of ordered packing in a normally amorphous material. The higher the crystallinity in a type of polyethylene, the more rigid the sample is, and the lower the level of crystallinity in a polyethylene sample, the more flexible the plastic. A low level of crystallinity causes the sample to have a lower density. Low-density polyethylene (LDPE) is utilized in such things as plastic storage bags. A high level of crystallinity causes the sample to have a higher density. High-density polyethylene (HDPE) is found in milk jugs and water bottles.

Polyethylene terephthalate (PET or PETE) is a thermoplastic polymer formed by condensation reactions. PET in the fibrous form is utilized in the manufacture of clothing in the form of polyester. The most common and easily identifiable use of PET is in plastic soda and water bottles.

Polypropylene, another common thermoplastic, is formed by addition reactions and is made up of propylene units. Polypropylene is utilized in kitchenware, appliances, food packaging, and other containers. With a density intermediate between those of LDPE and HDPE, polypropylene has improved flexibility over these substances and a high stability under heat. Medical equipment that needs to be sterilized is often made out of polypropylene.

Polystyrene is a thermoplastic polymer made from addition reactions of styrene monomers. In the solid form, polystyrene makes a hard, rigid plastic that is often used for such items as CD cases. Polystyrene in an expanded form is most often encountered in local restaurants, cafeterias, or shipping boxes. Air is passed through the polystyrene, creating foam, most commonly known as Styrofoam. This expanded polymer has a high insulation value and is used in homes and coolers as an insulator. Styrofoam is also protective and can be used to provide cushioning in the form of packing peanuts used to package equipment such as computers.

Polyvinyl chloride, commonly known as PVC, is another thermoplastic polymer that results from addition reactions of vinyl monomers. PVC is one of the most commonly used plastics in the world, although there are concerns about the environmental and health risks of its use. The most familiar use of PVC is in the building industry. PVC pipes are used in plumbing and are significantly less expensive to purchase as well as install than copper pipes. Rigid PVC, also known as vinyl, is used in siding for houses.

Polyurethane is a polymer formed by condensation reactions that can be utilized in such applications as furniture upholstery foam, Spandex materials, and water-protective coatings for wood. Polyvinyl

The development of polymers, such as the polyethylene of which these bottles are made, has revolutionized the world. *(David R. Frazier/Photo Researchers, Inc.)*

to recycle. Elastomers are polymers that are rubbery and regain their original shape after being stretched or bent. Elastomers can also be made into fibers such as nylon and polyester and are utilized for multiple commercial uses.

Examples of commercial polymers include such common substances as polyethylene, polyethylene terephthalate, polypropylene, polystyrene, polyvinyl chloride, polyurethane, and nylon. Organic chemists and polymer scientists developed each of these polymers in order to fill the need for a commercial product.

The thermoplastic polyethylene is a representative example of a class of polymers known as homopolymers. They are made up of identical monomers in repeating units. The ethylene units of polyethylene are connected by an addition reaction. Polyethylene formation is catalyzed by a free radical reaction on the double bond of the ethylene molecule, causing the previously bonded electrons to form new bonds with new ethylene molecules. Billions of pounds of polyethylene are produced every year. Types of polychloride can be produced in a rigid or in a more flexible form.

Nylon, developed by DuPont, is a thermoplastic polymer formed through condensation reactions and is utilized mostly in the form of fibers. Nylon is formed from a reaction consisting of a diamine, a molecule that has -NH$_2$ groups at both ends, and a diacid, a molecule that has -COOH groups at each end. The formation of the nitrogen-carbon bond occurs via the elimination of a water molecule. Originally thought of as simply an inexpensive replacement for silk, nylon now is used in numerous types of products such as parachutes, carpet fibers, fishing line, and women's hosiery.

Synthetic polymers often have substances known as plasticizers with lower molecular mass that interfere with the polymers' ability to crystallize and therefore maintain the plastic's pliability. Examples of this include the difference between polyvinyl chloride without the addition of plasticizers (hard drain pipes) and with plasticizers (soft dog toys).

Natural rubber is derived from the bark of the *Hevea brasiliensis* tree (rubber tree). The product, a polymer of isoprene (C_5H_8), as it is isolated from the tree is a suspension. This type of rubber is soft and reactive, so it is not commercially useful. This changed in the 1930s, when Charles Goodyear developed a process, known as vulcanization, that stabilizes the structure of the rubber through heat and treatment with sulfur to cross-link the natural rubber. By cross-linking a small percentage of the double bonds in the polymer, the rubber becomes resilient and flexible.

RECYCLING PLASTICS

Thermoplastic plastics are easily recycled, while thermosetting plastics are not. The chemical processes that take place to produce thermosetting polymers make them difficult to convert into other products. Thermoplastic plastics are melted down and cast into other shapes and forms. Recycling capabilities differ by region of the country and from city to city. The types of polymers that are able to be recycled are designated by numeric categories as shown in the table Plastic Types. Plastics numbered 1 and 2 are recycled in nearly all municipalities. Higher-numbered plastics are only recycled in certain areas.

POLYMER SCIENCE

Polymer science is a unique branch of chemistry in that it is truly economy driven. Very little basic research is being done at present in the field of polymers. The production of compounds that are intended for the marketplace is the main emphasis of polymer science in research and development divisions of corporations. Theoretical as well as laboratory training is critical in polymer science. Education and training in the field requires an in-depth and comprehensive study of organic chemistry. Many scientists in the field believe that a degree in organic chemistry should be required for employment, but degrees in polymer science with a heavy emphasis on organic chemistry are becoming more and more common.

The type of degree required to be a polymer scientist depends upon the type of job desired. Degrees in polymer science range from four-year bachelor of science degrees to research-based master of science degrees, and Ph.D. degrees. In general, laboratory technician positions require a bachelor's or master's degree. Supervisory roles require either a master's degree or a Ph.D. Full-time academic positions generally require a Ph.D., and many require additional postdoctoral work in the field.

Polymer science positions are as diverse as the number of products on the market today. Polymer scientists can work on product development for any area of the medical field, drug development, environmental issues, the energy field, automotive, clothing and textiles, food, and household goods, just to name a few. The space and aeronautics industry employs

PLASTIC TYPES

Number	Abbreviation	Polymer
1	PET	Polyethylene teraphthalate
2	HDPE	High-density polyethylene
3	V	Polyvinyl chloride (PVC)
4	LDPE	Low-density polyethylene
5	PP	Polypropylene
6	PS	Polystyrene

polymer scientists to create useful products for their applications. Polymer scientists can hold positions in academia or business, as well as governmental jobs. It is beneficial to have flexibility and a broad base of training when pursuing such a career. More than any other field of chemistry, polymer science is dependent on the economy. As companies downsize and cut research and development divisions, polymer scientists may have difficulty finding jobs. When the economy is strong, however, polymer science is a strong field.

See also BIOCHEMISTRY; CARBOHYDRATES; NUCLEIC ACIDS; ORGANIC CHEMISTRY; PROTEINS.

FURTHER READING

Ebewele, Robert O. *Polymer Science and Technology*. London: CRC Press, 2000.

power   The rate of performance of work (i.e., work done per unit time) as well as the rate of transfer or conversion of energy (which is energy per unit time) is called power. The SI unit of power, which is a scalar quantity, is the watt (W), equivalent to a joule per second (J/s). Other common SI related

units of power are the kilowatt ($1 \text{ kW} = 10^3 \text{ W}$), the megawatt ($1 \text{ MW} = 10^6 \text{ W}$), and the gigawatt ($1 \text{ GW} = 10^9 \text{ W}$). A common non-SI unit of power is the horsepower (hp), now used almost exclusively in the United States. The value of one horsepower is

$$1 \text{ hp} = 746 \text{ W}$$
$$= 0.746 \text{ kW}$$

James Watt, the 18–19th-century Scottish inventor for whom the power unit watt is named, measured the rate of work of a large draft horse and formalized the value of the horsepower. In the British system of units, used mostly in the United States, the unit of work or energy is the foot-pound (ft·lb), and the unit of power is accordingly the foot-pound per second (ft·lb/s) or foot-pound per minute (ft·lb/min). Then

$$1 \text{ hp} = 550 \text{ ft·lb/s}$$
$$= 33,000 \text{ ft·lb/min}$$

and

$$1 \text{ ft·lb/s} = 1.36 \text{ W}$$

A familiar example of power rating is the 100-W lightbulb, which converts electrical energy into heat

The "horsepower" as a unit of power was derived from the rate at which a large draft horse can perform work, here pulling a tree out of a forest. *(Jürgen Schulzki/Alamy)*

These incandescent lightbulbs consume electric energy at the rates of—from smallest bulb to largest—11 watts (W) 100 W, 300 W, and 620 W. The 620-W bulb serves in aviation warning fixtures. *(Photos courtesy of Royal Philips Electronics)*

and light energy at the rate of 100 joules per second. Automobile manufacturers specify the power of their engines in horsepower for the U.S. market and in kilowatts for the European market. Another familiar use of a unit of power is in the kilowatt-hour (kWh) unit of electric energy, as it is sold to us by the electricity company. This unit is the amount of energy consumed during one hour at the rate of one kilowatt. The kilowatt-hour relates to the joule by

$$1 \text{ kWh} = 3.6 \times 10^6 \text{ J} = 3.6 \text{ MJ}$$

If the amount of work being performed or the energy being transferred during time interval Δt is ΔW or ΔE, respectively, then the average power is defined as

$$P_{av} = \frac{\Delta W}{\Delta t} \text{ or } \frac{\Delta E}{\Delta t}$$

Instantaneous power P is defined as the limit of the average power as Δt goes to 0, so that

$$P = \frac{dW}{dt} \text{ or } \frac{dE}{dt}$$

If a force **F** (a vector), in newtons (N), is acting on a body whose instantaneous velocity (also a vector) is **v,** in meters per second (m/s), the instantaneous power of the work the force is performing on the body is given by the scalar product

$$P = \mathbf{F} \cdot \mathbf{v}$$

or, in terms of magnitudes,

$$P = Fv \cos \theta$$

where F and v denote the magnitudes of the force and velocity, respectively, and θ is the smaller angle (less than 180°) between the two vectors.

Thus, for example, one can find the thrust force of a 20-horsepower boat engine as it propels a boat at full throttle through the water at 50 feet per second (15.2 m/s). The data are

$$v = 15.2 \text{ m/s,}$$

$$\theta = 0, \quad \cos \theta = 1,$$

$$P = 20 \text{ hp} = (20 \text{ hp})\left(\frac{746 \text{ W}}{1 \text{ hp}}\right) = 1.49 \times 10^4 \text{ W}$$

From the magnitude equation one obtains for the thrust force

$$F = \frac{P}{v \cos \theta}$$

$$= \frac{1.49 \times 10^4 \text{W}}{(15.2 \text{ m/s}) \times 1}$$

$$= 980 \text{ N}$$

So the thrust force of the boat engine under the given conditions is 980 newtons. Or, for another example, the instantaneous power of the work being done by a force of magnitude three newtons ($F = 3$ N) acting on a body whose instantaneous speed is five meters per second ($v = 5$ m/s), when the force is acting at an angle of 35 degrees ($\theta = 35°$) to the direction of the body's velocity, is

$$P = Fv \cos \theta$$

$$= (3 \text{ N})(5 \text{ m/s})\cos 35°$$

$$= 12 \text{ W}$$

Similarly, if a torque τ, in newton-meters (N·m), acts on a body with respect to an axis about which the body rotates with instantaneous angular speed Ω, in radians per second (rad/s), the instantaneous power of the work being performed by the torque is

$$P = \tau\Omega$$

As another example, if a battery whose terminal voltage is V, in volts (V), delivers electric current i, in amperes (A), to a circuit, the instantaneous power being delivered to the circuit is

$$P = Vi$$

In numbers, the instantaneous power that a 12-volt ($V = 12$ V) battery delivers, when it supplies a current of 90 amperes ($i = 90$ A), is

$$P = Vi$$

$$= (12 \text{ V})(90 \text{ A})$$

$$= 1,080 \text{ W}$$

$$\approx 1.1 \text{ kW}$$

See also ELECTRICITY; ENERGY AND WORK; FORCE; ROTATIONAL MOTION; SPEED AND VELOCITY.

FURTHER READING

Young, Hugh D., and Roger A. Freedman. *University Physics*, 12th ed. San Francisco: Addison Wesley, 2007.

pressure   This quantity is related to forces distributed and acting over surfaces and is the force acting perpendicularly on a surface per unit area of surface. Pressure is a scalar quantity, whose SI unit is the pascal (Pa), equivalent to newton per square meter (N/m²). Other common units for pressure are the atmosphere (atm); millimeter of mercury (mm Hg), also called torr (torr); inch of mercury (in Hg); pound per square inch (lb/in² or psi); and bar. These relate to the pascal as follows:

$$1 \text{ atm} = 1.013 \times 10^5 \text{ Pa}$$

$$1 \text{ mm Hg (torr)} = 1.333 \times 10^2 \text{ Pa}$$

$$1 \text{ in Hg} = 3.386 \times 10^3 \text{ Pa}$$

$$1 \text{ lb/in}^2 \text{ (psi)} = 6.895 \times 10^3 \text{ Pa}$$

$$1 \text{ bar} = 1.000 \times 10^5 \text{ Pa}$$

The standard atmosphere, 1 atm, which is a nominal approximation to the actual pressure of Earth's atmosphere at sea level, relates to the other units in this way:

$$1 \text{ atm} = 1.013 \times 10^5 \text{ Pa}$$

$$= 7.60 \times 10^2 \text{ mm Hg (torr)}$$

$$= 29.92 \text{ in Hg}$$

$$= 14.70 \text{ lb/in}^2 \text{ (psi)}$$

$$= 1.013 \text{ bar}$$

Devices for measuring pressure are called pressure gauges and, for measuring atmospheric pressure, barometers.

Pressure is relevant to fluids (i.e., to liquids and gases), which exert pressure on the walls of their container as well as on any object immersed in them. If the fluid pressure at a point on a surface that is in contact with the fluid is denoted p, then the magnitude of the force acting on an infinitesimal area of surface dA due to the pressure at that point is given by

$$dF = p\, dA$$

Here dF denotes the infinitesimal magnitude of force, in newtons (N), and dA is in square meters (m²). The direction of the force due to the pressure is perpendicular to the surface and pointing outward from the fluid toward the surface. The discussion is assuming positive pressure, which is the usual case. Negative pressure has an inward-pulling effect, with the surface being sucked toward the fluid. To take negative pressure into account, the previous relation might better be written

$$dF = |p|\, dA$$

The vector form of this equation, correct for positive and negative pressures, is

$$d\mathbf{F} = -p\, d\mathbf{A}$$

where $d\mathbf{A}$ is a vector whose magnitude is dA and whose direction is perpendicular to the surface at the point in question and pointing from the surface inward toward the fluid. The vector $d\mathbf{F}$ denotes the infinitesimal force resulting from the action of the pressure.

The total force on a finite area of surface is obtained by integrating this relation over the area upon which the pressure acts. For the simple situation in which the pressure is uniform and the surface is flat, the magnitude of the force, F, on the surface area is

$$F = |p|A$$

with A denoting the area. This relation can be put in the form

$$|p| = \frac{F}{A}$$

This shows, for instance, that large pressures can result from small forces, if they act over sufficiently small areas, and that large forces can cause small pressures, if the areas over which the forces act are sufficiently large.

As an example, find the pressure acting on the flat, horizontal floor of a rectangular Jacuzzi, when the Jacuzzi contains water to a depth of $d = 0.80$ m. Take into account the density of water $\rho = 1.0 \times 10^3$ kg/m^3 and the acceleration due to gravity $g = 9.8$ m/s^2. For the solution, denote the length and width of the Jacuzzi l and w, respectively, although these quantities will not affect the result. The volume of water is $V = lwd$. The mass of water is $m = \rho V$. The weight of water is $W = mg$, which is also the magnitude of the force F on the Jacuzzi's floor, whose area is $A = lw$. The pressure on the floor is

$$
\begin{aligned}
|p| &= \frac{F}{A} \\
&= \frac{W}{lw} \\
&= \frac{mg}{lw} \\
&= \frac{\rho V g}{lw} \\
&= \frac{\rho lwdg}{lw} \\
&= \rho dg \\
&= (1.0 \times 10^3 \text{ kg/m}^3)(0.80 \text{ m})(9.8 \text{ m/s}^2) \\
&= 7.8 \times 10^3 \text{ Pa}
\end{aligned}
$$

A SCUBA (for self-contained underwater breathing apparatus) diver carries an air supply at high pressure in the tank on his or her back. If the same amount of air were at atmospheric pressure, it would take up a much larger volume. *(DJ Mattaar, 2008, used under license from Shutterstock, Inc.)*

So the pressure caused by the water on the Jacuzzi's floor is 7.8 × 10³ Pa.

A perpendicular pressure force is the only force a fluid at rest exerts on a surface with which it is in contact. A moving fluid exerts also a tangential force on a surface (i.e., a force parallel to the surface) due to viscosity.

A commonly encountered example of pressure is the tire pressure of vehicles. In the United States, tire pressure is designated in the unit of pound per square inch, or psi. What a tire pressure gauge shows, however, is not the actual pressure of the air in the tire, but rather the difference between that pressure and the pressure of the air surrounding the tire, the atmospheric pressure. The term for pressure that is specified relative to the external pressure is *gauge pressure*. The actual pressure, which is the gauge pressure plus atmospheric pressure, is called *absolute pressure*.

Another example of pressure in daily life, at least for weather fans, is barometric pressure, as announced by meteorologists. This is the actual pressure of the atmosphere and in the United States is commonly announced in inches, as in "the barometric pressure is 29.7 inches," which is short for *29.7 inches of mercury*.

The pressure units "millimeter of mercury" (mm Hg) and "inch of mercury" (in Hg) are based on the way atmospheric pressure is measured with a mercury barometer, which gives a very accurate reading. A transparent tube, closed at one end, is completely filled with mercury. The open end of the tube is then stopped with a finger, the tube is inverted, the stopped end inserted in a bowl of mercury, and the finger withdrawn. The tube should be long enough, longer than 30 inches (760 mm), so that the column of mercury in it drops a little, leaving a vacuum in the space above it. The column of mercury is then supported by the pressure

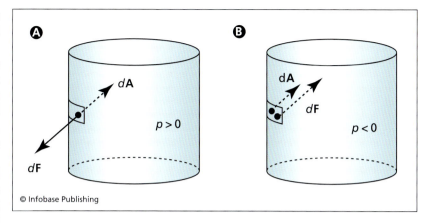

The pressure in a fluid causes a force on a surface with which the fluid is in contact. The force is perpendicular to the surface and points away from the fluid for positive pressure and into the fluid for negative pressure. The infinitesimal magnitude of the force on an infinitesimal element of area dA is $dF = |p|\, dA$, where p denotes the pressure. The vector relation among pressure, surface area, and force is $d\mathbf{F} = -p\, d\mathbf{A}$, where $d\mathbf{F}$ represents the infinitesimal force vector and $d\mathbf{A}$ is the vector of magnitude dA that is perpendicular to the area element and pointing toward the fluid. The situation is depicted for fluid in a container, showing the direction of force on an element of inner wall surface for positive fluid pressure (a) and for negative pressure (b).

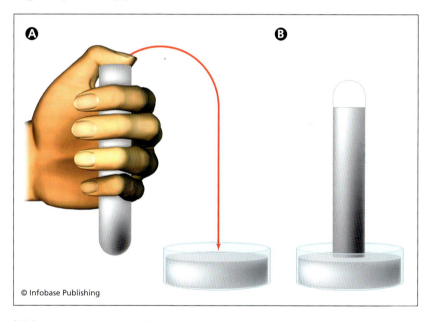

(a) A transparent tube, closed at one end, is completely filled with mercury. The open end of the tube is then stopped with a finger, the tube is inverted, the stopped end inserted in a bowl of mercury, and the finger withdrawn. The result is shown in (b). The tube should be long enough so that the column of mercury in it drops a little, leaving a vacuum in the space above it. The column of mercury is then supported by the pressure of the atmosphere, transmitted to the bottom of the column by the mercury in the bowl. The height of the mercury column, from the surface of the mercury in the bowl to the top of the column, then serves to measure atmospheric pressure.

of the atmosphere, transmitted to the bottom of the column by the mercury in the bowl. The height of the mercury column, from the surface of the mercury in the bowl to the top of the column, then serves to measure atmospheric pressure. This is the length that designates the pressure in inches or millimeters. When the weather forecaster announces a high atmospheric pressure, usually accompanying nice weather, a higher-than-average air pressure pushes on the top of the mercury in the bowl. This pressure is transferred to the base of the mercury column and raises the column to a height greater than 760 mm (which is the height that corresponds to standard atmospheric pressure).

See also FLUID MECHANICS; FORCE; VECTORS AND SCALARS.

FURTHER READING
Young, Hugh D., and Roger A. Freedman. *University Physics*, 12th ed. San Francisco: Addison Wesley, 2007.

Priestley, Joseph (1733–1804) English *Chemist*
Joseph Priestley was an 18th-century English chemist. He is best known for his discovery of the element oxygen. Priestley also invented carbonated beverages, delineated the process of photosynthesis, invented apparatuses that improved the efficiency of gas collection before the word *gas* even existed, and discovered 10 new gases himself.

Joseph Priestley was an 18th-century English chemist who is best known for his discovery of the element oxygen. *(Sheila Terry/Photo Researchers, Inc.)*

EARLY YEARS AND EDUCATION
Joseph Priestley was born on March 13, 1733, in Fieldhead, England. He was the oldest of six children born to Jonas Priestley (a cloth merchant) and his first wife, Mary Swift. Priestley's mother died during the birth of her sixth child in as many years, and at age nine Joseph was sent to live with his aunt, Sarah Priestley Keighley, until adulthood. Here he was raised in a strict religious household.

By the time he was a teenager, Joseph showed signs of being a prodigy. He had already mastered several languages and was tutored privately in algebra, geometry, and Newtonian mechanics, but he felt called to the ministry. Priestley was not a member of the Church of England and therefore could not enter Oxford or Cambridge University. At the age of 19, Priestley became the first student to enroll in the dissenting academy at Daventry. While there, he enjoyed the inquiry-based learning atmosphere and spent lots of time formulating his thoughts concerning different religious doctrines. Priestley left the academy in 1755.

BEGINS TEACHING CAREER
Priestley left the academy in 1755 and took a position as an assistant minister at Needham Market in Suffolk. As Priestley's personal religious beliefs matured,

his congregation and senior minister became uncomfortable with his viewpoints, in particular, his denial of the Trinity, or the union of the Father, the Son, and the Holy Ghost. When he had fulfilled the obligations of his term, he eagerly accepted a position in 1758 as a minister at Nantwich, in Cheshire. To supplement his income there, he opened a successful school for the girls and boys in his congregation. His own interests in natural philosophy increased, and he purchased an air pump and an electrical machine that his students used for experiments.

His reputation as a teacher grew, and in 1761 Priestley was invited to join the faculty at the dissenting academy at Warrington, in the county of Lancashire. There, he began to write and publish successful texts on language, grammar, and education and conducted electrical experiments. Priestley married Mary Wilkinson on June 23, 1762. In 1764, he was awarded a doctor of laws degree from the University of Edinburgh for his studies on education. Priestley remained at Warrington for six years, until the income was no longer enough to support his growing family.

While traveling to London in 1766, Priestley met Benjamin Franklin, an American statesman who was representing the colonies in discussions with the

British government. Franklin was a respected scientist who had made the famous discovery that lightning was an electrical phenomenon. Priestley took advantage of this opportunity to discuss his electrical experiments with Franklin. Franklin encouraged Priestley to write his *History and Present State of Electricity, with Original Experiments,* a text that gave an up-to-date accounting of all related research. Franklin even assisted Priestley in obtaining reference material. One new observation from Priestley's own experiments was that carbon, a nonmetal, could conduct electricity. He reported his deduction that the electrical attraction between bodies had an inverse-square relationship to the distance between them. He also recorded the first description of an oscillatory discharge, the same principle used in wireless telegraphy by the Italian inventor Guglielmo Marconi. Even before the book, which was published in 1767, reached the shops, news of Priestley's experiments had spread among other English scientists. In 1766, he was elected a fellow of the Royal Society on the basis of his electrical work, quite an honor for someone with no formal training in science. In 1767, he moved his family to Leeds and took over the Presbyterian parish at Mill-Hill Chapel.

CARBON DIOXIDE AND SODA WATER

Although Priestley had spent most of his life as a minister and a tutor, he still had a great love of scientific research. At the time of Priestley, the chemistry of gases was not understood. The term *gas* was not even established yet, and the term used for the study of gases was *airs*. Priestley began his interest in this field by studying the gases formed above fermentation vats in a brewery. His scientific observations demonstrated that the gas found above the vats was denser than air and was able to put out a flame. This gas is now known to be carbon dioxide.

Priestley found that if this gas was allowed to dissolve in water, it created soda water. During attempts to create soda water experimentally, he discovered that heating limestone or heating chalk with water and hydrochloric acid created the same gas he had seen at the brewery. Priestley's work led to his publication of *Directions for Impregnating Water with Fixed Air* in 1772. Although others later took commercial advantage of soda water, Priestley never attempted to market his discovery.

Priestley's experimentation with "airs" led to the refinement of the apparatus known as a pneumatic trough. A reaction vessel was connected by a tube to a bottle inverted in a tub of liquid, either mercury or water. When the reaction in the vessel produced a gas, it traveled through the tube into the bottle; if the gas is less dense than air, it will displace the water in the collection bottle, be collected, and be measured.

Priestley demonstrated his technique for the production of soda water to the Royal Society of London and even to the British navy in hopes that it would help prevent scurvy. Although soda water had no effect on scurvy, the Royal Society awarded him their highest scientific honor, the Copley Medal, in 1773.

DISCOVERY OF OXYGEN

In 1773, William Fitzmaurice Petty, the second earl of Shelburne, hired Priestley as his librarian, literary companion, and supervisor of his sons' education. Lord Shelburne promised him not only a generous salary, but a well-equipped laboratory and an extra stipend to purchase chemicals and supplies for experiments. Taking this position also meant moving his family to Calne, in Wiltshire. Priestley spent the summers with his family and the winters with Lord Shelburne in London. His years with Shelburne, from 1773 to 1780, were by far his most productive in the advancement of chemistry.

In August 1774, Priestley burned mercurius calcinatus (red mercuric oxide) using a large magnifying lens. He collected the resultant gas, using his usual method of passing it through mercury into an inverted bottle in the pneumatic trough. Shiny globules of elemental mercury were left behind. He collected three bottles full of the gas released. Because he had a lighted candle nearby, he held it to the gas in one bottle, and the flame burned brighter. He took a glowing ember of wood and held it to the second bottle, and it immediately burst into flames. Even more remarkably, a mouse could live entrapped in a bottle with this air longer than with ordinary air.

At the time, scientists widely accepted the "phlogiston theory" to explain how materials burned. They believed that a substance called phlogiston was responsible for allowing things to burn. If a substance had a lot of phlogiston, it burned easily. If a substance had little phlogiston, then it was more resistant to burning. When all the phlogiston had left a substance, burning would cease. The fact that a candle flame would be extinguished if kept under a jar was explained as the air's becoming saturated with phlogiston, so that it could absorb no more, and the flame would die. Because the new air that Priestley had extracted from mercurius calcinatus allowed the candle and the ember to burn more brightly, he reasoned that the air produced had little to no phlogiston present in it. Because of this, it was able to suck the phlogiston out of the candle and the wood much more readily. He called this special type of air "dephlogisticated." Priestley had discovered oxygen although his commitment to the phlogiston theory held back his true understanding of the chemistry of oxygen. He inhaled some himself and found that it made him feel light and easy. He predicted that breathing this new air would be a good medical treatment for people with respiratory problems, yet he worried that if it were used by healthy persons, it might cause one to die too soon. He wrote up his results the next year and sent them to the Royal Society in March 1775.

WORK ON PHOTOSYNTHESIS

Priestley also performed multiple experiments to learn about how plants breathed. Experimenters knew that a mouse placed in a bottle would use up all of the "good air" and soon die. When the air left behind in the mouse experiment was tested with a flame, the flame was immediately extinguished. This told Priestley that the gas being produced was similar to that which he used to make soda water.

In order to determine whether plants would react similarly, Priestley placed some mint in inverted bottles kept in a tub of water to prevent any outside air from entering the plant's environment. Mint was a good specimen for examining this phenomenon because the sprigs could live in water. He set up the bottles outside in his garden, where, surprisingly, the plants survived for weeks. To test the air left behind, he placed a lighted candle inside a bottle in which the mint had been growing and the flames became brighter and lasted longer. Clearly this experiment showed Priestley that the gas being produced by the plant was not the same gas being produced with the mouse. When he placed a mouse and a sprig of mint together in a bottle from which all of the "good air" had been removed by a burning candle, both the mint and the mouse survived just fine. Priestley determined that plants are able to take in bad air and release good air. The same phenomenon was being studied at the time by the Dutch physician Jan Ingenhousz.

Priestley continued to study this process. He filled several bottles with water and inverted them over bowls containing water. Some of the bottles had green pond scum (probably algae) in the water, and some did not. All of them were placed outside in the sunlight. By the end of the day, the water had been displaced in the bottles over the scummy water but not in the bottles over the plain water. A gas had been produced in the presence of the pond scum. When he placed a glowing ember in the gas and it burst into flames, he realized that the green plants (or algae) had produced oxygen. Priestley wondered whether the fact that green plants required sunlight to survive was related to their oxygen-producing capability. He repeated the experiment, but this time he put some of the bottles in the dark. No oxygen was produced, demonstrating that light was required. This well-researched process is termed photosynthesis.

DISCOVERY OF OTHER TYPES OF "AIRS"

Prior to Priestley, only three different airs had been identified: ordinary common air, fixed air (carbon dioxide), and inflammable air (hydrogen gas). Ordinary air was thought to be an element and, therefore, impossible to break down any further. Today scientists know that air is made up of several types of gases, but primarily nitrogen (78 percent) and oxygen (21 percent). Fueled by his successes with fixed air, Priestley decided to see what other airs he might produce.

Using a pneumatic trough and several new methods he developed, Priestley was able to produce and identify what are now called nitric oxide, nitrogen dioxide, nitrous oxide (laughing gas), ammonia, hydrogen chloride, sulfur dioxide, silicon tetrafluoride, nitrogen, and carbon monoxide. Some of these experiments were presented in 1772 to the Royal Society in his paper "Observations on Different Kinds of Air." Other discoveries were published in his six-volume series entitled *Experiments and Observations on Different Kinds of Air and Other Branches of Natural Philosophy*, published during the period 1774–86. Most of the work presented in his *Airs* series was performed during Priestley's years with Shelburne. Because of his advancements in the field of pneumatic chemistry, Priestley was elected to the French Academy of Sciences.

PHLOGISTON THEORY CHALLENGED

During fall 1774, Priestley accompanied Lord Shelburne to Europe. One night they dined with other famous scientists including the French chemist Antoine Lavoisier, who was much younger than Priestley but already very respected in the field. He asked Priestley about his current experiments, and Priestley openly shared his exciting discovery of dephlogisticated air. The other scientists were impressed and peppered him with questions, but Lavoisier just listened silently. Unbeknown to Priestley, Lavoisier was already trying to incorporate this new knowledge into a set of experiments he would perform over the next few months. Lavoisier later repeated Priestley's experiments and presented his own results to the French Academy of Sciences in April 1775 without giving any credit to Priestley for his intellectual contribution. Lavoisier called the dephlogisticated air *oxygen*, from the Greek word *oxys*, which means "sharp" (like an acid), and *gen*, which means "to be born." He also showed that ordinary air is made up of approximately 20 percent oxygen. He went even further by using this information to debunk the entire phlogiston theory.

Lavoisier challenged, "If phlogiston is released when something is burned, then why does the weight of such elements as magnesium increase after burning?" Lavoisier correctly hypothesized that burning, or combustion, resulted from the combination of a substance with oxygen. Candle flames were extinguished in enclosed spaces because the oxygen was used up. Mice died after a while under a jar for the same reason. Priestley could not accept this—he believed that phlogiston had a quality called levity, a sort of negative weight. He thought the explanation of phlogiston saturation was sufficient to explain why candle flames burned out after a period of time. Another problem with the phlogiston theory was that phlogiston had never been isolated. To this, Priestley responded that neither gravity nor electricity nor magnetism had been isolated and that phlogiston resembled a power more than a substance.

More concerned with seeking truth than fame, Priestley did not make a fuss after Lavoisier tried to steal the credit for the discovery of dephlogisticated air, or oxygen. He believed that the importance was the discovery itself, and that benefits could be derived no matter who received credit for it.

THE PRIESTLEYS IN AMERICA

The Priestley family had grown to a total of four children, and they wanted to stay in England. That was not to be. Joseph Priestley's controversial religious beliefs eventually caused them to move from England. Priestley had defended the rights of Americans to break away from England, he sympathized with the French revolutionaries, and he had consistently chipped away at the heart of the nation's prescribed religion. His own religious views were unpopular, he was shunned by his former Royal Society colleagues, and he and his wife no longer felt welcome in their motherland. They made arrangements to emigrate from England to the United States, where their three sons had moved a few years earlier.

Joseph and Mary Priestley set sail for America in April 1794. Though he was practically chased out of England, the United States warmly greeted Priestley. The couple stayed in New York for a few weeks and then moved on to Philadelphia. Priestley had become a member of the American Philosophical Society (founded by Ben Franklin) and wanted to visit some of the members there, though his old friend and confidante, Ben Franklin, had died in 1790. The Priestleys moved in with their son Joseph until their own house was completed. A laboratory was set up after the surprise arrival of several pieces of equipment sent by Josiah Wedgwood and others from England. One and one-half years after their arrival, their youngest son, Henry, died of pneumonia. Nine months later, Mary died.

Though most of his time was spent puttering in his lab, writing religious texts, and corresponding with colleagues including Thomas Jefferson, Benjamin Rush, and John Adams, Priestley continued to keep up with new developments in chemistry. He hoped to determine the amount of phlogiston in various metals. In 1799, he discovered yet another new gas, carbon monoxide, by heating coal in a small amount of air. Though poisonous, this gas has many industrial uses. He continued to defend the phlogiston theory passionately despite the fact that it had been demolished, ironically, as a result of his own discovery (oxygen). His last scientific paper was one titled "The Doctrine of Phlogiston Established," though, of course, it was not. Years before, Lavoisier had finally vindicated Priestley for his discovery of oxygen, but from that discovery, Lavoisier had given birth to a new revolution in chemistry, one that had no place for phlogiston.

Priestley's health began to weaken in 1801. By 1803, he was mostly bedridden. On February 6, 1804, he died with his son by his side, in Northumberland, Pennsylvania. To recognize Joseph Priestley's distinguished services to chemistry, the American Chemical Society established the Priestley Medal in 1922, honoring the father of pneumatic (gas) chemistry and the discoverer of oxygen. In his lifetime, Joseph Priestley completed more than 134 publications for future theologians, historians, linguists, philosophers, and scientists to contemplate.

See also CAVENDISH, HENRY; LAVOISIER, ANTOINE-LAURENT; METABOLISM; PHOTOSYNTHESIS.

FURTHER READING

Chemical Heritage Foundation. "Joseph Priestley." Available online. URL: http://www.chemheritage.org/classroom/chemach/forerunners/priestley.html. Accessed July 25, 2008.

Horvitz, Leslie Alan. *Eureka! Scientific Breakthroughs That Changed the World.* New York: John Wiley & Sons, 2002.

Schofield, Robert E. *The Enlightenment of Joseph Priestley: A Study of His Life and Work from 1733 to 1773.* University Park: Pennsylvania State University Press, 1997.

proteins Proteins, one of four types of biomolecules found within all living organisms, are the workhorses of cells. Made from different combinations of 20 building blocks called amino acids, these biomolecules exist as single polymers or complexes of multiple polymers that act in concert to perform a particular task. Proteins play many roles within a single cell and between cells by acting as enzymes, serving as structural support, providing immune protection to multicellular organisms, generating and transmitting nerve responses, regulating growth and development, and transporting ions, small molecules, and other proteins.

AMINO ACIDS

Many biological molecules contain not only carbon, hydrogen, and oxygen, but also nitrogen. Nitrogen fixation is the process used by plants to convert atmospheric nitrogen (N_2) to ammonia (NH_3), the form of nitrogen that can be used in the synthesis of amino acids, molecules that contain nitrogen. Amino acids contain a central carbon referred to as the alpha (α) carbon, bonded to a hydrogen atom, an amino group (-NH_3), a carboxyl group (-COOH), and one of 20 side chain groups (R groups) that give the biomolecule its specific characteristics. The peptide backbone can be represented as N-C-C with the nitrogen from the amino group followed by the α carbon and finally the carboxyl carbon. Because the α carbon is surrounded by four different side groups, each amino acid has two optical isomers, or stereoisomers, with the exception of glycine, which has a single hydrogen atom as its R group. In general, stereoisomers may be right-handed or left-handed, yet most amino acids associated with biological organisms are the left-handed, or L, form. Right-handed isomers exist in some sea-dwelling creatures, such as cone snails, and the peptidoglycan cell walls of bacteria. Occasionally, posttranslational modification of the amino acid aspartic acid to its right-handed form occurs in aging proteins.

The chemical nature of the 20 side chains dictates the behavior of each amino acid found in proteins. Eight amino acids are hydrophobic (water hating) in nature; they are alanine, cysteine, glycine, isoleucine, leucine, methionine, proline, and valine. Glycine is the smallest amino acid with a single hydrogen atom as its R group; consequently, it imposes no steric constraints on a protein's structure, unlike larger amino acids. Alanine, with only a methyl (-CH_3) group, is abundant and small enough that it can be found on the interior or exterior of a protein. Cysteine contains a sulfhydryl (-SH) group that can react with adjacent sulfhydryl groups on the interior of a protein to form disulfide bridges, which serve to strengthen the three-dimensional structure of a protein. Methionine is a larger hydrophobic amino acid that also contains a sulfur atom, and it is the first amino acid incorporated into every new polypeptide during translation, although it is sometimes subsequently removed. With long aliphatic side chains consisting of only carbon and hydrogen, isoleucine, leucine, and valine position themselves in the interior of proteins away from the aqueous environment of a cell. Proline is a unique hydrophobic amino acid because its R group circles back to form a covalent linkage with the peptide backbone. This conformation results in a kink along the peptide backbone and disrupts the ordered folding of a polypeptide. The presence of proline in a polypep-

tide sequence is common only in collagen, the main protein of connective tissue in humans.

Three hydrophobic amino acids are composed of side groups that include aromatic rings: phenylalanine, tryptophan, and tyrosine. The large, rigid rings on their side chains make these amino acids the largest of the 20, and they tend to orient themselves toward the interior of folded proteins. All three of these amino acids serve as precursors to an interesting array of products. Phenylalanine is used to make aspartyl-phenylalanine-1-methyl ester, the artificial sweetener more commonly known as aspartame, and L-dihydroxyphenylalanine (L-Dopa), a drug used in the treatment of Parkinson's disease. The body uses trytophan, which contains a double-ringed structure, to produce serotonin, and tyrosine serves as a precursor for melanin, epinephrine, and thyroid hormones.

Amino acids with positively (basic) or negatively (acidic) charged side chains also exist. The acidic amino acids are aspartic acid and glutamic acid, and the basic amino acids are arginine, lysine, and histidine. Because of their charged side chains, these amino acids tend to be on the outer surface of a protein, although oppositely charged amino acids pair up occasionally in the interior portion of certain proteins. With their strong negative charges, aspartic acid and glutamic acid make a protein highly water soluble when found on its outer surface. These two amino acids bind to positive ions and molecules, and they often fix a metal ion to its position within an enzyme. Both lysine and arginine have long, positively charged side chains that promote their interaction with negatively charged molecules like deoxyribonucleic acid (DNA); thus, nucleosomes, which are DNA-binding proteins, contain a large number of lysine and arginine residues. Histidine is a third basic amino acid that undergoes conformational changes when exposed to acidic environments, and the cell uses these changes as means to regulate proteins in organelles such as lysosomes. Histidine is a fairly rare amino acid, and very few of these residues can induce large-scale changes in protein structure.

Other hydrophilic, or water-loving, amino acids include asparagine, glutamine, serine, and threonine. Asparagine and glutamine are the neutralized forms of the two acidic amino acids, aspartic acid and glutamic acid. Both serine and threonine have a short R group that has a hydroxyl (-OH) group with a hydrogen atom that is easily removed, making these amino acids frequent hydrogen donors in enzymes. All four of these hydrophilic amino acids are usually located on the outer region of proteins, where they interact with the aqueous solution.

Humans cannot synthesize eight of these 20 amino acids, so they must obtain them through diet. These amino acids, called essential amino acids, are

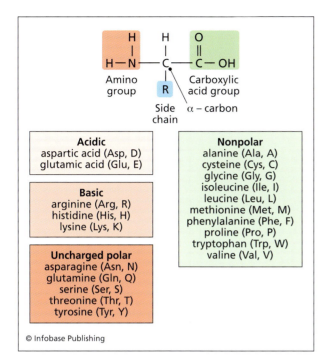

A generic amino acid is depicted at the top. The 20 amino acids used in proteins, along with their standard three-letter and one-letter abbreviations, are grouped by their chemical properties.

isoleucine, leucine, lysine, methionine, phenylalanine, threonine, tryptophan, and valine. Children must also obtain arginine and histidine through diet because their metabolic pathways necessary for synthesis of these amino acids are not developed fully. Obvious sources for these amino acids include high-protein foods such as meat, eggs, and cheese. Vegetarians can obtain the essential amino acids through consuming a wide variety of plant products.

Individuals with certain metabolic disorders have to avoid particular amino acids. Phenylketonuria (PKU) is a genetic disorder that prevents phenylalanine metabolism. Individuals with PKU must abstain from consuming any food that might contain phenylalanine—meat, nuts, dairy products, and starchy foods like potatoes and corn. If left untreated, PKU results in brain damage and progressive mental retardation due to an accumulation of phenylalanine and its by-products. Performing a simple blood test shortly after birth can determine whether or not an individual has this disorder.

PEPTIDES AND POLYPEPTIDES

Proteins are strings of amino acids connected together in a specific way dictated by the genetic code embedded in another biomolecule, DNA. Proteins are synthesized during the process called translation. In the nucleus of the cell, transcription generates a copy of the DNA, in the form of a molecule of messenger ribonucleic acid (mRNA), containing instructions for building a particular protein. The mRNA leaves the nucleus to associate with ribosomes, organelles involved in the translation process. During this process, ribosomes hold free amino acids in close proximity to one another, permitting a condensation reaction to occur between the carboxylic acid of one amino acid and the amino group of a second amino acid, resulting in the formation of a peptide bond and release of a water molecule. The rigid bond that forms between these adjacent amino acids behaves as a double covalent bond, yet a large degree of rotational freedom exists on each side, permitting folding to take place. One reads the amino acid sequence of a growing protein beginning at the end with the free amino group of the first amino acid until reaching the carboxylic acid end of the last amino acid incorporated. In some cases, after translation, a protein undergoes additional modifications that enhance its ability to perform its role, or a protein associates with other proteins to form a complex necessary for its function.

Four levels of protein structure exist: primary, secondary, tertiary, and quaternary. The most basic level of protein structure, the primary level, defines the amino acid sequence of a particular peptide, polypeptide, or protein. A peptide has 10 or fewer amino acids bound together, a polypeptide consists of 10–100 amino acids, and a protein is composed of 100 or more amino acids. Many times, peptides and polypeptides are linear and unstructured. Examples of these biomolecules include oxytocin and calcitonin, two peptides that are involved in hormone signaling and metabolism, respectively. A folded polypeptide, or protein, can have as many as 27,000 amino acids, for example, the muscle protein titin, with masses measured in kilodaltons.

The next level of protein structure, the secondary structure, results in portions of a newly synthesized protein folding into of one of two repeating structural motifs, the alpha (α) helix or the beta (β) sheet. The alpha helix looks like a right-handed coil with the amino acids spiraling up an invisible vertical axis securing the shape via hydrogen bonds between the NH and C=O groups of the peptide backbone at amino acids separated by four residues. The R groups splay out around this central core. The lengths of these helices vary from 40 angstroms (Å) to 1,000 Å, or encompass 90 to almost 2,500 amino acid residues. Alpha helices are the major structural motif in the oxygen-carrying proteins myoglobin and hemoglobin, but other proteins, such as chymotrypsin, a digestive enzyme, have no α helices in the folded state. The other secondary structural motif, the β-pleated sheet, resembles a sheet of extended polypeptide chains bound together with hydrogen bonds between the NH and C=O groups of adjacent backbones. The peptide backbone from one chain to the next can run parallel or antiparallel in a β sheet with two to five strands within a specific motif. Connective proteins, such as silk fibronin, often contain β sheets within their structure.

The third level of protein structure consists of the overall, or three-dimensional, shape of a single protein molecule. In addition to determining the secondary structure, a protein's R groups found along its backbone dictate the way a protein folds. Hydrophobic side chains bury themselves in the interior core of a globular protein. Charged amino acids engage in ionic interactions with oppositely charged molecules within or on the exterior of the protein. Disulfide bridges form between neighboring cysteine residues in the interior of proteins, and 11 of the 20 amino acids can form hydrogen bonds that add to the stability of the three-dimensional structure a protein assumes. In the case of proteins such as hemoglobin, another level of structure called the quaternary structure assembles multiple folded proteins together as subunits of functional complexes. Sometimes the shape, or conformation, of a protein will change after arriving at its tertiary or quaternary structure in order to perform its job. When a conformational change occurs in a protein, one structure is biologically active where the alternative is not; thus, conformational changes can regulate protein activity.

Most proteins belong to one of five main classes based on their structures and roles in the cell: soluble proteins, filamentous proteins, membrane-associated proteins, motor proteins, and proteins that protect other proteins. Soluble proteins, as the name suggests, are proteins that are soluble in the aqueous environment of the cell. Enzymes that catalyze biochemical reactions and antibodies that help fight infection are examples of this class of protein. Filamentous proteins perform structural functions, forming scaffolding and matrices within a cell or organism. Tubulin and actin are filamentous proteins that serve as structural proteins within the cell. Collagen and keratin are two filamentous proteins that provide the framework for skin, hair, and nails in multicellular organisms. Membrane-associated proteins traverse the lipid bilayer and permit the exchange of information and materials between the outside and inside of a cell. An example of a membrane-bound protein is the sodium-potassium pump that permits movement of these two ions across an amphipathic cell membrane. Membrane-associated proteins also include channels and receptors attached to the lipid bilayer. A fourth class of proteins, including motor proteins such as myosin and dynein, functions in the movement of organelles, cells, or organisms. Another type of protein protects other proteins from denaturation, or unfolding, during times of stress for a cell; they also serve as chaperones ensuring that newly synthesized proteins fold correctly and move to their proper location within the cell. Two examples of this type of protein are the heat shock proteins HSP70 and HSP90, named according to their molecular weights.

The study of proteins depends on many different techniques to elucidate information about the sequence, folding, structure, and function of these biomolecules. Protein chemists traditionally used a process called Edman degradation to determine the sequence of short peptides one amino acid at a time. Now they more commonly use recombinant DNA technology and decipher the amino acid sequence from nucleotide sequence information using the genetic code. Knowing the amino acid sequence allows protein biochemists to attempt to predict how the primary structure will then fold into a three-dimensional shape, and then surmise hypothetical functions. After isolating a crude protein, biochemists have several methodologies at their disposal to determine the size, charge, and binding affinities. A type of electrophoresis called sodium dodecyl sulfate (SDS) polyacrylamide gel electrophoresis uses detergent to denature (unfold) a protein of interest, separating mixtures of these biomolecules by size on a polyacrylamide matrix with larger proteins staying near the top and smaller proteins traveling toward the bottom of the gel. Gel filtration chromatography also separates protein mixtures by size. Both enzyme-linked immunoabsorbent assays (ELISA) and Western blots use protein-specific antibodies to detect and quantitate low levels of proteins of interest. Microscopy techniques, such as fluorescence microscopy, also use antibodies to detect the presence of proteins in cells. With a large amount of protein, X-ray crystallography can determine the three-dimensional structure of a protein, permitting its folding patterns, its relationship to other proteins, and its catalytic sites to be studied.

See also BIOCHEMISTRY; CARBOHYDRATES; CHROMATOGRAPHY; ENZYMES; LIPIDS; NUCLEIC ACIDS; PAULING, LINUS; PERUTZ, MAX.

FURTHER READING

Tanford, Charles, and Jacqueline Reynolds. *Nature's Robots: A History of Proteins.* Oxford: Oxford University Press, 2001.

Whitford, David. *Proteins: Structure and Function.* London: John Wiley & Sons, 2005.

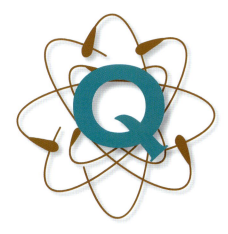

quantum mechanics Quantum mechanics is the realization of the principles of quantum physics as they apply to actual physical systems. This entry starts with a discussion of quantum physics, the most general and widely applicable understanding of nature that physicists have. Quantum physics deals with nature on a broad range of scales, both the larger scales that form the domain of classical physics and the smaller scales—the atomic and molecular scales and the scale of elementary particles—where classical physics does not apply.

CHARACTERISTICS OF QUANTUM PHYSICS

Quantum physics is characterized by the Planck constant, h = 6.62606876 × 10^{-34} joule-second (J·s), rounded to 6.63 × 10^{-34} J·s, which is nature's elementary unit of a physical quantity called action. The Planck constant is named for Max Planck, the 19–20th-century German physicist who introduced it and who is considered one of the founders of quantum physics. Some further characteristics of quantum physics, which apparently reflect fundamental properties of nature, are uncertainty, indeterminism, discontinuity and discreteness, nonlocality, wave-particle duality, and particle indistinguishability.

Uncertainty. Physical quantities do not generally possess sharp values. They do not even have *any* value until they are measured. They do not have values that are simply not known but become revealed by measurement; rather, they have no value at all until a measurement "forces" them to exhibit one and, thus, so to speak, endows them with a value. When repeated measurements of a physical quantity are performed on the same state of a system, the results are generally different, giving an uncertainty in the value of the quantity. In particular situations, however, the results can be the same, whereby the quantity has a sharp value. Certain pairs of physical quantities obey the Heisenberg uncertainty principle (discussed in more detail later), according to which the minimal uncertainty of one is inversely proportional to the minimal uncertainty of the other. Thus, if one such quantity possesses a sharp value, the other has maximal uncertainty and can show any allowed value upon measurement.

Indeterminism. Given an initial state of an isolated system, what evolves from that state at future times is only partially determined by the initial state. The system is described by a wave function, which evolves deterministically until the system spontaneously and suddenly undergoes a transition, such as radioactive decay, or until a measurement is performed on it. The occurrence of transitions and the values of physical quantities are not in general uniquely determined. Rather, the wave function gives the probabilities of transitions and the probabilities that the system's physical quantities, when measured, will have any of their allowed values. The probability, then, is determined, while indeterminism reigns with regard to the actual occurrence of transitions and to the values of quantities. From this state of affairs, it follows that quantum predictability exists only with respect to probabilities. The time of occurrence of a spontaneous transition or the actual value measured for some physical quantity is, in general, undetermined and therefore unpredictable.

Discontinuity and discreteness. An isolated physical system, when left to its own devices, evolves continuously in time for some duration. That continuity ends in discontinuity, when the system undergoes

Some devices representing modern technology. From left to right: a personal digital assistant (PDA), a mobile phone, and a laptop computer. They all rely for their operation on quantum effects in semiconductors. *(JUPITER IMAGES/Comstock Images/Alamy)*

spontaneous transition, sometimes called a quantum jump, or when a measurement is performed on it. An additional type of quantum discontinuity holds for certain physical quantities whose range of possible values, according to classical physics, is continuous. But according to quantum physics, those quantities possess a range of allowed values that can include, or possibly even consist wholly of, a set of discrete values.

As an example, the allowed energy values of the bound electron and proton that constitute a hydrogen atom form a discrete set and are given, in units of electron volts—where one electron volt (eV) equals 1.60×10^{-19} joule (J)—by

$$E_n = -\frac{13.6}{n^2} \text{ eV for } n = 1, 2, \ldots$$

Here n is the quantum number for energy and E_n is the energy of the nth energy level. The unbound electron and proton (i.e., a positive hydrogen ion, which is simply a proton, and an electron wandering off somewhere) can possess any positive value of energy. So, the allowed energy values of the electron-

proton system consist of both a discrete set of negative values and a continuous range of positive values. According to classical physics, on the other hand, the range of energy values for this system is completely continuous.

Nonlocality. The situations at different locations can be entangled, so that what happens at one affects what happens at the other in a manner that is not explainable in terms of an influence propagating from one to the other. This is expressed by the statement that a quantum state is nonlocal. Consider this example. Two particles might be emitted by a common source such that their spins are in opposite directions. When they are very far apart, the spin direction of one is measured. Then, within a time interval too short for a signal to pass from one location to the other, the spin direction of the other is measured. The spins will invariably exhibit opposite directions. This effect is considered a nonlocal quantum effect based on the fact that a particle does not possess a spin direction until it is measured. So the first measurement affects not only the particle whose spin direction is first measured, but the whole quantum state that involves both particles. The second-measured particle instantaneously "knows" the spin direction it must have in order for it to be opposite that of the first-measured one.

Wave-particle duality. All matter possesses both wave and particle characteristics. A wave is a propagating disturbance, possibly characterized by characteristics such as frequency and wavelength, that is spread out over space, so is not a localized entity. A particle, on the other hand, is localized—has a definite position—and is characterized, for example, by mass, velocity, position, and energy. Every wave possesses a particle aspect and every particle has a wave aspect. The photon and the phonon, for example, form the particle aspects of electromagnetic and acoustic waves, respectively. On the other hand, the electron and the proton, normally viewed as particles, are each related to corresponding waves. Which aspect manifests depends on the phenomenon being observed. When a wave is exchanging energy with some system, for instance, it does so in distinct units, so its particle aspect is apparent. When a beam of particles is split and rejoined, interference takes place; thus, the wave aspect is revealed.

Particle indistinguishability. Identical particles are fundamentally indistinguishable. That characteristic leads to the Pauli exclusion principle, stating that no more than a single fermion can occupy the same state, and to Bose-Einstein statistics and Fermi-Dirac statistics for many-particle systems of bosons and fermions, respectively.

Although quantum physics gives the best description of nature that physics currently possesses, classical physics can well approximate quantum physics, while being considerably simpler than it, in the appropriate domain. That domain, called the classical domain, is characterized by lengths, durations, and masses that are not too small, that are in general larger, say, than the atomic and molecular scales.

QUANTUM MECHANICS

Quantum mechanics, as mentioned, is the realization of the principles of quantum physics as they apply to actual physical systems. The following is a brief presentation of the ideas, concepts, and terminology that are involved in quantum mechanics.

In quantum mechanics, a physical system is described by a wave function Ψ, which specifies the state of the system at any time. The wave function is a complex function (i.e., its values are complex numbers) that depends on the system's generalized coordinates and on time t. Generalized coordinates are the set of physical quantities needed to specify a state. In the case of a system of particles, for instance, the generalized coordinates consist of all the coordinates of the individual particles. For the sake of simplicity, let us assume the system comprises a single particle, whose position is specified by the position vector **r**, which might be expressed in terms of its Cartesian coordinates (x, y, z). Then, the wave function is written as $\Psi(\mathbf{r}, t)$ or $\Psi(x, y, z, t)$. In a one-dimensional situation, the wave function is simply $\Psi(x, t)$.

The wave function contains all the information about the system. One most important property of the wave function is that it gives probabilities for the system. In the one-dimensional case, the probability that the particle is found in the interval between x and $x + dx$ at time t, denoted by $P(x, t)\,dx$, is given by the square of the absolute value of the wave function, $|\Psi(x, t)|^2 = \Psi(x, t)^*\Psi(x, t)$, where $\Psi(x, t)^*$ denotes the complex conjugate of $\Psi(x, t)$. The relation is

$$P(x, t)\,dx = |\Psi(x, t)|^2\,dx$$

From this, it follows that the probability for the particle to be found in the finite interval $a \le x \le b$ at time t is

$$\text{Probability} = \int_a^b |\Psi(x,t)|^2\,dx$$

In order for the wave function to serve in this capacity correctly, it must be normalized, in order to ensure that the probability for the particle to be anywhere is unity. The normalization condition for the wave function is then

$$\int_{-\infty}^{+\infty} |\Psi(x,t)|^2\,dx = 1$$

This condition is appropriately generalized in more general cases. Some uses of the wave function are described in the following.

PHYSICAL QUANTITIES

In quantum mechanics, physical quantities are represented by operators, which are mathematical expressions that "operate" on the wave function. One use of operators is this. If O_Q denotes the operator representing the physical quantity Q, then the average of the values of Q that are found by repeated measurements of Q when the system is in the state specified by $\Psi(x, t)$, called the expectation value of Q and denoted by $<Q>$, is given by

$$<Q> = \int_{-\infty}^{+\infty} \Psi(x,t)^*\, O_Q\Psi(x,t)\,dx$$

in the one-dimensional case. The operator corresponding to position x, for example, is simply multiplication by x. Thus, the expectation value of position is given by

$$<x> = \int_{-\infty}^{+\infty} \Psi(x,t)^*\, x\, \Psi(x,t)\,dx$$

The linear-momentum operator is given by $[h/(2\pi i)]\,(\partial/\partial x)$, where h denotes the Planck constant, referred to previously. So, the expectation value of the particle's momentum p is

$$<p> = \int_{-\infty}^{+\infty} \Psi^* \frac{h}{2\pi i} \frac{\partial \Psi}{\partial x}\, dx$$

$$= \frac{h}{2\pi i} \int_{-\infty}^{+\infty} \Psi^* \frac{\partial \Psi}{\partial x}\, dx$$

For another example, the operator corresponding to the particle's total energy E is $(ih/[2\pi])(\partial/\partial t)$, giving for the energy expectation value

$$<E> = \int_{-\infty}^{+\infty} \Psi^* \frac{ih}{2\pi} \frac{\partial \Psi}{\partial t}\, dx$$

$$= \frac{ih}{2\pi} \int_{-\infty}^{+\infty} \Psi^* \frac{\partial \Psi}{\partial t}\, dx$$

In general, in quantum mechanics, a physical quantity does not possess a sharp value for an arbitrary state of a system. As mentioned, the expectation value then gives the average of the various values that are found for the quantity when it is measured for the same state. The action of an operator O_Q, representing the physical quantity Q, on the wave function $\Psi(x, t)$ specifying a state might, however, result in the same wave function multiplied by a number. That number is just the value of the quantity Q, q, for that state, for which the quantity then possesses a sharp value. The formula describing this is

$$O_Q\, \Psi(x, t) = q\, \Psi(x, t)$$

In such a case, the wave function is called an eigenfunction of the operator O_Q with eigenvalue q, and the state specified by the wave function is called an eigenstate of the quantity Q, also with eigenvalue q. Then the expectation value of Q equals its sharp value q:

$$<Q> = \int_{-\infty}^{+\infty} \Psi(x,t)^* O_Q \Psi(x,t)\, dx$$

$$= \int_{-\infty}^{+\infty} \Psi(x,t)^* q \Psi(x,t)\, dx$$

$$= q \int_{-\infty}^{+\infty} \Psi(x,t)^* \Psi(x,t)\, dx$$

$$= q$$

The last equality follows from the normalization condition for the wave function.

HEISENBERG UNCERTAINTY PRINCIPLE

Named for its proposer, Werner Heisenberg, a 20th-century German physicist, the Heisenberg uncertainty principle states that certain pairs of physical quantities cannot simultaneously possess sharp values: that greater sharpness of one is at the expense of less sharpness of the other. The principle is expressed in terms of the uncertainty of the physical quantities, where uncertainty and sharpness are inversely related: greater uncertainty means less sharpness and less uncertainty signifies greater sharpness. Zero uncertainty is synonymous with perfect sharpness.

The uncertainty of a physical quantity Q is denoted ΔQ and is defined by

$$\Delta Q = \sqrt{<Q^2> - <Q>^2}$$

It expresses the spread of the values about the average of those values found by repeated measurements of Q, for whatever state of the system is being considered. If Q has a sharp value, its uncertainty is zero, $\Delta Q = 0$.

Denote such a pair of physical quantities by A and B and their respective uncertainties by ΔA and ΔB. Then, the Heisenberg uncertainty relation for A and B states that the product of ΔA and ΔB cannot be less than a certain amount, specifically

$$\Delta A\, \Delta B \geq \frac{h}{4\pi}$$

where h denotes the Planck constant, referred to earlier. Given the uncertainty of one of such a pair, say ΔA, the least possible uncertainty of the other is determined by

$$\Delta B \geq \frac{h}{4\pi \Delta A}$$

When one quantity of such a pair is sharp (i.e., its uncertainty is zero), the uncertainty of the other is infinite. That means its range of values is unlimited; it is as "unsharp" as possible.

The Heisenberg uncertainty principle is understood to concern the measured values of the physical quantities. It is not a matter of the system's always possessing a sharp value and the uncertainty's being in the physicist's knowledge of it. Rather, according to quantum physics, the system does not possess any value until a measurement "forces" it to take one. Sharpness means that repeated measurements on the system in the same state (or measurements on many identical systems in the same state) always result in the same value for the quantity. In a situation of nonzero uncertainty, however, such a set of measurements produces a range of values. That range is what uncertainty means.

The pairs of physical quantities that are subject to the uncertainty principle include every generalized coordinate of the system together with its corresponding generalized momentum. The generalized coordinates are the set of physical quantities needed to specify a state of the system. If we denote a generalized coordinate by X, its generalized momentum is the physical quantity represented by the operator $[h/(2\pi i)](\partial/\partial X)$. In the case of a system of particles, for instance, the generalized coordinates are simply the coordinates of the positions of all the particles, and the corresponding generalized momenta are the corresponding components of the particles' linear momentum. As an example, let us consider the x-coordinate of a particle's position x and the x-component of its linear momentum p_x, represented by the operator $[h/(2\pi i)](\partial/\partial x)$, as a pair of quantities for which the Heisenberg uncertainty relation holds. Their respective uncertainties are denoted Δx and Δp_x and are subject to the constraint

$$\Delta x \, \Delta p_x \geq \frac{h}{4\pi}$$

where Δx is in meters (m) and Δp_x in kilogram-meters per second (kg·m/s). Given the uncertainty of Δx, say, the least possible uncertainty of Δp_x is determined by

$$\Delta p_x \geq \frac{h}{4\pi \Delta x}$$

and similarly

$$\Delta x \geq \frac{h}{4\pi \Delta p_x}$$

When one of these two quantities is sharp, the uncertainty of the other is infinite. If the particle's position is precisely known, the particle's momentum is maximally uncertain. And, conversely, if the momentum is sharp, the particle has no location and can be anywhere.

An example of what that means in practice is this. A "quantum dot" is a structure in a semiconductor crystal that confines electrons very closely in all three dimensions. Its size is of the order of tens of nanometers. The location of an electron confined in a quantum dot is known with an uncertainty that equals the diameter of the quantum dot. The uncertainty of the component of the electron's momentum in any direction follows from the previous uncertainty relation. To illustrate this, take a diameter of 50 nanometers, $\Delta x = 50 \times 10^{-9}$ m. Then, the uncertainty relation gives for Δp_x

$$\Delta p_x \geq \frac{6.63 \times 10^{-34}}{4\pi \left(50 \times 10^{-9}\right)}$$

$$= 1.06 \times 10^{-27} \text{ kg·m/s}$$

The momentum of electrons in a quantum dot is quantized (see the section Quantization): that is, the allowed values of electron momentum form a discrete set. When an electron possesses one of the allowed values, there remains the uncertainty that it might be moving with that magnitude of momentum in, say, either the x direction or the $-x$ direction. The uncertainty is then twice the magnitude of the momentum, so the uncertainty is minimal when the momentum is minimal, that is, when the electron has the least allowed momentum. Minimal uncertainty is given by the uncertainty relation, when the $\geq$ relation is taken as equality. Thus, the uncertainty relation determines that the momentum of an electron in its slowest allowed state for this example is of the order of 10^{-27} kg·m/s. And, in particular, no electron confined to a quantum dot can be at rest.

An uncertainty relation also holds for energy and time in quantum mechanics, although time is not considered a physical quantity, but rather a parameter of evolution. The form of this uncertainty relation is

$$\Delta t \, \Delta E \geq \frac{h}{4\pi}$$

where ΔE denotes the uncertainty of energy, in joules (J), and Δt is in seconds (s). The significance of Δt is the duration, or lifetime, of the state of the system whose energy is E. The longer the state lives, the sharper its energy can be, and the briefer the state's lifetime, the more uncertain its energy. So stable states of atoms and nuclei, for example, and other states with long lifetimes possess sharp, well-defined values of energy. On the other hand, the energies of very unstable atomic and nuclear states can be quite indefinite and become less and less sharp for shorter and shorter lifetimes.

SCHRÖDINGER EQUATION
The wave function is a function of time. Its temporal evolution is described by the Schrödinger equation, named for its discoverer, the 19–20th-century Austrian physicist Erwin Schrödinger. For a single particle whose potential energy is a function of position, $U(\mathbf{r})$, the Schrödinger equation for the wave function $\Psi(\mathbf{r}, t)$ takes the form

$$\frac{ih}{2\pi} \frac{\partial \Psi}{\partial t} = -\frac{h^2}{8\pi^2 m} \nabla^2 \Psi + U(\mathbf{r})\Psi$$

[This is actually an operator equation expressing the fact that the total energy equals the sum of the kinetic energy and the potential energy for the state specified by $\Psi(\mathbf{r}, t)$.] Expressed in terms of coordinates, the equation takes the form

$$\frac{ih}{2\pi}\frac{\partial\Psi}{\partial t} = -\frac{h^2}{8\pi^2 m}\left(\frac{\partial^2}{\partial x^2}+\frac{\partial^2}{\partial y^2}+\frac{\partial^2}{\partial z^2}\right)\Psi + U(x,y,z)\Psi$$

In the one-dimensional case, the Schrödinger equation for the wave function $\Psi(x, t)$ is

$$\frac{ih}{2\pi}\frac{\partial\Psi}{\partial t} = -\frac{h^2}{8\pi^2 m}\frac{\partial^2\Psi}{\partial x^2} + U(x)\Psi$$

The Schrödinger equation allows physicists to predict the results of atomic processes and to calculate the properties of atomic and molecular systems.

QUANTIZATION

Quantization is the achievement of understanding of a physical system in terms of quantum physics, rather than of classical physics, and according to the rules of quantum mechanics. One aspect of quantization is that some physical quantities possess a discrete set of possible values rather than a continuous range of values. In classical physics, a harmonic oscillator, for example, may have any value of energy from zero up. The quantized harmonic oscillator, on the other hand, may only possess energy values given by the formula

$$E_n = \left(n + \tfrac{1}{2}\right)hf \quad \text{for } n = 0, 1, \ldots$$

where E_n denotes the value of the energy, in joules (J), for the nth energy level; f represents the classical frequency of the oscillator, in hertz (Hz); and h is the Planck constant, referred to already. (Note that the value 0 is not allowed for the energy.) The number n is referred to as a quantum number, the energy quantum number in this case.

An additional example of quantization is the system comprising a proton and an electron bound by the force of their electric attraction, that is, the hydrogen atom. As shown, its energy levels, in electron volts, are

$$E_n = -\frac{13.6}{n^2}\,\text{eV} \quad \text{for } n = 1, 2, \ldots$$

That leads to the common expression that energy is quantized for the harmonic oscillator and the hydrogen atom. Similarly, one says that any physical quantity is quantized when quantum mechanics

allows it only a discrete set of possible values, whereas in classical physics, it is a continuous variable. Angular momentum serves as another example of such a quantity. In classical physics, angular momentum can have any value. But in the quantum domain, it is quantized, as its component in any direction is restricted to integer multiples of $h/(2\pi)$, that is, to $mh/(2\pi)$, where $m = 0, \pm1, \pm2, \ldots$ is the corresponding quantum number. This set of allowed values for the component of a particular angular momentum in any direction depends on the magnitude of the angular momentum, which is also quantized and can have only values $\sqrt{\ell(\ell+1)}\,h/(2\pi)$, where $\ell = 0, 1, 2, \ldots$ is the quantum number for angular momentum. For a particular value of ℓ, the allowed values of the component in any direction are given by

$$m = \ell, \ell - 1, \ldots, 0, \ldots, -(\ell - 1), -\ell$$

so that the allowed values themselves are

$$\ell\frac{h}{2\pi}, \ (\ell-1)\frac{h}{2\pi}, \ldots, 0, \ldots, -(\ell-1)\frac{h}{2\pi}, \ -\ell\frac{h}{2\pi}$$

The angular momentum that is associated with the spin of elementary particles is similarly, but not identically, quantized. Its component in any direction can only possess a value that is a half-integer multiple of $h/(2\pi)$, or $(n/2)[h/2\pi]$, where $n = 0, \pm1, \pm2, \ldots$. The exact set of values for a particular type of particle is determined by the spin of that particle type. For a particle of spin s (i.e., whose spin quantum number is $s = 0, 1/2, 1, 3/2, \ldots$), the allowed values of the component of its spin angular momentum in any direction are $m_s h/(2\pi)$, where

$$m_s = s, s - 1, \ldots, -(s - 1), -s$$

giving for the allowed values themselves

$$s\frac{h}{2\pi}, \ (s-1)\frac{h}{2\pi}, \ldots, -(s-1)\frac{h}{2\pi}, \ -s\frac{h}{2\pi}$$

The allowed values of the spin component of an electron, for instance, which is a spin-½ particle, are $h/(4\pi)$ and $-h/(4\pi)$.

An additional aspect of quantization is that some physical quantities do not possess sharp values but, rather, have probabilities for each of their various allowed values to be detected when a measurement is performed on the system. In certain situations, the quantized harmonic oscillator, for example, might possess a definite value for its energy, one of its E_n. In other situations, however, it might not have a sharp value of energy, but only a set of probabilities for each of the values E_n to be found when the system's energy is measured and a single value is determined. And even when the quantized harmonic oscillator does have a definite value for its energy, its position is not sharp. Its position is not restricted and may have any value, but it does not have a *definite* value until a position measurement "forces" it to take one, according to the system's probabilities for position. This characteristic of the quantum world was discussed earlier.

A further aspect of quantization is that certain pairs of physical quantities are mutually exclusive, according to the Heisenberg uncertainty principle, as discussed.

QUANTUM FIELD THEORY

The application of quantum mechanics to the understanding of fields is termed quantum field theory. The underlying idea of quantum field theory is that all interactions are described by fields that possess a particle aspect and all matter particles are described by fields, such that the principles of quantum physics and the rules of quantum mechanics are obeyed.

Quantum electrodynamics (QED) is the very successful quantization of the electromagnetic field, its interaction with matter, and the electromagnetic interaction of matter with matter. According to QED, the particle aspect of the electromagnetic field is the photon; electromagnetic waves are viewed as flows of photons; the interaction of the electromagnetic field with matter is understood as the emission, absorption, and scattering of photons by matter; and the electromagnetic interaction of matter with matter is comprehended in terms of exchange of photons. The electromagnetic field also possesses a classical theory, embodied in Maxwell's equations, named for the 19th-century Scottish physicist James Clerk Maxwell.

The strong interaction and the weak interaction do not possess classical theories, and quantum field theory very successfully deals with them. The interaction field of the strong interaction is the gluon field, of which gluons form the particle aspect. Quarks are the elementary particles most directly affected by the gluon field, and they are described by their corresponding fields. The quantum field theory of the strong interaction is called quantum chromodynamics (QCD). The particle aspect of the weak-interaction field comprises the intermediate vector bosons, which are named and denoted W^+, W^-, and Z^0.

While the gravitational field possesses a classical theory, which is the general theory of relativity, its quantization is problematic and has not yet been successfully achieved. This appears to be due to the fact that, while the other fields seem merely to play their various roles in the arena of space-time, so to speak, the gravitational field is intimately linked to space-time, according to general relativity, so that its quantization would seem to involve the quantization of space-time itself. Although it has not yet been experimentally detected, the graviton is the putative particle aspect of the gravitational field.

HISTORY OF QUANTUM PHYSICS AND QUANTUM MECHANICS

By the end of the 19th century, physicists viewed their field as essentially complete, clear, and beautiful, but with two "small" exceptions, referred to as two clouds in the clear sky. One problem was the negative result of the Michelson-Morley experiment, which seemed to indicate that the "ether" that was assumed to form the material medium carrying electromagnetic waves did not, in fact, exist. The other "cloud," called the blackbody spectrum problem, was that the energy distribution among the frequencies of the electromagnetic radiation emitted by a radiating body was not explainable by the physics of the time. Both "clouds" served as seeds from which the two revolutions of modern physics sprouted in the early 20th century. The German-Swiss-American physicist Albert Einstein's special theory of relativity explained the negative result of the Michelson-Morley experiment, and the solution of the blackbody spectrum problem by the German physicist Max Planck led to the development of quantum physics.

Quantum physics describes nature and explains natural phenomena with amazing accuracy. It is the best and most-encompassing physics that physicists have at present. The following is a brief review of some of the events and people in the history of quantum physics and quantum mechanics:

- In order to explain the blackbody spectrum, Max Planck proposed in 1900 that electromagnetic energy could be emitted only in quantized form, that is, in multiples of the elementary unit of energy hf, where h denotes the Planck constant, presented earlier, and f is the frequency of the electromagnetic wave. Planck was awarded the 1918 Nobel Prize in physics "in recognition of the services he rendered to the advancement of physics by his discovery of energy quanta."
- In 1905, Albert Einstein explained the photoelectric effect, that light shining on certain metals causes electrons to be emitted, by postulating that light consists of individual particles (photons). In 1921, Einstein was awarded the Nobel Prize in physics, not for relativity, for which he is most widely known, but "for his services to theoretical physics, and especially for his discovery of the law of the photoelectric effect."

- The New Zealand-British physicist Ernest Rutherford discovered in 1907 the structure of atoms: a relatively tiny positively charged nucleus at the center, with negative charge distributed around it. Rutherford, too, was awarded a Nobel Prize, in 1908, but in chemistry, not in physics. And his prize was not for the discovery of atomic structure, but rather "for his investigations into the disintegration of the elements, and the chemistry of radioactive substances."
- In 1913, the Danish physicist Niels Bohr proposed his quantum model of the hydrogen atom, to explain the light emission spectrum of hydrogen. Bohr was awarded the Nobel Prize in physics in 1922 "for his services in the investigation of the structure of atoms and of the radiation emanating from them."
- The French physicist Louis de Broglie proposed in 1923 that not only do waves have a particle aspect (see the discussion of Planck and Einstein), but particles, such as electrons, also possess a wave nature. (The electron microscope is based on the wave character of electrons.) De Broglie was awarded the 1929 Nobel Prize in physics "for his discovery of the wave nature of electrons."
- In 1925, the Austrian-Swiss physicist Wolfgang Pauli presented the "Pauli exclusion principle," which states that no two fermions (particles with spin 1/2, 3/2, etc.) can exist in the same quantum state. This principle is essential for understanding the electron configurations of atoms. Pauli was awarded the Nobel Prize in physics in 1945 "for the discovery of the Exclusion Principle, also called the Pauli Principle."
- On the basis of de Broglie's idea of electron waves, in 1926, the Austrian physicist Erwin Schrödinger formulated an equation, presented earlier in this article, to describe the distribution of electrons in space and the evolution of the distribution in time. He shared the 1933 Nobel Prize in physics with Paul Dirac "for the discovery of new productive forms of atomic theory."
- The British physicist Paul Dirac proposed, in 1928, an equation to describe elementary particles of spin ½, such as electrons, consistent with both the principles of quantum physics and the special theory of relativity. The Dirac equation predicted the existence of antiparticles. As mentioned, Dirac shared the 1933 Nobel Prize in physics with Erwin Schrödinger.
- During the rest of the 20th century, much progress was achieved in the application of quantum physics to fields, called quantum field theory. Application to the electromagnetic field produced quantum electrodynamics (QED), which—is an amazingly accurate field theory—indeed, the most accurate theory in existence. Leaders in this endeavor were the Japanese physicist Sin-Itiro Tomonaga and the American physicists Julian Schwinger and Richard Feynman, who shared the 1965 Nobel Prize in physics "for their fundamental work in quantum electrodynamics, with profound consequences for the physics of elementary particles."
- Further developments in quantum field theory were recognized by the 1979 Nobel Prize in physics, awarded to the American physicists Sheldon Glashow and Steven Weinberg and the Pakistani physicist Abdus Salam "for their contributions to the theory of the unified weak and electromagnetic interactions between elementary particles, including inter alia the prediction of the weak neutral current." Additionally, the Dutch physicists Gerardus 't Hooft and Martinus Veltman were awarded the 1999 Nobel Prize in physics "for elucidating the quantum structure of electroweak interactions in physics."

See also ACOUSTICS; ATOMIC STRUCTURE; BOHR, NIELS; BOSE-EINSTEIN STATISTICS; BROGLIE, LOUIS DE; CLASSICAL PHYSICS; DUALITY OF NATURE; EINSTEIN, ALBERT; ELECTRICITY; ELECTROMAGNETIC WAVES; ELECTROMAGNETISM; ENERGY AND WORK; FERMI-DIRAC STATISTICS; FEYNMAN, RICHARD; GENERAL RELATIVITY; GRAVITY; HARMONIC MOTION; HEISENBERG, WERNER; MASS; MATTER AND ANTIMATTER; MOMENTUM AND COLLISIONS; PARTICLE PHYSICS; PAULI, WOLFGANG; PHOTOELECTRIC EFFECT; PLANCK, MAX; RADIOACTIVITY; ROTATIONAL MOTION; RUTHERFORD, SIR ERNEST; SCHRÖDINGER, ERWIN; SPECIAL RELATIVITY; SPEED AND VELOCITY; WAVES.

FURTHER READING

Rae, Alastair. *Quantum Physics: A Beginner's Guide.* Oxford: Oneworld, 2006.

———. *Quantum Physics: Illusion or Reality?* 2nd ed. Cambridge: Cambridge University Press, 2004.

Serway, Raymond A., Clement J. Moses, and Curt A. Moyer. *Modern Physics,* 3rd ed. Belmont, Calif.: Thomson Brooks/Cole, 2004.

Styer, Daniel F. *The Strange World of Quantum Mechanics.* Cambridge: Cambridge University Press, 2000.

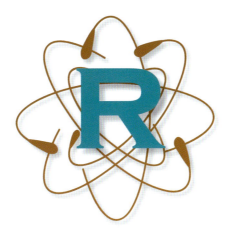

radical reactions A unique type of chemical reaction, a radical reaction, involves atoms and molecules known as radicals or free radicals that contain lone unpaired electrons. The removal of an electron from an atom or molecule can form a free radical, as can the addition of an electron. Radicals are very reactive species that often undergo spontaneous reactions and can be utilized in chain reactions, reactions in which the starting material is constantly regenerated allowing the reaction to continue. Valuable players in chemistry, radicals also have a valuable role in the manufacture of plastics and are relevant to processes such as aging. Understanding the mechanisms of free radical reactions provides useful information regarding their functions.

FORMATION OF RADICAL REACTIONS

Standard chemical reactions involving bond cleavage are known as heterolytic cleavages. In this common form of bond breaking, one of the species participating in the bond assumes both of the electrons from the broken bond. Heterolytic cleavage leads to the formation of ions. This cleavage causes the formation of a positively charged cation from the atom (known as a nucleophile) that gave up the electron and a negatively charged anion from the atom (known as an electrophile) that received the extra electron. Curved arrows represent the direction of the movement of the electrons from the bond to the atom that gains the electrons.

The most common method for creating free radicals is through a related process called homolytic cleavage. In homolytic cleavage, a bond that involves two shared electrons breaks, with each of the participating atoms receiving one of the shared electrons. This creates two atoms that each have a lone unpaired electron. Fishhook arrows (arrows with only half a head) are used to represent the direction of a single electron that is moving. This process only occurs in nonpolar compounds because both of the atoms involved in the bonds have to have an equal claim to the electrons, meaning both of the species in the bond must have similar electronegativities (the tendency for an atom to attract the electrons in a bond). This homolytic cleavage results in the formation of free radicals, which are written using a dot beside the symbol of the atom. The notations for the free radical of chlorine and a methyl free radical are shown.

chlorine free radical: $Cl\bullet$

methyl free radical: $CH_3\bullet$

In the free radical of the methyl group, the dot representing the unshared electron truly belongs to the carbon atom of the methyl group. Generally, the dot is written after the radical species, as shown.

DISCOVERY OF FREE RADICALS

The discovery of free radicals occurred in carbon compounds. This discovery was met with incredulous skepticism. Most scientists believed that the creation of unpaired electrons in organic compounds was not possible. The work of a Russian immigrant, Moses Gomberg, changed this way of thinking. Gomberg was born in Russia in 1866, and after harsh treatment his family immigrated to the United States in 1884. Through his own hard work, in 1886, Gomberg entered the University of Michigan, where he completed his bachelor's, master's, and a doctor of science degree and where he eventually became a professor. Near the turn of the century, Gomberg was studying the reaction of triphenylmethyl halides with metals, research that led to his discovery of free radical carbon compounds. Gomberg reported his findings on the formation of the triphenylmethyl radical ($Ph_3C\bullet$) in the *Journal of the American Chemical Society* in 1900. The American Chemical Society honored Gomberg's discovery of organic free radicals on the 100th anniversary of the event, on June 25, 2000, with the National Historical Chemical Landmark in a ceremony at the University of Michigan. Several decades after the discovery of free radicals passed before chemists fully embraced their existence. They are now widely used in the manufacture of plastics and rubber and are critical to many other biochemical, medical, and agricultural processes.

EXAMPLES OF FREE RADICALS

Several examples of common free radicals are discussed in the following. They include chlorine, reactive oxygen species (ROS) that include oxygen and peroxide, and carbon compounds (discussed later).

Chlorine is representative of atoms that create free radicals when heated or subjected to strong light. The chlorine molecule (Cl_2) contains two chlorine atoms. Upon treatment with ultraviolet radiation, the chlorine molecule breaks down through homolytic cleavage to create two chlorine free radicals, $Cl\bullet$. This radical is involved in chain radical reactions and in the production of plastics. The free radical nature of chlorine compounds also has applications to chlorofluorocarbons and their effect on the ozone layer. When these compounds enter the stratosphere and are subjected to strong ultraviolet (UV) radiation, they lead to the destruction of the ozone layer.

Reactive oxygen species (ROS) include such free radicals as oxygen and peroxide. These substances can be damaging to biological systems by damaging DNA and other components of cells such as cellular membranes. The ability of oxygen to form free radicals stems from oxygen's two unshared electrons in its valence shell. When electrons are added to oxygen, several different ROS can form, including a superoxide anion, peroxide, and a hydroxyl radical. A superoxide anion contains one additional electron and is represented by the symbol $O_2^-\bullet$. A peroxide species contains two additional electrons and is represented by the symbol $O_3^-\bullet$. The hydroxyl radical forms from an oxygen covalently bonded to a hydrogen atom. It contains one additional electron and is represented by the symbol $\bullet OH$. A variety of biological processes and external stimuli such as ultraviolet radiation can induce the formation of reactive oxygen species.

CHAIN REACTIONS

Chain reactions are divided into three separate phases, known as initiation, propagation, and termination. Two types of these chain reactions are the reaction of methane with bromine and of methane with chlorine in the presence of UV light. This gentle treatment of these two compounds still manages to create dramatic effects. The mechanisms of both of these reactions are identical and are known as photochemical reactions. The process of adding a halogen (such as chlorine or bromine) to an organic compound is known as halogenation. The products of these types of reactions are haloalkanes.

The first stage of the reaction, the initiation, requires the treatment of Br_2 with UV light according to the following reaction, which creates two free radicals:

$$Br_2 \rightarrow 2Br\bullet$$

The ultraviolet light source is no longer required after the initiation step. Each of these bromine free radicals has the ability to continue the reaction by reacting with methane during the following chain propagation steps:

$$2Br\bullet + CH_4 \rightarrow CH_3\bullet + HBr$$
$$CH_3\bullet + Br_2 \rightarrow Br\bullet + CH_3Br$$

The first propagation step utilizes the bromine free radical to react with methane to create a methyl free radical and hydrogen bromide. The methyl free radical can then react with additional bromine to recreate the bromine free radical and methylbromide (a haloalkane). This bromine free radical can then also react with methane to continue the chain.

The reaction will continue until the chain termination steps occur. In order to terminate the chain reaction, two radicals must react to form an atom or molecule. This can occur by having two bromine free radicals, two methyl free radicals, or one bromine free radical with one methyl free radical.

$$2Br\bullet \rightarrow Br_2$$
$$Br\bullet + CH_3\bullet \rightarrow CH_3Br$$
$$CH_3\bullet + CH_3\bullet \rightarrow CH_3CH_3$$

In each of these cases, the unshared electron of each of the radicals forms one-half of the bond between the two radicals, and the reaction is terminated.

BOND DISSOCIATION ENERGIES

The energy involved in bond forming and breaking can be calculated and determined by the bond dissociation energy, a measure of the amount of energy that is stored in the covalent linkages and that is released upon breakage. When a nonpolar covalent compound goes through homolytic cleavage, there are no ions created from the breakage. Therefore, the two atoms are not attracted to each other, and the bond dissociation energy directly represents the amount of energy necessary to pull apart the two atoms by breaking the bond. When comparing the bond dissociation energies of bonds, trends start to appear.

ORGANIC FREE RADICALS

Organic free radicals are carbon-containing compounds. Carbon has four valence electrons and is capable of forming four covalent bonds with other atoms. When a free radical forms with a carbon atom, there are one unshared electron in its valence shell and three other covalent bonds. According to the American Chemical Society, organic free radicals, once formed, react with other species in one of several ways.

- A free radical can donate its unshared electron to another molecule.
- A free radical can steal an electron from another molecule in order to make itself more stable.
- The free radical can also remove a whole group of atoms from the molecule it interacts with.
- A free radical can add itself to the new molecule, creating a new compound.

Organic compounds differ in the stability of their free radicals. The less stable the free radical, the shorter time that it exists or the less likely it is to form in the first place. When comparing the bond dissociation energies of various carbon compounds, the order of the stability of their free radicals is as follows, from the most stable to the least stable:

- Tertiary (a carbon attached to three other carbons)
- Secondary (central carbon attached to two other carbons)
- Primary (central carbons attached to one other carbon)
- Methyl (contains only one carbon attached to three hydrogens)

APPLICATIONS OF FREE RADICALS

One biological implication of free radicals relates to the aging process. Research shows that the formation of oxygen free radicals causes damage to DNA. The primary impact of oxygen free radicals occurs in the cellular organelle called the mitochondria. The mitochondria are the site of cellular respiration. The electron transport chain passes electrons from the electron carriers (NADH and $FADH_2$) formed during glycolysis and the citric acid cycle to the ultimate electron acceptor molecular oxygen (O_2). When the electron transport chain runs, there is the possibility for the electrons to be added inappropriately to water, creating oxygen free radicals $O_2\bullet$. These radicals can detrimentally affect the mitochondrial DNA and damage the organelle. As this mitochondrial damage builds up, aging occurs, and the cell eventually dies. The addition of antioxidants can help to remove these oxygen free radicals and prevent the DNA damage. Vitamins such as vitamin A, C, and E have been shown to eliminate free radicals. Whether these antioxidants reach the mitochondria to reduce the oxygen free radicals found there is not clear.

Ionizing radiation, such as from X-rays, can cause the removal of electrons from an atom or molecule, producing a radical. For example, if water is irradiated, the products formed are shown as follows:

$$H_2O \rightarrow H_2O^+ + e^-$$

This charged water species forms by the removal of an electron. This

Bond	Energy (kJ/mol)	Bond	Energy (kJ/mol)
H – H	436	N – N	160
C – H	413	N = O	631
N – H	393	N ≡ N	941
P – H	297	N – O	201
C – C	347	N – P	297
C – O	358	O – H	464
C – N	305	O – S	265
C – Cl	397	O – Cl	269
C = C	607	O – O	204
C = O	805	C – F	552
O = O	498	C – S	259

© Infobase Publishing

Radicals are atoms with unpaired electrons. When the bonds between the atoms are broken, the electrons are distributed evenly between the two elements. The amount of energy given off by breaking bonds is known as bond dissociation energy and is given in this table for some common bonds, in units of kilojoules per mole (kJ/mol).

positively charged H_2O molecule then reacts with another water molecule to produce the following:

$$H_2O^+ + H_2O \rightarrow H_3O^+ + OH\bullet$$

Organic free radicals are responsible for the production of polymers, including polyethylene and other plastics. A polymer is a high-molecular-weight substance made up of a chain of smaller monomers in a repeating structure that produces a tough and durable product. Examples of commercial polymers include such common substances as polyethylene, polyethylene terephthalate, polypropylene, polystyrene, polyvinyl chloride, polyurethane, and nylon. Each of these polymers has been developed by organic chemists and polymer scientists to fill the need for a commercial product.

Polyethylene is a thermoplastic and is an example of a class of polymers known as homopolymers. They are made up of identical monomers in repeating units. The ethylene units of polyethylene are connected by a radical addition reaction. Polyethylene formation occurs by a free radical reaction on the double bond of the ethylene molecule, causing the previously bonded electrons to form new bonds with new ethylene molecules. Billions of pounds of polyethylene are produced every year. Types of polyethylene can be used in packaging and in plastic soda, water, and juice bottles.

See also CHEMICAL REACTIONS; CITRIC ACID CYCLE; ELECTRON TRANSPORT SYSTEM; ORGANIC CHEMISTRY; POLYMERS.

FURTHER READING

Baskin, Steven, and Harry Salem. *Oxidants, Antioxidants and Free Radicals.* Washington, D.C.: Taylor & Francis, 1997.
Parsons, A. F. *An Introduction to Free Radical Chemistry.* Boston: Blackwell Science, 2000.

radioactivity Radioactivity is the spontaneous decay of unstable nuclei with the concomitant release of energy and particles. The nucleus of an atom is made up of positively charged protons and electrically neutral neutrons that together are known as nucleons. The ratio of protons to neutrons determines whether or not a particular version of a nucleus (nuclide) will be stable or not. Nuclear stability requires a balance in two forces—the attractive nuclear force holding the nucleus together and the repulsive electrostatic force of protons pushing the nucleus apart. When these two forces are unbalanced, the nucleus will decay. A plot of the number of neutrons in a nuclide (a nucleus with a specific number of protons and neutrons) versus the number of protons reveals its stability. Nuclides that fall within the zone of stability on the graph are stable. If a nuclide has too many protons, it will fall below the zone of stability and will need to release a positively charged particle to become stable. If the nuclide has too many neutrons, it will undergo beta decay (release an electron or a positron) to become stable.

All nuclides with atomic numbers greater than 84 are radioactive and will spontaneously decay. Nuclides with even numbers of protons and neutrons have the most stable nuclei. When nuclei are small, a 1:1 ratio of protons to neutrons will confer stability, but as the nuclei become larger, more neutrons are needed to balance the repulsive effects of the protons. Certain numbers of protons make the nucleus especially stable: 2, 8, 20, 28, 50, 82, and 126.

Radioactivity is a random process. There is no way to predict when any particular unstable nucleus will decay, but it is possible to describe radioactivity in terms of the rate of decay, that is, the fraction of a sample of a radioactive substance that will decay in any time interval. In fact, for the same substance, the same fraction will decay in the same interval. The most common way of characterizing the rate of decay is in terms of half-life, the time required for one half of the sample to decay. So, say that some radioactive isotope has a half-life of two years. At the end of two years, half of the sample will have decayed. Wait another two years, and half of the remaining undecayed sample will have decayed, leaving one-quarter of the original sample undecayed. And so on.

In 1896, a French physicist named Henri Becquerel first discovered spontaneous radioactivity, an achievement for which he received the Nobel Prize in physics in 1903. Becquerel learned that uranium gave off energy that could expose photographic film. Initially, Becquerel believed that the uranium was absorbing the energy from the sunlight and then reemitting the energy in another form. However, Becquerel placed uranium in a dark drawer beside unexposed film and was surprised to find that the film had been exposed, demonstrating that the uranium itself was the source of the energy rays.

Much of what is now known about radioactivity can be credited to the hard work of two married physicists, Marie and Pierre Curie, who dedicated their lives to the study of these energetic emissions. The Curies investigated the radioactive properties of uranium by studying pitchblende, the unrefined ore from which uranium was obtained. The Curies observed that the pitchblende yielded hundreds of times more radioactivity than pure uranium. This led to their refinement of the pitchblende and the discovery of two new radioactive elements, which they named polonium (Po) and radium (Ra). Marie Curie determined that the decay of a radioactive element could be measured and predicted, setting the stage for the determination of half-lives in later studies. The Curies together received the Nobel Prize in physics in 1903, shared with Becquerel, for their work with radioactivity, and Marie received a second Nobel Prize in chemistry in 1911 for the discovery of radium and polonium. Pierre was killed in an accident in the early 1900s. Their daughter, Irène, followed in her parents' footsteps and continued studies into radioactivity and atomic structure. She earned a Nobel Prize in chemistry, shared with her husband, Frédéric Joliot-Curie, in 1935 for their work on synthesizing new radioactive elements.

Three basic types of radioactive emissions are distinguished by the direction and magnitude of their deflection observed in a magnetic field: alpha particles, beta particles, and gamma rays. Alpha particles show a slight deflection in a magnetic field in one direction, beta particles show considerable deflection in the opposite direction, while gamma rays show no deflection. Each type of radiation has its own characteristic energy, penetrating power, and properties.

Alpha radiation occurs when an element loses two protons and two neutrons, essentially a helium nucleus. The nucleus that is left behind is lighter, the atomic number has been reduced by 2, and the atomic mass number has been reduced by 4. The resulting nucleus is much more stable than the initial radioactive nucleus. Relative to other forms of radioactivity, alpha particles are large and slow moving. Alpha radiation exerts its effects over a short range and can be blocked with a barrier as simple as a piece of paper. Because alpha radiation interacts strongly with matter, it cannot travel far but can still do serious damage to the cells and tissues with which it has contact. Alpha radiation is most detrimental if it is ingested. Large nuclides commonly release an alpha particle; for example, thorium decays by alpha particle emission to produce radium by the following reaction:

$$^{230}_{90}\text{Th} \rightarrow\, ^{4}_{2}\text{He} + ^{226}_{88}\text{Ra}$$

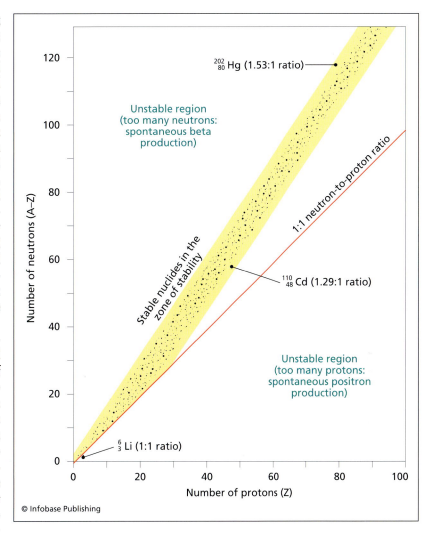

The ratio of protons to neutrons in a nuclide determines its stability or type of radioactive decay. The band running from corner to corner represents the "zone of nuclide stability" in this plot of neutron number versus proton number. Nuclides falling within this zone are stable, while those possessing too many neutrons or protons undergo radioactive decay.

Beta radiation is divided into two categories known as beta positive radiation (β^+), or positron production, and beta negative radiation (β^-), or just beta particle production. Beta radiation results in the release of one of two sets of particles, a positron with a neutrino (β^+ radiation) or an electron with an antineutrino (β^- radiation). A positron is a positively charged particle with the mass of an electron. Neutrinos and antineutrinos are fundamental particles that were first proposed by Wolfgang Pauli in 1931 in an attempt to explain why nuclear reactions did not seem to follow the law of conservation of energy. The amount of energy released was not high enough, and Pauli suggested that additional energy was leaving in small, undetectable, massless, neutral particles. β^- radiation results when a nucleus that contains too many neutrons converts a neutron into

RADIATION THERAPY: DEVELOPMENTS AND DISCOVERIES

by Lisa Quinn Gothard

Radiation therapy has been used for the treatment of cancer since 1896, when a patient with an advanced stage of breast cancer was treated in Chicago. The patient's chest wall was exposed to radiation, and the tumor size was dramatically reduced. In radiation therapy, radiation damages the deoxyribonucleic acid (DNA) of rapidly dividing tumor cells, causing them to die. In addition to destroying the cancer cells, the radiation prevents the tumor from developing the vascular tissues necessary to support tumor growth. The major problem with this type of treatment is that even when it is targeted to a specific tissue, normal healthy cells can also be killed. Optimization of this type of treatment is critical to the effective treatment of cancer, and advances in radiation therapy have increased its success rate.

Radiation therapy utilized for cancer treatments includes external radiation therapy, internal radiation therapy, and systemic radiation therapy. The most common type of radiation therapy is external radiation therapy, which utilizes X-rays or gamma rays from an exter-

nal source. The patient can receive this type of radiation therapy on an outpatient basis. External radiation therapy is used for all types of cancer treatments.

Different types of radiation sources are used as an energy source for external radiation therapy, including X-rays, gamma rays, and particle beam radiation. X-rays and gamma rays are both forms of electromagnetic radiation. The wavelength of X-rays is shorter than that of ultraviolet light, and gamma rays have even shorter wavelengths. The short wavelengths of X-rays translate into X-rays' having higher energy than ultraviolet light, and gamma rays have even higher energy than that. The penetrating power of these two types of radiation allows them to reach any part of the body from the outside. Another energy source for external radiation does not utilize electromagnetic radiation but instead uses a particle beam made of subatomic particles. This type of radiation therapy does not have the penetration power of X-rays or gamma rays. Because of this, it is most effective when used on cancer cells that are close to the surface.

Internal radiation therapy, also known as brachytherapy, involves the implanta-

tion of a source of radiation inside the body, either inside or very near the tumor. Internal radiation therapy generally requires a hospital stay, as the patient needs to be isolated from the public, as his or her body contains a radioactive source. Nurses and care providers need to take precautions, such as shielding themselves and disposing of biohazardous wastes properly.

Systemic radiation treatment involves ingestion of the radiation source (usually iodine 125 or iodine 131). This type of radiation treatment is often used for the treatment of thyroid cancers. The patient exudes radiation after this treatment and therefore requires special care during the course of treatment.

As advances in targeting radiation therapy have come about, other research has focused on methods to prevent damage to healthy cells during radiation treatment. Molecular biologists and chemists have worked together to determine ways to prevent the damage to healthy tissues while still killing the cancer cells. Two experimental systems have been shown to protect mice and potentially humans from the damaging effects of radiation. These new developments include the dis-

a proton and releases an electron and an antineutrino in an attempt to stabilize. The atomic number of the isotope is increased in this type of radiation. β^+ radiation (positron production) results when a nucleus that contains too many protons stabilizes by converting a proton to a neutron, thus decreasing the atomic number of the isotope. During creation of the neutron, the nucleus releases a positron and a neutrino. Unlike alpha radiation, which has fixed values for the energy given off, beta radiation exhibits a wide range of released energy quantities. Beta radiation has more penetrating power than alpha radiation but much less than gamma rays. A layer of aluminum can block beta radiation, which also does the most damage if ingested.

Gamma radiation, or gamma rays (γ), usually accompanies alpha and beta radiation. A form of electromagnetic radiation, gamma radiation does not

change the charge, mass number, or atomic number of an isotope. A nucleus often remains unstable after releasing alpha or beta radiation, and the accompanying loss of energy in the form of gamma rays enables the nucleus to restabilize. Gamma rays travel the farthest and are the most penetrating form of all radiation. Only a layer of lead can sufficiently block gamma rays. The production of gamma rays as a result of alpha radiation from uranium is shown in the following:

$$^{238}_{92}\text{U} \rightarrow {}^{4}_{2}\text{He} + {}^{234}_{90}\text{Th} + 2\gamma$$

Another type of radiation, known as electron capture, occurs when a nucleus is unstable because it has too many protons, so it captures one of its inner energy level electrons and uses it to turn a proton into a neutron. The release of gamma rays accompa-

covery of a protein known as flagellin that forms the filament in bacterial flagella. When administered prior to radiation treatment, flagellin protects healthy cells from death by apoptosis. A second development is using gene therapy to administer a transgene whose product is able to protect the cells from the damaging effects of radiation.

The American cancer researcher Andrei Gudkov demonstrated that mice that had been injected with a synthetic form of the protein flagellin survived potentially lethal doses of radiation. All of the mice that were given flagellin resisted the damage caused by the radiation, while all of the control mice were killed by the radiation dose. Gudkov's group followed up this work by synthesizing a drug, CBLB502 (Protectan), to mimic the function of flagellin, and they determined that it offered protection against radiation damage. The drug works by preventing the process of apoptosis (programmed cell death) in those cells that are not meant to be killed by the radiation. Amazingly, all tests performed by scientists on cancer cell lines and tumors in mice have shown that the cancerous cells are not protected by the flagellin-based drug. The potential of this medication to be used as protection for those undergoing radiation therapies for cancer treatment is tremendous. Admin-

istration of a dose of Protectan prior to radiation treatment could prevent the damage of healthy cells in the patient's body. In addition, the U.S. Department of Defense has contracted with Cleveland BioLabs to research this medication for first responders including paramedics and military personnel in the event of a nuclear disaster. Studies need to be completed to determine whether Protectan is safe for human use.

Another approach to radiation protection involves gene therapy, the insertion of a copy of a gene that will lead to the production of a useful gene product. The American scientist Joel Greenberger demonstrated that in mice gene therapy conferred resistance to the radical damage caused by radiation treatment. The transgene, manganese superoxide dismutase plasmid liposome (MnSOD-PL), causes the cell to produce MnSOD. This gene product helps protect the cells from free radicals formed by the ionizing radiation treatment. Scientists found that the body metabolizes and removes the transgene within 72 hours, and therefore the production of the protective enzyme is only for the short term. The therapy would need to be repeated for longer-term radiation exposures.

The ability to prevent damage to healthy cells and tissues during radia-

tion therapy will lead to more successful outcomes of treatment for cancers. Safety will be improved, and negative side effects such as hair loss and nausea may be reduced or eliminated. Another immediate implication of this type of therapy would be as a prophylactic measure in the event of a nuclear war or a meltdown of a nuclear power plant. Allowing access to these potential medications by first responders as well as those in direct line of exposure could prevent the damage normally caused by radiation exposure. New advances in radiation therapy have the potential to create a safer method of cancer treatment with the added benefit of being able to protect humans in time of emergency nuclear exposure.

FURTHER READING

Department of Health and Human Services, National Institutes of Health. U.S. National Library of Medicine: Medline Plus. "Radiation Therapy." Available online. URL: http://www.nlm.nih.gov/medlineplus/radiationtherapy.html. Last updated April 15, 2008.

National Cancer Institute. U.S. National Institutes of Health. "Radiation Therapy for Cancer: Questions and Answers." Available online. URL: http://www.cancer.gov/cancertopics/factsheet/therapy/radiation. Accessed April 20, 2008.

nies the conversion of the proton to the neutron. The atomic mass number of the nuclide does not change, since the total number of nucleons remains the same, but the atomic number of the nuclide decreases by 1 as a result of the loss of a proton. The following reaction demonstrates the process of electron capture:

$$^{73}_{33}\text{As} + ^{0}_{-1}\text{e} \rightarrow ^{73}_{32}\text{Ge} + \gamma$$

As mentioned, the length of time required for half of a quantity of a particular radioactive isotope to decay is known as its half-life ($t_{1/2}$). The values for half-life vary from seconds to thousands of years, depending on the stability of the radioactive nuclide. If a given sample contains 100 grams of a nuclide, at the end of one half-life, the sample will contain 50 grams. After two half-lives, 25 grams

will remain, and after three half-lives, 12.5 grams will remain, and so on. Every half-life decreases the amount of nuclides by a factor of 2 from the starting value (N_0). The formula for calculating half-life is

$$t_{1/2} = 0.693/k$$

where k is the rate constant, expressing the fraction that decays in a unit of time, for decay of the specific nuclide. For example, knowing that protactinium 234 has a half-life of 6.75 hours, one can determine the rate constant using the equation for half-life.

$$6.75 \text{ hours} = 0.693/k$$
$$k = 0.693/(6.75 \text{ hr})$$
$$k = 0.103/\text{hr}$$

Alternatively, if the rate constant for protactinium 234 is known to be 0.103 per hour, one can calculate the half-life using the same formula.

$$t_{1/2} = 0.693/k$$

$$t_{1/2} = 0.693/(0.103/hr)$$

$$t_{1/2} = 6.75 \text{ hr}$$

Radioactive phenomena have many useful applications: devices such as smoke detectors, luggage inspection techniques, a broad range of medical procedures and treatments, energy production, archaeological methods for determining the age of remains, and weaponry. In the medical field, radioactivity is used to treat cancers, and iodine 131 is used to detect and treat thyroid disorders. Many nuclides such as thallium 201 and phosphorus 32 can be used as tracers to visualize internal structures better in order to diagnose and treat human diseases. Some power plants use the radioactive isotope uranium 235 as a fuel in nuclear fission reactions to generate electricity. Plutonium 238 is used as a source of power for spacecraft. The use of radioactive dating methods help scientists discover the age of items recovered in archaeological digs. Nuclear weapons also depend on the force and power provided by radioactivity.

Negative effects of radioactivity have been well documented. Radiation effects on the human body can cause sickness and even death. Radiation can also cause genetic defects that can be passed to future generations. The extent of the damage differs, depending on the energy of the radiation, the type of radiation, the length of exposure time, and the distance from the radioactive source during exposure.

Radioactivity is measured in several units, including the roentgen (R). One roentgen is equal to 2.58×10^{-4} curie/kilogram, where a curie (Ci) is equivalent to 3.7×10^{10} decays per second. The energy of the radiation is measured in radiation absorbed dose (rad). One rad is equal to 0.01 joule per kilogram of tissue.

When measuring the biological damage created by radioactivity, one calculates the roentgen equivalent for man (rem), calculated by the following formula:

$$1 \text{ rem} = \text{rads} \times \text{RBE}$$

where RBE is a factor known as the relative biological effectiveness for that type of radiation.

The disposal of radioactive waste is a serious growing environmental problem. The radioactive substances, if not disposed of properly, could contaminate water supplies and air, causing sickness in those people who live nearby. Radioactivity with a long half-life will affect the environment for many years after its disposal.

See also ATOMIC STRUCTURE; CURIE, MARIE; ENERGY AND WORK; FISSION; ISOTOPES; MATTER AND ANTIMATTER; NUCLEAR CHEMISTRY; NUCLEAR PHYSICS.

FURTHER READING
Hobson, Art. *Physics: Concepts and Connections*, 4th ed. New York: Prentice Hall, 2007.

Ramsay, Sir William (1852–1916) English Chemist

Sir William Ramsay was an English chemist who is best known for his discovery of the family of elements known as the noble gases. Ramsay used apparent discrepancies in the masses of nitrogen in the atmosphere to determine that there must be an unreactive element that existed in the air. Using this line of research, Ramsay discovered argon in 1894. He also discovered helium in 1895, and neon, krypton, and xenon in 1898. Ramsay received the Nobel Prize in chemistry in 1904 "in recognition of his services in the discovery of the inert gaseous elements in air, and his determination of their place in the periodic system."

EARLY YEARS AND EDUCATION

William Ramsay was born on October 2, 1852, in Glasgow, Scotland, to William and Catherine Ramsay. He inherited his love of science and chemistry from his family. Both of his grandfathers had a scientific background. His mother's father was a physician, and his paternal grandfather was a chemical manufacturer.

Ramsay's progress in school was accelerated. When he was only 14 years old, he entered the University of Glasgow, where he demonstrated a significant interest in science. In 1870, he enrolled at the University at Tübingen, and he received a Ph.D. for his work in organic chemistry in 1872. After completion of his doctorate, Ramsay accepted his first teaching position, at Anderson College in Glasgow. In 1880, he became a professor of chemistry at University College in Bristol. After seven years in Bristol, in 1887 Ramsay took a position as chair of the chemistry department at the University College in London (UCL), where he remained until he retired in 1913.

DISCOVERY OF NOBLE GASES

Ramsay began his work at UCL by studying a phenomenon earlier reported by the English physicist Lord Rayleigh (John William Strutt). Rayleigh later received the Nobel Prize in physics in 1904 for his studies of the differences between atmospheric nitrogen and experimentally derived nitrogen. There was a significant difference in mass between these two forms of nitrogen. Ramsay took up this line of experimentation and sought to determine the difference between these forms of nitrogen. He was able to demonstrate that the atmospheric nitrogen Rayleigh studied was truly a mixture of gases including N_2 and an as of yet undiscovered element, which Ramsay named *argon*. Ramsay and Rayleigh presented their findings in a paper to the Royal Society on January 31, 1895.

Ramsay followed up his research on argon by pursuing a question proposed by the work of the American physicist William Hillebrand. When certain rocks were heated in the presence of sulfuric acid, they emitted an unknown gas. Ramsay endeavored to identify this gas. He studied the spectral lines given off by heating the gas and found that it did not correspond to any known element on Earth. He then recognized a connection between the unknown spectral lines seen when observing the Sun and the gas found by Hillebrand. The gas that gave off these spectral lines had previously been named *helium*. Ramsay was the first to demonstrate that helium existed naturally on Earth.

Ramsay's discoveries of the other noble gases depended significantly on his belief in periodic law. The regular repeating patterns of properties seen across rows (periods) of the periodic table convinced Ramsay that the 18th group of the periodic table must contain additional unknown elements that had chemical properties similar to those of helium and argon. His understanding of these repeating periodic properties and the octet rule led Ramsay to discover neon, krypton, and xenon by 1898.

THE NOBLE GASES

Ramsay's experimental work eventually led to the discovery of all of the noble gases except radon. The noble gases fall into what is now the final group of representative elements (Group VIIIA, or IUPAC Group 18), known as the noble gases. This group contains helium, neon, argon, krypton, xenon, and radon. The noble gases have a filled outer energy level with two electrons in the *s* sublevel and six in the *p* sublevel. Since they have a full outer shell, these elements do not gain or lose electrons and are therefore considered unreactive.

The noble gases have the highest ionization energies of all of the elements in the periodic table, with helium having the highest first ionization energy. The ionization energy decreases as one proceeds down a group of the periodic table. The farther down the table, the more energy levels shielding the nucleus from the valence electrons, and the harder it is for the protons to hold on to the valence electrons. This

Sir William Ramsay was an English chemist best known for his discovery of the nonreactive family of the periodic table known as the noble gases. *(Visual Arts Library [London]/Alamy)*

helps explain why it was later found that the noble gases of higher mass can undergo reactions under specific reaction conditions as discussed later.

After hydrogen, helium is the second most abundant element in the universe. Helium is a colorless, odorless, tasteless gas having an atomic number of 2, meaning that it has two protons in its nucleus and two electrons in its stable atomic form. Two isotopes of helium exist, helium 3 and helium 4; helium 3 contains one neutron while helium 4 contains two. Both isotopes of helium are stable. Because the two electrons found in helium fill the first energy level completely, helium is unreactive. Since helium is unreactive, it will not combust and is not flammable. This makes helium a fine choice for applications such as helium balloons. Helium is found in natural gas deposits, which consist of up to 7 percent helium. Helium is constantly lost to space, but since radioactive alpha particles are essentially helium atoms, it can be replenished by alpha-radioactive decay. Helium has the lowest melting point and boiling point of any element and can be used as a coolant for nuclear power plants.

Neon is a colorless gas with an atomic number of 10. Neon has three stable isotopes, neon 20, neon 21, and neon 22. Neon 20, the most prevalent one, contains 10 neutrons, neon 21 has 11, and neon 22 has 12 neutrons. Two electrons fill the first energy level of neon, and eight fill the second energy level, giving neon its unreactive electron configuration. Neon is a noble gas that does not form any stable compounds, and it glows with a red-orange color in a discharge tube. A natural component of the atmosphere, neon can be isolated by distillation. The main uses for neon include the manufacture of neon signs, and gas lasers and as a cryogenic refrigerant.

Argon was named by Ramsay in 1898 for the Greek word "inactive." Argon is an odorless gas with an atomic number of 18 and a mass number of its most common isotope of 40. Argon has six known isotopes, from argon 36 to argon 41. Three of these isotopes are stable: argon 36, argon 38, and argon 40. Argon 36 has 18 neutrons, argon 38 has 20 neutrons, and argon 40 has 22 neutrons. The electron configuration of argons's 18 electrons fills the first energy level with two electrons, and both the second and third energy levels with eight electrons each. Argon is found in electric lightbulbs, as it does not react with the filament but is able to conduct heat away from the filament, extending the life of the bulb. Argon can be used to prevent oxidation of reactions with metals. Since argon is unreactive with oxygen, it forms a layer of insulation around the metal so oxygen cannot enter. Argon is found in the atmosphere and is isolated from air by distillation.

Krypton was named by Ramsay in 1898 for the Greek word *kryptos,* meaning "hidden." Krypton is a gas with an atomic number of 36 and a mass number of 84 for its most common isotope. Krypton has 15 known isotopes, six of these stable isotopes: krypton 78, krypton 80, krypton 82, krypton 83, krypton 84, and krypton 86. These isotopes contain 42, 44, 46, 47, 48, and 50 neutrons, respectively. Krypton's electron configuration has two electrons in the first energy level, eight electrons in the second, 18 electrons in the third energy level, and a full fourth energy level containing eight electrons. This complete octet in the outer energy level of krypton gives it the unreactive nature characteristic of noble gases. Krypton is found in the Earth's atmosphere and is removed from air by distillation. The primary uses of krypton are in lighting.

Ramsay discovered another element in 1898 and named it *xenon,* meaning "stranger." The atomic number of xenon is 54, and the mass number of its most common isotope is 131. Xenon has 21 known isotopes; nine of these are stable isotopes: xenon 124, xenon 126, xenon 128, xenon 129, xenon 130, xenon 131, xenon 132, xenon 134, and xenon 136. These isotopes have 70, 72, 74, 75, 76, 77, 78, 80, and 82 neutrons, respectively. The electron configuration of xenon has two electrons in the first energy level, eight electrons in the second, 18 electrons in the third, 18 electrons in the fourth energy level, and eight electrons completely in its valence shell in the fifth energy level. Xenon's chemical unreactivity stems from the complete octet in its valence shell. Xenon is present in the Earth's atmosphere and can be removed from the air by distillation. The size of xenon makes it more likely than the smaller noble gases to undergo reactions.

In contrast to the other noble gases, radon was discovered in 1898 by Friedrich Ernst Dorn. Radon is different from the other noble gases in that it is not normally present in the Earth's atmosphere. Radon is an odorless, tasteless, radioactive noble gas with an atomic number of 86. The mass number of its most common isotope is approximately 222. Radon is a short-lived radioisotope that has seven known isotopes that range in half-life from 0.6 millisecond to 3.82 days. None of the isotopes of radon is stable.

As radium that is naturally present in rocks decays and releases radon, the gas can seep into homes through basements and foundations. Radon exposure is the number one cause of lung cancer in nonsmokers and is second only to smoking as a cause of lung cancer in smokers. The Environmental Protection Agency (EPA) estimates that radon exposure causes 14,000–20,000 deaths in the United States per year, primarily from lung cancer. Specific testing for radon is the only way to know whether this gas contaminates a structure. Radon levels above four picoCuries per liter (pCi/L) require mitigation. Simple measures such as sealing cracks in the floor and walls and ventilating basements can help reduce radon levels. Venting of the gases from below the foundation is known as subslab depressurization. This method removes the radon gas before it can enter the home. Only qualified radon contractors should perform these repairs.

COMPOUNDS FORMED BY NOBLE GASES

Until the 1960s, it was thought that noble gases were completely inert. Although the gases are unreactive, it is possible for them to form compounds under very specific reaction conditions. The larger the noble gas, xenon and radon, the lower its ionization energy, so the more easily it gives up its valence electrons. Xenon and radon are not very reactive; however, compounds of these elements have been formed, especially with fluorine and oxygen, two reactive elements.

CONTRIBUTIONS OF RAMSAY

William Ramsay was a chemist who began his career studying organic compounds but who is best known for his contributions to inorganic chemistry with the discovery of an entire family of elements, the group of the periodic table known as the noble gases. The unreactivity of the noble gases made them difficult to find initially. After the discovery of argon, Ramsay was able to use the periodic trends set up by Mendeleyev to begin looking for the other elements in the group. Ramsay received many awards and honors throughout his life, including knighthood in 1902 and the Nobel Prize in chemistry in 1904. William Ramsay died at High Wycombe, Buckinghamshire, on July 23, 1916, leaving behind his wife, Margaret, and one son and one daughter.

See also ELECTRON CONFIGURATIONS; PERIODIC TABLE OF THE ELEMENTS; PERIODIC TRENDS.

FURTHER READING

The Nobel Foundation. "The Nobel Prize in Chemistry 1904." Available online. URL: http://nobelprize. org/nobel_prizes/chemistry/laureates/1904/. Accessed April 12, 2008.
"Sir William Ramsay: His Life and Works at UCL." Available online. URL: http://www.ucl.ac.uk/museumstudies/websites06/das/index.htm. Accessed April 12, 2008.

rate laws/reaction rates   The speed at which a chemical reaction takes place is known as the reaction rate. Many chemical reactions take place on the order of pico- (10^{-12}) or femto- (10^{-15}) seconds, while others take place on the order of thousands of years. The study of the speed at which chemical reactions occur is known as chemical kinetics. After determining the type of chemical reaction that will take place and what products will be formed, one must verify the rate at which this reaction will occur. Rate laws are the physical laws that govern the speed at which reactants are converted into products. They can be used to establish the speed of the reaction and, therefore, the amount of product that will accumulate after a given period.

Many factors can affect the rate of a reaction: concentrations of the reactants, temperature of the reaction, surface area of the reactants, and whether or not a catalyst is present. Increasing the concentration of reactants in the reaction vessel without changing the volume of the mixture will lead to more collisions between the particles and, therefore, a faster reaction rate. Increasing the pressure on a given concentration of reactant will also increase the kinetic energy, because if the temperature and number of moles remain constant, then the volume will decrease, leading to an increased number of collisions due to the molecules' running into each other more often in the enclosed space. The rate of the reaction increases when the concentration of reactant increases, because of the increased likelihood that two of the reactant particles will collide. A burning fire will glow red when the amount of oxygen supplied to the burning wood increases (by blowing on it or using bellows), causing the combustion reaction to occur more quickly and leading to a roaring fire. The concentration of oxygen present increases the rate of the reaction. The rate law demonstrates how the rate of the reaction is dependent on concentration of reactants. The following reaction

$$A + B \rightarrow C$$

given a rate law of

$$\text{Rate} = k[A][B]$$

where k is a rate constant specific for that reaction, and the concentrations of A and B are for a given time point, demonstrates that as the concentration of either A or B is increased, the rate of the reaction increases proportionally.

Kinetic energy is the energy of motion. The higher the level of kinetic energy the mixture has, the higher the temperature, as temperature is simply the measure of the average kinetic energy of the particles. In order to react, the reactants must first bump into each other or collide. These collisions happen naturally and spontaneously; however, they may not take place in a timely manner. The rates at which these collisions occur depend on the kinetic energy of the reacting molecules. The number of collisions increases as the kinetic energy increases. This is evident in many chemical reactions that take place more quickly at warmer temperatures. For example, many metabolic functions and enzyme catalyzed reactions in the human body take place quickly at 98.6°F (37°C) but slow at lower temperatures. Bacterial metabolism and, therefore, growth also increase at warmer temperatures. This explains why refrigerating dairy items helps delay spoilage, and why the leftovers from a restaurant need to be refrigerated immediately to prevent the bacteria left on them from multiplying and spoiling the food.

The increased temperature of a reaction leads to an increased number of collisions between the particles and helps the reactants overcome the activation energy of the reaction, the energy that must be supplied in order to allow the reaction to proceed. The activation energy is much like the bump at the top of a ski slope. In order to commit to the reaction, even if it is a downhill spontaneous reaction, an initial input of energy is required to get the process going. The higher temperature of the reaction, which is due to increased kinetic energy of the particles, supplies this activation energy and allows the reaction to proceed. Therefore, if those particles have a higher kinetic energy, then they also have a higher number of collisions and, therefore, a faster rate of reaction. Kinetic molecular theory states that the higher the temperature of the particles, the more collisions take place between the particles. The Arrhenius equation also demonstrates the relationship between the rate of the reaction and the temperature.

$$k = Ae^{-E_a / RT}$$

where k is the rate coefficient of the reaction, T is the absolute temperature, A is the frequency factor of the reaction (based on the order of the reaction, the exponent that the concentration is raised to in the rate law), E_a is the activation energy of the reaction, and R is the ideal gas constant, which has units of L·kPa/K·mol. When the activation energy is given in joules per molecules instead of joules per mole, the ideal gas constant is replaced with the Boltzmann constant (1.3807×10^{-23} J/K).

The rate of a chemical reaction depends on the surface area of the reactants in reactions involving solid and liquid reactants. The larger the surface area of the reactants in the reaction, the more quickly the reaction will proceed. The collision of large particles with small surface area allows smaller surface area with which the particles can react, leading to a slower reaction rate. When a chemical reactant is ground into a powder, the rate of reaction will be quicker than if the same chemical were a single solid crystal because there is more surface area to react.

Another factor that affects the rate of a chemical reaction is the presence of a catalyst. A catalyst is a substance that speeds up the rate of reaction without itself being changed by the reaction. A catalyst cannot cause reactions that would not normally occur, and a catalyst cannot change the equilibrium point of a reaction, but it does help the reaction reach equilibrium faster. A catalyst functions mainly by reducing the activation energy of the reaction, allowing the reaction to occur more quickly because more collisions will be likely to achieve this activation energy. Most biological reactions require protein catalysts known as enzymes. An enzyme is specific for the reaction that it catalyzes and often only recognizes one or a handful of target molecules known as substrates. Enzymes have binding sites for the reactants and an active site in which the reaction takes place. The active site can be located in the binding site or in a different location on the enzyme. By placing the reactants in close proximity and in the proper ori-

entation, a catalyst decreases the amount of energy required to initiate progression of the reaction and thus decrease the amount of time required for the reaction to occur. Many industries depend on catalysts to speed reactions used in the manufacture of chemicals or other products.

A reaction rate indicates how quickly a reaction proceeds and can be experimentally determined by measuring either the amount of reactant consumed or the amount of product synthesized over a certain period. Both of these methods are valid for calculating the speed of a reaction, and which method an investigator uses usually depends on the ease of measurement of the reactants and products. If it is simpler and more reliable to measure the disappearance of a reactant, then the rate will be calculated this way; if it is more reliable and simpler to measure the production of a product, then the reaction rate will be measured that way. When calculating the rate of a reaction, one can use the average reaction rate or the instantaneous reaction rate.

The average reaction rate for the reaction of A → B can be calculated according to the following formula:

$$\text{average rate} = \frac{-\Delta \text{moles A}}{\Delta t} \text{ or } \frac{\Delta \text{moles B}}{\Delta t}$$

The change in time (Δt) can be any defined time frame, and the amount of reactant A consumed (ΔmolesA) or product B produced (ΔmolesB) equals the difference in the amount at the beginning and the end of this period. If measuring the amount of ΔA, the value will be negative, since the amount of A decreases as the reaction proceeds. Since the rate is a positive value, the formula requires a negative sign. If measuring ΔB, the value for average rate will be positive, since the amount of product B increases as the reaction proceeds.

To calculate instantaneous reaction rates, the change in the amount of reactant A or product B is calculated at a given time point rather than over a time interval. In order to calculate the change in reactant or product, a graph is made of the reaction progress. The instantaneous rate is calculated using the tangent line that touches the reaction curve at the point of interest. The calculated slope of that tangent line represents the reaction rate at that exact time point.

All reaction rates are simply dependent on the number of collisions that take place in the reaction vessel. Regardless of the frequency of collisions, a catalyst cannot force a reaction to proceed unless it is thermodynamically possible, meaning that the energy of the products would be lower than the energy of the reactants. Even if they have a high number of collisions, if the compounds cannot chemically react with one another, the increased temperature, pressure, or concentration will have no effect. However, if two reactants are reactive toward one another, the probability of a reaction will increase with increased temperature, pressure, surface area, or concentration.

Most reactions can occur in both directions: they are reversible. In a reversible reaction, if reactants A and B combine to form product C, then product C can also break down to form reactants A and B according to the following equation:

$$A + B \rightleftharpoons C$$

Measuring the amount of each of the reactants or product will give information about the speed at which the forward or reverse reactions are taking place. The concept of equilibrium is important to the study of chemical kinetics. Equilibrium occurs when a reversible reaction achieves a balanced state between product formation and reactant consumption. When a reaction occurs and forms products, and then the products can react in reverse to produce reactants once again, there must be a state of equilibrium at which the rates of the two reactions are equal. At equilibrium, it appears that no net reaction is taking place, because the forward reaction occurs at the same rate as the reverse reaction, and there is no net change in the amount of reactants or products. The equilibrium point of a reaction is fixed and cannot be changed without altering the factors in the experiment. In biological systems, the equilibrium is referred to as dynamic equilibrium. This means that the reactions are still taking place: the reactants are being turned into products and the products are being reformed into reactants even though no net changes occur.

The equilibrium point of a reversible chemical reaction has a fixed point unless the reaction is changed in some way. The French chemist Henri-Louis Le Chatelier first described this phenomenon, called Le Chatelier's principle, which states that when a system at equilibrium is put under stress such as a change in pressure, temperature, or concentration of a reactant or product, then the equilibrium point of the system will shift to adjust to the change and reestablish the equilibrium. Increasing the concentration of a reactant will favor the forward reactions, while increased concentrations of the product by buildup of the product will favor the reverse reaction. Often the removal of a product will allow for the reaction to continue in the forward direction, while the buildup of a product will favor the reverse direction.

See also CHEMICAL REACTIONS; CONCENTRATION; EQUILIBRIUM.

FURTHER READING
Connors, Kenneth A. *Chemical Kinetics: The Study of Reaction Rates in Solution.* New York: VCH, 1990.

reactions in aqueous solution   A solution is a homogeneous mixture of two or more types of substances. The substance that does the dissolving is known as the solvent and is present in the solution in the greatest amount. The solute is the substance that is dissolved and is present in the solution in a lesser amount than the solvent. Aqueous solutions utilize water as a solvent and serve special roles in biological and chemical systems. The importance of aqueous solutions necessitates that reactions that take place in this type of solution be given special consideration. Reactions in aqueous solutions are dictated by the properties of the reactants as well as the properties of the water molecules themselves.

Water is a compound made up of one oxygen atom covalently bonded to two hydrogen atoms. The oxygen atom has a higher electronegativity (attraction for an electron that is shared between two bonded atoms) than the hydrogen atoms; the bond formed between them shares the electrons unequally, creating a polar covalent bond. The result of the polar covalent bond is that the oxygen atom of water has a slightly negative charge and the hydrogen atoms have slightly positive charges. This polarity causes the most structurally important feature of water, the formation of the intermolecular attraction known as hydrogen bonds. The attraction of the water molecules to one another confers on water several unique physical properties such as cohesion, the ability to moderate temperature, a denser liquid than solid state, and its universality as a solvent, which is due to hydrogen bonding between polar water molecules.

The physical property known as cohesion, the attraction of like molecules for each other, is responsible for such processes as the transport of water and nutrients from the roots of a tree to the leaves through transpiration. The water molecules from the base of the plant are pulled upward by the evaporation of water molecules from the leaves of the plant. The hydrogen bonds between the polar water molecules attract the water molecules to one another, pulling them up the plant. The physical property of water known as surface tension is also due to the hydrogen bond patterns of water. This skinlike attraction of water molecules to each other on the surface of the liquid allows such physical phenomenon as waters striders' walking on water and rocks' skipping across the surface of water. As long as the downward force exerted by the object is not greater than the surface tension of the liquid, it will stay on the top of the water.

The hydrogen bonding of water is also responsible for the moderation of Earth's temperatures. Water has a specific heat (amount of energy required to raise one gram of the substance one degree Celsius) that is higher than that of other compounds with similar structures. Because of the high amount of hydrogen bonding in water, the substance can absorb a large amount of heat before its temperature rises, as the energy goes into breaking the hydrogen bonds. Water also holds heat longer than other substances before its temperature lowers. This causes large bodies of water to moderate the temperature of the land. Areas of the planet that do not have bodies of water surrounding them have much more extreme temperature differences without water to moderate the temperature. Comparing the temperature difference between San Diego, California, and Topeka, Kansas, demonstrates the difference a large body of water can cause in a climate. The average daily temperature in the coastal city of San Diego varies from 57°F to 71°F (14–22°C) over the course of an entire year, and Topeka's average daily temperature has summer and winter daily temperature averaging 27–79°F (-3-26°C).

The hydrogen bonding of water creates a unique feature of water—a lower density in the solid state than in the liquid state. Unlike most substances, water expands when it freezes into ice, which is characterized by a very rigid structure. Ice forms when single water molecules participate in four hydrogen bonds, leaving a large volume of unoccupied space existing within that structure, making is less dense than the liquid form, in which each water molecule participates in only two to three hydrogen bonds. Water's density has biological as well as environmental implications, as Earth's oceans do not freeze solid. The ice floats on the surface of the body of water and acts as an insulating barrier to the water and life below.

The function of water as a universal solvent demonstrates water's great ability to dissolve a wide variety of compounds. Nearly all ionic compounds are soluble, at least to some extent, in water. When an ionic compound has contact with water, the cations (positive ions) and anions (negative ions) separate from one another and become surrounded by water molecules. The slightly negative oxygen end of the water molecules surrounds the cations, while the slightly positive hydrogen end of the water molecules surrounds the anions. This successfully separates the ions from one another and prevents their reassociation. Evaporation of the water frees the ions to associate and reform the ionic compounds.

FACTORS AFFECTING SOLUBILITY

Solubility is a measure of the amount of solute that can dissolve in a given volume of solvent at a given temperature. The rate of solubility and the maximal amount of substance that is soluble in a certain solute are controlled by many factors, including temperature, amount of solvent, concentration of solute, surface area of solute, pH of solution, and pressure of the system.

Temperature plays a large role when determining the maximal amount of solute that can dissolve in a solution. When the solute is a solid, as the tempera-

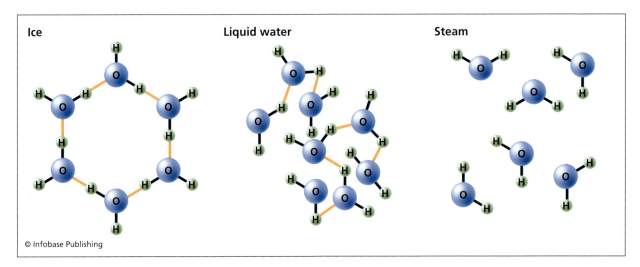

Water exists in three states: solid, liquid, and vapor. The solid form of water is ice, and the polar water molecules have a rigid yet open structure that causes ice to be less dense than liquid water. Solid ice has approximately four hydrogen bonds to each water molecule (shown in red). Liquid water has approximately two to three hydrogen bonds to each water molecule, while water vapor (steam) has few if any hydrogen bonds between water molecules.

ture increases, so does the solubility of the solute. The lower the temperature, the lower the solubility of the solute. Increasing the kinetic energy of the solvent particles increases the amount of solute particles that can be dissolved. When the solute is a gas and the solvent is a liquid, the solubility decreases as the temperature increases, as in carbonated water. When the water is cold, the amount of dissolved carbon dioxide (the carbonation) is high. As the beverage warms, the amount of gas dissolved in the water decreases and causes the soda to lose its carbonation, or become flat. The increased kinetic energy of the liquid solvent forces the gas molecules out of solution.

Another factor that affects the solubility of a substance is the amount of solvent. As the volume of solvent increases, the mass of solute that can be dissolved also increases. Solubility values are reported per volume or mass of solvent to prevent misunderstandings due to differences in solvent amount. The density of water is one gram (g) per one milliliter (mL), such that the volume and mass of water as the solvent are the same value. In other words, 500 g of water is equivalent to 500 mL of water.

The total surface area of the solute influences the amount that can be dissolved in a given amount of solvent. A solute with a greater surface area will dissolve more quickly than one with less surface area. If a solid is in a crystalline form, it will have a lower solubility than if it has been ground into a powder with greater surface area. Another process that increases solubility of a substance in water is stirring. The concentration of solute will affect the solubility. When added in small quantities with stirring, the solute will dissolve more easily than if added all at once. One reason is increased kinetic energy. Another reason is that solute concentration also affects solubility—mixing prevents the local solute concentration from becoming too high and thus inhibiting further solvation.

The pH of a solution can affect the amount of a solute that will dissolve if the ions produced are acidic or basic. Basic ions produced from a solute have a higher solubility as the acid concentration of the solution increases. Just as lowering the pH increases the solubility of basic ions, raising the pH increases the solubility of acidic ions.

When the solute is a gas dissolved in a liquid, increasing the pressure of the gas in the surroundings increases the solubility of the gas in the system. This pressure effect on solubility explains why patients who have breathing difficulties are placed in oxygen tents or given oxygen. The higher the concentration of inhaled oxygen, the higher the oxygen pressure and the more oxygen that can be dissolved in the blood. The same phenomenon is seen in carbonated beverages. As the pressure is released from a bottle,

the gas escapes because of the lower carbon dioxide pressure outside the bottle. This lowered pressure decreases the solubility of carbon dioxide in the bottle, and, if left uncapped, the beverage eventually becomes flat.

When dissolved in water, many ionic compounds conduct an electrical current; they are known as electrolytes. Insoluble ionic compounds are poor electrolytes, while very soluble ionic compounds are good electrolytes. Covalent compounds are poor electrolytes, as they do not dissociate into ions when in aqueous solutions.

COMMON ION EFFECT

When an aqueous solution contains multiple solutes, the concentration of one of the solutes affects the equilibrium of the other solutes. Significant examples of this are found in the human body and in seawater, which are made up of predominantly water and have many solutes dissolved at the same time. The effect of one ion on the equilibrium of another related ion is known as the common ion effect. Calculation of the common ion effect on equilibrium is observed when a weak acid is in solution with a salt of that acid. The equilibrium of acetic acid would be measured by the equilibrium constant of the acid, K_a, according to the following formula:

$$CH_3COOH \rightleftharpoons CH_3COO^- + H^+$$

$$K_a = [CH_3COO^-][H^+]/[CH_3COOH]$$

When multiple solutes are present in the same aqueous solution, ions that are present in both equilibrium reactions can affect one another. For example, if sodium acetate were also present in this aqueous solution, it would have the following equilibrium equation:

$$NaCH_3COO \rightleftharpoons Na^+ + CH_3COO^-$$

$$K_{eq} = [Na^+][CH_3COO^-]/[NaCH_3COO]$$

Notice that the acetate ion (CH3COO$^-$) is present in both equilibrium equations. The alteration of one of the equilibriums will affect one another. Sodium acetate dissociates completely upon addition to water, so the concentration of intact sodium acetate is negligible. The addition of sodium acetate to the solution ensures that more acetate ion will be produced. The production of additional acetate ion will push the equilibrium of acetic acid to the left, toward the formation of acetic acid, because of LeChatelier's principle. This demonstrates the complexity of aqueous solutions that have multiple solutes dissolved simultaneously, with the equilibrium of one affecting the other.

HETEROGENEOUS AQUEOUS SOLUTIONS

Often the substances in an aqueous solution are not in the same phase, producing a heterogeneous situation. A precipitation reaction is a type of chemical reaction in which one or more of the products formed are insoluble and fall out of solution as a precipitate. Precipitation reactions play an important role in the purification and isolation of particular species. If the solubility of the compounds is known, the reaction can be manipulated such that only the product of interest precipitates and the unwanted compounds stay in solution. The desired compound is isolated from solution, dried, and further purified if necessary. When reactants are known, prediction of precipitates as well as soluble compounds based on the chemical makeup of the reactants and products is possible. General precipitate predictions can be based on the guidelines represented in the following table.

These solubility rules predict which substances will be soluble and which will precipitate but do not give a quantitative sense of their solubility. Calculation of the solubility product constant K_{sp} gives this quantitative look at how much of a solute will dissolve in aqueous solution at a given temperature. The solubility-product constant is the equilibrium constant between the solid ionic compound and its ions in an aqueous solution. Following a solubility-product constant example for calcium hydroxide gives the following reaction:

$$Ca(OH)_2 \leftrightharpoons Ca^{2+} + 2OH^-$$

$$K_{sp} = [Ca^{2+}][OH^-]^2$$

The solubility-product constant is equal to the concentration of the ions raised to the coefficient of their ion in the balanced dissociation reaction. The K_{sp} of calcium hydroxide in water at 77°F (25°C) is 5.5 × 10^{-6}, a low value, demonstrating that a small quantity of the calcium hydroxide will dissolve in water. The qualitative information from the solubility table that stated hydroxides were insoluble supports this quantitative information. Solubility represents the concentrations of solute that can dissolve in a solvent at a given temperature. As shown in the example in the section Common Ion Effect, the amount of a substance changes, given the presence of other ions in the solution. The solubility product constant is a fixed value for a given substance at a given temperature.

Precipitation can be predicted by comparing the product of the ion concentrations (Q) with the K_{sp} of the equilibrium reaction. When Q is greater than K_{sp}, the solute will precipitate. When $Q = K_{sp}$, the solution is at equilibrium, and when Q is less than K_{sp}, the solute will dissolve until the solution is saturated.

Understanding the solubility product constants and the precipitation of given solutes allows for qualitative analysis of aqueous solutions containing multiple solutes. This process sets up conditions for precipitation of certain metal ions and confirms the presence of these ions after precipitation. Examples that can be determined through this type of qualitative analysis include insoluble chlorides, acid-insoluble sulfide, base-insoluble sulfides and hydroxides, and insoluble phosphates.

See also ANALYTICAL CHEMISTRY; ATMOSPHERIC AND ENVIRONMENTAL CHEMISTRY; CHEMICAL REACTIONS; EQUILIBRIUM; RATE LAWS/REACTION RATES.

FURTHER READING

Hein, Morris, and Susan Arena. *Foundations of College Chemistry,* 11th ed. Indianapolis, Ind.: Wiley, 2004.

representing structures/molecular models

The relationship between the structure and function of a molecule is a reappearing theme throughout all of science. Understanding the structure of a molecule helps explain how it functions. The phrases "form fits function" and "structure equals function" explain that the way a molecule works depends on the way it is put together. Destruction of the shape of a molecule will render it nonfunctional. Scientists use

SOLUBILITY RULES

Ion Present	Soluble/ Insoluble	Significant Exceptions
Nitrate	Soluble	None
Sulfate	Soluble	Ag^+, Pb^{2+}, Hg_2^{2+}, Ca^{2+}, Sr^{2+}, Ba^{2+}
Acetates	Soluble	None
Group IA and NH_4^+	Soluble	None
Chlorides, bromides, iodides	Soluble	Ag^+, Pb^{2+}, Hg_2^{2+}
Phosphates, carbonates, sulfides, hydroxides, chromates	Insoluble	Group IA and NH_4^+ compounds

drawings and models of molecules to demonstrate the attachments between atoms as well as the three-dimensional shape of molecules.

Molecules can be represented in two and three dimensions, depending on the type of visualization desired. Two-dimensional representations of a molecule are useful because they can easily be done on paper, although they do not show the actual structure of the molecule in space. Three-dimensional models that utilize model kits are useful because they represent the true structure of the molecule in three-dimensional space, but they become cumbersome if the molecule is too large. Recent developments in computer modeling have created the best of both worlds with ease of manipulation and no size restraints, as well as clear representations of the three-dimensional structure. New learners sometimes have difficulty visualizing three-dimensional shape from a computer-generated model on a screen, however, so two- and three-dimensional model kits still play a role in textbooks and chemistry instruction.

REPRESENTING THE TWO-DIMENSIONAL SHAPE OF A MOLECULE

The three-dimensional shape of molecules often needs to be represented in two dimensions so it can be written down and understood. Many methods exist for this, including Lewis structures, stereochemical diagrams, Fischer projections, and Newman projections.

In Lewis structures, all of the valence electrons of an atom are shown, and the bonds formed between atoms in a molecule are represented by shared electron pairs (usually as dots) in the plane of the paper. Single covalent bonds are shown by drawing one pair of shared electrons between the atoms, double covalent bonds by drawing two pairs, and triple covalent bonds by showing three pairs. Often the shared electron pairs are simplified by using one, two, or three horizontal lines between the bonded atoms. Structural isomers can be easily represented by this method. Lewis structures are useful for determining the relationship between the atoms in a molecule but do not depict the true three-dimensional shape of a molecule, especially a large molecule.

Stereochemical diagrams are used to represent the geometric arrangement of the atoms. They are an improvement on Lewis structures, because the three-dimensional arrangement of the atoms is represented in this type of structure. Wedges represent atoms that project out of the plane of the paper, straight lines represent atoms in the same plane as the central atom, and dashed wedges indicate atoms that project into the plane of the paper.

Fischer projections were created by the German chemist Emil Fischer. They represent a quicker way to draw stereochemical diagrams; however, they are not as representative of the actual structure of the molecule. In a Fischer projection, the horizontal bonds are assumed to project out of the page, while the vertical bonds project into the page. Fischer projections are widely used to represent the three-dimensional structure of carbohydrates. An asymmetric carbon atom (a carbon with four unique substituents attached) is represented with four lines extending from it, with the carbon in the center of the cross. Molecules that have more than one carbon atom are generally oriented such that the carbon chain is vertical on the page. This method of representation allows for stereoisomers of sugar molecules such as glucose and galactose to be readily identified and compared.

The conformations, spatial relationship between atoms due to rotation around a single bond, of certain

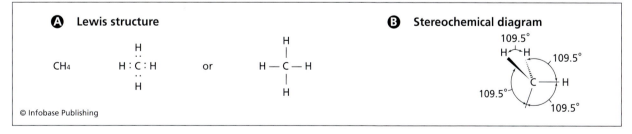

(a) Lewis structures show the bonding arrangements of atoms in a molecule. The valence electrons are shown as dots and two electrons between atoms represent a single covalent bond, as in methane, CH_4. Lewis structures are sometimes shown with straight lines representing covalent bonds. (b) Stereochemical diagrams show the three-dimensional structure of a compound. Straight lines represent bonds that are in the plane of the paper. Dashed wedges represent bonds that are projecting into the paper, and filled-in wedges represent bonds projecting out of the paper. Methane is shown as an example.

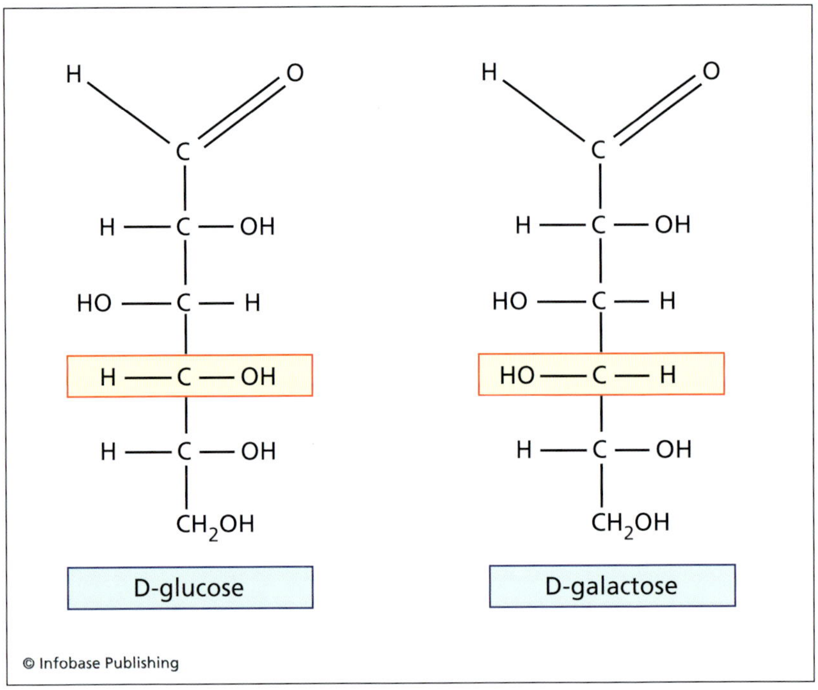

A Fischer projection represents the shape of a molecule with the carbon chain written vertically in the plane of the paper. This method of representing structures shows the difference between stereoisomers such as glucose and galactose. The only difference between these two sugar molecules is the location of the -OH group.

organic molecules often need to be represented on paper. The most common method of representing these conformations uses Newman projection formulas developed by the American chemist M. S. Newman. In a Newman projection, instead of drawing the carbon-carbon backbone vertically on the paper, as in a Fischer projection, the perspective is down the carbon-carbon bond. The front carbon is represented by a point, and the back carbon is represented by a circle. This enables one to compare rotation around that single carbon-carbon bond between molecules.

REPRESENTING THE THREE-DIMENSIONAL SHAPE OF A MOLECULE

Many types of molecular models exist to demonstrate the shape of a molecule. For decades, the molecular models have been the most definitive method of representing the structure of a molecule. Two-dimensional drawings do not give the true shape of a molecule, so chemists developed three-dimensional models.

A ball-and-stick model represents the bonding attachments between the atoms as well as the three-dimensional arrangements of the atoms in a molecule. The atoms are represented with balls of different colors, and long and short sticks or springs are used as bonds. Single covalent bonds are longer than double and triple bonds, so the length of the

stick indicates the type of covalent bond. Just as a multiple bond alters the final conformation of a molecule, shortening a stick keeps the atomic balls closer together and changes the three-dimensional shape of the molecular model. Ball-and-stick models are useful for demonstrating to chemistry students how atomic arrangements make up a larger molecule. This type of model is easily manipulated and easily rotated to demonstrate concepts such as isomers, bond angles, and three-dimensional shape. Some drawbacks of ball-and-stick models are that they are cumbersome to build for very large molecules, and the size of the balls representing the atoms does not accurately represent the true size of the atoms relative to one another.

A second type of molecular model is the space-filling model, in which each part of the model that represents an atom is the appropriate size relative to other model atoms. The amount of space taken up by an atom gives a true representation of the three-dimensional structure of the molecule. This model type is the most effective at demonstrating the lack of space within a large molecule. All of the atoms in a molecule are represented in a space-filling model. The van der Waals radius of an atom represents how close together two atoms can get to one another. Visualizing the bonding differences in space-filling models is more difficult for novices, but such models are more useful for allowing experienced scientists to see the relative size of the atoms. The colors of the atoms in a space-filling model are always the same, simplifying the comparison of different models.

Sulfur:	Yellow
Oxygen:	Red
Carbon:	Black
Hydrogen:	White
Phosphorus:	Purple
Nitrogen:	Blue

These types of models are useful when visualizing the structure of very large molecules such as pro-

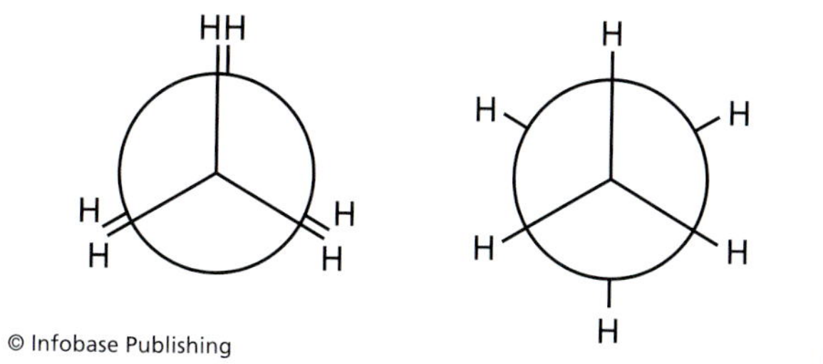

© Infobase Publishing

In a Newman projection the three-dimensional structure of a compound is represented by projecting the carbon chain into the paper. A point represents the carbon atom in front, while the carbon atom in the back is represented by a circle. Two three-dimensional versions of ethane are shown here.

teins, especially to highlight changes in structure between two different conformations of the same protein.

Both ball-and-stick models and space-filling molecular models become very complicated and over-detailed when every atom is represented in a large molecule, such as large biological molecules. Scientists have developed simpler methods of representing large molecules that do not involve the visualization of every atom in the molecule; rather, they show the backbones and overall shape of the molecule. These types of models include backbone models and ribbon diagrams.

Backbone models only show the backbone structure of the molecule, such as the sugar-phosphate backbone of DNA or the polypeptide backbone of a large protein. This method of depicting molecules is very useful to discern overall trends in the molecule's shape.

Ribbon diagrams are molecular representations that highlight secondary structures such as alpha-helices and beta-sheets in large protein molecules. Important parts of the molecule, such as specific amino acids in the binding site of a protein, are represented specifically, while the rest is shown in the general shape.

MOLECULAR MODELING WITH COMPUTERS

In the 21st century, the term *molecular modeling* means a process that is performed with a computer program. A computer has the capability to represent the same molecule by multiple methods, including showing the ribbon diagram, the ball-and-stick model, as well as the space-filling model all on the same screen. Molecular modeling programs used to be complex and required large computer systems to run. Presently, molecular modeling programs can be run on personal computers with ease. The ability of

a computer program to represent the three-dimensional shape of a molecule in a short period greatly speeds the identification and determinations that are made for large molecular structures.

One can also rotate the models on the computer screen and even model what conformational changes would occur if changes in the molecules arrangements were made. For instance, one can demonstrate the effect on the three-dimensional structure of a protein when one amino acid is changed. The folding properties that are altered are quickly represented, and the two models of the protein (the original and the modified) can be compared.

Molecular mechanics is a method that allows for the evaluation of the stability of different conformations of a molecule. Computer modeling programs take this into account to determine the most likely three-dimensional arrangement of the atoms involved. Concepts such as the amount of strain placed on a molecule's structure relative to the most stable form can be calculated as strain energy (E_s).

$$E_s = E_{\text{bond stretching}} + E_{\text{angle bending}} + E_{\text{torsional}} + E_{\text{van der Waals}}$$

where E_s is the total strain energy, $E_{\text{bond stretching}}$ is from bond distances deviating from their ideal values, $E_{\text{angle bending}}$ represents the deviation of the bond angle from ideal values, $E_{\text{torsional}}$ represents the amount of torsional strain on the molecule, and $E_{\text{van der Waals}}$ represents the interactions between atoms that are placed in close proximity to one another without bonding. The computer molecular mechanics programs determine the most likely combination to minimize the total strain energy on the molecule.

Many molecular modeling and molecular graphic programs are available, as well as large molecular modeling groups that are able to perform the molecular modeling for scientists. Collections of molecular modeling databases are available for structural comparison. The National Center for Biotechnology Information (NCBI), sponsored by the National Institutes of Health (NIH) and the National Library of Medicine (NLM), maintains many molecular modeling databases. Examples include the Molecular Modeling Database (MMDB), which contains three-dimensional information on macromolecules; the Vector Alignment Search Tool (PubVast), which enables the comparison of protein chains; PubChem, which provides information on small moles; and the Conserved Domain Database (CCD), for identification of conserved domains in proteins.

See also COVALENT COMPOUNDS; IONIC COMPOUNDS; ISOMERS.

FURTHER READING

Berg, Jeremy M., John L. Tymoczko, and Lubert Stryer. *Biochemistry*, 6th ed. New York: W. H. Freeman, 2007.

NIH: Center for Molecular Modeling home page. Available online. URL: http://cmm.cit.nih.gov/. Accessed July 25, 2008.

ribozymes (catalytic RNA)   Ribozymes or catalytic RNA are molecules of ribonucleic acid (RNA) that perform enzymatic roles in living organisms. This newly discovered potential of RNA has changed the traditional thinking of biochemists. The functioning of all living things is determined by the genetic information that is written within the deoxyribonucleic acid (DNA) code in every cell. The traditional understanding that held firm from the 1950s to the late 1980s was that the genetic code inherent in the DNA was transcribed into messenger RNA, which was then used as a template for the translation of that molecular information into a functional protein. Thus the flow was as follows:

$$DNA \rightarrow RNA \rightarrow proteins$$

Ribosomes carry out the process of translation by assisting in the formation of peptide bonds between adjoining amino acids, according to the sequence of nucleotides in the RNA (which was determined by the nucleotide sequence of the DNA).

Biochemists traditionally considered the sole function of the messenger RNA to be a simple carrier molecule of the information from the DNA to the protein synthesis machinery. As the so-called workhorses of the cell, proteins are inarguably important, but the importance of the RNA was considered secondary. Biochemists generally agreed that the two other types of RNA, transfer RNA and ribosomal RNA, played a more active and therefore significant role in protein synthesis. Transfer RNA delivers the correct amino acids to the ribosome, and ribosomal RNA makes up part of the ribosomal machinery that forms the proteins. The discovery of the catalytic abilities of RNA shined new light on the function of the RNA molecules. RNA is no longer considered a secondary molecule that only has passive value in the production of proteins. Scientists now recognize the function of RNA in biochemical catalysis and in the regulation of cellular function to be extremely important.

DNA AND RNA STRUCTURE

The monomers that make up DNA are known as nucleotides. Every nucleotide contains the following three components:

- a deoxyribose sugar (ribose that is missing an oxygen atom at the 2′ carbon atom)
- a phosphate group (initially a triphosphate consisting of three bonded phosphate groups, but becoming a monophosphate after the cleavage of two of the phosphates during phosphodiester bond formation, when the nucleotide is incorporated into a molecule of DNA)
- a nitrogenous base (adenine, guanine, cytosine, or thymine)

DNA is a polymer of nucleotides that are connected by phosphodiester bonds between the 3′ hydroxyl group of the deoxyribose sugar and the 5′ phosphate group of the adjacent nucleotide. The formation of these bonds creates a sugar-phosphate backbone that contains a sequence within the order of their bases. The nitrogenous bases that are found on one strand of a DNA molecule are hydrogen bonded to the nitrogenous bases on the opposite strand. This creates a double-helix structure that is characteristic of DNA.

RNA is a nucleic acid polymer that is created by the process of transcription, during which DNA serves as a template for copying. The main differences between RNA and DNA are listed:

- The nucleotides of RNA contain ribose rather than deoxyribose.
- RNA is generally single-stranded, although it may adopt secondary structures by the formation of hydrogen bonds between nucleotides within a single strand.
- The nitrogenous bases used in RNA include adenine, guanine, cytosine, and uracil (rather than thymine).

PROTEINS AND ENZYMES

Proteins are considered the workhorses of the cell. They perform numerous functions necessary for many aspects of life: they function as structural proteins, immune proteins, transport proteins, and catalytic proteins known as enzymes. Proteins are responsible for many cellular functions (all proteins are not enzymes), and biochemists long believed that all enzymes were proteins. An enzyme is a catalyst that speeds up the rate of a biochemical reaction without itself being consumed or altered in the reaction. Within any living thing, millions of chemical reactions are simultaneously occurring. These reactions would not occur within a biologically significant time frame without the help of enzymes. Protein enzymes have binding sites for the reactants or substrates and an active site where the reaction is carried out. When the substrate binds to the active site of an enzyme, a complex known as the enzyme-substrate complex is formed. This allows the substrate to be cleaved or to react with another substrate. Enzymes can only increase the rate at which a reaction occurs; they cannot cause a reaction to occur if it is not energetically favorable. The protein enzyme has amino acids that are specifically able to bind to the substrate in its binding site and amino acids that have specific catalytic properties in the active site. For decades, biochemists thought protein enzymes were the only type of enzymes.

DISCOVERY OF CATALYTIC RNA

The Nobel Prize in chemistry in 1989 was given to two scientists who changed the way biochemists thought about the traditional role of RNA. Sidney Altman and Thomas Cech were recognized for their contributions to the discovery of the important functions of RNA and their recognition of the different roles that RNA plays in living things. Altman discovered the ability of a molecule known as RNase P to cleave or activate transfer RNA (tRNA) molecules. Cech demonstrated the ability of a RNA molecule to self-splice. Both of these discoveries demonstrated for the first time that RNA acted as something other than a messenger molecule between DNA and protein. RNA played a significant and active role in the functioning of living things.

SIDNEY ALTMAN

Sidney Altman is a Canadian-born chemist who trained and worked the majority of his life in the United States. Altman was born into a modest family on May 7, 1939, and his career in science was largely influenced by the political events of the time. The Manhattan Project and the development of the atomic bomb had a large impact on him as a young boy and led to his interest in science, especially physics. Altman's family, although not rich, understood the importance of education. He attended college in the United States, studying physics at the Massachusetts Institute of Technology (MIT). This is where Altman was first introduced to the field of molecular biology, the study of life processes at the molecular level. He began his graduate work in physics at Columbia University but later transferred to the University of Colorado at Boulder, where he worked on the insertion of compounds (intercalation) between the bases of a DNA molecule. This work in biophysics continued as he earned a Ph.D. from the University of Colorado in 1967. Altman then worked on the replication of the T4 bacteriophage (a virus that infects bacteria). After completion of his degree he began work as a postdoctoral fellow at Harvard

Sidney Altman is a Canadian-born chemist who discovered the first catalytic ribonucleic acid (RNA) molecule, ribonuclease P (RNase P). *(Michael Marsland/Yale University)*

University, where he stayed from 1967 to 1969. He then took a position at the Medical Research Council Laboratory in Molecular Biology in Cambridge, England, from 1969 to 1971 in the group of the South African–born British molecular biologist Sydney Brenner and the English molecular biologist Francis Crick (the codiscoverer of the double-helix structure of DNA). Here Altman began his work on the catalytic activity of RNase P, an enzymatic activity that catalyzes the cleavage of RNA. In 1971, he accepted a faculty position at Yale University, where he has remained ever since.

Altman first examined RNase P as a visiting research fellow while studying the intercalation of alcidine dyes into DNA of *Escherichia coli* tRNA genes. While many mutants of tRNA molecules existed, the outcome of Altman's work was unique. He found that an enzymatic activity converted the mutant tRNA molecules back to the wild type. The identity of the enzyme, which they named RNase P, was elusive and, surprisingly, continued work demonstrated that this function might be conferred by a nucleic acid rather than a protein. Altman's work later showed that RNase P contained both protein and RNA subunits and that both were required for function of the enzyme. RNase P was shown to be essential in vivo for the synthesis of all tRNAs. Altman had discovered one of the first known catalytic RNA molecules.

Altman presently holds the position of Sterling Professor of Molecular, Cellular, and Developmental Biology as well as professor of chemistry, biophysical chemistry, and organic chemistry at Yale. He continues to research the functions of RNase P in bacterial and mammalian cells and is also investigating the possibility of using RNase P to activate genes in mammalian tissues.

THOMAS CECH

Tom Cech was born on December 8, 1947, in Chicago, Illinois. Cech's father was a doctor, and their household was steeped in scientific topics and discussions as he grew up in Iowa City, Iowa. After graduating from Grinnell College with a degree in chemistry in 1970, he moved to Berkeley with his wife and received a Ph.D. in chemistry from the University of California in 1975. Cech undertook postdoctoral work at the Massachusetts Institute of technology (MIT) in Cambridge, Massachusetts. After his postdoctoral work, Cech accepted a position at the University of Colorado at Boulder.

At UC Boulder, Cech carried out his Nobel Prize–winning research using a simple pond organism known as *Tetrahymena*. It was in this ciliated protozoon that Cech discovered the ability of RNA to self-splice. Splicing is the specific cleavage of a RNA molecule in order to remove unnecessary nucleotide sequences to form a functional RNA. Cech's laboratory determined that the RNA found in this organism was capable of splicing in the absence of any detectable protein. As protein was previously thought always to be responsible for such catalytic activity, it was difficult for his lab to demonstrate that the RNA was truly acting alone. Proving that there was not some undetectable amount of protein present and responsible for the splicing of the RNA was virtually impossible, so Cech changed his tactics and eventually developed in vitro experiments that had no possibility of being contaminated by cellular proteins. He first demonstrated the catalytic activity of RNA using the in vitro system. Replication of the experiments by others in different organisms only furthered the concept of catalytic RNA put forward by Cech.

Cech presently serves as president of the Howard Hughes Medical Institute in addition to maintaining an active lab at the University of Colorado at Boulder. His present research interests still revolve around the catalytic properties of RNA. He is researching the structure and function of these RNAs as well as the structure and function of telomerase, an enzyme responsible for preserving the ends of chromosomes.

HUMAN GENOME PROJECT

Beginning in 1999, a joint group of government researchers sponsored by the U.S. Department of Energy began an enormous endeavor of sequencing the entire human genome. The joint cooperative created automated technology that simplified the sequencing process in order to speed up the process of DNA sequencing. The Human Genome Project was successfully completed in 2003 with some surprising results. Of the 3 billion bases in the entire human genome, researchers only found approximately 20,000 to 25,000 genes. This was a surprisingly low number, considering that significantly less complicated organisms do not have a correspondingly significant number of genes.

The rest of the genome that did not code for protein was considered "junk" DNA and was, until the work of Altman and Cech, not considered of much importance. Now, because of their work and the discovery of the catalytic and regulatory properties of RNA, it has become evident that although there are only 25,000 protein-coding genes, there are many thousands of genes that code for only RNA. This has dramatic implications for the study of molecular biology and biochemistry. Understanding that the part of the genome previously considered nonessential for all of these years may actually be involved in controlling the genome by regulating which genes are expressed will allow those who conduct future research in this area to understand the regulation of genes more comprehensively.

EVOLUTION AND RIBOZYMES

The discovery of catalytic RNA has shed new light on the controversy surrounding the origin of life. The question of which came first—the chicken or the egg or, in this case, nucleic acid or protein—has been constantly applied to hypotheses concerning the origin of life. DNA contains the hereditary information to form all of the proteins needed for life; however, proteins are required for the replication and transcription of DNA. How could proteins be made from something that required proteins themselves to be formed? There was no end to the circular chain of questioning. The discovery of catalytic RNA added a new wrinkle to the question. RNA could have served as the carrier of hereditary information as well as provided the catalytic ability—a potential solution to this problem. The discoveries of ribozymes by Cech

Thomas Cech is an American chemist who first demonstrated the ability of RNA to self-splice in *Tetrahymena,* a simple pond organism. *(University of Colorado)*

and Altman have opened up new possibilities in the understanding of the origins of life.

See also BIOCHEMISTRY; CRICK, FRANCIS; DNA REPLICATION AND REPAIR; ENZYMES; NUCLEIC ACIDS; RATE LAWS/REACTION RATES.

FURTHER READING

Morris, Kevin V., ed. *RNA and the Regulation of Gene Expression: A Hidden Layer of Complexity.* Norfolk, England: Caister Academic Press, 2008.

The Nobel Foundation. "The Nobel Prize in Chemistry 1989." Available online. URL: http://nobelprize.org/nobel_prizes/chemistry/laureates/1989/illpres/what.html. Accessed July 25, 2008.

rotational motion   The change of spatial orientation taking place over time is called rotation. Rotational motion is one type of motion that a body can undergo; the other is translation, which is change of position. It is convenient and common to consider separately the motion of a body's center of mass and the body's rotation about its center of mass. If the center of mass moves, then the body is undergoing translational motion. Otherwise, it is not. The most general motion of a body involves simultaneous motion of its center of mass and rotation about the center of mass. As an example, consider the motion of the Moon, which revolves around Earth while keeping its same side facing Earth. So, as its center of mass performs nearly circular translational motion around Earth with a period of about a month, the Moon simultaneously rotates about its center of mass at exactly the same rate. Similarly, Earth is traveling through space in elliptical translational motion about the Sun, with a period of a year, while also rotating about its axis, with a period of about a day (exactly one sidereal day, to be precise). In the simplest kind of rotational motion, a body rotates about a line of fixed direction passing through its center of mass, a fixed axis of rotation. Then all particles of the body undergo circular motion about that line. That is what the Moon and Earth are doing to a reasonable approximation.

Consider rotational motion of a body about an axis of fixed direction. The body's orientation about the axis is specified by a single angle that is in general a function of time, $\theta(t)$. The SI unit for angle is the radian (rad), but angles are also specified in degrees (°). One complete turn, 360°, equals 2π rad, so

$$1 \text{ rad} = \frac{360°}{2\pi} = \frac{180°}{\pi} = 57.30°$$

$$1° = \frac{2\pi}{360} \text{ rad} = \frac{\pi}{180} \text{ rad} = 1.745 \times 10^{-2} \text{ rad}$$

The time derivative of this function gives the instantaneous angular speed $\omega(t)$,

$$\omega(t) = \frac{d\theta(t)}{dt}$$

The SI unit of angular speed is radian per second (rad/s). Other commonly used units include degrees per second (°/s) and revolutions per minute (rev/min, or rpm), where

$$1 \text{ rpm} = 2\pi \text{ rad/min} = \frac{2\pi}{60} \text{ rad/s}$$

$$= \frac{\pi}{30} \text{ rad/s} = 0.1047 \text{ rad/s}$$

Angular speed can also be expressed in terms of a vector, the angular velocity vector $\boldsymbol{\omega}(t)$. The magnitude of $\boldsymbol{\omega}(t)$ is $\omega(t)$, and its direction is the direction of the rotation axis, so that if the right thumb points in its direction, then the curled fingers of that hand indicate the sense of rotation. The average angular speed ω_{av} for the interval between times t_1 and t_2 is defined as the angular displacement of the body during the interval, $\theta(t_2) - \theta(t_1)$, divided by the time interval:

$$\omega_{av} = \frac{\theta(t_2) - \theta(t_1)}{t_2 - t_1}$$

As an example, say that at time $t_1 = 0.20$ s the valve stem of a rotating car wheel has the orientation $\pi/6$ rad (30°) with respect to the upward vertical, so

$$\theta(t_1) = \theta(0.20 \text{ s})$$

$$= \frac{\pi}{6} \text{ rad}$$

The wheel continues turning so that by time $t_2 = 1.30$ s it has completed two full turns and a bit more, and at this time the valve stem makes the angle of $\pi/3$ rad (60°) with respect to the upward vertical. Because of the two complete turns, the orientation at t_2 is

$$\theta(t_2) = \theta(1.30 \text{ s})$$

$$= \left(2\pi + \frac{\pi}{3}\right) \text{ rad}$$

$$= \frac{7\pi}{3} \text{ rad}$$

The displacement during the time interval from t_1 to t_2 is

Some amusement park rides, such as the carousel and the Ferris wheel at Cypress Gardens Adventure Park in Winter Haven, Florida, involve large-scale rotational motion. *(© Jeff Greenberg/The Image Works)*

$$\theta(t_2) - \theta(t_1) = \theta(1.30 \text{ s}) - \theta(0.20 \text{ s})$$

$$= \left(\frac{7\pi}{3} - \frac{\pi}{6}\right) \text{ rad}$$

$$= \frac{13\pi}{6} \text{ rad}$$

and the value of the time interval is

$$t_2 - t_1 = (1.30 - 0.20) \text{ s} = 1.10 \text{ s}$$

So, the average angular speed of the wheel during the given time interval is

$$\omega_{\text{av}} = \frac{\theta(t_2) - \theta(t_1)}{t_2 - t_1}$$

$$= \frac{13\pi / 6 \text{ rad}}{1.10 \text{ s}}$$

$$= 6.19 \text{ rad/s}$$

$$= 356° / \text{s}$$

The angular acceleration $\alpha(t)$, in radians per second per second (rad/s²), is the time derivative of the angular speed and, accordingly, the second time derivative of the orientation angle:

$$\alpha(t) = \frac{d\omega(t)}{dt}$$

$$= \frac{d^2\theta(t)}{dt^2}$$

The average angular acceleration α_{av} for the interval between times t_1 and t_2 is defined as

$$\alpha_{\text{av}} = \frac{\omega(t_2) - \omega(t_1)}{t_2 - t_1}$$

SIMPLE ROTATIONAL MOTIONS

The simplest rotational motion about a fixed axis is the state of rotational rest, a state of constant, time-independent orientation. Then

$$\theta(t) = \theta_0$$

where θ_0 denotes the body's constant orientation angle. This is rotation at constant zero angular speed:

$$\omega(t) = 0$$

The next simplest rotational motion about a fixed axis is one with constant nonzero angular speed:

$$\omega(t) = \omega_0$$

where $\omega_0 \neq 0$ represents the body's constant angular speed. Then, the orientation angle as a function of time is given by

$$\theta(t) = \theta_0 + \omega_0 t$$

where θ_0 denotes the orientation angle at time $t = 0$. This is rotation at constant zero angular acceleration

$$\alpha(t) = 0$$

The next simplest rotation after that is rotation at constant nonzero angular acceleration:

$$\alpha(t) = \alpha_0$$

with $\alpha_0 \neq 0$ denoting the value of the constant angular acceleration. The time dependence of the angular speed is then

$$\omega(t) = \omega_0 + \alpha_0 t$$

where ω_0 now denotes the value of the angular speed at time $t = 0$. The orientation angle is given as a function of time by

$$\theta(t) = \theta_0 + \omega_0 t + \frac{1}{2}\alpha_0 t^2$$

where θ_0 now denotes the orientation angle at time $t = 0$.

As a numerical example, call the orientation of a rotating carousel 0 at time $t = 0$, $\theta_0 = 0$. At the same time, the carousel has an instantaneous angular speed of 20 degrees per second, or $\omega_0 = \pi/9$ rad/s, and a constant angular acceleration of one degree per second per second, which is $\alpha_0 = \pi/180$ rad/s^2. What is the carousel's orientation after 20 seconds, at time $t = 20$ s? Substituting the data in the formula for orientation angle gives

$$\theta(10\ \text{s}) = 0 + \left(\frac{\pi}{9}\ \text{rad/s}\right)(20\ \text{s}) + \frac{1}{2}\left(\frac{\pi}{180}\ \text{rad/s}^2\right)(20\ \text{s})^2$$

$$= 3.33\pi\ \text{rad}$$

Since every 2π rad of orientation indicates a full turn, this must be subtracted from the result as many times as possible, leaving a positive number. In the present case, 2π rad can be subtracted from 3.33π rad only

once, leaving 1.33π rad. So during the 20-second time interval, the carousel rotates more than once, but not twice, and at the end of the interval has the orientation of 1.33π rad (239°).

TORQUE

Also called the moment of force, torque is a vector quantity that expresses the rotating, or twisting, capability of a force with respect to some point. Let force **F** act on a particle or body, and denote by **r** the position vector of the point of application of the force with respect to some reference point O. So, **r** is the spatial vector extending from reference point O to the point of application of the force. Then the torque of the force with respect to O, $\boldsymbol{\tau}_O$, is given by the vector product

$$\boldsymbol{\tau}_O = \mathbf{r} \times \mathbf{F}$$

Here force is in newtons (N), **r** is in meters (m), and torque is in newton-meters (N·m). Note that the SI unit of energy or work, the joule (J), is equivalent to the newton-meter. But torque and energy are very different quantities, and the unit of torque, the newton-meter, is *not* called a joule.

The magnitude of a force's torque, τ_O, is given by

$$\tau_O = rF \sin \phi$$

where r and F denote the magnitudes of **r** and **F**, respectively, and ϕ is the smaller angle (less than 180°) between **r** and **F**. The direction of the vector $\boldsymbol{\tau}_O$ is perpendicular to the plane in which the two vectors **r** and **F** lie. Its sense is such that if the base of **F** is attached to the base of **r** and the base of $\boldsymbol{\tau}_O$ joined to those bases, then from the vantage point of the tip of $\boldsymbol{\tau}_O$, looking back at the plane of **r** and **F**, the rotation from **r** to **F** through the smaller angle between them (ϕ, the angle less than 180°) is counterclockwise. Equivalently, if one imagines grasping the vector $\boldsymbol{\tau}_O$ with the four fingers of the right hand so that the thumb is pointing in the direction of $\boldsymbol{\tau}_O$, the curving four fingers indicate the sense of rotation from **r** to **F** through the smaller angle between them. The torque $\boldsymbol{\tau}_O$ measures the capacity of the force **F** to cause a rotation about an axis passing through point O in the direction of $\boldsymbol{\tau}_O$.

In many practical cases, a body is constrained to rotate about an axis that is fixed in a given direction, and the force is perpendicular to that direction, such as in the application of a wrench to a bolt or a screwdriver to a screw. Then the magnitude of the force's torque for rotation about that axis equals the product of the magnitude of the force and the perpendicular distance (which is also the shortest

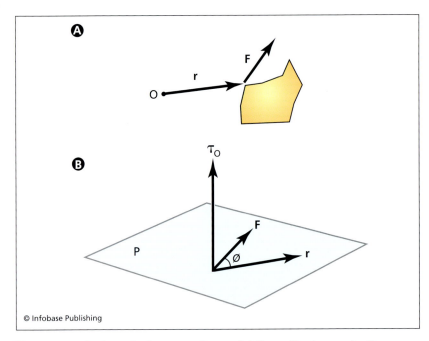

The torque of a force in the general case. (a) Force **F** acts on a body. **r** denotes the position vector of the force's point of application with respect to reference point O. The torque of the force with respect to O is the vector $\tau_O = \mathbf{r} \times \mathbf{F}$. (b) The relation among vectors **r**, **F**, and τ_O. Plane **P** contains vectors **r** and **F**. Vector τ_O is perpendicular to **P** in the sense shown. The smaller angle (less than 180°) between **r** and **F** is denoted ϕ. The magnitude of the torque equals $rF \sin \phi$, where r and F denote the magnitudes of **r** and **F**, respectively.

distance) $r_\perp$ between the force's line of action and the axis, called the moment arm, or lever arm, of the force:

$$\tau_O = r_\perp F$$

This can be expressed in terms of the position vector of the force's point of application with respect to the axis, that is, taking reference point O to lie on the axis, as near to the application point as possible. Then the position vector **r** is perpendicular to the axis and its magnitude r is the perpendicular distance of the application point from the axis. The torque's magnitude then equals

$$\tau_O = rF \sin \phi$$

where, as defined, ϕ is the smaller angle (less than 180°) between **r** and **F**. The direction of the torque vector is parallel to the axis and in the sense indicated by the right-hand thumb when the right hand grasps the axis with the fingers curving in the sense of rotation that the force tends to impart to the body.

As an example of these considerations, try pushing on a door, preferably one with an automatic closing mechanism. If the door is equipped with a closing mechanism, push to open the door. In any case, push perpendicularly to the surface of the door and at various distances from the hinges. Try to keep the magnitude of the pushing force more or less constant. The conclusion should be clear: the rotational effectiveness of the pushes—their torque—is least near the hinges, is greatest at the edge away from the hinges, and increases as the point of push moves from the former position to the latter. That is consistent with the previous discussion of constant magnitude of force F and force perpendicular to the axis, $\phi = 90°$, $\sin \phi = 1$. Then $\tau_O = rF$, so the torque is proportional to the distance from the axis.

For an example of a real-world application, consider a wrench being used to tighten or loosen a nut on a bolt. The force applied to the wrench, of magnitude F, is perpendicular to the rotation axis, which is the axis of the bolt. So again $\phi = 90°$, $\sin \phi = 1$, and $\tau_O = rF$. This time, r is the length of the wrench, or more accurately, the distance from the axis of the bolt to the center of the hand pushing or pulling the wrench. A longer wrench, with the same applied force, can produce greater torque.

MOMENT OF INERTIA

The moment of inertia, I, of a rigid body consisting of a number of point particles, with respect to any axis of rotation, is

$$I = \sum m_i r_i^2$$

In this formula, m_i denotes the mass of the ith particle, in kilograms (kg); r_i is the particle's perpendicular distance from the axis, in meters (m); and the summation is taken over all the particles constituting the body. For continuous bodies, the summation is replaced by integration. For a single point particle, the moment of inertia is simply

$$I = mr^2$$

The SI unit of moment of inertia is kilogram-meter² (kg·m²).

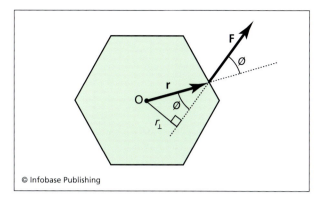

© Infobase Publishing

The torque of a force that is perpendicular to a rotation axis. The rotation axis is perpendicular to the plane of the figure. Its point of intersection with the plane of the figure is denoted O. Force **F** acting on the body lies in the figure plane and is thus perpendicular to the axis. **r** denotes the position vector of the force's point of application with respect to O. The smaller angle (less than 180°) between **r** and **F** is denoted ϕ. The force's moment arm, or lever arm, which is the perpendicular distance between the force's line of action and the axis, is $r_\perp$. The magnitude of the force's torque with respect to the axis equals $\tau_O = r_\perp F = rF \sin \phi$, where r and F denote the magnitudes of **r** and **F**, respectively. The torque vector's direction is perpendicular to the figure plane and, in this example, points out of the page.

Moment of inertia is the rotational analog of mass. The following three examples show that. Whereas the magnitude of linear momentum, p, in kilogram-meter per second (kg·m/s) of a body of mass m, in kilograms (kg), and speed v, in meters per second (m/s), is given by

$$p = mv$$

the magnitude of a body's angular momentum (discussed later), L, is

$$L = I\omega$$

where ω denotes the body's angular speed, as explained earlier.

In addition, the rotational version of Newton's second law of motion is

$$\tau = I\frac{d\omega}{dt}$$

or more generally

$$\tau = \frac{dL}{dt}$$

where τ denotes the magnitude of the torque acting on a rigid body and $d\omega/dt$ is the magnitude of the body's resulting angular acceleration. The translational version of Newton's second law of motion is

$$F = m\frac{dv}{dt}$$

or more generally

$$F = \frac{dp}{dt}$$

where F denotes the magnitude of the force acting on a body and dv/dt is the magnitude of the body's acceleration. The analogy here should be clear. In both cases, the translational and the rotational, the mass and the moment of inertia measure the body's inertia, its resistance to changes in its translational velocity and its angular velocity, respectively.

Further, while the translational kinetic energy of a body of mass m moving at speed v is given, in joules (J), by the expression $\frac{1}{2}mv^2$, the rotational kinetic energy of a body that is rotating at angular speed ω about an axis, with respect to which the body's moment of inertia is I, is given by the expression $\frac{1}{2}I\omega^2$. The sum of the two expressions gives the total kinetic energy of a body that is undergoing both translational and rotational motions:

$$\text{Total kinetic energy} = \frac{1}{2}mv^2 + \frac{1}{2}I\omega^2$$

The radius of gyration of a rigid body with respect to an axis is the distance from the rotation axis at which a point particle of the same mass as the body would have the same moment of inertia as does the body. In other words, a point particle of the body's mass that is located at the radius of gyration is rotationally equivalent to the body itself. If we denote the radius of gyration by r_{gyr} in meters (m), we have

$$r_{\text{gyr}} = \sqrt{\frac{I}{m}}$$

where I denotes the body's moment of inertia and m its mass.

Moment of inertia obeys the parallel-axis theorem, which relates a body's moment of inertia with respect to any axis to its moment of inertia with respect to a parallel axis through its center of mass. If a body's moment of inertia with respect to an axis through its center of mass is I_0, then its moment of inertia I, with respect to a parallel axis at distance d, in meters (m), from the center-of-mass axis, is given by

$$I = I_0 + md^2$$

where m denotes the body's mass.

ANGULAR MOMENTUM

In the physics of rotational motion, angular momentum plays a role that is analogous to the role played by linear momentum in translational (linear) motion. The angular momentum of a point particle is a vector quantity **L** given by

$$\mathbf{L} = \mathbf{r} \times \mathbf{p}$$

where the SI unit of **L** is kilogram-meter^2 per second (kg·m^2/s) and **r** is the particle's position vector, in meters (m). The vector **p** represents the particle's linear momentum, in kilogram-meters per second (kg·m/s),

$$\mathbf{p} = m\mathbf{v}$$

where m is the particle's mass, in kilograms (kg), and the vector **v** is its velocity, in meters per second (m/s). Note that angular momentum depends on the coordinate origin chosen, since the position vector is origin dependent, while linear momentum does not. Thus, angular momentum can only be defined with respect to some point. The angular momentum vector **L** is perpendicular to the plane in which the two vectors **r** and **p** lie. Its sense is such that if the base of **p** is attached to the base of **r** and the base of **L** joined to those bases, then from the vantage point of the tip of **L**, looking back at the plane of **r** and **p**, the rotation from **r** to **p** through the smaller angle between them (the angle less than 180°) is counterclockwise. The magnitude of the angular momentum, L, is

$$L = rp \sin \phi$$

where r and p are the magnitudes of vectors **r** and **p**, respectively, and ϕ is the smaller angle (less than 180°) between **r** and **p**. Since $r \sin \phi$ is the perpendicular distance of the origin from the line of the linear momentum vector, L

can be thought of as the product of the magnitude of the linear momentum and the perpendicular distance of the origin from the line of the linear momentum. Alternatively, since $p \sin \phi$ is the component of the linear momentum perpendicular to the position vector, L can be thought of as the product of the distance of the particle from the origin and the component of its linear momentum perpendicular to its position vector.

Another way of viewing angular momentum is through the relation

$$\mathbf{L} = I\omega$$

where

$$\omega = \frac{\mathbf{r} \times \mathbf{v}}{r^2}$$

is the angular velocity vector of the particle, in radians per second (rad/s), and $I = mr^2$ is the particle's moment of inertia as defined. (This follows directly from the earlier definition of L.) Then the angular momentum of a point particle is seen to be moment of inertia times the angular velocity, by analogy to

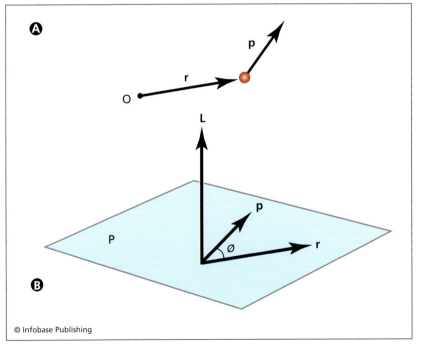

© Infobase Publishing

(a) The angular momentum of a particle that is moving with linear momentum **p** and whose position vector with respect to the origin O is **r** is **L** = **r** × **p**. (b) The relation among vectors **r**, **p**, and **L** is shown. Plane P contains **r** and **p**. Vector **L** is perpendicular to P in the sense indicated. The smaller angle (less than 180°) between **r** and **p** is denoted ϕ. The magnitude of the angular momentum, L, equals $rp \sin \phi$, where r and p denote the magnitudes of **r** and **p**, respectively.

linear momentum as mass times linear velocity, with moment of inertia serving as the analog of mass.

The total angular momentum of a system of point particles with respect to some point is the vector sum of the individual angular momenta with respect to the same point. The magnitude of the total angular momentum of a rigid body consisting of a number of particles, with respect to any axis of rotation, is then

$$L = I\omega$$

where ω is the magnitude of the angular velocity of the body, the body's angular speed, about the axis and I, the body's moment of inertia, is given by

$$I = \sum m_i r_i^2$$

as presented earlier. For continuous bodies the summation is replaced by integration.

Angular momentum obeys a conservation law, which states that in the absence of a net external torque acting on a system, the total angular momentum of the system remains constant in time. Athletes and dancers, for example, make use of conservation of angular momentum to control their angular speed. Consider a figure skater spinning on the tip of her skate. The torque that is exerted on her by the ice is quite small, so her angular momentum is approximately conserved, at least for short enough times. In order to increase her angular speed while spinning, the skater reduces her moment of inertia by drawing in her outstretched arms. That results in a noticeable increase in angular speed.

At the quantum level, angular momentum, including spin, is quantized: that is, it takes values that are only an integer or half-integer multiple of $h/(2\pi)$, where h denotes the Planck constant, whose value is $6.62606876 \times 10^{-34}$ joule-second (J·s), rounded to 6.63×10^{-34} J·s. Only spin makes use of the half-integer multiples, while all other angular momenta are quantized in integer multiples.

CIRCULAR MOTION

As the name straightforwardly implies, circular motion is motion of a particle along a circular path (i.e., a path of constant radius confined to a plane).

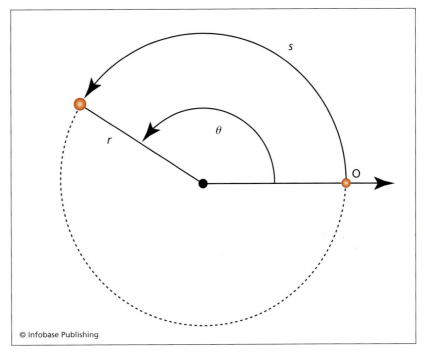

The particle is moving in a circle of radius *r*. The particle's angular displacement θ (in radians) from the indicated reference direction relates to the length of circular arc *s*, from the reference point O to the particle's position, by $\theta = s/r$.

One can view such motion as translation and as a kind of rotation. Every point of a body that is rotating about an axis passing through it is undergoing circular motion. Since the direction of such motion is continuously changing, whether or not the speed is changing, the particle is undergoing continuous acceleration, called centripetal acceleration. This acceleration is directed toward the center of the circular path. Its magnitude a_c is related to the particle's speed v and the (constant) radius of the circular path r by

$$a_c = \frac{v^2}{r}$$

Acceleration is in units of meter per second per second (m/s²), speed is in meters per second (m/s), and radius is in meters (m).

The centripetal acceleration is brought about by some centripetal force, a center directed force, whose magnitude F_c, according to Newton's second law of motion, is

$$F_c = ma_c = \frac{mv^2}{r}$$

where m denotes the mass of the particle. The force is in newtons (N) and the mass in kilograms (kg).

If the particle's speed is changing as well, its acceleration has an additional component, a tangential component a_t, perpendicular to the centripetal acceleration and given by

$$a_t = \frac{dv}{dt}$$

Thus, the magnitude of the total acceleration a of a particle in circular motion is

$$a = \sqrt{a_c^2 + a_t^2}$$
$$= \sqrt{\left(\frac{v^2}{r}\right)^2 + \left(\frac{dv}{dt}\right)^2}$$

Circular motion is also described by a number of angular quantities. Angular displacement θ is the position of the particle in terms of its angle around the circular path, as measured from some reference direction. Unless the particle is at rest, this quantity is a function of time. It relates to the length of the circular arc s from the corresponding reference point on the circle to the particle's position

$$\theta = \frac{s}{r} \quad \text{or} \quad s = r\theta$$

where θ is in radians (rad) and s is in meters (m). Angular speed ω, in radians per second (rad/s), is defined as

$$\omega(t) = \frac{d\theta(t)}{dt}$$

which relates it to the speed by

$$\omega = \frac{d\theta}{dt}$$
$$= \frac{d}{dt}\left(\frac{s}{r}\right)$$
$$= \frac{1}{r}\frac{ds}{dt}$$
$$= \frac{v}{r}$$

or more simply,

$$v = r\omega$$

Angular acceleration α, defined as

$$\alpha(t) = \frac{d\omega(t)}{dt}$$

relates to the tangential acceleration by

$$\alpha = \frac{d\omega}{dt}$$
$$= \frac{d}{dt}\left(\frac{v}{r}\right)$$
$$= \frac{1}{r}\frac{dv}{dt}$$
$$= \frac{a_t}{r}$$

or, more simply, as

$$a_t = r\alpha$$

The unit of angular acceleration is radian per second per second (rad/s²). In summary, the relations between the corresponding translational and angular quantities are

$$s = r\theta$$
$$v = r\omega$$
$$a_t = r\alpha$$

In terms of these angular quantities, the magnitude of the centripetal acceleration is

$$a_c = r\omega^2$$

the magnitude of the centripetal force is

$$F_c = mr\omega^2$$

and the kinetic energy, in joules (J), is

$$E_k = \tfrac{1}{2}mv^2$$
$$= \tfrac{1}{2}m(r\omega)^2$$
$$= \tfrac{1}{2}mr^2\omega^2$$

Circular motion at constant speed is periodic motion, with period T (the time of one revolution, which is the circumference divided by the speed) in seconds (s)

$$T = \frac{2\pi r}{v}$$
$$= \frac{2\pi r}{r\omega}$$
$$= \frac{2\pi}{\omega}$$

and frequency f in hertz (Hz)

$$f = \frac{1}{T}$$

$$= \frac{1}{2\pi/\omega}$$

$$= \frac{\omega}{2\pi}$$

See also ACCELERATION; CENTER OF MASS; CONSERVATION LAWS; ENERGY AND WORK; FORCE; MASS; MOMENTUM AND COLLISIONS; MOTION; NEWTON, SIR ISAAC; QUANTUM MECHANICS; SPEED AND VELOCITY; VECTORS AND SCALARS.

FURTHER READING

Serway, Raymond A., Clement J. Moses, and Curt A. Moyer. *Modern Physics,* 3rd ed. Belmont, Calif.: Thomson Brooks/Cole, 2004.

Young, Hugh D., and Roger A. Freedman. *University Physics,* 12th ed. San Francisco: Addison Wesley, 2007.

Rowland, F. Sherwood (1927–) American Chemist

F. Sherwood Rowland is an American chemist who has spent the majority of his life studying the chemistry of radioactive compounds. In the early 1970s, Rowland began exploring the atmospheric and environmental implications of a class of compounds known as chlorofluorocarbons (CFCs). In 1974, Rowland, along with Mario Molina, then his postdoctoral fellow at the University of California, Irvine, discovered the link between such compounds and the destruction of the ozone layer. For their work on CFCs, the pair was given the 1955 Nobel Prize in chemistry, shared with the Dutch chemist Paul Crutzen.

EDUCATION

F. Sherwood Rowland was born on June 28, 1927, in Delaware, Ohio. Rowland's father was a professor at Ohio Wesleyan University, and Rowland grew up in a house full of books. Rowland was a bright student and his schooling was quite accelerated. He graduated from high school at the age of 16 and entered Ohio Wesleyan University. Rowland's studies were interrupted by his decision to enlist in the navy. After serving in the navy, Rowland returned to school at the age of 21 and worked toward a triple major in mathematics, physics, and chemistry. Upon receiving his bachelor's degree in 1948, Rowland continued his education at the University of Chicago as a doctoral student. His research at this time was on radioactivity and tritium radioactive tracers. Rowland's dissertation was on the production of radioactive bromine atoms using a cyclotron. He completed the requirements for a Ph.D. in 1952, the same year he

F. Sherwood Rowland is an American chemist who contributed significantly to the discovery of the link between chlorofluorocarbons (CFCs) and destruction of the ozone layer. *(AP Images)*

married Joan Lundberg. The couple moved to Princeton University, where Rowland accepted a position as a chemistry professor.

Rowland explored the field of tritium chemistry while working at the Brookhaven National Laboratory during his time at Princeton. In 1956, he took a position in the chemistry department of the University of Kansas.

EXPLORATIONS IN ATMOSPHERIC CHEMISTRY

In 1964, Rowland relocated to the Irvine campus of University of California, where he continued his work on various aspects of radiochemistry. In the early 1970s, he turned his attention to issues of environmental chemistry and eventually determined the fate of the chemical compounds known as chlorofluorocarbons in the atmosphere. Although scientists at the time viewed this compound as completely inert and therefore of no consequence to environmental issues, Rowland had trouble believing that the compound would remain that way while in the upper atmosphere.

Throughout his graduate, postdoctoral, and professorial careers, Rowland was lucky enough to receive nearly constant funding from the Atomic

Energy Commission (AEC). The AEC was rightfully interested in his work on CFCs and continued to support Rowland's work financially throughout the 1970s and 1980s, until 1994.

The question of which student in his laboratory would work on the atmospheric chemistry project was answered when a newly graduated postdoctoral fellow named Mario Molina joined Rowland's laboratory and took on the project. They later received the Nobel Prize in chemistry for this work.

CFCS AND THE OZONE LAYER

CFCs are synthetic compounds originally developed in the 1920s and 1930s as a safer alternative to refrigerants and coolants of the time. Their market expanded until the 1980s, as CFCs were considered safe compounds that were nontoxic, nonreactive, and nonflammable. CFCs were utilized in refrigeration (both home and commercial) under the common name Freon, aerosol propellants, insulation (foam-type), and Styrofoam.

An ozone molecule (O_3) is made up of three covalently bonded oxygen atoms. Ozone at ground level is a major component of smog, especially surrounding large cities. The presence of ozone in the lower portions of the stratosphere, the second level of Earth's atmosphere, plays an important role in protection from ultraviolet radiation reaching Earth's surface. Ozone is formed when molecular oxygen (O_2) is struck with solar energy according to the following reactions:

$$O_2 + sunlight \rightarrow O + O$$

$$O + O_2 \rightarrow O_3$$

At the time Molina and Rowland began working on atmospheric ozone, scientists still believed that CFCs were essentially inert substances with no harmful effects. Rowland and Molina set out to determine what happened to these molecules once they were released. This project afforded Molina the opportunity to apply his knowledge of physical chemistry to the new field of atmospheric chemistry. Molina obtained very few positive results when he attempted reacting CFCs with compounds in the troposphere, the level of the atmosphere closest to ground level. Only after he began to explore the possibility of their passing into the stratosphere did the potential downside of CFCs become apparent.

Molina and Rowland showed that the reaction between chlorofluorocarbons and ozone caused the destruction of the ozone molecule. The method of destruction of the ozone layer by CFCs does not involve the entire CFC molecule. Upon being subjected to intense solar energy at the higher levels of the atmosphere, the chlorofluorocarbons break down, leaving primarily reactive chlorine (Cl). The chlorine reacts with a molecule of ozone and then atomic oxygen (O) to produce two molecules of molecular oxygen according to the following reactions:

$$Cl + O_3 \rightarrow ClO + O_2$$

$$ClO + O \rightarrow Cl + O_2$$

As is shown in the reactions, chlorine first reacts with ozone to create one molecule of molecular oxygen (O_2) and one molecule of chlorine monoxide (ClO), which then reacts with an atom of oxygen to recreate a free chlorine atom and form an additional oxygen molecule. The net result of these reactions is that O_3 and O are converted into two molecules of O_2. Because chlorine is not consumed in the reaction, it can continue reacting with ozone, which becomes depleted.

EFFECTS OF DAMAGE TO THE OZONE LAYER

The loss of the ozone layer or even the depletion of the ozone layer has dramatic effects on the human population. The ozone layer acts as a protective blanket that eliminates the majority of the Sun's ultraviolet rays. Without it, life on Earth would cease to exist. Destruction of the ozone layer could lead to increases in skin cancer and damage to crops, wildlife, and habitats. CFCs also impact the environment by trapping solar energy in Earth's atmosphere, contributing to the greenhouse effect, the phenomenon in which greenhouse gases in the atmosphere prevent solar energy reflected from Earth's surface from escaping, leading to a rise in global temperature. CFCs are greenhouse gases and are more effective greenhouse gases than the more commonly discussed carbon dioxide.

Molina and Rowland published a paper in *Nature* in 1974, relating the destruction of the ozone layer to CFCs. The initial reaction to their findings was skepticism. Scientists understood the information, but not much was done to reduce the use of CFCs because many believed that it would never cause a dramatic effect. Business and political interest groups did not believe that such a critical component of the economy as CFCs could be eliminated. Others had difficulty believing that the continued use of a compound with such a safe history could have such negative consequences.

Several countries did begin eliminating the production of CFCs in aerosols in 1978. Serious environmental implications of the reaction between CFCs and the ozone began to appear in 1985, when the English scientist Joseph Farman identified a large hole in the ozone layer over Antarctica. The hole encompassed an area larger than the area of the United States. This dramatic finding, coupled with the solid understanding that CFCs were directly linked to the production of this hole, spurred governments worldwide to action. In 1985, an international treaty known as the Montreal Protocol was passed to reduce the number of CFCs produced and attempt to phase out their production by 1996. Developing countries were given extended deadlines. The Environmental Protection Agency succeeded in phasing out CFCs in the United States in 2000. CFCs have already proven to be an environmental hazard by their destruction of the ozone layer of Earth's atmosphere. They have been taken off the market, and current laws prohibit the release of CFCs from older appliances into the atmosphere.

Scientific studies by the Environmental Protection Agency show that the rate of destruction appears to be decreasing. The CFCs that have already been released have an incredibly long life span in the stratosphere and remain capable of causing damage for up to 100 years after they have been released. Despite this, the elimination of CFCs will eventually have a significant effect on preventing further loss of the ozone layer.

PRESENT RESEARCH INTERESTS

F. Sherwood Rowland is a great thinker, an adventurous man, and a scientist not afraid to take on a new challenge. The nature of his work on the atmospheric effects of CFCs was far removed from his work on radiochemistry, yet he embraced it and the implications of the work with both hands. Rowland, Molina, and Crutzen have changed the course of environmental history by discovering the link between aerosol CFCs and the damage to the ozone layer.

After the work in the 1970s, Rowland's research group altered their research focus from radiochemistry to atmospheric and environmental chemistry nearly exclusively. Rowland is still a member of the chemistry department at the University of California, Irvine, where he is a Bren Research Professor of Chemistry and Earth System Science. His research group is studying the atmospheric and environmental chemistry of remote locations of the Pacific, areas with high deforestation, and areas with high pollution.

See also AIR POLLUTION (OUTDOOR/INDOOR); ATMOSPHERIC AND ENVIRONMENTAL CHEMISTRY; CRUTZEN, PAUL; GREEN CHEMISTRY; GREENHOUSE EFFECT; MOLINA, MARIO.

FURTHER READING

Hoffman, Matthew J. *Ozone Depletion and Climate Change: Constructing a Global Response.* Albany: State University of New York Press, 2005.

Molina, Mario J., and F. Sherwood Rowland. "Stratospheric Sink for Chlorofluoromethanes: Chlorine Atomic-Catalysed Destruction of Ozone." *Nature* 249 (1974): 810–814.

The Nobel Foundation. "The Nobel Prize in Chemistry 1995." Available online. URL: http://nobelprize.org/nobel_prizes/chemistry/laureates/1995/. Accessed April 10, 2008.

The Rowland-Blake Group, UC Irvine, Chemistry. Available online. URL: http://www.physsci.uci.edu/~rowlandblake/. Accessed April 10, 2008.

Rutherford, Sir Ernest (1871–1937) New Zealander *Physical Chemist* Ernest Rutherford profoundly impacted the studies of atomic structure and radioactivity through his research and theories. After discovering alpha and beta rays emitted by uranium and earning the Nobel Prize in chemistry in 1908, Rutherford collaborated with Frederick Soddy to formulate the transformation theory of radioactivity. Rutherford later elucidated the nuclear structure of the atom by assimilating years of experimental results into a theoretical model that is still accepted today.

Born August 30, 1871, in Nelson, New Zealand, Ernest Rutherford was the fourth of 12 children born to James, a wheelwright, and Martha, a schoolteacher. Ernest received his early education in New Zealand in government schools. When he

Sir Ernest Rutherford was an experimentalist who studied natural and induced radioactivity and discovered that atoms possess a nucleus. He was awarded the 1908 Nobel Prize in chemistry. *(HIP/Art Resource, NY)*

was 16, Ernest entered a nearby secondary school called Nelson Collegiate School, and then, in 1889, the University of New Zealand, where he enrolled at Canterbury College to study mathematics and physical science, earning an M.A. in 1893 with first-class honors in both disciplines. While Ernest was at Canterbury College, he studied the effects of magnetization of iron by high-frequency discharges. To measure the changes to iron, he created a device that could detect wireless wave signals (i.e., electromagnetic waves), and he took this device with him in 1895 to Cambridge University's Cavendish Laboratory to continue his studies of electromagnetic phenomena with Sir J. J. Thomson, who was the director of the laboratory at that time.

Rutherford spent a great deal of energy increasing the range and sensitivity of the electromagnetic wave detector, and, consequently, the instrument was sensitive enough to detect signals more than half a mile away. Thomson was impressed, and he asked Rutherford to collaborate with him to study the effects of X-ray radiation on ion formation in gases. During this time, Rutherford quantified the speeds of the ions being released, the rate of recombination of ions, and the differences in ion formation in different gases after exposure to X-rays. Rutherford then examined the effects of ultraviolet light and uranium radiation on ion formation.

In 1898, Rutherford reported that uranium emitted two types of particles, alpha and beta particles. When Rutherford bombarded thin aluminum foil sheets in the laboratory with uranium radiation, he observed that alpha particles, which have a positive charge, were easily absorbed, and beta particles, which have a negative charge, penetrated deeper. Rutherford spent the next 40 years characterizing these particles.

During this time, McGill University in Montreal offered him a position in physics. He expanded his studies of radioactivity to include another radioactive element, thorium, after arriving in Montreal. Rutherford detected a radioactive gas that he called an emanation being released by thorium. All of the known radioactive elements appeared to produce emanations that were initially highly radioactive yet lost activity over time. As the rate of this decrease of activity was unique for each element, Rutherford used this information to distinguish one element from another. Rutherford recruited a chemist, Frederick Soddy, to help with these studies. Rutherford and Soddy separated the most active emanating product, which they referred to as thorium X, from thorium. They observed, over the course of a few weeks, that thorium X consistently lost its radioactivity, while thorium seemed to regain its original activity. A correlation between the loss of activity by the emanation

and the regeneration of activity by the radioactive element led to the theory of transformation in 1902. Rutherford and Soddy proposed the existence of parent radioactive atoms that decay at specific rates, generating daughter atoms and emanations that form a decay series of atoms until a stable end product is reached. If the parent atom is isolated from the daughter atoms, it regenerates its activity, so that additional daughter atoms can be produced and the process can continue. Daughter atoms are unable to regenerate activity; thus their activity diminishes over a specific period. A great deal of experimental evidence, including the discovery and assignment of a number of new radioactive elements that fell within the proposed decay series, supported this theory, which suggested radioactive phenomena were atomic rather than molecular in nature.

After nine years at McGill, Rutherford accepted a position at the University of Manchester in 1907. He continued his research on the properties of radioactivity, specifically alpha particles. Rutherford set out to study atomic structure with Hans Geiger, who later created a device to count the number of alpha particles by electronic means. He directed massive alpha particles made from helium atoms missing their two electrons at thin pieces of gold foil. Earlier, Thomson proposed the plum pudding model, in which all of the positive charges are distributed evenly throughout an atom, and negatively charged electrons are dispersed as raisins are in pudding. If this model were correct, Rutherford predicted that most of the alpha particles in his studies would pass through the gold atoms in the foil. The majority of the atoms passed through the foil as he had hypothesized, but a small fraction of the alpha particles did not pass through the gold foil. This fraction of particles deflected at large angles away from the foil and, in some cases, deflected straight back toward the source of the particles. Rutherford concluded that the plum pudding model was invalid, and he proposed that an atom has a small dense core, or nucleus, of positively charged particles that makes up most of an atom's mass, and that negatively charged electrons float around in the remaining space. In 1912, Niels Bohr went on to suggest that the electrons circled the nucleus in orbitals much as planets circle the Sun, and he applied Max Planck's quantum theory to Rutherford's model. This model for the structure of the atom remains valid today, and Rutherford's hand in its development is considered his most significant contribution to the scientific community.

Rutherford published several books including *Radioactivity* (1904), which was the first textbook on the subject; *Radiation from Radioactive Substances* (1919, 1930) with James Chadwick and C. D. Ellis; and *The Newer Alchemy* (1937). An inspiring mentor, he also advised and directed many future Nobel Prize winners, including Niels Bohr, James Chadwick, Patrick Blackett, John Cockcroft, and Ernest Walton. Rutherford married Mary Newton in 1900, and they had one child, Eileen, who married the renowned physicist R. H. Fowler. In 1914, Rutherford was knighted, and he became First Baron Rutherford of Nelson, New Zealand, and Cambridge in 1931. He died in Cambridge on October 19, 1937. His ashes were placed beside Lord Kelvin's in the nave of Westminster Abbey.

See also ATOMIC STRUCTURE; BOHR, NIELS; NUCLEAR CHEMISTRY; NUCLEAR PHYSICS; PLANCK, MAX; RADIOACTIVITY.

FURTHER READING

Heilbron, J. L. *Ernest Rutherford: And the Explosion of Atoms (Oxford Portraits in Science)*. New York: Oxford University Press, 2003.

The Nobel Foundation. "The Nobel Prize in Chemistry 1908." Available online. URL: http://nobelprize.org/chemistry/laureates/1908. Accessed July 25, 2008.

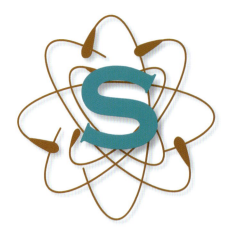

Schrödinger, Erwin (1887–1961) Austrian *Physicist* Statistical mechanics, thermodynamics, color and its perception, the general theory of relativity, unified field theory, particle physics, philosophy, the history of science, molecular genetics, and, especially, quantum mechanics—the multifaceted genius Erwin Schrödinger studied and contributed to all these fields in the early 20th century and earned half of the 1933 Nobel Prize in physics for his major accomplishment in quantum mechanics, the Schrödinger equation. He even invented a cat that possessed the amazing property of being alive and dead at the same time. The proverbial nine lives of an ordinary cat pale in comparison.

SCHRÖDINGER'S LIFE

Erwin Rudolf Josef Alexander Schrödinger was born on August 12, 1887, in Vienna, Austria, to Rudolf and Emily Schrödinger. Rudolf was a talented and broadly educated man who had studied chemistry, Italian painting, and botany and had published papers in botany. At the time of Erwin's birth, Rudolf owned and managed a linoleum factory. Little is reported about Erwin's mother, Emily. She was the daughter of Alexander Bauer, Rudolf's chemistry professor at the Vienna Technical College. Emily was half English on her mother's side.

Until age 10, Erwin was schooled at home, where he learned to speak both German and English, as both languages were spoken there. In 1898, he entered gymnasium (equivalent to high school in the United States). There he excelled in all fields but disliked the classes that required rote memorization. For example, Erwin loved physics, mathematics, and Latin grammar, which are all strictly logical, while he hated memorizing dates and facts. He grasped the physics and mathematics material right away, without doing any homework, and was easily able to solve problems in them immediately after class. During Erwin's eight years in gymnasium, he developed and broadened his cultural interests. He graduated from gymnasium in 1906 at age 18 and in the same year entered the University of Vienna.

Schrödinger attended the University of Vienna for four years, during 1906–10. It attests to the academic level of his gymnasium in that period that at the university he studied such advanced physics subjects as analytical mechanics, electromagnetic theory, optics, statistical mechanics, and thermodynamics. Schrödinger studied advanced mathematics as well, including group theory, differential equations, theory of complex functions, projective geometry, and statistics. He was especially influenced by the theoretical physics lectures given by Fritz Hasenöhrl, who was the successor to Ludwig Boltzmann, the statistical mechanics pioneer. In 1910, after only four years of university study, Schrödinger graduated with a doctorate, awarded for his theoretical dissertation, "On the Conduction of Electricity on the Surface of Insulators in Moist Air."

From 1910 to 1914, Schrödinger remained at the University of Vienna performing research and publishing papers. World War I broke out in 1914 and lasted until 1918. During this period, Schrödinger served in the Austrian army on various fronts as artillery officer. Even then, he managed to make progress in his theoretical research and publish a number of additional papers. In 1917, Schrödinger was transferred back to Vienna to serve there, and after the war ended in 1918, he remained in Vienna until 1920, doing research and publishing papers. In 1920, at the age of 33, Schrödinger married Annemarie (Anny) Bertel, age 24, with whom he had become engaged the previous year. He could not find a sufficiently remunerative job in Vienna, so he took a position in Jena, Germany, and the couple moved there in 1920. Very soon, they moved again, to Stuttgart, Germany, and then, in 1921, to Zurich, Switzerland, where Schrödinger took a position at the University of Zurich, which he held for six years, until 1927.

Zurich proved to be an excellent location for Schrödinger. The academic atmosphere was intellectually stimulating, and Schrödinger associated closely with talented colleagues, including especially Hermann Weyl, who helped him develop his mathematical thinking and skills. In Zurich, Schrödinger made his most important contributions to physics.

In 1927, Schrödinger visited the University of Wisconsin, Madison, to give a series of lectures in January and February. He succeeded brilliantly and was offered a professorship there, but he declined. He had learned that he was a leading candidate to succeed Max Planck in the chair of theoretical physics at the University of Berlin, in the capital of Germany. Later in the year, he was offered the position and accepted. He and Anny moved to Berlin in 1927, and Schrödinger became immersed in an atmosphere of lively discussion, debate, and research with a group of eminent colleagues, among them Albert Einstein. This idyll continued until 1933, when Adolf Hitler and the Nazi Party came to power in Germany and established the persecution of Jews as national policy. Although he was not Jewish and not in personal danger, Schrödinger decided he could not continue living in Germany in such a situation, and he and Anny left Berlin for two visiting positions during 1933–36: first in Oxford, England, and then in Princeton, New Jersey. Schrödinger and Paul Dirac were awarded the 1933 Nobel Prize in physics for their work in quantum mechanics, or, to quote the Nobel Prize committee, "for the discovery of new productive forms of atomic theory." Schrödinger's part in this discovery consisted of his equation, called "the Schrödinger equation," which described the behavior and properties of quantum systems such as atoms. In 1934, Princeton University offered him a position, which he did not accept.

Schrödinger missed his native country and culture. In 1936, after much deliberation over the political situation in neighboring Germany, he decided to return to Austria to take a position at the University of Graz that had been offered him. Schrödinger's decision turned out to have been unfortunate, however, when in 1938 Germany annexed Austria and sent in an occupation force. Schrödinger immediately found himself in trouble with the new authorities, since they considered his departure from Germany five years earlier an unfriendly act toward the Nazi

Erwin Schrödinger was a theoretical physicist and one of the founders of quantum physics. Famous for his equation for the quantum wave function, known as the Schrödinger equation, he shared the 1933 Nobel Prize in physics. *(Photograph by Francis Simon, courtesy AIP Emilio Segrè Visual Archives)*

regime. The administration of the University of Graz, which was renamed Adolf Hitler University, fired Schrödinger, and the political authorities forbade him to leave the country and required him to find a job in industry. He and Anny immediately fled Austria for Rome. From there, Schrödinger returned to Oxford for a while and then went to the University of Ghent, in Belgium. In 1939, he moved to Ireland to take the position of director of the School for Theoretical Physics at the newly founded Institute for Advance Studies in Dublin. Schrödinger stayed at the institute until he retired in 1955. He and Anny then returned to Vienna. Schrödinger died in Vienna of tuberculosis on January 4, 1961, at age 73.

SCHRÖDINGER'S WORK

Besides quantum mechanics, the field in which his work merited him half a Nobel Prize, Schrödinger had many other interests. He was a broad and deep thinker. While working at the University of Vienna, after earning his doctorate in 1910 and until the war started in 1914, Schrödinger published several papers, the most important of which concerned statistical mechanics. During his 1914–18 military service, he managed to continue his research and publish a few more papers. While in Vienna during 1918–20, Schrödinger investigated the phenomena of color and color vision. This work resulted in three publications in 1920, two in *Annalen der Phyisk* and one in *Zeitschrift für Physik*. He also studied the statistical properties of radioactive decay, the mechanical behavior of crystal lattices, and quantum mechanics.

During 1920–21, while on the move from place to place, Schrödinger continued his study of color vision. The next six years, from 1921 to 1927, in Zurich were Schrödinger's most fruitful period. He studied and published in the fields of the mechanics of crystal lattices (investigating the specific heats of solids, in particular), atomic spectra, color vision, general relativity, and quantum mechanics. During this time, in 1926, he made his most important contribution to physics by discovering the equation that bears his name, the Schrödinger equation, for which he shared the 1933 Nobel Prize in physics.

The Schrödinger equation is a partial differential equation, that is, an equation involving partial derivatives, for an unknown function called the wave function, which is commonly denoted Ψ or ψ (capital or lowercase Greek psi). The wave function is a function of position and time and describes the physical system completely. When one solves the Schrödinger equation for the wave function of a particular system under some set of conditions, one is then able to know the properties of the system and predict its future behavior. These predictions are, in general, of a probabilistic nature. For example, the Schrödinger equation might reveal the allowed energy states that some system, such as a hydrogen atom, might possess. The equation also gives the probabilities of actually finding the system in any of those states when the system's energy is measured. Similarly, the equation might be used to show where some particle might be located, where it is forbidden to be, and the probability of finding it at any allowed location. For another example, the Schrödinger equation gives the probability for an initially undecayed radioactive (unstable) nucleus to decay at any time.

In 1935, Schrödinger published an essay, "The Present Situation in Quantum Mechanics," in which he discussed the achievements and difficulties of quantum mechanics as they appeared to him at that time. In the essay, in order to clarify one of the problems of quantum theory, Schrödinger introduced to the world his famous cat, now known as "Schrödinger's cat." To start the presentation of the problem, consider a radioactive nucleus. It can exist in an undecayed state or in a decayed state. In classical physics and intuitively, these are the only two states available to the nucleus. But according to quantum mechanics, the nucleus can also exist in what is known as a superposition state, a state in which it is in a sense in both states, undecayed and decayed, at the same time. In such a state, the nucleus possesses some probability of being found in the undecayed state, if it is observed, and some probability of appearing in the decayed state. But until it is observed, its existence is one of simultaneous undecay and decay.

Return to the example of Schrödinger's equation as applied to such a radioactive nucleus. When the nucleus starts from an undecayed state, the solution of the equation, the wave function of the nucleus, describes the nucleus's evolution in time. It describes the state of the nucleus as starting out as a pure undecayed state, then evolving into a superposition of undecayed and decayed states, and, after sufficient time has passed, reaching a pure decayed state. The superposition state evolves such that it shows a continually decreasing probability of the nucleus's being found in its undecayed state, if observed, and a continually increasing probability of its appearing in its decayed state. Eventually, the former probability becomes 0 and the latter 100 percent. Although such behavior has no counterpart in classical physics and goes against human intuitions, it is one of the characteristics of quantum physics and one needs to and can get used to it.

But Schrödinger revealed a problematic aspect of such quantum behavior by setting up a thought experiment. Imagine, he proposed, a sealed box that initially contains a live cat, a radioactive substance, and a mechanism that, when it detects the radioactive decay of the substance, breaks a vial of poison gas and kills the cat. The cat, substance, and mechanism form a single system. Classically, this system possesses two states: state A, in which the substance has not decayed and the cat is alive, and another state, D, with decayed substance and dead cat. But the quantum wave function, the solution of Schrödinger's equation for the system, describes the system as starting in state A, evolving into a superposition of states A and D, and eventually, after a sufficiently long time, winding up in state D. Whenever an observer opens the box, he discovers a cat that is either alive or dead. But as long as the box is sealed, the cat is in a superposition state: it is neither purely alive nor purely dead; it is both alive and dead at the same time. The problem here is that real-world cats never exist in such states. They are either alive or dead, but never both. The difficulty that Schrödinger was showing is this: as long as quantum behavior is confined to the atomic and subatomic worlds, it works superbly and people must and can accept it. But quantum mechanics should also apply to the macroscopic world, the world of cars, cakes, and cats. Yet the macroscopic world does not exhibit quantum behavior. Schrödinger was on to something very deep, as this problem is still under investigation.

While at the Dublin Institute of Advanced Study during 1939–55, Schrödinger continued his work and publication. He studied electromagnetism and relativity and tried his hand, unsuccessfully, at a unified field theory, an idea that Einstein was working on at the time. Schrödinger studied the philosophy and science of the ancient Greeks and in 1954 published his book *Nature and the Greeks*. He also devoted thought to theoretical biology, resulting in his book *What Is Life?*, published in 1944. This book laid the theoretical foundation for the idea of a genetic code and inspired Francis Crick and James Watson to discover that code in the structure of deoxyribonucleic acid (DNA), the discovery for which they shared with Maurice Wilkins the 1962 Nobel Prize in physiology or medicine.

After his retirement in 1955 and return to his native Vienna, Schrödinger summarized his life philosophy and worldview in his final book, *Meine Weltansicht* (*My World View*), published in 1961.

SCHRÖDINGER'S LEGACY

Erwin Schrödinger was a person of many interests, into which he was capable of delving deeply. His many contributions to human culture were mostly in physics, but also in such near and far fields as color perception, biology, history, and philosophy. Schrödinger inspired others to develop ideas and make their own contributions in various fields. His most important achievement was the Schrödinger equation, for which he shared the Nobel Prize. The idea for which he is best known beyond the physics and chemistry communities is surely Schrödinger's famous alive-and-dead cat, which continues to haunt the thoughts of physicists and philosophers to this very day.

See also ATOMIC STRUCTURE; BROGLIE, LOUIS DE; CRICK, FRANCIS; EINSTEIN, ALBERT; ELECTROMAGNETISM; ENERGY AND WORK; GENERAL RELATIVITY; HEAT AND THERMODYNAMICS; OPTICS; PLANCK, MAX; QUANTUM MECHANICS; RADIOACTIVITY; STATISTICAL MECHANICS; WATSON, JAMES; WILKINS, MAURICE.

FURTHER READING

Cropper, William H. *Great Physicists: The Life and Times of Leading Physicists from Galileo to Hawking*. New York: Oxford University Press, 2001.

Gribbin, John. *In Search of Schrödinger's Cat: Quantum Physics and Reality*. New York: Bantam, 1984.

———. *Schrödinger's Kittens and the Search for Reality: Solving the Quantum Mysteries*. New York: Back Bay, 1996.

Heathcote, Niels Hugh de Vaudrey. *Nobel Prize Winners in Physics 1901–1950*. Freeport, N.Y.: Books for Libraries Press, 1953.

Moore, Walter J. *Schrödinger: Life and Thought*. Cambridge: Cambridge University Press, 1992.

The Nobel Foundation. "The Nobel Prize in Physics 1933." Available online. URL: http://nobelprize.org/physics/laureates/1933/. Accessed February 20, 2008.

Schrödinger, Erwin. *My View of the World*. Woodbridge, Conn.: Ox Bow, 1982.

———. *Nature and the Greeks* and *Science and Humanism*. Cambridge: Cambridge University Press, 1996.

———. *What Is Life?* with *Mind and Matter* and *Autobiographical Sketches*. Cambridge: Cambridge University Press, 1992.

Segrè, Emilio. *From X-Rays to Quarks: Modern Physicists and Their Discoveries*. San Francisco: W.H. Freeman, 1980.

scientific method　　The scientific method is an organized series of steps that scientists use to solve a problem. The first step of the scientific method involves observation and recognition of the question to be solved. In order to define the problem clearly, the appropriate background research must be completed. Understanding the present level of knowledge on the topic allows for preparation of a logical hypothesis and prevents duplication of previously performed experiments.

HYPOTHESIS

After becoming familiar with the background knowledge related to a particular subject, a scientist formulates a logical hypothesis. A hypothesis is often called an educated guess, though the term *guess* is misleading, since a hypothesis is a proposed answer to the problem based on the background research conducted. One critical characteristic of a valid hypothesis is testability. If crop circles appear in a farmer's field, many people believe that the possible solution to the problem (hypothesis) is that aliens are responsible for the formations. Testing whether aliens are responsible is not scientifically possible, and therefore scientists may propose a hypothesis that these formations were created by humans. They would then collect data and document their findings in attempts to support this alternate solution.

EXPERIMENTAL DESIGN

Designing a good experiment follows the development of a testable hypothesis. Experimental design can involve many required steps, but one can only test a single variable at a time. A variable is something that changes during the experiment. Scientific experiments involve two types of variables, the independent (manipulated) variable and the dependent (responding) variable. The independent variable is the factor in the experiment that the experimenter changes and that might affect other variables. For example, someone examining factors that affect the speed of a car on a ramp might alter conditions such as ramp height and ramp material. The dependent variables are the factors that the experimenter does not change but that might vary in response to the independent variable. The dependent variable is what the scientist observes or measures in the experiment. In the ramp experiment, dependent variables that could be measured include car speed or time required to travel the entire length of the ramp.

The experimental design must also include appropriate controls. The use of controls is twofold, meaning the experimental subjects should include an experimental group and an equivalent control group. The experimental group is the sample that the experimenter subjects to the treatment or independent variable being studied. The control group includes similar subjects that do not receive treatment. In a single-blind experiment, the experimenter does not know what treatment the subjects receive. In a double-blind experiment, neither the experimenter nor the subject knows the treatment subjects receive. One way to carry out a blinded experiment is by using a placebo, a treatment that looks identical but that does not contain the active ingredient being tested (the independent variable). The subjects in the control group receive the placebo, whereas the subjects in the experimental group receive the real treatment. Performing double-blind studies reduces bias of both the experimenter and the subjects.

With the exception of the independent variable, all the factors in the experiment for both the control and experimental groups must be kept constant in order to determine whether the independent variable causes any observed changes in the dependent variable. These controls require that everything between the experimental and control groups with the exception of the independent variable be kept the same. If the ramp height is the independent variable, then the experimental and control groups must both have the same size car, same length ramp, and same type of ramp material, and any other possible differences need to be minimized in order to ensure that the independent variable is responsible for the observed effect.

DATA AND RESULTS

Completion of the experiment involves collecting data and making observations, particularly with respect to the dependent variable. Observations are simply the facts that can be obtained through use of the five senses regarding the experimental and control groups. Interpretation is not part of observation; the experimenter simply records the data, completely and objectively. In order to prevent bias, the experimenter must record all of the data, not just that which appears to fit the hypothesis or seems important at the time. Record keeping should include both quantitative (involving numbers and units) and qualitative (involving observations and descriptions) factors of the experiment. These notes are maintained in a logbook that is dated in order to document the progress of the experiment. Experimental observations should be repeatable and are not subject to interpretation. If a car moved down the ramp in 25 seconds in the control group and only 15 seconds in the experimental group, that is a measurable fact. One could expect to obtain similar results upon repeating the experiment in the same manner.

ANALYSIS AND CONCLUSIONS

The actual analysis of the results takes place during the next phase of the scientific method, drawing conclusions. At this time, the scientist studies the data, compares the results, makes inferences (assumptions based on previous knowledge), and proposes possible models for the application of the results. The scientist decides whether the results support the hypothesis or not. If the data do support the hypothesis, further experiments, based on this new information, are performed. If the data do not support the hypothesis, the hypothesis is revised and a new experiment based on the new hypothesis planned. Unlike the results and data collected in the experiment, the conclusions are subject to interpretation by the scientist performing the experiment. Scientists often vary in their confidence that the experimental results support the conclusions found in the experiment. Different background experiences can also lead to alternative interpretations of the results.

Statistical analysis determines how strongly the experimental results support or refute the hypothesis. A properly designed experiment allows for the comparison of the responses of the experimental and control groups. When large differences exist between the groups or when small differences are observed consistently, statistical analysis supports the results and conclusions; they are considered statistically significant. One common type of statistical analysis involves using a two-sample Student's t-test procedure to compare the average results from the experimental group to the control. This type of test involves computation of the likelihood that the two groups have equal average results. If this likelihood of equality is less than a predetermined threshold, called alpha, then one can conclude that the means are different and the hypothesis is supported. This alpha threshold is typically 5 percent. If the results

THEORIES: NOT MERE SPECULATIONS!

by Joe Rosen, Ph.D.

"It's only a theory!" Everybody has heard this put-down, or maybe even declared it oneself on occasion. It means: do not take the idea too seriously; it is only a speculation; or, it is merely a hypothesis. To give one example, this is the all-too-common creationist/intelligent-designist reaction to Charles Darwin's theory of biological evolution. "It's only a theory; it's not fact." However, Darwin's theory, together with such theories as Albert Einstein's theories of relativity, are *not* mere speculations, *not* off-the-cuff hypotheses. They are well-founded, solidly confirmed explanations of very many facts and phenomena of nature. They offer consistent, unifying, and far-reaching explanations of the way things are and the way they happen. They can predict new effects and thus be retested over and over again.

What then is the problem? Simply, the term *theory* is used in different ways in everyday conversation and in science. In the language of scientists, a theory is an *explanation,* no more and no less. It might indeed be a hypothesis, such as a grand unified theory (GUT) of elementary particles and their interactions. It might even form a speculation, such as any of various ideas about why the expansion of the universe seems to be accelerating. Many theories do start their lives as plausible, or even implausible, hypotheses. But a theory might offer a unifying explanation of a broad range of phenomena and might, furthermore, predict new phenomena and thus be testable. And such a theory might continually pass test after test after test. It then becomes accepted as *the* theory. Such in fact are Darwin's and Einstein's theories, each in its field of application.

But whether a scientific theory is speculative, hypothetical, promising, competing, or reigning, it always forms an attempt to explain. It always proposes a way of understanding what is known. Moreover, to be taken seriously, it should also predict what is not yet known. That makes it falsifiable and thus testable. Correct predictions confirm a theory. Sufficiently many such successes can make it accepted. A single failure, however, causes it to crash. Then a revision might be called for, or possibly a new theory altogether.

Do successful theories live forever? Not ordinarily, at least not so far. And at least not in the field of physics. While the basic ideas of Darwin's theory of evolution, for instance, might possibly maintain their validity for all time, that does not seem to be the case in physics. What has been happening is that as theories are tested under increasingly extreme conditions, they invariably reveal their limitations. They then become subsumed under more general theories of broader applicability. Sir Isaac Newton's laws of motion proved inaccurate for speeds that are a large fraction of the speed of light. Einstein's special theory of relativity fixed that problem. Newton's law of gravitation could not correctly deal with strong gravitational fields and high speeds. Einstein's general theory of relativity gave the correct treatment and, moreover, extended the limits of validity of the special theory. As for limitations on the general theory of relativity, it does seem that the theory is not compatible with quantum theory, itself another very successful physics theory. Physicists commonly assume that both Einstein's general theory of relativity and quantum theory will eventually be subsumed under a very general theory, which, it is hoped, will offer deep insight into space, time, and the quantum aspect of nature.

Here is an archetypal example of a physics theory and how it develops. Johannes Kepler carefully studied Tycho Brahe's observations of the motions in the sky of the Sun and of the planets that were then known—Mercury, Venus, Mars, Jupiter, and Saturn. As a result, he developed three laws that correctly describe the motions of the planets, including that of planet Earth, around the Sun. Kepler's laws, however, do not form a theory, since they do not, and were not intended to, explain anything. What they do, though, is show that the motions of the six planets are not independent. Their motions present particular cases of certain general laws of planetary motion. Kepler's laws have proven valid for the additional planets that were discovered—Uranus, Neptune, and Pluto (since demoted from planethood)—for the other objects orbiting the Sun, and also for the systems of moons around the multimoon planets.

Being an excellent physicist, Newton did not just accept the laws that Kepler had discovered but inquired into their cause. *Why does nature behave in just this way?* He looked for an explanation, a *theory,* for Kepler's laws. And he found a theory, in the form of his own three laws of motion and law of gravitation. This theory explains not only Kepler's laws in the sky, but also a host of other phenomena, including, very importantly and usefully, mechanical effects on Earth. His theory was so successful that it lasted some 200 years, until its limitations became apparent. Newton's theory is still the theory used for everyday life, such as for designing buildings and vehicles and for athletic activities, including baseball and pole vaulting. But in principle, Newton's theory has been subsumed under Einstein's theories and quantum theory.

Of course, one cannot change the way *theory* is commonly used in ordinary conversation. But one *can* be cognizant of the very different meaning of the term in science. That way one can avoid the absurdity of labeling as mere hypotheses such monumental achievements of science as Darwin's and Einstein's theories.

FURTHER READING

Feynman, Richard. *The Character of Physical Law.* Cambridge, Mass.: M.I.T. Press, 1965.

Hatton, John, and Paul B. Plouffe. *Science and Its Ways of Knowing.* Upper Saddle River, N.J.: Prentice Hall, 1997.

Wilson, Edward O. *Consilience: The Unity of Knowledge.* New York: Vintage, 1998.

are not statistically significant, that is, if the computed value does not reach the alpha threshold, then either a revision of the hypothesis is necessary or the experiment may need to be repeated using a larger sample size. Consideration of the analysis phase is critical at the experimental design phase to ensure all of the necessary control and treatment options are included so that one can draw reliable conclusions in the end. Improper or incomplete experimental design opens the door for the refutation of statistically supported decisions.

After determining whether the results support the hypothesis of the experiment, the scientist publishes the results in peer-reviewed journals in order to share the information with other scientists. Sharing information allows others to repeat and support the results and conclusions or to point out problems with the experimental design, results, or conclusions.

SCIENTIFIC THEORIES AND LAWS

The term *theory* is often misused by the general population to mean a guess or possible explanation, much like a hypothesis. A scientific theory involves much more than that and can often be used to predict outcomes in new experimental systems. After experimental models or conclusions survive the scrutiny of multiple tests, repeated results, and many scientists drawing the same conclusion, they evolve into a scientific theory. For example, Albert Einstein's theory of relativity has been clearly supported by multiple researchers, since he first proposed it in 1905.

Scientific laws *describe* how things happen and can be used to predict future events, while scientific theories *explain* scientific laws, give rise to laws, and can thus possibly be used for prediction as well. Unlike human laws, such as speed limits, scientific laws cannot be broken. A car has the physical ability to travel faster than the speed limit; the driver makes a decision to maintain a certain speed. The law of gravity represents a scientific law and cannot be broken. An apple simply cannot choose to fall upward.

FURTHER READING

Gauch, Hugh G., Jr. *Scientific Method in Practice*. New York: Cambridge University Press, 2003.

separating mixtures A mixture is a physical association between two or more substances. Mixtures can be classified as either heterogeneous or homogeneous, depending on whether or not the mixture has a uniform consistency. Homogeneous mixtures, also called solutions, are the same throughout. Solutions are made up of a solute, the substance being dissolved, and a solvent, the substance in which the solute is dissolved. The average size of a solute particle in a solution is less than 3.9×10^{-8} inch (1 nm). A mixture with a nonuniform composition is known as a heterogeneous mixture and is classified into categories based on the size of the particles found in the mixture. When the particle size exceeds 100 nm, the mixture is called a suspension, and the particles will settle out upon standing. Sand stirred up in water is an example of a suspension. One must continue to stir the mixture or else the sand will settle to the bottom of the container. Heterogeneous mixtures with intermediate particle size between approximately 1 nm and 100 nm are called colloids. Examples of colloids include egg whites, marshmallows, mayonnaise, and fog.

Separating mixtures into their component parts is often desirable in order to use the individual substances. Separation techniques capitalize on the physical properties of the components involved. Any characteristic that distinguishes a particular component from the others in the mixture can be used for separation. Several physical properties that are useful in separating the components include the size, color, density, magnetic properties, solubility, boiling point, charge, and affinity for other molecules.

The most straightforward method of separation is simple physical separation done by hand or with a tool. Manual separation of large particles in the mixture can take place using forceps and physically picking out the pieces of interest. Separating pebbles from a sand mixture would easily be accomplished by hand. Magnetic components can be separated from the nonmagnetic components of a mixture by using a magnet. A large-scale example of this is a recycling center with steel and aluminum cans. Most cities and towns collect aluminum and steel cans for recycling in the same dumpster. This creates a mixture of metal types. The nonmagnetic aluminum cans are separated from the magnetic steel cans through use of a large magnet. Capitalizing on the different magnetic properties of the two components allows the recycling company to use fewer dumpsters and makes it easier for the community to recycle. Mixture components can also be isolated on the basis of density. Placing the mixture in water causes any component that is less dense than water to float and be isolated from the surface of the water. Components that are denser than water will sink to the bottom and then can be separated by filtration.

Filtration works by separating the components in the mixture by size. If the mixture is not already in a liquid phase, the first step is to suspend it in water. One must choose filter paper with a pore size that is smaller than the particles one is trying to remove and large enough to allow the other mixture components to pass. The liquid sample is poured over the filter paper, and the filtrate (liquid that passes through the filter) is collected. Any solid particles that are too large to fit through the pores of the paper will be retained by the filter paper. Further treatment of the filtrate can be performed to isolate any remaining components, or, alternatively, if the component of interest is on the filter paper, then it can be dried and tested for purity. Filtration appears in many separation methods as a preliminary step to remove larger particles prior to finer separation techniques.

One can take advantage of differences in the solubility of mixture components to facilitate the separation process. If one of the components in a mixture dissolves in water and the others will not, then separation of that component can be achieved by combining the mixture with water. After the component of interest dissolves, the liquid can be decanted off and separated from the insoluble components in the mixture. The soluble component can then be separated from the water by evaporating off the water. When a large amount of water is present, heating speeds the removal of liquid from the sample. After the water evaporates, the solid component, which can be further purified or studied, will be left behind.

Distillation separates components on the basis of differences in boiling points. The production of petroleum products and alcoholic beverages utilizes this process, which works well when a mixture contains a nonvolatile solid and a volatile liquid. For example, distillation purifies water by removing salts that are present. Components with similar boiling points are difficult to separate by distillation. To use distillation effectively, the boiling point of the individual components or, minimally, the boiling point of the substance one is trying to isolate must be known. Distillation removes impurities from a mixture that contains solids dissolved in a liquid and also is used to remove certain liquids from a mixture containing more than one liquid.

Several types of distillation procedures exist; the most basic are simple distillation and fractional distillation. Simple distillation uses an apparatus that includes a boiling flask, condenser, and collecting flask. The sample of the mixture is heated; when the boiling point of the first substance is reached, that substance will evaporate and move up into the condenser that is surrounded by cooling water. The water cools the evaporating substance; it condenses again and can be collected in the collecting flask. Fractional distillation differs from simple distillation because fractional distillation separates components that have similar boiling points. In fractional distillation, samples or fractions that contain different mixture components are collected. The procedure is similar to simple distillation, including boiling the sample, vaporizing the liquid, and condensing the

vapor. In fractional distillation, the sample is then vaporized and condensed again, allowing for the separation of liquids that have closely related boiling points.

Chromatography is a class of procedures that involve passing the mixture over a medium, either paper, a solid column, or a gas column, and separating the sample by size and its attraction for the column or paper. There are many types of chromatography, including size-exclusion chromatography, ion-exchange chromatography, affinity chromatography, and gas chromatography. In size-exclusion chromatography, the mixture is passed over a column made up of porous beads. The size of the components in the mixture determines the speed at which they move through the column. The larger particles are excluded from the smaller pores in the column; therefore, they have a shorter distance to travel in the column and will pass through first. The smaller particles will travel through the column, moving in and out of the pores of the beads along the way, taking a longer time to pass through the entire column. Samples called fractions collected as the sample flows out of the bottom of the column contain the mixture components separated by size, with the earlier samples containing the larger particles.

In ion-exchange chromatography, the mixture is passed over a charged column that interacts with the oppositely charged components in the mixture. If the column is negatively charged, then the positively charged components will be attracted and bind to the column. The negatively charged particles will be repelled and pass right through the column, efficiently separating the charged particles. The positively charged particles can then be eluted, removed from the column by passing a solution over the column that competes for binding and releases the component for collection or study. Ion-exchange chromatography is the basis for separating the ions from hard water in a water purifier.

Affinity chromatography is a separation technique based on the attraction of the mixture components for molecules affixed to the column. Known substrates for enzyme molecules can be attached to the column, and an enzyme sample can be passed over the column. The corresponding enzyme-substrate interactions occur, and the enzyme is attached to the column. The enzyme is then eluted from the column by breaking that interaction.

Gas chromatography (GC) is useful in the separation of liquid-liquid mixtures that include at least one volatile liquid. The samples are injected into the GC and instantly raised to a temperature above their boiling point to flash-vaporize the sample. The sample then moves with the help of a carrier gas over the column and is separated on the basis of size and attraction for the column, and the fractions are collected at the end and analyzed for purity.

Centrifugation is a class of separation techniques that separate the mixture on the basis of differential molecular weights and density by rapidly spinning the sample. Centrifugal force is not actually a force but a function of the inertia of the mixture components. While it appears that the particles are being forced away from the center of rotation, the particles are simply attempting to continue their straight-line motion in one direction. The force pushing against the particles is created by the outside of the rotor and causes the particles to move in a circle. Centrifugation aids in the isolation of cellular components by cell biologists and allows for the removal of unwanted components in a mixture, as well as the collection of desired components. A centrifuge spins the samples at a high rate of speed, which causes the heaviest particles to settle to the bottom of the tube as a pellet. The liquid portion of the sample, the supernatant, contains the less-dense particles. The supernatant can be removed for further separation and study, while the pellet may be resuspended in a buffer or water and studied.

Dialysis is a separation technique based on size and shape of the molecule and the ability of the molecule to move through the pores of a semipermeable membrane. A liquid mixture is placed in dialysis tubing and suspended in a buffer of different concentration. The components of the mixture that are smaller than the pores of the semipermeable membrane will diffuse through the tubing and dissolve in the buffer until reaching equilibrium. The particles that are too large to leave the dialysis tubing will stay inside. This is a useful method to change the concentration of a buffer in a mixture without taking the solute molecules out of solution. Dialysis is a common laboratory technique and is the same principle applied to the medical field in kidney dialysis.

See also CENTRIFUGATION; CHROMOTOGRAPHY.

FURTHER READING

Wilbraham, Antony B., Dennis D. Staley, Michael S. Matta, and Edward L. Waterman. *Chemistry*. New York: Prentice Hall, 2005.

simultaneity   The occurrence of two or more events at the same time is termed simultaneity. It is of particular concern in Albert Einstein's special theory of relativity, since the theory shows that simultaneity is observer dependent and, thus, relative (according to the meaning of the term *relative* in the special theory of relativity). Two events might possess the relation that a light signal cannot connect them, that if such a signal were to be emitted at one event, it would

reach the location of the second event only after that event occurred. Such events are said to have a spacelike separation. For such pairs of events, and only for such pairs, some observers might measure their times of occurrence and find the events to be simultaneous. Other observers, in motion with respect to those, find that one of the events occurs before the second. Depending on the direction of an observer's motion, either event might be the earlier. So, for events that possess spacelike separation, their temporal order, including their simultaneity, is observer dependent and, thus, relative.

The following imaginary scenario shows how the relativity of simultaneity comes about. It is based on one of the fundamental postulates of the special theory of relativity, that the speed of light in vacuum has the same value, $c = 2.99792458 \times 10^8$ meters per second (m/s), for all observers that are moving with respect to each other at constant velocity (i.e., at constant speed in a straight line). Let a train car be moving past a station platform at constant speed, lower than c. An observer sits in the center of the car. On the platform, another observer has arranged electric contacts and circuits so that when the ends of the car simultaneously pass and close two sets of contacts, two simultaneous light flashes are produced, one at each end of the car. When that happens, the platform observer sees light pulses from the flashes rushing toward each other and meeting in the middle, where the car observer was sitting at the instant of contact. But in the meantime, the car observer moves some distance. As a result, the light pulse from the front of the car reaches and passes her before meeting its partner where the middle of the car was when the pulses were generated, and the pulse from the rear catches up with the car observer only after meeting and passing its partner from the front. So, the platform observer sees the car observer detecting the light pulse from the front of the car before detecting the one from the rear and understands how that comes about.

Now take the view of the car observer. She sees the light pulse from the front of the car first, followed by the pulse from the rear. She knows that both pulses traveled the same distance—half the length of the car. She also knows that both light pulses traveled at the same speed, c. This latter fact is crucial, as it is where the special theory of relativity enters the scenario. The car observer, thus, knows that both pulses were in transit for the same time interval (the same distance divided by the same speed). Since she detected the pulse from the front before she detected the one from the rear, the pulse from the front had to be produced before the pulse from the rear. Here is the relativity of simultaneity. According to the platform observer, both flashes are simultaneous, while

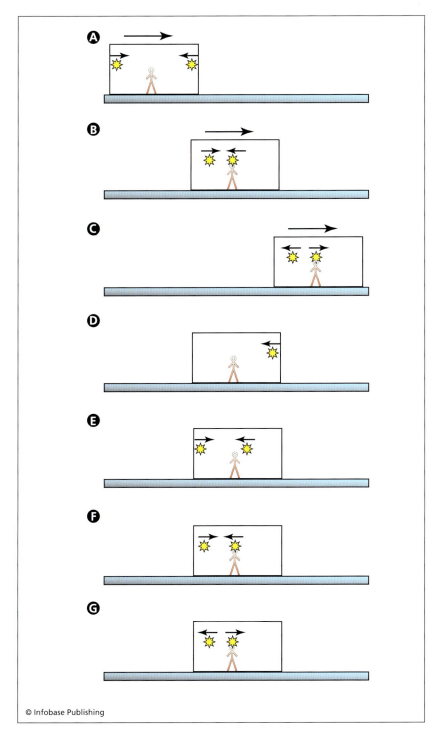

Simultaneity is relative (i.e., observer dependent), according to the special theory of relativity. A train car, moving to the right with an observer at its center, passes a station platform. As observed from the platform: (a) Two light pulses leave the ends of the car simultaneously. (b) The pulse from the front of the car reaches the car observer first, since she is moving toward it. (c) After the pulses pass each other, the one from the rear of the car overtakes the car observer. As the car observer sees it: (d) A light pulse leaves the front of the car. (e) Later a light pulse leaves the rear of the car. (f) The pulse from the front reaches the observer. (g) Then the pulse from the rear reaches the observer, after the pulses have passed each other. The light pulses' leaving the ends of the car occurs simultaneously for a platform observer, while for the car observer the pulse from the front leaves before the pulse from the rear does. Both observers agree that the pulse from the front reaches the car observer before the one from the rear, but this fact is interpreted differently by the observers. (The figures are not drawn to scale.)

© Infobase Publishing

for the car observer the one in front occurs before the one in the rear.

As a general rule, if one observer detects two events as occurring simultaneously, then a second observer, in motion along the line connecting the locations of the events, detects the event at the foremost location as occurring before the event at the rearmost location. The relativity of simultaneity lies at the root of the effects of length contraction and time dilation in the special theory of relativity.

See also EINSTEIN, ALBERT; SPECIAL RELATIVITY; SPEED AND VELOCITY; TIME.

FURTHER READING

Jammer, Max. *Concepts of Simultaneity: From Antiquity to Einstein and Beyond.* Baltimore: Johns Hopkins University Press, 2006.

Serway, Raymond A., Clement J. Moses, and Curt A. Moyer. *Modern Physics,* 3rd ed. Belmont, Calif.: Thomson Brooks/Cole, 2004.

solid phase peptide synthesis Solid phase peptide synthesis is a method of preparing proteins, polypeptides (long chains of amino acids), or smaller peptides by linking monomers together by creating peptide bonds. Proteins are one of four basic types of macromolecules important to biochemistry, along with lipids, carbohydrates, and nucleic acids. These polymers of amino acids are responsible for most of the cellular activity in a living organism. Having roles in movement, structure, catalysis of reactions, defense against invading organisms, cellular signaling, and cell division, proteins are the most diverse of all the macromolecules and are often considered the workhorses of the cell. Development of the technical capability of scientists to produce desired peptides has a far-reaching effect, as peptides are now known to act as hormones, neurotransmitters, signaling molecules, toxins, and antibiotics. Solid phase peptide synthesis is a reliable method for producing polypeptides.

PROTEINS AND PEPTIDES

Protein polymers are made up of monomers known as amino acids. In most animals, 20 different amino acids make up all of the known proteins. All amino acids have the same basic structure with the difference being in the side chain of the amino acid (-R). Each amino acid contains a central carbon (the α carbon), which has an amino group (-NH$_3$), a carboxyl group (-COOH), a hydrogen (-H), and an R group. The 20 different R groups confer the chemical characteristics on the individual amino acids. The side chains vary from a single hydrogen atom (in the amino acid glycine) to large ring structures (in the amino acid phenylalanine) and acidic and basic side chains (in aspartic acid and glutamic acid).

When forming a peptide bond, these amino acids are joined through a dehydration reaction (a bond formed with the removal of water) between the carboxyl group of the first amino acid and the amino group of the next amino acid, creating a peptide bond. Each peptide and polypeptide has an amino terminus (an end with a free amino group) and a carboxyl terminus (an end with a free carboxylic acid group). The reactivity, polarity, and size of the R groups that are attached on the polypeptide chain help control the ways the protein folds and functions. The peptide backbone can be represented as N-C-C with the nitrogen from the amino group followed by the α carbon and finally the carboxyl carbon. Because the α carbon is surrounded by four different side groups, each amino acid has two optical isomers, or stereoisomers, with the exception of glycine, which has a single hydrogen atom as its R group. In general, stereoisomers may be right-handed or left-handed, yet most amino acids associated with biological organisms are the left-handed, or L, form.

DEVELOPMENT OF SOLID PHASE PROTEIN SYNTHESIS

Solid phase protein synthesis requires anchoring the carboxyl terminus of an amino acid to a resin. The growing peptide is then formed from the carboxyl to amino terminus with the addition of sequential amino acids. This process was developed by the American chemist Bruce Merrifield. His interest in chemistry began when he was still in high school. Merrifield attended graduate school in the biochemistry department of UCLA and completed his doctorate in 1949. While performing his dissertation research and his subsequent work at the Rockefeller Institute of Medical Research on peptide growth factors, he recognized the need for a reliable method to synthesize proteins and peptides for research purposes. Prior to the work of Merrifield, the synthesis of even dipeptides and tripeptides was still a complicated process. Once the peptides became larger, the technical difficulties in their synthesis made the process ineffective.

Merrifield built on the work that was known at the time, including peptide synthesis methods developed by the American chemist Vincent du Vigneaud, who received the Nobel Prize in chemistry in 1955 for his synthesis of oxytocin, a peptide consisting of nine amino acids. The main obstacle to synthesis of larger peptides was that after the addition of each amino acid, the product would have to be isolated from the rest of the materials in the reaction. This could be a complex and time-consuming process. Merrifield developed the solid phase protein synthesis method, in 1959, in which the peptide being produced was immobilized to a resin. This allowed the sample to be washed to remove the excess reactants and any by-products while keeping the peptide product secure, creating very high yields from each synthesis step. Efficient peptide synthesis was a reality, and the process could even be automated. Merrifield's first goal was to synthesize an enzyme, and he succeeded in synthesizing ribonuclease A. Merrifield's work and the development of a method that allowed scientists for the first time to synthesize large peptides earned him the Nobel Prize in chemistry in 1984.

Advantages of solid phase peptide synthesis over previous methods included the following:

- All of the reactions can take place together rather than purifying between and transferring vessels.

- Since no purification is necessary, only washing, the losses that are routinely seen in crystallization or other purification techniques do not occur.
- Reactions can be forced to completion, thereby increasing yield.

RESINS

The basis of solid phase protein synthesis involves the linking of an amino acid to an insoluble yet porous resin. There are hundreds of types of resins available, but the similarity between them is that they are porous and swell upon the addition of liquids. A common type of resin was developed by Merrifield out of spherical beads made of styrene and a cross-linking agent. These original beads can swell to up to 200 times their original size in the presence of an organic solvent such as dichloromethane or dimethylformamide, which swells the resin and consequently increases the rate of the reaction. The coupling reactions occur within the swelled resin and solvent, allowing access to more of the growing peptide chain.

PROTECTION

Protection is necessary for two areas of the amino acids being used in the synthesis. The side chains of the amino acids being used must be protected throughout the synthesis process. The Merrifield method of protection includes benzyl groups, or substituted benzyl groups. Side chain protection in the Fmoc (9-fluorenylmethoxycarbonyl) system described later is carried out by the addition of *tert*-butyl groups that can be removed with trifluoroacetic acid (TFA).

The amino end of the first amino acid must also be protected during the synthesis process. This protection can be carried out by several methods. The original Merrifield method utilized a *t*-butoxycarbonyl group (Boc), but now the most common is the Fmoc group added to the amino terminus. The amino end of each subsequent amino acid must also be protected in the same way. This protection must be reversible, in order for the next amino acid to be attached to that amino terminus. The addition of TFA removes the Boc without disrupting the side chain–protecting groups or disrupting the anchoring to the resin. As the process of solid phase peptide synthesis improved, chemists found that repetitive use of TFA could cause premature cleavage of the resin attachment or the side chain blocks. The use of Fmoc, which is removed using a base that has no effect on the side chains or the resin attachment (instead of Boc), improved the process. The Fmoc

process is a milder process and has been shown to have higher purity and yield.

ACTIVATION AND COUPLING

Activation is required for the amino acids being added to the growing peptide chain during solid phase peptide synthesis. The Merrifield method of activation includes such chemicals as dicyclohexylcarbodiimide, esters, or anhydrides. Once the incoming amino acid is activated, it can be added to the growing peptide chain. Coupling of the activated amino acid to the amino acid attached to the resin enables the peptide chain to grow. In order to couple the two, the Fmoc protective group on the first amino acid must be removed and the activated amino acid is added. The coupling process can be repeated as many times as necessary to achieve the desired peptide length.

DEPROTECTION

Once the peptide is complete, deprotection is required to remove the protective groups from the amino acid side chains. The Fmoc group from the amino terminus must also be removed, and the entire newly synthesized peptide must be removed from the resin attached at the carboxyl terminus. The most common method of deprotecting Boc-protected amino acids includes treatment with strong concentrations of acid such as hydrofluoric acid. Fmoc-protected groups can be removed with a base, and the *tert*-butyl attachments can be removed using TFA.

The illustration of solid phase peptide synthesis depicts a complete series of solid phase peptide synthesis. The first step is to bind the carboxyl terminus of the final amino acid in the chain to a resin. The amino terminus of this amino acid is blocked with a Fmoc group. Next, the second amino acid is treated with an activator that allows it to become part of the chain. This activated amino acid is coupled to the first amino acid that has been deprotected by the removal of the Fmoc group, creating a peptide bond, resulting in a dipeptide. This process is continually repeated until all of the required amino acids have been added. Once the chain has reached the required length, the final Fmoc group blocking the amino terminus of the peptide is removed. The final steps also require the removal of the protecting group from each of the R groups on the amino acids of the peptide. The peptide then needs to be cleaved from the resin in order to generate the free peptide.

The structure of peptides and proteins determines their function. Care must be taken when performing solid phase peptide synthesis to ensure that complete functional peptides are produced. Termination of the peptide prematurely will result in small, improperly functioning peptides. Deletion of even one amino acid can alter the structure and function of any peptide or protein. The addition of an incorrect amino acid can also dramatically affect the peptide or protein. Pure reagents and resins help prevent such problems, as does ensuring that the reaction goes to completion. Use of an automated system overcomes many

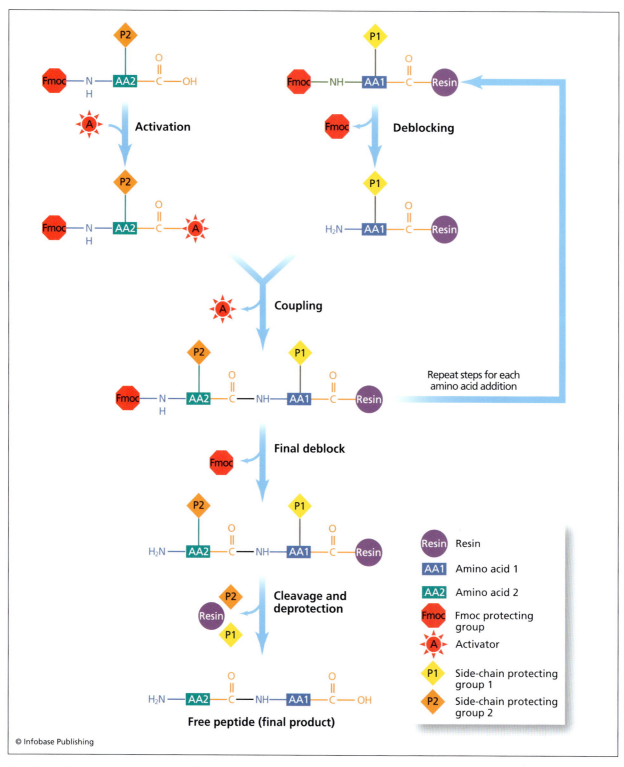

The formation of synthesized peptides involves the addition of the carboxyl group of the first amino acid to a stable resin, represented by AA1. The blocking agent on the amino terminus of this amino acid is removed and the first amino acid is coupled to a second amino acid, AA2. This process continues until all of the necessary amino acids are added to the chain. The final blocking agent is removed from the amino groups, and then the completed peptide is removed from the resin.

of these problems. Peptide synthesizers perform all of these tasks automatically, greatly increasing the purity and yield of peptide produced.

APPLICATIONS OF SOLID PHASE PEPTIDE SYNTHESIS

The synthesis of specific peptides is a growing field. After Merrifield's optimization of peptide synthesis and his production of ribonuclease A, several other scientists began using his technique to produce such peptides as epidermal growth factor (EGF), vasopressin derivatives, and, most commonly, antigen epitopes, in order to study the immunogenic response to antigens. Improvements in the technology and increased demand for synthesized peptides have greatly expanded the use of this process. Scientists have developed modified amino acids that differ from the traditional 20 amino acids and can create peptides with physical and chemical properties that differ from the true peptide, such as solubility. Using peptide synthesis methods and modified amino acids, scientists can create peptides with properties that make them useful in the laboratory and clinical settings.

After the inception of solid phase peptide synthesis, scientists quickly recognized that the process of anchoring a monomer, then creating a polymer was not restricted to peptides. Scientists have developed similar techniques to create polynucleotides (polymers of the building blocks contained in nucleic acids such as DNA and RNA) and polysaccharides (polymers of sugars or carbohydrates). These polymers are useful for research as well as for clinical applications.

See also BIOCHEMISTRY; PROTEINS.

FURTHER READING

Chan, Weng C., and Peter D. White. *Fmoc Solid Phase Peptide Synthesis: A Practical Approach.* New York: Oxford University Press, 2000.

solid state chemistry Solid state chemistry is a branch of chemistry involved in the study of the physical and chemical properties of substances in the solid state. All matter exists in one of four states, as determined by the interactions of the particles in the substance: solids, liquids, gas/vapor, and plasma. When the particles have little or no interaction between them, they are said to be gas or vapor. The term *gas* is used for substances that are normally a gas at room temperature, and the term *vapor* is used to describe a substance in the gaseous state that is normally a liquid at room temperature. The attractive forces between the atoms or molecules in a liquid are stronger than those seen in a gas. The attractive forces within a solid are stronger still.

The shape and volume differ among the states of matter. Matter that is in the gas phase has no definite shape or definite volume. Liquids have a definite volume, although the shape of a liquid depends on the shape of the container. Solids have both a definite shape and a definite volume. The motion of the particles also distinguishes the states of matter. The particles in a solid move very slowly in a vibrating manner. Particles in a liquid can flow past one another. Gaseous particles move in constant random motion and are able to move in all directions. Plasma is a distinct state of matter that exists when a gas is ionized by either electricity or extremely high temperatures such as that found in stars. This ionized form of a gas can carry an electric charge and, while not very common on Earth, is the most common form of matter elsewhere in the universe.

INTERMOLECULAR FORCES

The attractive forces between atoms and molecules in the solid state hold the molecules together. Intermolecular forces are forces that hold molecules together but that are not true bonds. Several types of attractions between molecules contribute to their shape and properties. Although intermolecular forces are not as strong as true ionic or covalent bonds, they are still significant forces in the structure and function of matter. Important types of intermolecular forces include dipole-dipole interactions, dispersion forces, and hydrogen bonds.

A dipole is a separation of charge within a molecule due to the unequal sharing of electrons. When electrons are shared in covalent bonds and one of the elements is more electronegative (attracted to shared electrons in a bond) than the other, a separation of charge exists and the electron spends more time in the vicinity of the more electronegative element, giving it a slightly negative charge. When a molecule has this separation of charge, it is known as a polar molecule. Attraction between polar molecules due to the formation of these dipoles creates a relatively strong intermolecular force known as a dipole-dipole interaction.

One specific type of this intermolecular force is hydrogen bonding. A hydrogen atom attached to a more electronegative element in a covalent bond will always have a slightly positive charge and be able to interact with the slightly negative atoms on other polar molecules. Water is a perfect example of a hydrogen-bonding substance, as the attraction of water molecules for one another confers many of the properties of water. The boiling point and melting point of water are higher than they would be if the molecules did not have strong intermolecular forces. Solid water is unique in that it is the only substance that expands as it cools. This expansion is due to the hydrogen bonds that form between the water molecules in ice. This expansion also explains why ice is less dense than water. Hydrogen bonds are responsible for many biochemical phenomena, as well. The three-dimensional structure of proteins is influenced by the hydrogen bonding of polar amino acids in the proteins. The double helix found in DNA is stabilized by the hydrogen bonding between the nitrogenous bases (adenine, thymine, guanine, and cytosine) of the nucleotide subunits that compose DNA molecules.

Putting a polar molecule in the vicinity of a nonpolar molecule can create what is known as an induced-dipole in the nonpolar molecule. An induced dipole is created when the slight charge of a polar molecule is placed in close proximity with a nonpolar molecule, causing the electrons of the nonpolar molecule to move toward the other side of the molecule as they are repelled from the neighboring electrons. This sets up a dipole in a molecule that normally would not have one. The interaction between a polar molecule and a nonpolar molecule in this way is known as a dipole-induced dipole interaction.

When two nonpolar molecules are in close proximity, it is possible for their electron arrangements to become synchronized, creating an induced dipole in the atoms or molecules. This induced dipole can then create an induced dipole in neighboring molecules to create a weak intermolecular force known as a dispersion force or London force.

MELTING POINTS

The melting point of a solid is a physical characteristic that can be used to identify the solid. Melting points vary from low temperatures to extremely high temperatures. The melting point of steel varies, depending on its composition, but on average steel melts at temperatures near 2,500°F (1,370°C). Tungsten is the metal with the highest melting point, at 6,192°F (3,422°C), while francium has a melting point of only 81°F (27°C). Determination of the melting point of a substance can be used as one key to determining the makeup of a solid as well as the purity of a solid. When a solid changes to a liquid during the process of melting, the kinetic energy of the particles increases. This causes the temperature to rise, as temperature is simply the measure of the average kinetic energy of the particles. The stronger the intermolecular forces between the atoms and molecules in a solid, the higher the temperature required to melt the solid. Substances that have very weak intermolecular forces melt and vaporize at temperatures lower than those with strong intermolecular forces.

CRYSTALLINE SOLIDS

Many solids exist in the form of crystals, ordered regular arrangements of the atoms in a solid. Crystals have characteristic structures and are very rigid. Scientists originally studied crystals in terms of their macroscopic geometry. To decipher the arrangement of the atoms, crystallographers measured the angles of the crystal face relative to theoretical reference points (crystallographic axes) and established the symmetry of the crystal being studied. The different types of crystals are made up of a fundamental unit called a unit cell, which combines with other unit cells to form the crystal. The unit cell is the smallest part of the crystal structure that contains all of the attributes of the crystal. Crystal structures can be divided into nine categories: cubic, body-centered cubic, face-centered cubic, hexagonal, monoclinic, triclinic, orthorhombic, rhombohedral, and tetragonal.

Cubic crystal structures fall into three different arrangements of atoms in a unit cell. These are the simple cubic, face-centered cubic, and body-centered cubic structure. The unit cell of a simple cubic arrangement displays one of each atom at the corners of each face. The body-centered cubic arrangement has four atoms at each corner of the cubic unit cell and one atom in the center of the unit cell. The face-centered cubic arrangement has one atom at each corner of each face and one atom in the center of each face, leaving the center of the cube empty.

AMORPHOUS SOLIDS

Any solid that does not form an organized crystal lattice structure is known as an amorphous solid. These types of solids include such substances as plastics and glass. These substances existed as liquids that were then cooled in a nonspecific manner. Amorphous solids do not have specific melting points, as the melting of their nonspecifically ordered structure is not consistent. The Nobel Prize in physics for 1977 was shared by three scientists who studied the properties of amorphous solids: the American physicist Philip Warren Anderson, the British physicist Sir Nevill F. Mott, and the American physicist John Hasbrouck van Vleck. Anderson and van Vleck worked primarily on the magnetic properties of solids. Mott received the Nobel Prize for his discoveries on the electrical properties of amorphous solids. He determined the chemical properties that led amorphous solids to be good semiconductors.

METALLIC SOLIDS

Solids formed between metal atoms are unique in their type of bonding. Atoms within metallic solids have what is known as delocalized electrons in their valence shells, meaning the electrons do not strictly stay with the atom they belong to. By definition, metal atoms have three or fewer valence electrons. In order to complete their valence shell, these atoms would have to share up to seven electrons with other atoms, but this is not a plausible bonding mechanism. Delocalized electrons form what is known as a "sea of electrons" surrounding the metal atoms and allowing all of the metal atoms to utilize one another's valence electrons to complete their outer shell. This type of bonding takes place between atoms of all types of metals, including copper, as well as within alloys, such as brass. Because of the mobility of the valence electrons within metals, they are good conductors of electricity.

TECHNIQUES

The study of solids and crystal structures involves common chemistry structural techniques such as X-ray crystallography and nuclear magnetic resonance (NMR). Solid state chemists study the atomic arrangements in the solid as well as the intermolecular arrangements between the solids.

X-ray diffraction or X-ray crystallography is a technique used to understand the material and molecular structure of a substance. Crystallography now depends on the analysis of diffraction patterns (the way rays are scattered by the molecules they hit) that emerge from a closely packed, highly organized lattice, or crystal, targeted by a beam of short waves. Scientists most frequently use X-rays, a form of electromagnetic radiation, as the beam source, although neutrons and electrons (two kinds of subatomic particles possessing wavelike properties) are alternative sources. X-ray crystallography is useful for studying microscopic entities or molecules at the atomic level. X-rays have shorter wavelengths than the atoms being studied, making it possible to differentiate atomic positions relative to each other in a molecule.

NMR is a technique that reveals the location and arrangements of atoms in a molecule on the basis of their spin in a magnetic field when they are bombarded with radio waves. The atoms spin according to the way they line up in the magnetic field. Once the atoms reach equilibrium in the magnetic field, the samples are subjected to radio waves of particular frequencies. When the radio frequency matches the frequency of the precession (change in the direction of rotation around an axis) of the atom, the atom changes its direction of spin. Energy will be absorbed from the radio waves if the atom changes from the low-energy state to the high-energy state. Energy will be given off by the atom to the radio waves if the change is from high-energy state to low-energy state. NMR measures the amount of energy that must be gained or lost in order to change the direction of spin of the nucleus. The peaks present in the NMR display demonstrate the frequency of energy given off or taken up by the atom and thus information about the location of an atom and its neighbors.

See also COVALENT COMPOUNDS; NUCLEAR MAGNETIC RESONANCE (NMR); PROTEINS; STATES OF MATTER; X-RAY CRYSTALLOGRAPHY.

FURTHER READING
Smart, Leslie, and Elaine Moore. *Solid State Chemistry: An Introduction.* 3rd ed. London: Taylor and Francis, 2005.

special relativity Albert Einstein's special theory of relativity, called "special relativity" for short, which he proposed in 1905, deals with the fundamental concepts of space and time and the way distances and time intervals are measured and relate to each other. Special relativity is a theory about theories, what is called a supertheory, in that it lays down conditions that other physics theories must fulfill. Two basic postulates define special relativity: (1) The laws of physics are the same for all observers that move at constant velocity with respect to each other. In other words, every pair of observers who are in relative motion in a straight line and at constant speed discover the same laws of physics. (2) One such law is that light (and all other electromagnetic radiation as well) propagates in a vacuum at the speed of $c = 2.99792458 \times 10^8$ meters per second (m/s), or 3.00×10^8 m/s to a reasonable approximation, called "the speed of light in vacuum." This means that all observers, whatever their state of motion, must detect that in vacuum light propagates at the same speed c.

The result of the Michelson-Morley experiment, performed in 1887 by the American physicist Albert Abraham Michelson and American physicist and chemist Edward William Morley, supported the validity of the second postulate. The experiment was an attempt to compare the speed of light as measured in two different directions with respect to the motion of Earth. A positive result would have served as evidence for the existence of the ether, a hypothesized material medium for carrying electromagnetic waves, including light. Michelson and Morley detected no difference, however, proving that there is no ether, that light propagates through empty space (vacuum), and that the speed of light in vacuum is independent of the motion of the observer.

Defining the concepts of *relative* and *absolute* will facilitate further discussion. In the present context, *relative* means observer dependent. If different observers can validly claim different values for the same physical quantity, then that quantity is relative. On the other hand, an absolute quantity is one that has the same value for all observers. The second postulate of the special theory of relativity states that the speed of light in vacuum is an absolute quantity.

Note that in the special theory of relativity the terms *relative* and *absolute* mean no more and no less than observer dependent and observer independent, respectively. There is no claim whatsoever that "everything is relative." Note that the basic postulates of the theory concern absoluteness rather than relativity.

The first postulate implies the nonexistence of an absolute, observer-independent reference frame for constant-velocity motion. In other words, a state of absolute rest is a meaningless concept. This idea was not new with Einstein and is included in Newton's laws of motion, which predate Einstein by some 200 years. The second postulate was revolutionary. It runs counter to everyday experience with the composition of velocities. As an example, assume an observer has a beam of light going past her at speed *c* and has another observer passing her at speed $0.9c$ in the direction of the light beam. According to Newtonian mechanics and everyday experience with bodies in relative motion, the second observer should observe the light beam passing him at speed $c - 0.9c = 0.1c$. But the second postulate states that the second observer, too, will observe speed *c* for the light beam. In other words, the velocity of light is different from that of material bodies and holds a unique status in the special theory of relativity. In Newtonian mechanics, the velocity of light is relative, as are all velocities, while in the special theory of relativity it is absolute.

The differences between relativistic physics and Newtonian physics become manifest at speeds that are a considerable fraction of the speed of light and become extreme as speeds approach the speed of light. For speeds that are sufficiently small compared to the speed of light, relativistic physics becomes indistinguishable from Newtonian physics. This limit is called the nonrelativistic limit.

The special theory of relativity views matters in terms of *events* in space-time. An event is an occurrence at some location at some instant. Space-time, also referred to as Minkowski space (after Hermann Minkowski, a 19–20th-century German mathematician and physicist) in the context of the special theory of relativity, is a four-dimensional abstract space consisting of three spatial dimensions and one temporal one. An event is specified by its location—represented by three coordinates with respect to three spatial axes, say *x*, *y*, and *z*—and by its instant of occurrence—represented by a single coordinate with respect to a single temporal axis, *t*. So an event is represented by a point in space-time, specified by its four coordinates (*x*, *y*, *z*, *t*). A reference frame, or observer, is defined by a specific *x*, *y*, *z* coordinate system along with a resting clock ticking off the time *t*. The subject of space-time is addressed in more detail later in this article.

When reference frames, or observers, are in relative motion, their respective coordinate systems, along with their attached clocks, move with respect to each other. The most general mathematical relations between an event's coordinates in any coordinate system and the *same* event's coordinates in any coordinate system moving at constant velocity relative to the first one are given, in the special theory of relativity, by what are known as Poincaré transformations, after the 19–20th-century French mathematician Jules-Henri Poincaré. Certain more restricted transformations, called Lorentz transformations, named for Hendrik Antoon Lorentz, a 19–20th-century Dutch physicist, are very useful.

As an example of a Lorentz transformation, let coordinates (*x*, *y*, *z*) and time variable *t* be those used by observer O to designate locations and times in her reference frame, while (*x'*, *y'*, *z'*) and *t'* are used by observer O' in his. Let the reference frame of observer O' be moving at constant speed *v* with respect to the reference frame of observer O, in O's positive *x* direction. To fulfill the special condition for a Lorentz transformation, let the respective coordinate axes of O and O' instantaneously coincide when the observers' clocks at their coordinate origins both read 0. Physically, this is no limitation, since whenever one reference frame is in constant-speed rectilinear motion with respect to another, their coordinate axes can always be displaced and rotated and their clocks can always be reset so as to make this the case. Then, if an event is observed to occur at position (*x*, *y*, *z*) and at time *t* with respect to the reference frame of O, and if the *same* event is observed by O' at position (*x'*, *y'*, *z'*) and time *t'* with respect to his reference frame, the Lorentz transformation relating the variables is

$$x' = \frac{x - vt}{\sqrt{1 - v^2/c^2}}$$

$$y' = y$$

$$z' = z$$

$$t' = \frac{t - vx/c^2}{\sqrt{1 - v^2/c^2}}$$

or, inversely,

$$x = \frac{x' + vt'}{\sqrt{1 - v^2/c^2}}$$

$$y = y'$$

$$z = z'$$

$$t = \frac{t' + vx'/c^2}{\sqrt{1 - v^2/c^2}}$$

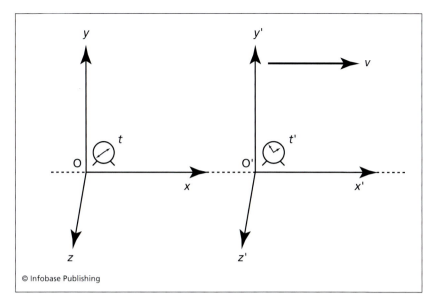

Lorentz transformations relate the coordinates and time variables of two inertial reference frames. The figure shows two such reference frames, O and O', with coordinates (x, y, z) and (x', y', z') and time variables t and t', respectively. O' is moving at constant speed v, which may approach the speed of light, in the positive x direction with respect to O, and their coordinate axes coincide at some time, which is taken to be t = t' = 0.

with c denoting the speed of light in vacuum. All spatial coordinates are in meters (m), all times in seconds (s), and v is in meters per second (m/s).

Note in particular that the observers' clock readings for the same event are related in a way that depends on the location of the event and on the speed v. Note also that as v/c approaches 0, called the nonrelativistic limit, both $\sqrt{1 - v^2/c^2}$ approaches 1 and the x- and x' dependence drops out of the time transformation. Thus, for speeds that are small compared to the speed of light, the transformations reduce to those that are familiar from Newtonian mechanics.

These transformations mix spatial and temporal coordinates. As a result, a pair of events that is observed as simultaneous (i.e., occurring at the same time) by one observer might not appear so to another observer. Other observers, depending on their motion, might measure one event as taking place before the other or the other before the one. In other words, simultaneity is relative. In Newtonian mechanics, on the other hand, simultaneity is absolute.

RELATIVE QUANTITIES

Another result of special relativity is that length is relative: different observers will measure the same object as having different values for its length. Let an observer for whom the object is at rest measure length L_0, called the object's rest length. Then an observer moving at speed v in the direction of the length (i.e.,

an observer for whom the object is moving longitudinally at speed v) measures a shorter length L, where

$$L = L_0 \sqrt{1 - v^2/c^2}$$

Here L and L_0 are expressed in the same unit of length and v is in meters per second (m/s). Note that in the limit of v approaching c, L contracts down to 0. At the other extreme, in the nonrelativistic limit (v small compared to c), L is indistinguishable from L_0. The length contraction effect is mutual, as an object at rest for the second observer is observed by the first to be contracted. While length is relative in the special theory of relativity, length is absolute in Newtonian mechanics.

Time intervals are relative as well. The same clock runs at different rates for different observers. Assume that a clock (or heart, etc.) is ticking off time intervals of duration T_0, as measured by an observer for whom the clock is at rest. Then an observer moving at speed v relative to the first one (i.e., an observer for whom the clock is moving at speed v) measures longer time intervals T for the same ticks (or heartbeats), where

$$T = \frac{T_0}{\sqrt{1 - v^2/c^2}}$$

with the two time intervals expressed in the same unit. In other words, a moving clock is measured as a slower clock. This effect is called time dilation. Similarly, a moving organism is observed to live longer than if it were at rest, and unstable elementary particles have longer lifetimes (or half-lives) when they decay in flight than when they are at rest. Note that as v approaches c, T tends to infinity, and the clock is observed to have almost stopped, the organism is in quasi-suspended animation, and the unstable particles almost never decay. In the nonrelativistic limit (v/c approaches 0), time dilation is negligible, reflecting the fact that in Newtonian mechanics time intervals are absolute. Time dilation is a mutual effect: each of two observers moving relative to each other measures the other's clock as running slower. This leads to what is known as the twin paradox, which is discussed later in this article.

Even mass is not immune to relativistic effects, as the mass of the same body is measured by different observers as having different values. Let m_0 denote the mass of a body for an observer for whom the body is at rest. This is called the rest mass of the body. An observer moving at speed v with respect to the first observer, who therefore observes the body moving at speed v, measures a greater mass m, where

$$m = \frac{m_0}{\sqrt{1 - v^2/c^2}}$$

with m and m_0 expressed in the same unit. Note that the mass of a moving body increases with its speed and approaches infinity as the speed approaches the speed of light. This means that a body's inertia increases with its speed and approaches infinity as v approaches c. In the nonrelativistic limit, the rest mass adequately represents the body's mass. In Newtonian mechanics, on the other hand, mass is absolute.

Relativistic linear momentum has the form

$$\mathbf{p} = m\mathbf{v} = \frac{m_0\mathbf{v}}{\sqrt{1 - v^2/c^2}}$$

where $\mathbf{p}$ and $\mathbf{v}$ designate the momentum and velocity vectors, respectively, and v is the magnitude of $\mathbf{v}$. The SI unit of momentum is kilogram-meter per second (kg·m/s), and the masses are in kilograms (kg). As v approaches c, the momentum approaches infinity, while in the nonrelativistic limit, the momentum attains the Newtonian form.

MASS AND ENERGY

The special theory of relativity assigns an energy E to a body:

$$E = mc^2$$

$$= \frac{m_0 c^2}{\sqrt{1 - v^2/c^2}}$$

where E is in joules (J). Note, very importantly, that even a body at rest possesses energy, called its rest energy

$$E_0 = m_0 c^2$$

So, every mass is assigned an equivalent energy, its mass energy. This energy can be tapped and made use of through the nuclear processes of fission and fusion. Conversely, every energy E possesses a mass equivalent m, whose value is given by

$$m = \frac{E}{c^2}$$

So, energy behaves as mass: it possesses inertia and is affected by the gravitational field.

A body's kinetic energy is its total energy less its rest energy, or

$$\text{Kinetic energy} = E - E_0$$

$$= m_0 c^2 \left(\frac{1}{\sqrt{1 - v^2/c^2}} - 1 \right)$$

In the nonrelativistic limit, this expression becomes the Newtonian expression for kinetic energy ½$m_0 v^2$.

A notable relativistic effect is that the speed of light is a limiting speed, in the sense that bodies cannot be accelerated to the speed of light and cannot move at that speed. The impossibility of accelerating a body to the speed of light can be understood in terms of the body's inertia approaching infinity as its speed approaches c, as seen earlier, so the accelerating effect of a force acting on the body approaches 0. The speed of light is a limiting speed for the transfer of information and for the propagation of causal effects, as well. Only light, and presumably also gravitational waves, propagate at the speed of light. Accordingly, photons (the quantum particles of light) and gravitons (the putative quantum particles of gravitational waves) move at that speed also and only at that speed.

RELATIVISTIC MECHANICS

Relativistic mechanics relates to the way velocities combine in the special theory of relativity. Let there be two observers O and O', such that O' is moving in a straight line at constant speed v with respect to O. Assume a body is moving in the same direction as O' and with speed u as observed by observer O, and let u be greater than v for simplicity of discussion. Denote by u' the body's speed as observed by observer O'. In Newtonian physics, these relative speeds combine by simple addition as

$$u = v + u'$$

which conforms with our everyday experience. If this relation were true for high speeds, speeds could potentially exceed the speed of light. For instance, if observer O' is moving at speed $0.6c$ relative to O and the body is moving at $0.5c$ relative to O', then the body would be observed moving at $1.1c$ relative to O, faster than the speed of light. In the special theory of relativity, the formula for the addition of speeds is, instead,

$$u = \frac{v + u'}{1 + vu'/c^2}$$

In the nonrelativistic limit, this formula does indeed reduce to the Newtonian one, but for high speeds, its results are quite different. In the example, the body is observed moving only at speed $0.85c$ relative to observer O. In fact, no matter how high the speeds v and u' are, as long as they are less than c, the result of their relativistic addition, u, is always less than c. When observing a light beam rather than a moving body, so that $u' = c$, then no matter what the relative speed of the reference frames v is, the formula gives $u = c$. This result conforms with the second postulate of the special theory of relativity, that all observers measure speed c for light.

Newton's second law of motion is commonly expressed as

$$\mathbf{F} = m_0\mathbf{a} = m_0 \frac{d\mathbf{v}}{dt}$$

where $\mathbf{F}$ denotes the force vector in newtons (N), $\mathbf{a}$ the acceleration vector in meters per second per second (m/s²), and t the time in seconds (s). Newton's original formulation was in terms of momentum:

$$\mathbf{F} = \frac{d(m_0\mathbf{v})}{dt}$$

which is equivalent to the former when the mass does not change in time. The latter is the valid form of the law, even for cases when a body accrues or loses mass over time (such as a rocket). We use m_0 here rather than the usual notation m, since in Newtonian physics, the mass does not depend on the body's speed and is, therefore, the rest mass.

Newton's second law does not conform with the special theory of relativity. An indication of this is the result that, according to the formula, a constant force acting for sufficient time will accelerate a body to any speed whatsoever, even exceeding the speed of light. The relativistic form of Newton's second law is based on its momentum formulation, but expressed in terms of relativistic momentum:

$$\mathbf{F} = \frac{d\mathbf{p}}{dt}$$
$$= \frac{d(m\mathbf{v})}{dt}$$
$$= m_0 \frac{d}{dt} \frac{\mathbf{v}}{\sqrt{1 - v^2/c^2}}$$

Although Newtonian mechanics requires modification to conform with the special theory of relativity, as just demonstrated, the laws of electromagnetism, as expressed by Maxwell's equations (named for their discoverer, the 19th-century British physicist James Clerk Maxwell), require no modification at all. These laws were formulated well before Einstein's work, yet they turned out to be perfectly relativistic as they are. They fulfill the second postulate of the theory by predicting that all observers, whatever their velocities relative to each other, will indeed observe the same speed c for light in vacuum. They even predict the correct value of c.

SPACE-TIME

The four-dimensional merger of space and time is called space-time: a merger, but not a total blending, since the temporal dimension of space-time remains distinct from its three spatial dimensions. Space-time is the arena for events, each of which possesses a location, specified by three coordinates, and an instant, or time, of occurrence, whose specification requires a single coordinate—together four coordinates. These coordinates can be taken as (x, y, z, t), which combine an event's spatial coordinates (x, y, z) and its time coordinate t.

The coordinates of the same event can have different values in different reference frames, that is, as measured by different observers with respect to their respective coordinate systems, in space-time. The special theory of relativity involves reference frames in relative motion at constant velocity. The mathematical relations between the coordinates of an event with respect to a pair of such reference frames are given by Poincaré transformations or Lorentz transformations (as in the example found near the beginning of this article). Einstein's general theory of relativity allows arbitrary reference frames, and the corresponding transformations are accordingly much more general. The following considerations will be those of the special theory.

In analogy with the distance between points in space, every pair of events in space-time possesses an interval D, which is given by

$$D^2 = (x_2 - x_1)^2 + (y_2 - y_1)^2 + (z_2 - z_1)^2 - c^2(t_2 - t_1)^2$$

where (x_1, y_1, z_1, t_1) and (x_2, y_2, z_2, t_2) are the space-time coordinates of the two events in any single reference frame, that is, as observed by any single observer. Although the coordinates of the same event have different values with respect to different reference frames (i.e., as measured by different observers) in space-time, the interval between any two events is invariant for reference frames in relative constant-velocity motion. It is an absolute quantity. In other words, the interval between a pair of events does not change under Poincaré or Lorentz transformations. If unprimed and primed symbols denote coordinates with respect to two such reference frames, this invariance is expressed as

$$(x_2' - x_1')^2 + (y_2' - y_1')^2 + (z_2' - z_1')^2 - c^2(t_2' - t_1')^2 =$$
$$(x_2 - x_1)^2 + (y_2 - y_1)^2 + (z_2 - z_1)^2 - c^2(t_2 - t_1)^2$$

This invariance assures that the speed of light has the same value with respect to both reference frames.

STRUCTURE OF SPACE-TIME

Two events for which $D^2 < 0$ are said to possess a timelike interval, or timelike separation. This relation is absolute (i.e., independent of reference frame). For two such events, there are reference frames with respect to which both events occur at the same location, and only their times of occurrence differ. Such events are said to lie within each other's light cone. Of two such events, one occurs after the other (with the other occurring before the one). This past-future distinction, too, is absolute. The later-occurring event is said to lie within the future light cone of the earlier one, and the earlier within the past light cone of the later. Such a pair of events possesses the property that a particle emitted from the earlier event and moving slower than c can reach the later event. Equivalently, a photon, which travels at speed c, that is emitted from the earlier event reaches the location of the later event before the later event actually occurs.

Another possibility for two events is that $D^2 = 0$. They are then said to have a lightlike interval, or lightlike separation. Such a pair of events are also said to lie *on* (rather than *within*) each other's light cone. Here, too, the past-future distinction is absolute. The later event lies on the future light cone of the earlier event, and the earlier on the past light cone of the later. A pair of events possessing lightlike separation can be connected by a photon emitted from the earlier and arriving at the later.

The last case is that for two events $D^2 > 0$. Such a pair of events have a spacelike interval, or spacelike separation. They are said to lie outside each other's light cone.

Their time ordering is relative, that is, depends on the reference frame. With respect to certain reference frames, one event occurs before the other. In other frames, it is the other that takes place before the one. Reference frames also exist in which both events occur at the same time (i.e., they are simultaneous). In no reference frame do the events take place at the same location. No material particle or photon emitted from one event can reach the location of the other event until after the event occurs. So, no signal or information can pass between events possessing spacelike separation, and they cannot affect each other.

Space-time is not isotropic: in other words, its properties are different in different directions. It possesses what is called a light-cone structure. From any event E, taken as origin, there are directions within its

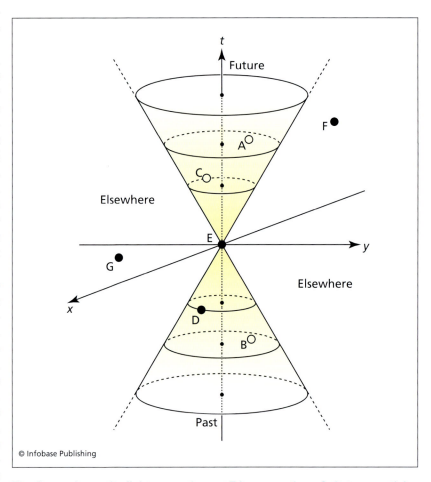

© Infobase Publishing

The figure shows the light cone of event E in space-time. Only two spatial dimensions, x (perpendicular to the page) and y (horizontal), are shown. The time axis, t, is vertical and forms the axis of the light cone. Events A and B lie within E's light cone, events C and D lie on the light cone, while events F and G lie outside the light cone (in the region labeled "elsewhere"). A and C are in E's absolute future; B and D are in its absolute past. F and G have no absolute temporal relation with E. E's separation from A and from B is timelike, from C and from D is lightlike, and from F and from G is spacelike.

future light cone, directions on its future light cone, directions outside its light cone, directions on its past light cone, and those that lie within its past light cone, and E's light cone is independent of reference frame. The different categories of direction possess different properties, as shown earlier. An event lying outside E's light cone occurs elsewhere from E and has no absolute (i.e., reference-frame-independent) temporal relation to E. Such events do not affect E; nor are they affected by E. On the other hand, an event lying within or on E's light cone can be said to occur elsewhen to E. That event occurs either in E's past or in E's future, and the difference is independent of reference frame. E can affect events in its future, and events in E's past can affect E.

WORLD LINES

The events that constitute the life of a massive particle form a continuous curve in space-time, called the particle's world line. Such a world line lies wholly within the light cone of any event on it. Massive particles that are created and then annihilated have their world lines starting at some event and ending at a later event, lying within the future light cone of the earlier event. The world line of a photon lies *on* the light cone of any of the events forming it. Processes involving a number of particles are represented by bundles of world lines. A body's existence appears as a dense skein of world lines, those of all its constituent particles. Collisions are represented by world lines converging toward each other, possibly actually intersecting each other, and diverging into the future.

TWIN PARADOX

The twin paradox is related to the relativistic effect of time dilation, which was discussed earlier. The effect is that a moving clock is measured as running more slowly than when it is at rest. The effect is not only for clocks, but for any process, including the aging of an organism: a moving heart is observed to beat more slowly than when it is at rest. Consider a pair of twins on Earth. One twin, the sister, takes a long round-trip in a very fast spaceship, while her

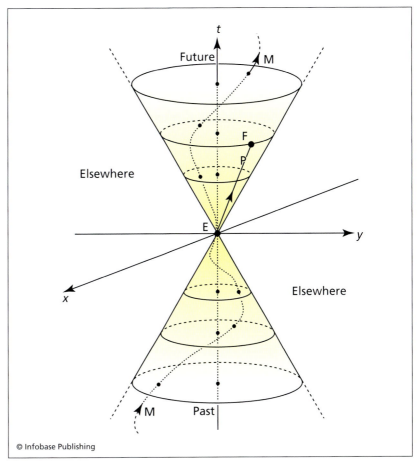

M denotes the world line of a massive particle in space-time. This world line lies wholly within the light cone of any event, E, in the life of the particle. Only two spatial dimensions, x (perpendicular to the page) and y (horizontal), are shown. The time axis, t, is vertical and forms the axis of E's light cone. The world line of a photon that is emitted at E and absorbed at event F lies on E's light cone and is labeled P.

twin brother remains on Earth. During the outward leg of the journey, the brother on Earth observes his traveling sister aging more slowly than he is. Then, on the return leg, the Earthbound twin again finds his sister aging more slowly than he. When the twin sister finally lands and steps off the space ship, she is indeed younger than her twin brother, who has aged more than she has during her absence.

The paradox, or actually *apparent* paradox, is as follows. Since time dilation is a mutual effect, the expectation is that after the trip, the twins will somehow find themselves at the same age. After all, their *relative* motion is what appears to matter. Each twin should have the same right to consider him- or herself to be at rest and the other one moving. Just as the Earthbound brother observes his sister aging more slowly than he, both while she is moving away and while she is returning, the spaceship-bound sister should see her brother aging more slowly than she,

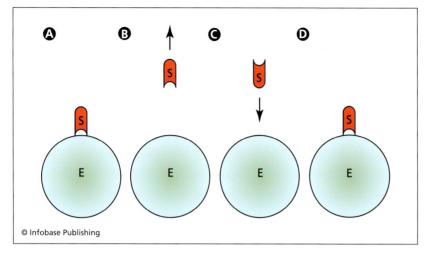

© Infobase Publishing

In the twin paradox, when the Earthbound twin greets his twin sister upon her return from a trip into space and back, she is younger than he is. Their different rates of aging can be attributed to the fact that during the trip, the Earthbound twin remains in a single inertial reference frame, while the traveling twin finds herself in four consecutive inertial reference frames and undergoes an acceleration from each frame to the next. (a) She and her spaceship S are in her first inertial reference frame, that of Earth E. Then she accelerates to (b) her second inertial reference frame, which is in constant-velocity motion away from Earth. Another acceleration turns her around and puts her in (c) her third inertial reference frame, moving toward Earth at constant velocity. Her third acceleration—a deceleration as viewed from Earth—takes her to rest on Earth and into (d) her fourth inertial reference frame.

both while Earth is moving away from the spaceship and while it is returning. The situation seems to be symmetric. Why, then, does one twin age more slowly than the other?

The paradox is only apparent, because, in fact, the twins are not in equivalent situations. While the Earthbound twin remains in the same inertial reference frame (i.e., a reference frame in which Newton's laws of motion are valid), that of Earth, throughout his sister's trip, the traveling twin finds herself in *four* successive inertial reference frames. She starts in her brother's. While she is on the outward leg, she is in her second. During her return, she is in the third. And upon landing, she returns to her initial reference frame, her fourth. She objectively differs from her brother in that she feels three acceleration forces, while the brother on Earth feels none. First, she accelerates to the outward speed. Then, she accelerates to turn the spaceship back toward Earth. And, finally, she decelerates for landing.

Under the simplifying assumption that the accelerations, in and of themselves, do not affect aging, the time dilation calculations can be carried out within the special theory of relativity. The result is that the traveler does indeed age more slowly. The involvement of accelerations requires the invocation of the general theory of relativity, though doing so does not

change the result. This effect has been tested and confirmed experimentally by synchronizing two ultraprecise atomic clocks and then taking one on a round-the-Earth trip by airplane, with the other remaining fixed on Earth.

STATUS OF SPECIAL RELATIVITY

Einstein's special theory of relativity is a very well confirmed theory. Physicists and engineers apply special relativity to the design and operation of such things as particle accelerators, the Global Positioning System (GPS), and nuclear devices. In the nonrelativistic limit, it reduces to Newtonian physics, which is adequate for most everyday applications. The special theory of relativity was generalized by Einstein to a theory of gravitation, as his general theory of relativity.

See also ACCELERATION; AC-CELERATORS; CLOCK; EINSTEIN, AL-BERT; ELECTROMAGNETIC WAVES; ELECTROMAGNETISM; ENERGY AND WORK; FISSION; FORCE; FUSION; GENERAL RELATIVITY; GRAVITY; MASS; MAXWELL, JAMES CLERK; MOMENTUM AND COLLISIONS; MOTION; NEWTON, SIR ISAAC; NUCLEAR PHYSICS; SPEED AND VELOCITY; TIME.

FURTHER READING
Mermin, N. David. *It's about Time: Understanding Einstein's Relativity.* Princeton, N.J.: Princeton University Press, 2005.

Serway, Raymond A., Clement J. Moses, and Curt A. Moyer. *Modern Physics,* 3rd ed. Belmont, Calif.: Thomson Brooks/Cole, 2004.

spectrophotometry Spectrophotometry is an analytical chemical method that is based on the absorption and emission of electromagnetic radiation by a substance. Electromagnetic radiation includes a large range of wavelengths and frequencies, from about 10^{-13} meter to 10 meters for wavelengths, corresponding to approximately 10^{22} hertz to 10^8 hertz for frequencies. This range is referred to as the electromagnetic spectrum and is arbitrarily divided into subranges, or bands, in order of increasing wavelength: gamma rays, X-rays, ultraviolet light, visible light, infrared rays, microwaves, and radio waves.

Quantum physics offers an alternate description of an electromagnetic wave in terms of a flow of particles, called photons. The energy of each photon is proportional to the frequency of the wave and is given by

$$E = hf$$

where E is the photon's energy in joules (J), f is the frequency of the wave in hertz (Hz), and h stands for Planck's constant, whose value is 6.63×10^{-34} J·s. Spectrophotometry makes use of the absorption and emission of electromagnetic radiation to identify elements in samples and to help determine, for example, concentration of an unknown sample.

ABSORPTION SPECTROPHOTOMETRY

In absorption spectrometry, a spectrophotometer passes a beam of light through a sample and detects the amount of light that is either absorbed by or transmitted through the sample. As the light passes through the sample, a certain percentage of photons of light will be absorbed by the sample, and only the remaining photons will reach the detector.

Absorption spectrophotometry is performed utilizing monochromatic light, light of a single wavelength. This is necessary because the samples will absorb different amounts of light of different wavelengths, depending on their chemical composition.

When beginning an experiment on a new sample using a spectrophotometer, one must first determine its absorbance spectrum by measuring the absorbance over multiple wavelengths. After the evaluation of an absorbance spectrum, the wavelength at which maximal absorbance occurs is known as λ_{max}. Utilizing λ_{max} absorbance reduces the error in the measurements and gives the strongest measurements.

In spectrophotometric methods, comparison of the reading sample to a standard is necessary. When measuring the transmittance through a certain substance that is dissolved in distilled water, using distilled water in the reference sample or blank enables one to measure the amount of change in transmittance relative to the distilled water consistently. Reference standards can be used containing stock samples of a known concentration of the sample being tested. This allows one to measure the transmittance relative to the known concentration, and a comparison can be made to the reference standard to obtain a quantitative measurement of a substance in the sample.

An absorption spectrophotometer has a light source (either visible light or ultra violet). The wavelength of the light passing through the spectrophotometer is isolated into individual wavelengths by diffraction gratings. The light of the appropriate wavelength passes through the sample cell, where it is either absorbed or transmitted through the sample. The light that is transmitted then reaches the detector, where it can be read directly. Many spectrophotometers are connected to computers that analyze the data as they are read.

The transmittance can be calculated as follows:

$$T = I_0/I$$

where T is the measure of the amount of light that passes through the sample, I_0 represents the value of the photons that pass through a blank sample, and I represents the amount of light that passes through the sample. Transmittance is generally the value that is directly read from the spectrophotometer.

Absorbance is a measure of how much energy from the light is absorbed by the sample. Absorbance can be calculated from the transmittance as follows:

$$A = -\log T$$

For example, if a sample has a transmittance of 68 percent at 400 nm, one can calculate its absorbance as follows:

$$A = -\log T$$
$$A = -\log (0.68)$$
$$A = 0.167$$

Once the absorbance of the sample is known, the concentration of the unknown sample can be calculated using a relationship known as Beer-Lambert law:

$$A = a_\lambda \times b \times c$$

where A stands for the absorbance; a_λ is an absorptivity coefficient, which is wavelength dependent; b is the path length, generally measured in centimeters; and c is the concentration of the sample.

EMISSION SPECTROPHOTOMETRY

One spectrophotometric method scientists use to identify an element in a sample being analyzed is based on the characteristic emission spectrum of that sample. An emission spectrum is obtained when a sample is energized in some way, such as by heating or with electricity, causing the electrons in the sample to become excited and to move up to higher energy levels. As these electrons fall back down to their ground state, they give off energy equivalent to the amount of energy absorbed from the excitation.

An understanding of emission spectra and how they related to atomic structure was originally determined by the Danish physicist Niels Bohr. While Bohr was developing his planetary model of the atom, he

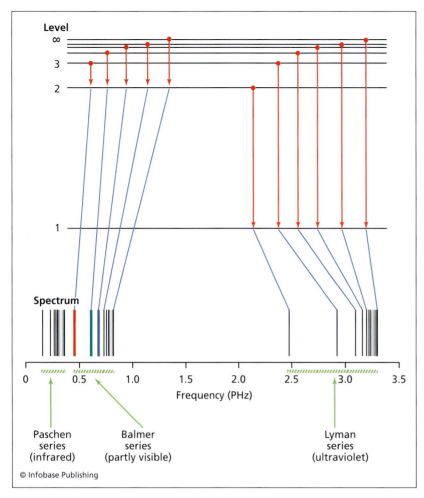

The energy given off from an excited hydrogen atom falls into three series: the Paschen series that has wavelengths in the infrared range, the Balmer series that falls into the visible light range, and the Lyman series that falls in the ultraviolet range.

he or she could not stay there indefinitely. As the electron falls back down to its ground state, it gives off energy proportional to the amount of energy it originally absorbed. The discrete bands seen in line spectra of an element such as hydrogen can be explained by the different energies (represented by the shorter wavelength with higher energy) being given off in the form of photons as the electrons returned to ground state.

Since the electron configurations of all elements are different, the number and location of the electrons being excited also differ. This gives a unique emission spectrum for each element. Scientists utilize emission spectra information to identify unknown substances. When an unknown sample's emission spectrum is compared to that of known elements and the line spectra are identical, the unknown sample can be identified as the known element.

Three types of emission spectra exist: the line spectrum, band spectrum, and continuous spectrum. When light is emitted at only certain wavelengths, the emission spectrum is known as a line spectrum. This type of emission spectrum is common when the sample is in a gaseous phase. A band spectrum has regions of wavelengths that are emitted, while a continuous spectrum represents all wavelengths of visible light. The Sun gives off a continuous spectrum. Solids that have been heated commonly emit continuous spectra.

The emission spectrum of hydrogen has wavelengths from the ultraviolet region, the visible region, and the infrared region of the electromagnetic spectrum. The different regions of the hydrogen spectrum are named for their discoverers.

- Lyman series: ultraviolet range
- Balmer series: ultraviolet and visible range
- Paschen series: infrared range
- Bracken series: infrared range

See also ATOMIC STRUCTURE; BOHR, NIELS; ELECTROMAGNETIC WAVES; ELECTROMAGNETISM; ELECTRON CONFIGURATIONS; MASS SPECTROMETRY.

developed a model that explained the emission spectrum of the hydrogen atom. Although Bohr's model of the atom was later updated, the basic principle of energy levels was still applicable and was essential to explaining the emission spectrum.

Bohr determined that the electrons moved around the nucleus of the atom in "orbits" that were certain distances from the nucleus. He also established that these electrons could not exist between orbits. To illustrate this, imagine a ladder. One can place his or her foot only on the rungs of the ladder and not between the rungs. As someone climbs a ladder, he or she gets higher and higher potential energy with the increase in height. The same can be said for electrons. The lowest energy state of the electron, which is the lowest energy level closest to the nucleus, is known as the ground state. Continuing the ladder analogy, if one were to become more energetic and reach higher rungs of the ladder, as an electron that has been excited would reach outer energy levels,

FURTHER READING

Gore, Michael G., ed. *Spectrophotometry and Spectrofluorimetry: A Practical Approach.* New York: Oxford University Press, 2000.

speed and velocity In the study of motion, speed and velocity play a major role. On the one hand, they are based on change of position, or displacement, over time. On the other hand, speed and velocity form the basis for acceleration.

DISPLACEMENT

Displacement is the separation between two points in space expressed as a vector. Specifically, the displacement of point B from point A is the vector whose tail is at A and whose head is at B. In other words, the displacement vector between B and A is a vector whose magnitude is the distance between A and B and whose direction is the direction from A to B.

The position vector associated with a point in space is the vector whose magnitude equals the point's distance from the origin of the coordinate system and whose direction is the direction of the point from the origin. In terms of the position vectors of points A and B, denoted by r_A and r_B, respectively, the displacement of B from A, which we denote by r_{BA}, is the vector difference,

$$r_{BA} = r_B - r_A$$

Alternatively, the position vector of point B is obtained by vectorially adding the displacement vector to the position vector of A:

$$r_B = r_A + r_{BA}$$

Note that whereas position vectors depend on the location of the coordinate origin, displacement vectors are origin independent.

VELOCITY

The time rate of displacement, that is, the change of position per unit time, is called velocity, or more precisely, instantaneous velocity. Like displacement, velocity is a vector quantity, possessing both magnitude and direction. In a formula, the instantaneous velocity v of a particle is

$$v = \frac{dr(t)}{dt}$$

Some racing cars can reach speeds of over 300 miles per hour (480 km/h.). Here is a scene from the Formula One Grand Prix of Brazil at Interlagos, near São Paolo. *(Gero Breloer/dpa/Landov)*

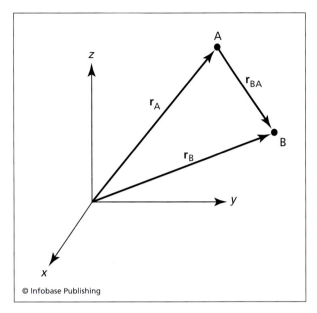

© Infobase Publishing

The displacement of point B from point A is specified by the displacement vector r_{BA}, pointing from A to B. This vector is the vector difference between r_B and r_A, the respective position vectors of points B and A: r_{BA} = r_B - r_A.

where **v** is in meters per second (m/s), and t is the time in seconds (s). The symbol $r(t)$ represents the position vector of the particle in meters (m) at time t, as described. Then, $d\mathbf{r}$ denotes the infinitesimal displacement vector, the infinitesimal change in **r**, of the particle during the infinitesimal time interval dt. Instantaneous speed v is the magnitude of the instantaneous velocity:

$$v = |\mathbf{v}| = \left|\frac{d\mathbf{r}}{dt}\right| = \frac{|d\mathbf{r}|}{dt}$$

The average velocity $\mathbf{v}_{av}$ over the time interval from t_1 to t_2 equals the net displacement during that time interval divided by the time interval. So,

$$\mathbf{v}_{av} = \frac{\mathbf{r}(t_2) - \mathbf{r}(t_1)}{t_2 - t_1}$$

where $\mathbf{r}(t_1)$ and $\mathbf{r}(t_2)$ denote the position vectors at times t_1 and t_2, respectively. One can calculate the instantaneous velocity from the average velocity by taking the limit of t_2 approaching t_1 (i.e., the limit of vanishing time interval). Note that average velocity depends only on the net displacement during the time interval and is independent of the motion of the object during that time interval.

In a simple situation, the motion of a body might be constrained to a straight line, say, in the x direc-

tion. In that case, we may drop the vector notation and use the formula

$$v = \frac{dx}{dt}$$

where v and x denote, respectively, the velocity and position in the x direction. If the velocity is constant (i.e., the position changes [increases or decreases] by equal increments during equal times intervals), the formula can be further simplified to

$$v = \frac{\Delta x}{\Delta t}$$

where Δx is the (positive or negative) change in position during time interval Δt.

Relative velocity is the velocity of a moving object as observed in a reference frame that is itself moving. The velocity of aircraft B as viewed from moving aircraft A, for instance, is the relative velocity of B with respect to A and is different from the velocity of B with respect to Earth. Let $\mathbf{v}_0$ denote the velocity of the reference frame and **v** the velocity of an object. Then, the relative velocity of the object with respect to the reference frame, $\mathbf{v}_{rel}$, is given by

$$\mathbf{v}_{rel} = \mathbf{v} - \mathbf{v}_0$$

In straight-line motion, that translates to

$$v_{rel} = v - v_0$$

As an example, take north as the positive x direction and let aircraft A be flying at 300 miles per hour (mph) due north and aircraft B flying in the same direction at 400 mph. How does B appear to the pilot of A? In terms of the symbols introduced, v = 400 mph and v_0 = 300 mph. The relative velocity of B with respect to A, v_{rel}, is then

$$v_{rel} = (400 \text{ mph}) - (300 \text{ mph})$$
$$= 100 \text{ mph}$$

The pilot of A sees B moving northward at 100 mph with respect to A. Now, reverse the problem: how does A appear to the pilot of B? In symbols now, v = 300 mph and v_0 = 400 mph. The relative velocity of A with respect to B is

$$v_{rel} = (300 \text{ mph}) - (400 \text{ mph})$$
$$= -100 \text{ mph}$$

So, B sees A moving southward at 100 mph.

Now let A be flying as before, but this time, let B be flying due south at 400 mph. How does B appear to A? In symbols, $v = -400$ mph and $v_0 = 300$ mph. The relative velocity of B with respect to A is

$$v_{rel} = (-400 \text{ mph}) - (300 \text{ mph})$$

$$= -700 \text{ mph}$$

The pilot of A sees B moving southward at 700 mph. Switch the point of view to ask how B sees A. In symbols, $v = 300$ mph and $v_0 = -400$ mph. A's relative velocity with respect to B is

$$v_{rel} = (300 \text{ mph}) - (-400 \text{ mph})$$

$$= 700 \text{ mph}$$

Relative to B, A is flying northward at 700 mph.

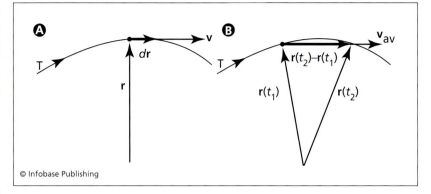

© Infobase Publishing

Instantaneous and average velocities for a particle moving along a trajectory T. (a) The instantaneous velocity of a particle with position vector **r** is **v** = d**r**/dt, where d**r** is the infinitesimal change in the particle's position vector during infinitesimal time dt. d**r** is in the direction of the trajectory at the particle's location, so **v** is tangent to the trajectory there. (b) The average velocity between times t_1 and t_2 is given by $\mathbf{v}_{av}$ = [**r**(t_2) − **r**(t_1)]/($t_2 - t_1$), where **r**(t) denotes the particle's position vector at time t. $\mathbf{v}_{av}$ is parallel to **r**(t_2) − **r**(t_1), the net displacement that the particle undergoes during the time interval.

SPEED

The magnitude of velocity is called speed, or instantaneous speed. It is a scalar quantity, and its SI unit is meter per second (m/s). If we denote by $s(t)$ the length of the path that a body follows as a function of time, t, then the body's instantaneous speed, $v(t)$, as a function of time is the time derivative of $s(t)$:

$$v(t) = \frac{ds(t)}{dt}$$

Here s is in meters (m) and t is in seconds (s). This is the same as the previous definition of instantaneous speed as the magnitude of the instantaneous velocity, since the magnitude of the infinitesimal displacement equals the infinitesimal path length,

$$|d\mathbf{r}| = ds$$

The average speed v_{av} is the ratio of the total distance traveled along the path of the motion s during a finite time interval Δt to the time interval,

$$v_{av} = \frac{s}{\Delta t}$$

Note that the average speed is not, in general, the same as the magnitude of the average velocity for the same time interval. While average velocity depends on the net displacement and is independent of the motion during the time interval, average speed depends on the total distance traveled during the time interval and, thus, depends on the motion. To help see the difference, imagine a swimmer swims six laps in a 50-meter pool. At the end of the swim, the swimmer finds herself at the same end of the pool where she started out, so her net displacement for the swim is 0. However, during her swim, she traveled a distance of 300 meters. Thus, her average velocity for the swim is 0, while her average speed equals 300 m divided by the time it took to swim that distance.

Relative speed is the speed of a body as measured with respect to a moving reference frame, such as a frame attached to another moving body. Relative speed is the magnitude of relative velocity. In the earlier, numerical example involving two aircraft, the relative speeds of A with respect to B and B with respect to A are both 100 mph, when both A and B are flying northward, and are both 700 mph, when they are flying in opposite directions.

See also ACCELERATION; MOTION; VECTORS AND SCALARS.

FURTHER READING
Young, Hugh D., and Roger A. Freedman. *University Physics*, 12th ed. San Francisco: Addison Wesley, 2007.

states of matter Matter is defined as anything that has mass and takes up space. Mass is the measure of the amount of material a body of matter contains. The more mass an object has, the more matter it contains. The particles of matter (atoms, molecules, ions, etc.) interact with each other in a way that depends

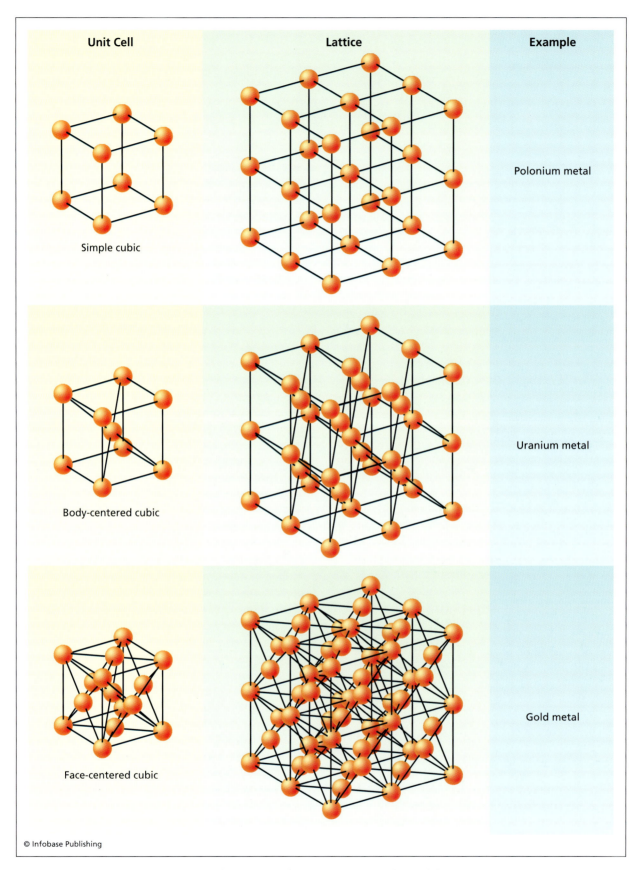

Unit Cell	Lattice	Example
Simple cubic		Polonium metal
Body-centered cubic		Uranium metal
Face-centered cubic		Gold metal

Three unit cells of cubic crystals are simple cubic, body-centered cubic, and face-centered cubic.

on the amount of energy the particles have and the attraction of the particles for each other. The energy of motion of particles is known as kinetic energy. The average random kinetic energy of the particles in a substance is proportional to its temperature. The higher the temperature of a substance, the more the particles move around, and, by definition, the more internal random kinetic energy it has.

Matter exists in one of four states, depending on the interactions of the particles of the matter: solid, liquid, gas/vapor, and plasma. Only three of these states, solid, liquid, and gas, are normally and stably present on Earth. Solid matter has a definite shape and a definite volume. The atoms or molecules that make up the solid matter have low kinetic energy. Their motion resembles vibrations more than actual movements, and the particles are strongly attracted to each other. Solids often assume a regularly ordered structure known as a crystal. The different types of crystals are composed of a fundamental unit called a unit cell, which combines with other unit cells to form the crystal. The unit cell is the smallest part of the crystal structure that contains all of the attributes of the crystal. Types of crystal structures include cubic, tetragonal, orthorhombic, rhombohedral, hexagonal, monoclinic, and triclinic.

Liquid is a state of matter that develops from a solid as the temperature (internal random kinetic energy) is raised and the particles vibrate more energetically and break their connections with the neighboring molecules or atoms. This leads to more space between the molecules and the ability to flow past each other. A liquid has a definite volume but does not have a definite shape; it assumes the shape of the container. The particles of a liquid have increased kinetic energy over the solid form of the same compound. Liquids can be volatile (easily evaporated) or not, depending on the strength of the intermolecular forces in the substance. Some substances, such as alcohol, evaporate very quickly, while other compounds, such as water, evaporate rather slowly.

When enough external energy is supplied, the molecules in a substance move rapidly enough to separate from their neighboring molecules in the substance. This state of matter is known as a gas or a vapor. Gases do not have a definite shape or a definite volume; they take the shape and volume of the container. The random kinetic energies of the particles in a gas are much higher than those of the liquid or solid form of the same compound. A substance is considered a gas if it normally exists as a gas at room temperature. A vapor is a gaseous form of a substance that is normally a liquid at room temperature. Water

is an example of a substance that is liquid at room temperature and that evaporates into a vapor.

Plasma is the fourth state of matter. Subjecting matter to extremely high temperatures, such as are present near the surface of the Sun, strips away the electrons, leaving positively charged ions behind. This state of matter, consisting of ions and electrons, shares some of the properties characteristic of gases, but, in contrast, electromagnetic fields and waves strongly affect plasmas, while they normally have no influence on gases, which consist of neutral particles.

The strength of the intermolecular forces (interactions among molecules of a substance) determines the degree of attraction of particles of matter, and this, in turn, determines the state of the matter. Though the strength of intermolecular forces is significantly lower than the strength of covalent bonds, the combined impact of all of the intermolecular forces in a substance dramatically affects the properties of that substance. Intermolecular forces include dipole-dipole forces, induced-dipole forces, and hydrogen bonds. In dipole-dipole interactions, the molecules involved are polar molecules, meaning they are formed by covalent bonding between two elements with a large difference in their electronegativities. The shared electrons in the covalent bond spend more time with the more electronegative atom (the electron "hog"), giving it a partial negative charge, and less time with the less electronegative atom, giving it a partial positive charge. This separation of charge creates the polar molecule. When many of these polar molecules interact with each other, the opposite partial charges between the negative ends and positive ends of different molecules attract one another.

An induced-dipole force is based on the phenomenon that at any given moment in time, a nonpolar molecule may have a temporary separation of charge due to the chance of the electron's location. When one molecule has a separation of charge, however brief, it can induce or cause electrons in a neighboring molecule to move away or toward the first molecule, creating a dipole in the second molecule that can then induce a dipole in a third molecule. The induced dipoles enable the interaction of these molecules and are an attractive force between molecules.

A common type of intermolecular force is the hydrogen bond, for which a more accurate name would be a *hydrogen interaction*. A hydrogen bond is formed between a polar molecule that has a highly electronegative atom covalently bonded to a hydrogen atom and a nitrogen, oxygen, or fluorine atom of a second molecule or at a distant location of the same molecule. When the shared electrons in that covalent bond spend more time with the more electronegative

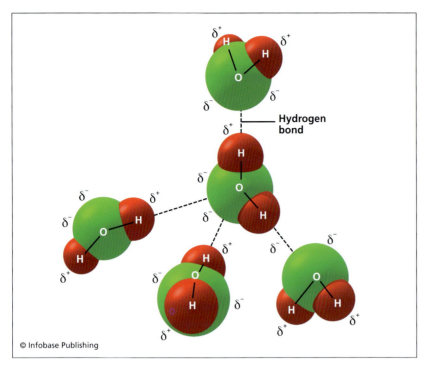

The very electronegative oxygen atom creates a water molecule that is polar. A positive hydrogen atom of one molecule and the negative oxygen atom of another attract each other. This attraction is the origin of the hydrogen bond between molecules.

element, usually oxygen or nitrogen, the hydrogen atom takes on the partial positive charge. This hydrogen can then interact with unpaired electrons on another molecule in the valence shell of a nitrogen, oxygen, or fluorine atom.

Hydrogen bonds confer unique properties on water, such as its high boiling point and surface tension. Water has a boiling point that is higher than that of molecules with similar formulas that cannot form hydrogen bonds. The strong interactions between the water molecules due to hydrogen bonds require the addition of a large amount of energy in order to boil water. Surface tension is caused by hydrogen bonds and creates a skinlike coating over the surface of a liquid. The interacting water molecules are strongly attracted to each other, and this attraction allows water striders to walk on water and rocks to skip on water. Hydrogen bonds are also responsible for the expansion of water upon freezing. In its liquid form, each water molecule interacts via hydrogen bonding with two to three other molecules of water, allowing them to pack closely to one another. Slowing the particles in the water by lowering the temperature causes more hydrogen bonds to form, up to four for each water molecule. This open, yet rigid structure is less dense than that of liquid water; thus, ice floats on water. The expansion of water upon freezing is responsible for the damage done to roads, bridges,

and concrete structures by the cycles of freezing and thawing that occur in the springtime. When the ice and snow melt, the liquid water flows into the cracks of the road surface. If the temperature cools again, that water freezes and expands, causing the road surface to break apart.

Hydrogen bonds are also involved in many biological structures, including deoxyribonucleic acid (DNA), ribonucleic acid (RNA), and proteins. Hydrogen bonds hold the bases of double-stranded nucleotides together.

Strong intermolecular forces draw molecules of a substance into close association. In order to change from one state to another state of higher energy, the random kinetic energy of the particles must be stronger than the intermolecular forces trying to hold the substance together. Each of the phase changes of matter is paired with a process that is its opposite. The changes between solids and liquids are known as fusion (melting) and solidification (freezing). Melting occurs when the kinetic energy of a solid object increases (such as by the input of heat), its particles break intermolecular interactions with their neighbors, and the phase changes to a liquid. Freezing occurs when the parti-

Sublimation is the physical process by which a solid transforms directly into a gas or vapor without passing through the liquid phase. Shown here is a common example of sublimation—solid dry ice subliming into carbon dioxide gas. *(Charles D. Winters/Photo Researchers, Inc.)*

cles of a liquid slow enough to interact strongly with their neighbors and form a solid. The phase changes between liquid and gas (vapor) are condensation and evaporation. Condensation occurs when the particles of a gas or vapor slow down, losing enough kinetic energy to allow the particles to interact with their neighbors to form a liquid. Evaporation is the process by which a liquid changes to a gas or vapor. As liquid particles move around more, they are able to "bump" into each other, and these collisions will cause particles to be hit hard enough to be ejected from of the surface of the liquid. These particles are then technically in the vapor phase and exert a force upward. Every liquid substance has a vapor pressure, the pressure exerted by particles escaping the liquid, which is characteristic of that compound. The boiling point of a liquid is the temperature at which the vapor pressure of the liquid equals the atmospheric pressure pushing down on the liquid. When the vapor pressure pushes up harder than the atmospheric pressure is pushing down, the particles of a liquid can escape from the surface and boil. The boiling point of water is dependent on the atmospheric pressure. At high elevations, atmospheric pressure is lower than at sea level; therefore, water will boil at a lower temperature because the kinetic energy of the water particles does not have to be as great in order for the particles to escape the surface of the liquid. The lowered boiling point of water necessitates that alterations be made when cooking or baking at higher elevations.

Two lesser-known phase changes are the changes that occur directly between a solid and a gas or vapor. Changing directly from a gas to a solid is known as deposition, a process exemplified by the condensation of water vapor in a cloud or a freezer directly into ice or snow. Sublimation is the change in phase directly from a solid to a gas. One example of sublimation is the conversion of solid carbon dioxide (dry ice) to carbon dioxide gas.

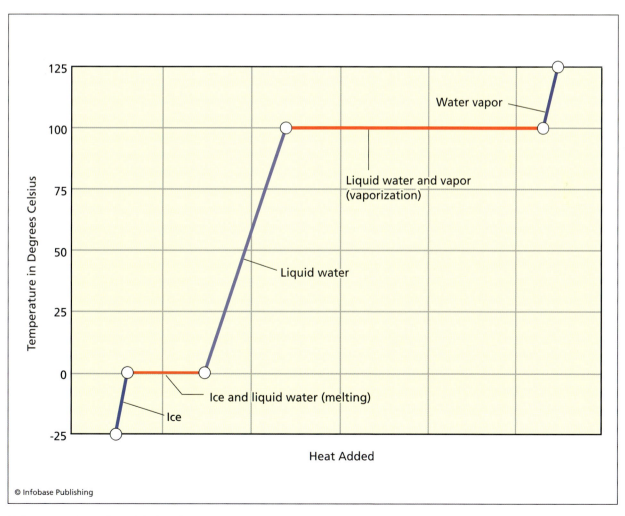

Using a graph such as this heating curve of water, one calculates in stages the amount of thermal energy required to melt or vaporize ice of given mass and temperature.

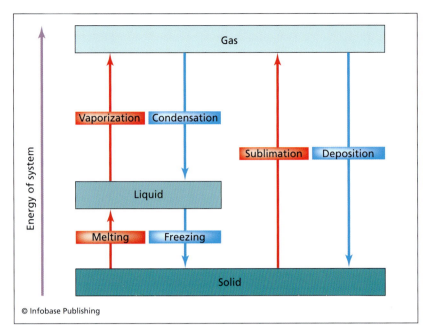

© Infobase Publishing

As the energy of the system increases, matter changes from a solid to a liquid to a gas.

A phase diagram for a compound demonstrates the temperature and pressure at which it exists in different states. The phase diagram of water reveals whether water will exist as a solid, liquid, or gas for given combinations of pressure and temperature. Reading, on the phase diagram, the temperature at standard atmospheric pressure (101.3 kPa) at which solid water changes to liquid water gives the normal melting point of the substance (32°F or 0°C). The phase diagram also shows the normal boiling point of water (212°F or 100°C) at the intersection of the liquid and vapor phases at 101.3 kPa. The triple point is the temperature and pressure at which the solid, liquid, and vapor phase all exist.

See also CALORIMETRY; ELECTROMAGNETIC WAVES; ELECTROMAGNETISM; ENERGY AND WORK; FORCE; HEAT AND THERMODYNAMICS.

FURTHER READING

Goldberg, David. *Schaum's Outline of Beginning Chemistry,* 3rd ed. New York: McGraw Hill, 2006.
Silberberg, Martin. *Chemistry: The Molecular Nature of Matter and Change,* 4th ed. New York: McGraw Hill, 2006.

statistical mechanics The field of statistical mechanics studies the statistical properties of physical systems containing large numbers of constituents. Examples of such systems are a gas (which comprises a very large number of atoms or molecules) and a galaxy (which contains billions, even thousands of

billions, of stars). Although the detailed evolution of such a system in terms of the behavior of every individual constituent might be traceable in principle, in practice the task is impossible. Even if it were possible, it might not be very useful. Statistical mechanics develops methods of deriving useful information about the system as a whole from limited knowledge about its constituents. In particular, this field explains the properties of bulk matter in terms of the properties and interactions of its constituent particles (atoms, ions, molecules). An example is kinetic theory, which explains properties of certain bulk matter in terms of the motion of its constituents. The kinetic theory of gases derives the properties of pressure, temperature, and quantity from the particle nature of gases. More generally, statistical mechanics offers a connection between thermodynamics, which deals with macroscopic properties of matter, and the underlying microscopic structure of matter and its properties.

Statistical mechanics allows for the testing of theories about the microscopic aspects of nature by means of macroscopic measurements. From assumptions about the interatomic forces in a solid and the nature of electrons, for instance, methods of statistical mechanics allow predictions of bulk properties such as heat capacity and electric and thermal conductivities. These properties can be measured and the predictions tested. In this manner, the microscopic assumptions can be evaluated.

See also BOSE-EINSTEIN STATISTICS; ELECTRICITY; FERMI-DIRAC STATISTICS; GAS LAWS; HEAT AND THERMODYNAMICS; MAXWELL-BOLTZMANN STATISTICS; MOTION; STATES OF MATTER.

FURTHER READING

Serway, Raymond A., Clement J. Moses, and Curt A. Moyer. *Modern Physics.* 3rd ed. Belmont, Calif.: Thomson Brooks/Cole, 2004.

stoichiometry Stoichiometry is the quantitative relationship between reactants and products in a chemical reaction. A balanced chemical equation is like a recipe for the relationships among the components in a reaction. The coefficients in a balanced chemical equation represent the relative quantities of each of the reactants and products involved in the chemical reaction. The coefficient can represent the number of atoms in the reaction, the number of molecules in the reaction, or the number of moles (amount of substance) in the reaction. To illustrate, in the following reaction, two molecules of hydrogen gas react with one molecule of oxygen gas to produce two molecules of water:

$$2H_2 + O_2 \rightarrow 2H_2O$$

Another valid and useful way to read this equation is that two moles of hydrogen gas react with one mole of oxygen gas to produce two moles of water.

In 1811, the Italian chemist Amedeo Avogadro defined the mole as the unit to describe the amount of a substance. One mole (abbreviated *mol*) equals 6.02 × 10^23 particles. The coefficients used in balanced chemical equations represent the molar relationships of the substances in the reactions. The law of conservation of matter states that matter can be neither created nor destroyed. This means that the same amount of atoms that go into a reaction as reactants must go out in some form as products. Balancing equations is necessary in order to know the precise value for the number of moles of reactant and product species. In this way, the balanced equation acts as the recipe that tells one how much of each reacting species is present in the reaction and how much of the products are formed in each reaction.

The mass of the substance being studied is the variable when determining how much of each reactant or product is present in a balanced reaction. Several different methods exist for discussing the mass of substances. The mass of an atom is determined by the number of protons plus the number of neutrons. The atomic mass unit (amu) is defined by giving the mass $(1.9924 \times 10^{-23}$ g) of the most common isotope of carbon, carbon 12, a value of 12 amu. Dividing the mass by 12 gives the value of a single amu.

$$1 \text{ amu} = (1.9924 \times 10^{-23} \text{ g})/12$$
$$= 1.66054 \times 10^{-24} \text{ g}$$

Average atomic mass (weighted average) is also known as atomic weight. In order to calculate the average atomic mass, one must consider the relative abundance of each isotope of that element. An isotope is a form of the element that has the same number of protons but a different number of neutrons. The characteristics of the elements are determined by the number of protons in the element, and the mass number of the element is the sum of the protons and neutrons. When calculating the average atomic mass, the relative frequency of the element is multiplied by the mass of that isotope. For instance, carbon has two different isotopes; carbon 12 represents 98.9

percent of all carbon and has a mass of 12 amu, and carbon 14 represents 1.1 percent of all carbon and has a mass of 13.003 amu. The calculation for determining the average atomic mass of carbon is shown.

$$(0.989)(12 \text{ amu}) + (0.011)(13.003 \text{ amu}) = 12.011 \text{ amu}$$

The mass of each ionic compound equals the sum of all of the masses of individual atoms in the compound. The subscripts in a chemical formula represent the number of atoms of each type of element in a molecule of that compound. The mass of the ionic compound is known as the formula mass and is based on the name of one unit of an ionic compound, the formula unit. If the compound is a covalent compound, one unit is known as a molecule. The mass of these compounds is determined by adding up the masses in amu of each of the atoms in the compound. The mass of covalent compounds is called the molecular mass.

Molar mass simplifies the atomic mass concept by describing how many grams are in each mole of the reactants and products. Molar mass is calculated by adding up all of the masses of the individual atoms regardless of whether a single atom, an ionic compound, or a covalent compound. This quantity is used when making conversions between grams and moles of a substance and has the units grams per mole (g/mol).

CONVERTING UNITS USING THE MOLE

The mole is the central unit when converting between other types of units, such as atoms, molecules, formula units, and grams of substances. In order to convert between moles and mass of a substance, one must first calculate the molar mass of the substance in question. The molar mass, reported in g/mol, is then used as a conversion factor to change the units of the substance. The unit conversion system known as dimensional analysis or the factor-label method facilitates these conversions. In dimensional analysis, the given information is placed on the first line, and a series of conversion factors that are equal to 1 are multiplied. The units in a dimensional analysis that are diagonal to one another in the grid can be cancelled, leaving the desired unit in the numerator. The answer in the dimensional analysis method is not calculated until all of the conversions are made. Consider the example.

How many moles of $CaCl_2$ are in 35.8 grams of $CaCl_2$?

1. Calculate the molar mass of $CaCl_2$ (Ca = 40.08 g/mol; Cl = 35.45 g/mol).

$$\text{molar mass of } CaCl_2 = 1(40.08 \text{ g/mol}) + 2(35.45 \text{ g/mol})$$
$$= 111.08 \text{ g/mol}$$

2. Set up a dimensional analysis using 111.08 g/mol as a conversion factor.

$$\frac{35.8 \text{ g } CaCl_2 \mid 1 \text{ mol } CaCl_2}{\mid 111.08 \text{ g } CaCl_2} = 0.3222 \text{ mol } CaCl_2$$

In order to determine the number of atoms or particles reacting from the number of moles, use Avogadro's number (6.022×10^{23} particles/mole) as a conversion factor. Unlike converting between grams and moles, converting between moles and particles uses 6.022×10^{23} particles/mole as the conversion factor regardless of the compound involved in the conversion. In the following example, the particles are formula units, since LiCl is an ionic compound.

How many formula units are in 2.5 moles of LiCl?

$$\frac{2.5 \text{ mol of LiCl} \mid 6.022 \times 10^{23} \text{ formula units LiCl}}{\mid 1 \text{ mol LiCl}} = 1.51 \times 10^{24} \text{ formula units LiCl}$$

A common conversion in chemistry involves converting a given amount of a compound in grams to particles. This two-step process must proceed through the calculation of the number of moles of substance before any other values can be compared. Converting from grams to particles first requires calculating the molar mass of the compound. After the units have been converted to moles, the number of particles can be calculated by using Avogadro's number, 6.022×10^{23} particles/mole.

How many formula units (the particles) are in 35.2 g of LiCl?

$$\frac{35.2 \text{ g LiCl} \mid 1 \text{ mol LiCl} \mid 6.022 \times 10^{23} \text{ formula units}}{\mid 42.39 \text{ g LiCl} \mid 1 \text{mol LiCl}} = 5.48 \times 10^{24} \text{ formula units LiCl}$$

Another conversion that can be made from the number of moles is the number of liters of a gas at standard temperature and pressure (STP). The conversion factor between these two units is 22.4 L of a gas, which is equal to one mole at STP. This means that if the mass, number of atoms, or volume of a gas is known, each of the other values can be calculated.

How many liters will 9.6×10^{28} molecules of hydrogen gas (H_2) occupy? (In this example, the particles are molecules, since H_2 is a covalent compound.)

$$\frac{9.6 \times 10^{28} \text{ molecules } H_2 \mid 1 \text{ mol } H_2 \mid 22.4 \text{ L } H_2}{\mid 6.022 \times 10^{23} \text{ molecules } H_2 \mid 1 \text{ mole } H_2} = 1.79 \times 10^9 \text{ L } H_2$$

Stoichiometric calculations allow for the determination of the theoretical amount of a product that can be formed. One can also calculate the precise amount of reagent necessary for a reaction to go to completion without wasting any reagents. These calculations are based on the molar relationships of the reaction components from the balanced chemical equation. In the following example, calculation of the product amount is done from a known amount of the reagent.

$$2Cl_2 + 2H_2O \rightarrow 4HCl + O_2$$

Given the preceding chemical equation, if 35 grams of water react with an excess of chlorine gas (Cl_2), what mass of hydrochloric acid (HCl) is produced? The first step in any of these problems is to calculate the number of moles of the given reactant or product. This involves the use of the molar mass of water to convert from grams to moles.

$$\text{molar mass of } H_2O = 2(1.01) + 1(15.99) = 18.01 \text{g/mol}$$

$$(35.0 \text{ g } H_2O) (1 \text{ mol}/18.01 \text{ g}) = 1.94 \text{ mol } H_2O$$

Once the value for the number of moles of water is known, the molar ratio between water and HCl, obtained from the coefficients in the balanced chemical equation, can be used to calculate the number of moles of HCl that can be produced.

$$1.94 \text{ mol } H_2O \times (4 \text{ mol HCl}/2 \text{ mol } H_2O) = 3.88 \text{ mol HCl produced}$$

The number of grams of HCl produced can then be calculated from 3.88 moles of HCl produced, using the molar mass of HCl.

$$\text{molar mass of HCl} = 1(1.01) + 1(35.45) = 36.46 \text{ g/mol}$$

$$(3.88 \text{ mol HCl}) (36.46 \text{ g HCl}/1 \text{mol HCl}) = 141 \text{grams HCl}$$

Therefore, the maximal amount of product (reported to the correct number of significant figures) is 141 grams of HCl. It is not always possible to reach this theoretical value experimentally. In order to determine how close the experimental value of product is to this theoretical value, one calculates the percent yield.

percent yield =
(actual yield/theoretical yield) × 100

In the example, if the actual yield is 132 g of HCl, the percent yield is given by the following calculation:

percent yield = (132 g/141 g) × 100

percent yield = 93.6%

Another valuable calculation that can be performed on the basis of formulas in a chemical equation is the percent composition. The percent composition of a compound shows the relative amount of each element present in a compound by mass. In order to calculate the percent composition of an element in a compound, one must first calculate the molar mass of the compound, divide the amount of grams of each element by the total, and multiply by 100. For example, the percent composition of calcium carbonate ($CaCO_3$) is calculated by first adding up the masses that each element contributes to one molecule of the compound.

Ca = 40.08 amu

C = 12.01 amu

O = (15.99) × 3 = 47.97 amu

Total mass of $CaCO_3$ = 100.06 amu

Calculating the percent composition of calcium carbonate requires dividing the amu contributed by each individual element by the total of 100.06 amu.

% Ca = (40.08/100.06) × 100 = 40.06%

% C = (12.01/100.06) × 100 = 12.00%

% O = (47.97/100.06) × 100 = 47.94%

Percent composition can be used as a conversion factor in stoichiometric calculations. When the amount of total compound is given, the amount of any of the elements in the compound can be determined.

How many grams of carbon (C) are in 35.0 g of calcium carbonate ($CaCO_3$)? The preceding calculation determined that the percent composition of carbon in $CaCO_3$ was 12 percent, meaning that 100 g of $CaCO_3$ would contain 12 g of carbon.

$$\frac{35.0 \text{ g } CaCO_3}{} \left| \frac{12.0 \text{ g carbon}}{100.0 \text{ g } CaCO_3} \right. = 4.20 \text{ g C}$$

Experimentally determined masses of reactants and products can be used to determine the formulas of an unknown compound. The empirical formula is the lowest whole number ratio between the atoms in a compound. Many compounds can have the same empirical formula, as C_2H_4 and C_6H_{12} each have the lowest whole number ratio between their elements as CH_2. In some cases, the empirical formula is identical to the molecular formula, the formula of the compound as it actually exists in nature. Examples such as water, H_2O, have the lowest whole number ratio between the atoms (2:1); thus, the empirical formula also serves as the molecular formula of the compound. In other cases, the empirical formula and the molecular formula are not the same. In order to calculate the molecular formula of a compound, one needs to know the molar mass of the compound and compare that to the molar mass of the empirical formula. The following problem demonstrates this situation.

Calculate the molecular formula for a compound with an empirical formula of CH_2 that has a molar mass of 28.06 g/mol.

1. The molar mass of the empirical formula CH_2 is 14.03 g/mol, based on the molar mass of one carbon atom (12.01 g/mol) and two hydrogen atoms 2(1.01) = 2.02.
2. Since 14.03 g/mol does not equal 28.06 g/mol, this empirical formula does not have the correct molar mass also to be the molecular formula of this compound.
3. Therefore, the empirical formula as a unit must be multiplied by a constant to equal the molar mass of 28.06 g/mol. The constant can be determined by dividing the molar mass by the empirical formula mass.

(28.06 g/mol)/(14.03 g/mol) = 2

2 (CH_2) = C_2H_4 = 28.06 g/mol

4. The molecular formula for this compound is C_2H_4 and is known as ethene.

See also ANALYTICAL CHEMISTRY; MEASUREMENT.

FURTHER READING
Chang, Raymond. *Chemistry,* 9th ed. Boston: McGraw Hill, 2006.

superconductivity The total lack of electric resistance that certain solid materials exhibit at sufficiently low temperatures, that is, at temperatures below their respective *critical temperatures,* is termed superconductivity. When an electric current is caused to flow in a material that is in the superconducting state, the current will continue to flow forever without diminishing, as far as is presently known. Another characteristic of the superconducting state is that it does not suffer the presence of a magnetic field within itself and even expels a preexisting field when the material is cooled to its superconducting state while immersed in a magnetic field. This is called the Meissner effect, named for the 19–20th-century German physicist Walther Meissner. The total cancellation of externally applied magnetic fields within the interior of a superconductor is accomplished by surface currents, which are electric currents flowing within a thin layer at the surface of the superconductor. Various laws of electromagnetism predict that a superconductor will maintain whatever magnetic field exists within it upon its becoming superconducting. The Meissner effect, however, shows that a superconductor goes beyond that, by completely canceling any interior magnetic field and then maintaining the

Because of the Meissner effect, a small disk magnet floats over a superconductor. This levitation is brought about by electric currents in the surface of the superconductor, which produce a magnetic field that both repels, and thus levitates, the floating magnet and completely cancels any magnetic field in the superconductor's interior. The superconductor is of the ceramic variety and is kept at a superconducting temperature by liquid nitrogen. *(David Parker/IMI/Univ. of Birmingham/Photo Researchers, Inc.)*

field-free state. In this way, the floating of a small magnet above a superconductor can be viewed as resulting from a "repulsion" of the magnet's magnetic field by the superconductor—as if the field lines behave as springs beneath the magnet and support the magnet above the impenetrable surface of the superconductor.

The strongest magnetic field in which a material can be a superconductor is called the material's critical field. An applied magnetic field that is greater than a material's critical field, even though it cannot penetrate the material in its superconducting state, nevertheless destroys the superconducting state. The value of the critical field depends on temperature, is greatest approaching 0 kelvin (K), and declines to 0 at the critical temperature. The critical field effectively limits the amount of current that can pass through a superconducting material. If the current is too great, the magnetic field it produces will surpass the critical field and suppress superconductivity.

Type I superconductors are certain pure metals: aluminum, mercury, and niobium, for example. They are characterized by critical temperatures below 10 K and possess relatively low critical fields, below about 0.2 tesla (T). Type II superconductors comprise various alloys and metallic compounds. Their critical temperatures are more or less in the 14–20 K range, and their critical fields are higher than 15 T. So type II superconductors are much more suitable than type I for the construction of superconducting electromagnets; they do not have to be cooled to such low temperatures and they can carry higher currents. Setting up and maintaining the field of an electromagnet normally require an investment of energy, but superconducting electromagnets only require energy investment for setting up the field, as long as the field remains below the critical field. A successful theory of superconductivity for types I and II, called the BCS theory (after John Bardeen, Leon Cooper, and John Schrieffer, all 20th-century American physicists), explains the effect as conduction by paired electrons, called Cooper pairs, whose existence involves the participation of the material's crystal structure.

A class of what are designated high-temperature superconductors, with critical temperatures ranging to above 130 K, has been discovered and is being expanded. These are metallic oxides in ceramic form and all contain copper. One importance of this class is that many of its members can be cooled to the superconducting state with liquid nitrogen, whose temperature is about 77 K and which is much cheaper than the liquid helium (at about 4 K) required for the cooling of superconductors of types I and II. High-temperature super-

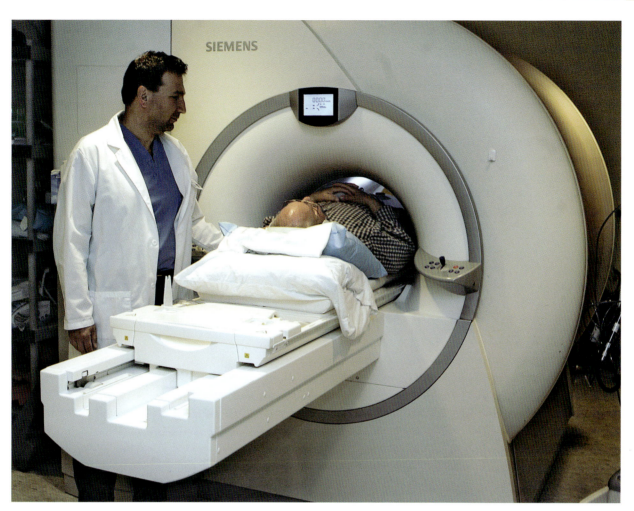

Magnetic resonance imaging (MRI) machines require high magnetic fields over large volumes. In order to achieve this, the machines are built with superconducting electromagnets, in which the coils are cooled to below their critical temperature by liquid helium or liquid nitrogen. Here a technician watches over a patient entering an MRI machine. *(AP Images)*

conductivity holds promise for many useful future applications, including loss-free electricity transmission, magnetic levitation, and liquid nitrogen–based superconducting electromagnets. The BCS theory does not seem applicable to high-temperature superconductivity, and no other explanation for it is currently available.

Superconductivity was discovered in 1911 by the Dutch physicist Heike Kamerlingh Onnes while he was investigating the electrical properties of metals at low temperatures. Kamerlingh Onnes was also the first to liquefy helium.

The best-known and most familiar application of superconductor technology is in the extrastrong electromagnets used in magnetic resonance imaging (MRI) scanners. The superconducting coils of such magnets allow the passage of high electric currents, which in turn produce the extrastrong magnetic fields needed for high-quality scans. Older MRI

devices require liquid helium for cooling the coils to their superconducting state, while newer machines make use of high-temperature superconductors and operate with liquid nitrogen.

See also ELECTRICITY; ENERGY AND WORK; HEAT AND THERMODYNAMICS; MAGNETISM.

FURTHER READING

The Nobel Foundation. "The Nobel Prize in Physics 2003." Available online. URL: http://nobelprize.org/nobel_prizes/physics/laureates/2003/public.html. Accessed July 25, 2008.

The Royal Society of Chemistry. The Superconductivity Web site. URL: http://www.chemsoc.org/exemplarchem/entries/igrant/main_noflash.html. Accessed July 25, 2008.

Serway, Raymond A., Clement J. Moses, and Curt A. Moyer. *Modern Physics*, 3rd ed. Belmont, Calif.: Thomson Brooks/Cole, 2004.

SUPERCONDUCTORS.ORG, a nonprofit, nonaffiliated Web site. URL: http://superconductors.org. Accessed July 25, 2008.

surface chemistry The study of the atomic and molecular interactions at the interface between two surfaces is known as surface chemistry. Research in this field allows scientists to get a closer look at reaction mechanisms, or the step-by-step processes by which reactions proceed. Traditional obstacles to studying reaction mechanisms stem from the fact that the reacting species interact with other substances in the mixture. Studying chemical reaction mechanisms is a complicated process but is simplified by first considering the reacting species in the gas form. If both reactants are in the gas phase, the gas molecules are only interacting with the individual molecules that are part of the reaction. When the reaction takes place in solution (usually water), the movement of the water and the other particles complicates the investigation of the mechanism of the reaction. Isolation of the reaction components is required in order to understand the mechanism of the reaction, and this is where surface chemistry plays a role. Surface chemistry takes place within heterogeneous mixtures known as colloids and in the process of heterogeneous catalysis. Training to study surface chemistry can begin with a degree in chemistry, analytical chemistry, or physical chemistry. Surface chemistry discoveries and advancements in the experimental methodology related to surface chemistry impact many areas of science, including the production of ammonia from atmospheric nitrogen to use in fertilizer, the development of the computer chip, corrosion resistance, and nanotechnology.

COLLOIDS

A mixture with no uniform composition is known as a heterogeneous mixture and is classified into categories based on the size of the particles found in the mixture. When the particle size exceeds 100 nanometers, the mixture is called a suspension, and the particles will settle out upon standing. Sand stirred up in water is an example of a suspension. One must continue to stir the mixture or else the sand will settle to the bottom of the container. Heterogeneous mixtures with intermediate particle size of approximately 1–100 nanometers are called colloids. Examples of colloids include egg whites, marshmallows, mayonnaise, and fog. Colloids are heterogeneous mixtures that can involve liquids dissolved in other liquids (known as emulsions), solids dissolved in a liquid (suspensions), or gases dissolved in a liquid (known as aerosols and foams). Studying surface chemistry allows scientists to understand colloids by determining what is occurring at the interphases of the states (the surface separating the two phases) from one another.

HETEROGENEOUS CATALYSIS

Catalysts are substances that speed up the rate of a reaction without themselves being consumed in the reaction. There are many types of catalysts, and they fall into two categories—homogeneous catalysts and heterogeneous catalysts. Homogeneous catalysts are those that exist in the same state as the reactants. For example, when the reaction is in solution, catalysts such as enzymes are also present in solution. Heterogeneous catalysis occurs when the state of the catalyst differs from that of the reactants. As it applies to surface chemistry, heterogeneous catalysis generally involves a solid catalyst (usually a metal) and a gaseous reactant. When a gas particle hits the surface of a solid, three potential scenarios exist: the particle can reflect off the surface without interacting; the particle can hit the surface and adhere to the surface (in some cases, being broken down to the atomic level); or the particle can react with another particle that has already adhered to the surface. In the second and third scenarios, the interaction of the reactant gas with the solid metal increases the speed at which the reaction occurs because the reacting molecule is broken down to the atomic level, making it available to carry out other chemical reactions. In the final scenario, the first reacting species adheres to the metal, forcing it to stay in place and facilitating interaction with the second gas molecule.

HISTORY AND ADVANCEMENTS IN SURFACE CHEMISTRY

Organic syntheses frequently perform hydrogenation (the addition of hydrogen atoms) to organic compounds. This process was recognized in 1912 with the first Nobel Prize for surface chemistry awarded to a French chemist, Paul Sabatier. He developed the basis for the hydrogenation method that is still in use today. The molecular hydrogen is first adsorbed to a metal surface, leading to the breakage of the molecular hydrogen into individual hydrogen atoms. These atoms are then available to react with an organic compound and add to the organic compound singly, producing a hydrogenated compound.

The German chemist Fritz Haber received the Nobel Prize in chemistry in 1918 for his development of a method for producing ammonia (NH_3) from atmospheric nitrogen (N_2). Because nitrogen is a critical component of the biomolecules proteins and nucleic acids, all living organisms require this nutrient. Molecular nitrogen makes up 78 percent of air. Despite its abundance, this form of nitrogen is not usable by plants or animals. The addition of nitrogen compounds to fertilizer has dramatically impacted crop production. Most plants can use nitrogen in the form of ammonia or in the form of nitrates (usually ammonium nitrate). Faber developed a method for synthetically producing ammonia, which living organisms can convert into useful compounds in the food chain. Chemists still use the Faber process of ammonia production with some modifications. The American chemist Irving Langmuir received the Nobel Prize in chemistry in 1932 for his developments in the field of surface chemistry and heterogeneous catalysis.

For the next several years, the stumbling block of keeping the reacting surfaces clean without having other molecules adsorb to these surfaces slowed the advancement of the field of surface chemistry. The 1960s introduced improvements in surface chemistry, including the ability to maintain a clean surface and to evaluate the process of the reaction as it was occurring on the surface. The discovery that if the surface is placed under a vacuum, other interacting particles could be reduced eliminated many of these complications.

The German chemist Gerhard Ertl is responsible for many of the aforementioned advances in surface chemistry. He won the 2007 Nobel Prize in chemistry for his contributions to the field. Ertl used a meticulous experimental approach to help describe the mechanism for several readily used examples of heterogeneous catalysis. He first demonstrated the arrangement of hydrogen atoms on the surface of such metals as platinum and lead. Using an experimental technique known as LEED (low-energy electron diffraction), Ertl demonstrated how the individual hydrogen atoms were singly arranged in a monolayer on the surface of the metal. This description of the arrangement of the hydrogen atoms enabled scientists to explain the proper mechanism for hydrogenation reactions.

Ertl was also responsible for demonstrating the precise reaction method of the ammonia-producing reaction developed by Haber that had been in use for a very long time. While people understood that the metal sped up the process of turning atmospheric molecular nitrogen into ammonia that could become useful in fertilizers, no evidence existed that explained the mechanism. Several different reaction mechanisms had been proposed, but no accurate experimental data supported these mechanisms. Ertl utilized his newly developed surface chemistry techniques to determine the mechanism. The reaction occurred in a series of steps in which the nitrogen and hydrogen gases are adsorbed onto the metal surface. Many doubted that nitrogen would dissociate into atomic nitrogen, a step that was necessary

for the reaction to occur, because of the strong triple bond joining the nitrogen atoms in molecular nitrogen. Ertl definitively confirmed several aspects of the proposed reaction mechanism. First, he used auger electron spectroscopy (AES) to demonstrate that, under high pressure, atomic nitrogen was clearly evident at the surface of the metal. This led the way for the determination of a reaction mechanism, as follows:

1. adsorption of molecular hydrogen to the surface of the metal

$$H_2 \rightarrow 2H_{adsorbed}$$

2. adsorption of molecular nitrogen to the surface of the metal

$$N_2 \rightarrow 2N_{adsorbed}$$

3. reaction of atomic nitrogen and atomic hydrogen

$$N_{adsorbed} + H_{adsorbed} \rightarrow NH_{adsorbed}$$

4. addition of another hydrogen atom to produce NH_2

$$NH_{adsorbed} + H_{adsorbed} \rightarrow NH_{2adsorbed}$$

5. addition of one more hydrogen atom

$$NH_{2adsorbed} + H_{adsorbed} \rightarrow NH_3$$

6. removal of ammonia from the metal

$$NH_{3adsorbed} \rightarrow NH_3$$

Determination of the mechanism for ammonia production was an economically important discovery. Fertilizers are a lucrative business, and by understanding the mechanism of the reaction, one can optimize the process. Surface chemistry studies allow a look at the reaction mechanism so scientists can ascertain how chemical reactions are taking place.

ACS DIVISION: COLLOID AND SURFACE CHEMISTRY

The American Chemical Society (ACS) has a branch known as the Colloid and Surface Chemistry division. Established in 1926, the division has been actively involved in expanding research in the field of surface chemistry. It is one the most active and growing divisions in ACS and has approximately 2,500 members. The mission of the division is the study of all ranges of reactions on every type of surface. They participate in the twice-yearly meetings of the American Chemical Society as well as holding yearly meetings of their own division.

See also ANALYTICAL CHEMISTRY; CHEMICAL REACTIONS; PHYSICAL CHEMISTRY; RATE LAWS/REACTION RATES.

FURTHER READING

McCash, Elaine M. *Surface Chemistry.* New York: Oxford University Press, 2001.

symmetry The possibility of making a change in a situation while some aspect of the situation remains unchanged is symmetry. As a common example, many animals, including humans, possess approximate bilateral symmetry. Consider the plane that bisects the animal's body front to back through the middle. With respect to that plane, the external appearance of the body has this balance: for every part on the right, there is a similar, matching part on the left (and vice versa). If the body were reflected through the bisecting plane, as if the plane were a two-sided plane mirror, the result would look very much like the original. Here we have the possibility of reflection as a change under which the body's external appearance remains unchanged (more or less).

Spatial symmetry is symmetry under spatial changes. Those include reflection (as in the example), rotation, displacement, and scale change. Here are some examples. A starfish possesses approximate fivefold rotation symmetry about the axis through its center and perpendicular to its plane, with respect to external appearance. It can be rotated through angles that are integer multiples of 360°/5 = 72°, and such rotations do not change its appearance. The laws of nature are symmetric under displacements, in that the same laws are valid at all locations in space (as

far as is presently known). Perform experiments here or perform them there: in both cases, one discovers the same laws of nature. Symmetry under change of scale is a property of certain natural phenomena and mathematically generated patterns. Such systems are called fractals. The shape of coastlines and the form of mountain ranges are approximately fractal, as they present more or less the same structure at different scales.

Besides spatial symmetry, there are many additional types of symmetry. Temporal symmetry involves changes in time, time ordering, or time intervals. As an example, the laws of nature also appear to be symmetric under time displacements (again, at least as far as we know): perform experiments now or at any other time, and one will find the same laws of nature. Spatiotemporal symmetry has to do with changes involving both space and time. An example is a change from one reference frame to another moving at constant velocity with respect to the first. These changes are expressed by Galilei transformations, Lorentz transformations, or Poincaré transformations. The laws of nature are found not to change under such transformations. Observers in relative straight-line motion at constant speed perform experiments and discover the same laws of nature. This symmetry forms an essential component of Albert Einstein's special theory of relativity. An even more general spatiotemporal symmetry lies at the heart of Einstein's general theory of relativity: observers in relative motion of any kind, including accelerated motion, find the same laws of nature.

Other changes, of more abstract character, are also useful in connection with symmetry. They include rearrangement (permutation) of parts or aspects of a system, changes of phase of wave functions, and mixing of fields, among others.

Symmetry of the laws of nature under some change is called an invariance principle. Invariance principles are related to conservation laws, to the fact that certain physical quantities remain constant in value over time. In addition to the invariance principles that underlie the theories of relativity, other invariance principles play major roles in physics. The theories of the strong force, electromagnetic force, weak force, and electroweak force are founded on invariance principles called gauge symmetries, which are symmetries under space-time-dependent changes that mix components of quantum fields.

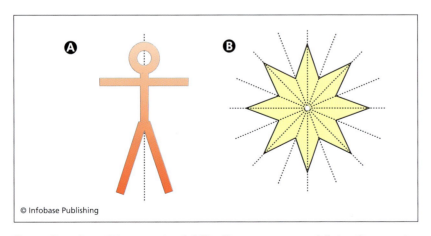

© Infobase Publishing

Examples of spatial symmetry. (a) The figure possesses bilateral symmetry (i.e., reflection symmetry) with respect to the dashed line. (b) Eightfold rotation symmetry about the figure's center is accompanied by bilateral symmetry with respect to each of the eight dashed lines.

Note that symmetry implies equivalence of parts, components, or aspects of a system. The more symmetries a system possesses, or the higher a system's degree of symmetry, the more equivalence it contains. Thus, symmetry correlates with uniformity. The higher the degree of symmetry, the higher the level of uniformity in a system. Compare a square plate and a circular disk with regard to their symmetry under rotations about their axis (through their center and perpendicular to their plane), for instance. The square plate maintains its appearance only under rotations through integer multiples of 360°/4 = 90°, while the circular disk is symmetric under rotation through any angle whatsoever. The disk possesses perfect rotational uniformity, while the uniformity of the square is severely limited.

Order correlates with distinguishability, inequivalence, nonuniformity. Its opposite, disorder, thus correlates with indistinguishability, equivalence, and uniformity. So, symmetry correlates with disorder. In this connection, let us compare the same substance in its crystalline state and its gaseous state. The crystal possesses a relatively high degree of order: the molecules are positioned only on the crystal lattice and not elsewhere (inhomogeneity), and the physical properties of the crystal are generally different in different directions (anisotropy). The crystal is characterized by a corresponding set of symmetries. The gas, on the other hand, is highly disordered: the molecules can be anywhere in the container (homogeneity), and all directions are equivalent with regard to the substance's physical properties (isotropy). As a result, the gas possesses a much higher degree of symmetry than does the crystal. Every symmetry of the crystal is also a symmetry of the gas, while the gas has many more additional symmetries.

The symmetry principle, also called Curie's symmetry principle or the Curie principle, for the 19–20th-century French physicist Pierre Curie, is of major importance in the application of symmetry. This principle states that the symmetry of a cause must appear in its effect. An alternative, equivalent statement of the principle is that any asymmetry (lack of symmetry) in an effect must also characterize the cause of the effect. So, an effect is at least as symmetric as its cause. Consider this example. The value of the index of refraction of a substance—as an effect—depends on the interaction of electromagnetic radiation, such as light, with the substance's constituent atoms, molecules, or ions—as the cause. By the symmetry principle, any anisotropy (dependence on direction) of the index of refraction, which is an asymmetry, must be found also in the distribution of the substance's constituents. When a substance is in a liquid state, its constituents are randomly and uniformly distributed, with no distinguished directions. Consequently, the index of refraction of liquids is isotropic (independent of direction). In a crystalline state of a substance, however, when the constituents are arrayed on a lattice, not all directions are equivalent. Then the index of refraction might very well have different values in different directions (an effect called birefringence).

See also ACCELERATION; CONSERVATION LAWS; CURIE, MARIE; EINSTEIN, ALBERT; ELECTROMAGNETIC WAVES; ELECTROMAGNETISM; GENERAL RELATIVITY; INVARIANCE PRINCIPLES; MOTION; OPTICS; PARTICLE PHYSICS; QUANTUM MECHANICS; SPECIAL RELATIVITY; SPEED AND VELOCITY; STATES OF MATTER.

FURTHER READING
Rosen, Joe. *Symmetry Discovered: Concepts and Applications in Nature and Science.* Mineola, N.Y.: Dover, 1998.

———. *Symmetry Rules: How Science and Nature are Founded on Symmetry.* Berlin: Springer, 2008.

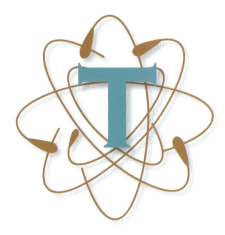

telescopes   A telescope is a device that allows for the visualization of distant objects by collecting electromagnetic radiation using reflecting mirrors or refracting lenses and focusing the rays on the focal plane or focal point to produce an image. Telescopes that utilize visible light, X-rays, infrared waves, and radio waves have been made. Objects in space emit different types and amounts of electromagnetic radiation. The electromagnetic waves travel at the speed of light, 1.86×10^5 mi/s (3.00×10^5 km/s) until they are picked up by a telescope. The most common types of telescopes are reflecting and refracting telescopes that use visible light.

The first reflecting telescope was developed by Sir Isaac Newton in the 1600s. Reflection is the return of light waves after hitting and bouncing off a surface. Reflecting telescopes use mirrors that are carefully formed surfaces that are coated with a reflective material to bounce light back in a controlled direction. Each reflecting telescope contains two: a primary mirror and a secondary mirror. The primary mirror is concave (indented in its center), sits in the back of the tube, and reflects the light to the secondary mirror, which then reflects the light onto the eyepiece to create an observable image for the viewer. If the telescope contains sensors, the secondary mirror reflects the light onto a focal plane, where the sensors can pick up the image.

In the early 1600s, Galileo Galilei developed a refracting telescope that used lenses to bend the light and collect it as an image. Refraction occurs when a wave bends as it passes through a medium. The primary lens of the telescope is convex (thicker in the middle than at the edges), and the second lens, located in the eyepiece, is concave (thinner in the middle than at the edges). Galileo studied the sky and

was the first to see the craters on the Moon and the moons of Jupiter. His observations using the refracting telescope led him to propose his theory that the

The Hale telescope of Palomar Observatory, located outside San Diego, California, is a reflecting telescope with a 200-inch mirror. *(Sandy Huffaker/Getty Images)*

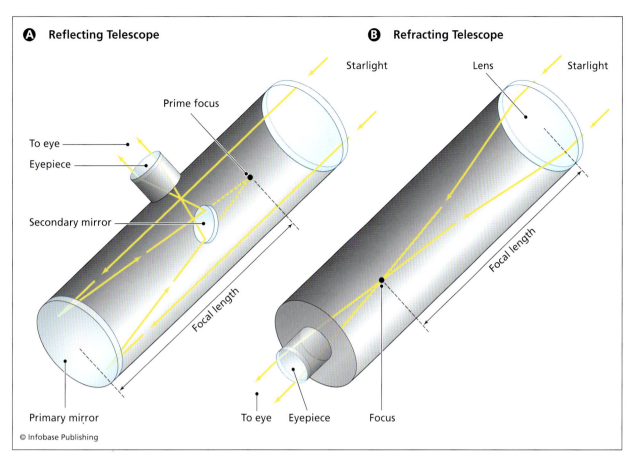

A Reflecting Telescope

B Refracting Telescope

Starlight

Lens

Starlight

Prime focus

To eye

Eyepiece

Secondary mirror

Focal length

Focal length

Primary mirror

To eye   Eyepiece   Focus

© Infobase Publishing

(a) Reflecting telescopes utilize mirrors, and (b) refracting telescopes utilize lenses to gather light and magnify the image.

Earth is not the center of the universe. In 1897, the world's largest refracting telescope was built in the Yerkes Observatory in Wisconsin.

Refracting telescopes use lenses to gather the light. They consist of a long tube that collects the light and holds the lenses at a fixed distance from each other. The objective or primary lens is at the light-collecting end of the telescope, and the eyepiece is where the viewer observes the image. The light passes through the objective lens (a convex lens), is refracted to a focal point, and then travels to the eyepiece, which can be either concave or convex, that allows one to look or, in the case of a refracting telescope, focus the light onto a focal plane.

Radio telescopes pick up radio waves from space, including from stars, galaxies, and other objects, with waves in the 100-m to 1-cm range. (Although scientists always report wavelengths in metric units, this range converts to approximately 330 ft to 0.4 in in customary units.) Radio telescopes require an antenna and a receiver. The National Radio Astronomy Observatory runs the world's most important radio telescopes, including the Very Large Array (VLA) in New Mexico, the Green Bank Telescope (GBT) in

West Virginia, the Atacama Large Millimeter Array (ALMA) in Chile, and the Very Long Baseline Array (VLBA), which comprises 10 radio telescopes situated in various North American locations.

Infrared telescopes detect radiation in the range of 1,000 nm to 700 nm. The *Infrared Astronomical Satellite (IRAS)* was a joint project of the United States, the Netherlands, and the United Kingdom. Construction began in 1975, and the *IRAS* was launched and used from January to November 1983. According to Ball Aerospace and Technology, the company responsible for building the *IRAS*, it was the first space-based, cryogenically cooled (by liquid helium) infrared telescope. *IRAS* was put into orbit in 1983 in order to survey the sky for infrared radiation. While in operation, the *IRAS* discovered many previously unobserved astronomical structures, including comet tails and dust bands.

X-ray telescopes are unique because they detect X-rays from extremely hot objects such as exploded stars even to temperatures up to millions of degrees Celsius. X-rays cannot penetrate the Earth's atmosphere; therefore, all X-ray telescopes are satellites. X-rays do not bounce off mirrors—they penetrate to an extent, so the mirrors in an X-ray telescope are nearly parallel to the path of the incoming X-ray. The most important X-ray observatory is NASA's *Chandra X-Ray Observatory,* which the space shuttle *Columbia* carried into space in 1999. Chandra has three parts: mirrors, scientific instruments, and the spacecraft. The mirrors collect the X-rays and send them to a focal point where scientific instruments detect the image. The spacecraft is responsible for housing the telescope, providing it power using solar panels, and communicating the results back down to Earth.

One of the most famous satellite telescopes to date is the reflecting, satellite *Hubble Space Telescope (HST)* or simply the *Hubble.* The *HST* is named for the American astronomer Edwin Hubble, who suggested the universe was expanding and discovered other galaxies using telescopes on Earth. The American theoretical physicist Lyman Spitzer first suggested building a telescope in space. In 1977, he gained congressional approval for the development of the *HST.* Perkin-Elmer developed the primary and secondary mirrors for the *HST,* and Lockheed Martin was in charge of building the telescope. The Space Telescope Science Institute, an organization founded by the National Aeronautics and Space Administration (NASA), manages and directs the *HST,* which

was put into operation in 1990 and was taken into space by the space shuttle *Discovery.* The *HST* is 43.6 ft (13.1 m) long and weighs approximately 24,500 lbs (11,100 kg). Since the *HST* functions in space, it must perform many tasks in addition to functioning as a telescope. The *HST* must supply its own power, which it accomplishes by using the solar arrays on the outside surface of the *HST.* This solar array provides power to run the *HST* when on the Sun side of the Earth and charges the battery supply to keep the *HST* running when the telescope moves to the shade side of the Earth. The communication equipment required for the *HST* is controlled by the movable antennae that can transmit data. The *HST* also needs to be steered. The control system contains reaction wheels that turn the *HST.* The fine guidance system (FGS) of the *HST* works in conjunction with the reaction wheels to ensure that the telescope stays on course.

The *HST* has provided key evidence for the existence of black holes and planetary disks, which are important in the formation of new planets. The *HST* orbits approximately 370 miles (600 km) above the surface of Earth and is able to complete one orbit around Earth in 97 minutes.

The major portion of the *HST* is the Optical Telescope Assembly, which includes the two mirrors

The *Hubble* telescope (*HST*) is one of the most famous satellite reflecting telescopes. This photo, taken by the STS-109 crew, shows the *Hubble* telescope returning to normal routine after five days of service and upgrade work, March 9, 2002. *(NASA Marshall Space Flight Center)*

This image of the Sombrero Galaxy (officially designated M104) was voted by astronauts to be the best picture taken by the Hubble telescope. This galaxy is 28 million light years from Earth, contains 800 billion stars, and is 50,000 light-years across. *(NASA Marshall Space Flight Center)*

and support trusses. The second part of the *HST* consists of filters that screen out light from other objects. The telescope contains a filter wheel that rotates to allow the correct filter to capture the beam of light. The third part of the operational system of the *HST* includes its light-sensitive detectors.

The *HST* works by allowing light to enter a tube with baffles that control which light passes in and out. The light hits a concave primary mirror that is 7.9 ft (2.4 m) in diameter and passes to a convex secondary mirror that is 12 in (0.3 m) in diameter. The convex secondary mirror reflects the light back through the primary mirror through a hole designed for this purpose. The reflected light is gathered on the approximately 8-in (20-cm) focal plane where the image is formed. The image is then transferred to the science instruments of the *HST*.

The *HST* presently contains six different science instruments, five of which are functional. These instruments include four cameras (one no longer in use), a spectrograph, and a sensor. Each of the cameras serves a distinct purpose. The Advanced Camera for Surveys (ACS) is the newest of the cameras, having been installed in 2002. The ACS is used for visualizing the weather on other planets and replaced the Faint Object Camera (FOC), a telephoto lens camera that was taken offline in 2002. The Widefield and Planetary Camera (WFPC2) is

the main camera for the *Hubble*. The Near Infrared Camera and Multi Object Spectrometer (NICMOS) is a heat sensor that can be used to study deep space.

The Space Telescope Imaging Spectrograph (STIS) separates light into its individual components by acting as a prism. The STIS suffered a power failure in 2004 and is presently not functioning. Many hope that future repair missions will restore the function of the STIS. The fine guidance sensors help to guide the position of the telescope and give it a frame of reference. The *HST* remains lined up even while it is moved by looking at stars and measuring their relative positions.

The mirrors of the *HST* are made of low-expansion glass with surfaces ground to a precision of 1/800,000 of an inch (0.03 μm). The glass must remain at a constant temperature of 70°F (21°C). The surfaces of the mirrors have a coating of aluminum 3/100,000 of an inch (0.76 μm) thick. A 1/1,000,000-in (0.025-μm) layer of magnesium fluoride makes the mirror more UV reflective.

After launching the *HST*, it was discovered that the primary mirror was not ground to the desired tolerance. The images returned from space were blurry because the light was being focused at several different locations rather than at one focal plane. All equipment that has been added subsequently to the

Hubble contains corrections that allow for a focused image even with the incorrect mirror. Bell Aerospace developed "eyeglasses for *Hubble*" that correct the defects in the primary mirror and is known as *COSTAR,* which stands for "corrective optics space telescope axial replacement."

The resolution or resolving power of a telescope determines how clear a picture is as well as how small an object can be seen by the observer through the eyepiece or a sensor at the focal plane. Astronomers describe resolving power using the term *arcseconds.*

1 degree = 60 arcminutes = 3,600 arcseconds

The resolving power of the human eye is approximately 60 arcseconds; the resolving power of a Galilean telescope is three arcseconds; the resolving power of large terrestrial telescopes is one arcsecond; and the *HST* has a resolving power of 0.05 arcsecond.

The Next Generation Space Telescopes (NGSTs) are being developed to replace the *HST* whenever it needs to be retired. The *James Webb Space Telescope (JWST)* is being developed and might be launched as soon as 2011. While the *HST* has a primary mirror diameter of 7.87 ft (2.4 m), the JWST design includes a primary mirror with a diameter of 26.2 ft (8 m), giving it nearly 10-fold greater power to see farther into deep space.

See also ELECTROMAGNETIC WAVES; GALILEI, GALILEO; NEWTON, SIR ISAAC; OPTICS; WAVES.

FURTHER READING

Seeds, Michael A. *Foundations of Astronomy,* 9th ed. Belmont, Calif.: Thomson Brooks/Cole, 2006.
Space Telescope Science Institute. "Hubble Site: The Telescope." Available online. URL: http://hubblesite.org/the_telescope. Accessed July 25, 2008.

textile chemistry Textile chemistry is the study of the chemistry and physics of textiles, dyes, and the technologies associated with the production of textiles into useful items. Textiles are used in materials, clothing, furniture, carpet, and thread. Textile chemistry can be subdivided into the chemistry of dyeing and finishing, the chemistry of polymers and fibers, and the chemistry of blending polymers. Looking around a room, one would find textiles in the carpets, drapes, upholstery, backpacks, coats, toys, linens, towels, clothing, and any other item that involves a thread or fabric. Textiles keep people warm, clothed and decorated and affect many aspects of everyday life.

Textile chemists have a wide variety of educational backgrounds. Many schools in the United States offer degrees in textile chemistry or another branch of textiles. Other related fields that feed into work as a textile chemist include polymer chemistry, chemical engineering, and other chemistry-based majors. Preparation for a career in textile chemistry requires coursework in organic chemistry and analytical chemistry, and experience in the manufacturing and production process as well as quality control and statistics is also useful. Some textile chemistry degree programs offer coordinated programs with international business degrees that allow the chemists to work in both the chemistry and the business aspects of the international textile market. Still other programs offer dual-degree programs in textile chemistry as well as design to increase the marketability and flexibility of their students. This allows the textile chemists to have a role in the design process. Some textile chemists pursue a doctoral degree to gain further expertise in the field.

Working as a textile chemist can lead to a wide variety of jobs. Textile chemists can work in research and development of new and innovative products in the textile fields. New materials, new fibers, and new dyes, as well as new manufacturing processes and technologies, allow for very diverse jobs in the research and development field. Many textile chemists work in the actual production process overseeing product quality and manufacturing. Others work in marketing, and still others in the design side of textiles. The diversity of jobs could include lab work, management and business, or design and marketing.

The study of textile chemistry includes the science of dyeing and finishing the fabrics and creating the fibers used to make the fabrics. People have dyed fabrics and fibers since ancient times, and until the 1800s, the process involved placing extracts from plants, berries, and other organic materials onto the fabric. Dramatic colors were not readily available, so the use of colored fabrics was considered a luxury and a sign of wealth. A multitude of colors can be obtained from natural dyes, which have a variety of sources, including leaves, twigs, stems, flowers, skins, hulls, wood, berries, seeds, or insects. The following table gives a sampling of natural dyes and their resulting colors.

William Henry Perkin produced the first commercially available synthetic dye in 1857 in England. He isolated a brilliant purple dye while working on the isolation of quinine in 1856 and marketed it as Mauveine the following year. Many synthetic dyes followed this discovery, including azo dyes, sulfur dyes, and Congo red, all of which are still used today. Disperse dyes were developed when the synthetic fibers that were being produced in the mid-1900s became available. Many of the older types of

NATURAL DYES

Common Name	Color
Alder	Yellow/ brown/ black
Apple	Yellow
Blackberry	Pink
Blackwillow	Red, brown
Bloodroot	Red
Buckthorn	Yellow, brown
Cherry (wild)	Pink, yellow, brown
Dahlia	Yellow bronze
Dyer's broom	Yellow
Elder	Yellow, gray
Groundnut	Purple, brown, pink
Indigo	Blue
Ivy	Yellow, green
Madder	Orange, red
Maple	Tan
Marigold	Yellow
Nettles	Beige, yellowy greens
Onion	Yellow, orange
Oak	Gold, brown
Privet	Yellow, green, red, purple
Ragwort	Yellow
Safflower	Yellow, red
Sloe-Blackthorn	Red, pink, brown
Weld (wild mignonette)	Olive green
Woad	Blue

of up to six colors onto a fabric. Most printing processes involve the use of pigments of low molecular weight rather than traditional dyes. This prevents the dyes from running to the unprinted area of the fabric. The pigments do not molecularly bind to the fabric so they must be processed with a copolymer binding agent that seals the pigment on the fabric. Technology has improved the efficiency of the textile printing industry as well as every other area of life in the 21st century, and many printing techniques now involve the use of computerized textile printers that are similar in concept to a home office ink jet printer. The print heads spray the drops of the pigments in the correct locations on the fabric to create the printed design. This allows designers to control fabric patterns down to the single drop of pigment and pixel level for more refined and reproducible printed fabrics.

Vat dyeing, the most common method of textile dyeing, involves the addition of the dye into a large vat of water and adding the fabric or the fiber. The solution is then brought to a boil and agitated in order to ensure proper covering of the fabric or fibers. The textile industry commonly employs vat dyeing because a large amount of fabric can be dyed at the same time. One drawback of this method is the production of a large amount of wastewater, which then becomes an environmental issue that needs to be addressed.

When dyeing, the stage at which the dye is applied can give dramatically different results. The application of dye before the fiber is spun is known as fiber dyeing. In package dyeing, the fiber is already spun and is dyed prior to its being turned into fabric. Piece dyeing occurs when the fabric is dyed after it is made. Most commercial dyes are water-soluble organic compounds. The dyes have an attraction for the fabric and are able to pass through the whole fiber or fabric to achieve complete coloration.

Any of four types of intermolecular forces help the dye to bind to the fibers of the fabric. These include hydrogen bonds, van der Waals forces, ionic bonds, and covalent bonds. The last two types of bonds are true chemical bonds between the atoms of the fibers and the dye. These are the strongest types of interactions and will lead to the longest-lasting coloration. Ionic bonds result from the electrostatic interaction between a positively charged cation and a negatively charged anion. A covalent bond is a chemical bond that results from the sharing of valence electrons between two different atoms. Covalent bonding leads to a strong dye-fiber interaction but is only functional in the dyeing process between a fiber and a certain type of dye known as a reactive dye.

dyes did not work on these new fabrics. Today more than 8,000 dyes are available, including an amazing array of color choices, fastness, and durability.

The development of textile printing allows for the creation of fabrics with different colored patterns and has increased the commercial capabilities of fabric production. In ancient times, patterned fabric was a result of painstaking embroidery or weaving. The development of the roller printing machine in 1783 by James Bell allowed for the simultaneous printing

Textile chemistry involves the production of innovative and versatile fibers. Fabrics are made of either natural components such as wool, silk, or cotton or synthesized fibers such as rayon, polyester, acrylic, and acetates. One of the fastest-growing areas of textiles today is the production of products known as intelligent textiles. These are synthetic textiles that are designed and engineered to fit a certain need or solve a particular problem. Some estimates put the intelligent textiles at as high as 38 percent of the textile market.

Many of these technologies were produced on the basis of knowledge gained from the research and development of special clothing for astronauts. Textile chemists and engineers utilized the technologies of space to solve everyday apparel problems. For example, upholstery manufacturing utilizes antibacterial materials, temperature-sensitive fabrics can alert the wearer to overheating or to a medical condition, and allergy-resistant materials help those with allergies. The now common Under Armour®

Jackets containing Lumalive fabric (a photonic textile), made by Philips, can display dynamic advertisements, graphics, and constantly changing color surfaces. *(Philips)*

technology allows one's clothing to breathe while its wearer sweats, yet it retains the insulating power necessary to keep athletes warm as well as dry when outdoors.

One current area of research in the field of textile chemistry is the creation of a line of pajamas that utilizes the sensory equipment used for astronauts in order to monitor heart rates and respiratory rates and then sounds an alarm when these values indicate dangerous conditions. Such clothing may help prevent deaths caused by sudden infant death syndrome (SIDS). Not only would this production save lives from SIDS, the monitoring and evaluation of the changes in heart rate and respiration could lead to a better understanding and prevention of SIDS in the future.

In 2001, *Time* magazine gave the best innovation of the year award to an Italian company for the development of textiles with a memory: that is, materials that would return to their original shape at body temperature, thus removing the need for ironing. Once again, drawing from alloys that were developed for use in space, textile chemists and engineers were able to produce new materials that simplify everyday life. This shape memory material is based on the concept that these materials have two forms—the original form and the temporary form—and they return to their original form upon heating. This technology also has very practical uses in the medical field in items such as sutures. Two of these alloys are nickel-titanium and copper (I)-zinc alloys. Shape memory materials can also offer protection against intense heat or cold by altering the stiffness or insulative powers of the materials on the basis of the conformation of the alloy in the material. Some clothing materials incorporate films inside that respond to the temperature of the outside material layer and change conformation to increase the air space and, thus, increase the insulation value of the garment. This conformational change reverses when the outer layer warms to a higher temperature.

Another area of interest in the textile chemistry industry is the improvement of the systems involved in checking color quality. Every year millions of yards of dyed fabrics are thrown away because they have inconsistent color in the dyeing process. This results in great financial loss to the company and creates environmental issues. In addition to the discarded materials, the massive amounts of contaminated dye solutions become waste. Solving this problem would be financially useful as well as environmentally responsible. A camera originally developed for use in space was developed that is able to scan the fabric at a speed of approximately 330 ft (100 m) per minute and pick up any variations in the dyeing process, including both shade differences and color differences. The first commercial version of these machines was developed by a joint venture in Europe including the European Space Association (ESA) and first sold in 2004.

A textile chemist plays a role in all aspects of the production process for fibers and fabrics. Textile chemists are involved in the production and development of new and creative fibers for specific applications. Current research also aims to continue increasing the efficiency and environmental friendliness of the dyeing process as well as the chemical design of fabric-specific dyes that allow for long-lasting and penetrating color for all types of fabrics. Textile chemists develop the improvements one sees in the textile industry every day.

See also INORGANIC CHEMISTRY; MATERIALS SCIENCE.

FURTHER READING

Collier, Billie J., and Helen Epps. *Textile Testing and Analysis.* Upper Saddle River, N.J.: Prentice Hall, 1999.
Collier, Billie J., Helen Epps, and Phyllis Tortora. *Understanding Textiles.* Upper Saddle River, N.J.: Prentice Hall, 2001.

theory of everything Since the beginning of time, people have attempted to explain natural phenomena and the world around them. How did the universe come into existence? What keeps the Moon associated with the Earth, and what prevents electrons from flying out of their atomic orbitals? How can light be both a wave and a particle? Why are some chemicals more reactive than others? Physicists seek simplicity and elegance in answers. A major goal of physics is to describe and explain physical phenomena in the fewest straightforward laws and theories possible. Laws grow in power and perceived beauty with each observation or event that they successfully describe or predict. Theories become more convincing and viewed as more beautiful the more laws they explain.

Physics is the study of matter and energy and their interactions with each other. Astronomical bodies exert forces on one another, but so do protons inside an atomic nucleus. Historically, physics has approached these two realms, one concerned with events occurring on the macroscopic level and the other concerned with the interactions at the submicroscopic level, from different perspectives. Classical physics explains phenomena that are readily observable, such as motion, heat and other forms of energy, sound, light, states of matter, electricity, and magnetism. Quantum physics deals mostly with phenomena that occur at or below the atomic level such

as nuclear decay and the nature and interactions of fundamental particles.

To reconcile the very large and the very small, scientists must explore the origin of the universe. According to present-day understanding, in the beginning, approximately 13.7 billion years ago, the entire universe consisted of a minuscule point of infinitely dense and hot matter. A tremendous explosion, termed the big bang, set in motion the events that led to the development of the universe as it is today with all its galaxies and galactic clusters, rapidly moving away from one another, becoming more distant as the universe continues to cool. Physicists generally hold the view that at the moment of the big bang, the four fundamental forces that govern the natural world and everything in it were unified in a single force that could be described by a single theory. The four fundamental forces are the gravitational force, the electromagnetic force, the strong nuclear force, and the weak force. An individual theory, each having its own sets of equations and numerical parameters, explains each of the four forces: Einstein's general theory of relativity explains gravitation; electromagnism is covered by Maxwell's equations; quantum chromodynamics (QCD) is the theory of the strong nuclear force; and the Standard Model deals with the weak force. In the eyes of a physicist, however, four separate theories are three too many, and throughout the 20th century and into the 21st century, physicists have been trying to unify these theories into one "simple" theory of everything (TOE). Electromagnetism and the weak force have been unified as the electroweak force. There are proposals under consideration for a grand unified theory (GUT) to unify the electoweak and strong forces. However, it remains a mystery even how to approach the challenge of unifying gravitation with the other three forces. Nevertheless, the draw of the unimaginable beauty of one single supertheory drives many physicists to continue their search for a theory of everything.

A BRIEF HISTORY OF CLASSICAL THEORY

At the end of the 19th century, laws proposed by Sir Isaac Newton in the 1600s still reigned, though they did not offer complete explanations for many observable phenomena. For example, Newton's law of gravitation stated that all objects with mass pulled at one another but failed to explain how this interaction occurred through space, including at the tremendous distances separating celestial bodies. How did the force exert itself without any physical connection between the two objects? The British physicist Michael Faraday first proposed the notion of a field, which is a modification of the region surrounding an object in which an effect exists, when describing electromagnetism. James Clerk Maxwell, a Scottish physicist, mathematically defined the laws of electromagnetism and showed that light was simply a disturbance in an electromagnetic field, in other words, an electromagnetic wave. These discoveries, though revolutionary, did not explain how the force acted over space. To address that, physicists imagined the existence of "ether," a hypothetical medium through which forces acted and waves traveled.

Newton's laws of motion depended on the idea that space and time were absolute and unchanging. One needed a reference frame to measure distances and speeds. This seems to follow common sense. If someone stands on the side of a racetrack, he or she can measure the speed at which cars pass by dividing the distance traveled by the time taken for the car to travel that distance. A fixed, unchanging backdrop seemed to be necessary to measure the motion of objects. Then in 1864, Maxwell showed that electromagnetic radiation traveled at a constant speed of 1.86×10^5 miles per second (3.00×10^8 m/s) and concluded that light itself was a form of electromagnetic radiation. But what would serve as the reference frame? Scientists assumed the speed of light must be constant relative to the ether, and therefore as Earth moved through the ether, measurements of the speed of light should give different results, depending on the direction of Earth's motion and the direction of the light beam. Two American scientists, Albert Michelson and Edward Morley, tested this idea in the late 1880s by measuring the speed of light when traveling parallel to Earth's motion around the Sun and perpendicular to Earth's motion. They did not detect any difference and thus concluded that the ether did not exist.

This was the state of knowledge when Albert Einstein entered the picture. In 1905, he published several landmark papers, including one on the special theory of relativity, a theory he developed by imagining what someone would see if she could travel at the speed of light. According to Newton, if someone were traveling at the speed of light alongside a beam of light, then the beam would appear to stand still, just as if one were traveling in a car alongside another car traveling at exactly the same speed. Maxwell's equations contradicted this scenario, since they showed that light was a moving wave and could not stand still. Einstein resolved this dilemma in his paper "On the Electrodynamics of Moving Bodies" by claiming that the speed of light was simply a property of the universe and was always the same for all observers. Measuring the speed of light required no reference frame, because it did not change with the motion of the observer. Furthermore, no moving object can reach the speed of light. If motion does not affect the speed of light, then time and space must be relative. The speed at which an object travels affects the rate at which time passes. Length, as measured in the direction of motion, is also affected by the motion. Mass increases as one approaches the speed of light, as well. In other words, lengths, time intervals, and masses are relative, whereas, according to Newton, they had been considered absolute. These phenomena of time dilation, length contraction, and mass increase are not readily observable, since nothing on Earth approaches the speed of light, and thus relativity is a difficult concept to grasp.

Later the same year, Einstein published another extension of special relativity, the famous equation

$$E = mc^2$$

indicating the equivalence of energy and mass. In this equation, E denotes energy in joules (J), m is the mass in kilograms (kg), and c is the standard symbol for the speed of light in vacuum, whose value is 3.00×10^8 meters per second (m/s). According to this equation, every mass—even of a body at rest—has an energy equivalent, and every quantity of energy possesses a mass equivalent. For example, in nuclear processes, the difference between the total mass of the interacting particles and the total mass of the resulting particles appears in the form of its energy equivalent according to Einstein's relation. That is the source of nuclear energy. On the other hand, since the energy of photons is equivalent to mass, photons are affected by gravity.

The special theory of relativity only dealt with motion at constant velocity, that is, at constant speed and in a straight line, hence the adjective *special*. In 1915, Einstein developed his general theory of relativity, which subsumed special relativity, but also incorporated accelerated motion. Einstein's general theory of relativity was a theory of gravitation. Whereas Newton imagined gravitation as a force that acted through space over time, Einstein described gravitation as a curvature of space-time, of the four-dimensional merger of space and time. All matter, including objects such as the Sun and the planet Earth, warped, or distorted, the space-time around them. Additionally, all matter followed the space-time analog of straight-line motion. That appeared in three-dimensional space as motion under the influence of gravity. Thus, matter served as a source of gravity by warping space-time and was affected by gravity by having its motion follow the curvature of space-time. This theory explained one observation that had long troubled astronomers—the fact that the planet Mercury did not follow a perfectly repeating elliptical pattern when orbiting the Sun; rather, its orbit had the form of a slowly rotating ellipse.

Physicists embraced Einstein's theory of general relativity, which gained further support as it predicted phenomena not yet observed, such as the bending of light as it passes close to the Sun and the warping of time by matter. These new phenomena were later observed and followed Einstein's predictions as laid out by his general theory of relativity.

Heralded as science's greatest genius, Einstein began probing what he perceived as physics's next greatest problem, trying to unify gravity with electromagnetism. The 20th-century revolution of quantum physics led most physicists in one direction, while Einstein remained committed to developing a unified theory for the remainder of his life. Though his formulations of special and then general relativity were steps toward a unified theory, he was unsuccessful in this endeavor.

Just as disturbances in the electromagnetic field create electromagnetic waves, Einstein proposed that accelerating masses would stretch and shrink space, creating gravitational waves. Because gravity is a weak force, such waves are very difficult to detect, though two American astrophysicists did receive the 1993 Nobel Prize in physics for their research exploring the concept of gravitational waves. They found that two stars rotating rapidly around one another slowed in their orbits, probably a result of energy loss due to emission of gravitational waves.

A BRIEF HISTORY OF QUANTUM THEORY

While Einstein struggled with cosmological problems, other physicists were dealing with additional issues that Newtonian physics did not adequately explain, importantly including the atom. When the ancient Greeks first proposed the concept of the atom in 500 B.C.E., they chose a name that reflected their belief about the particle's nature, *atomos,* which means "indivisible." In 1803, the British scientist John Dalton proposed the atomic theory, of which one major principle was that all matter is made of atoms, and that atoms were indivisible and indestructible. While the former part of this statement remains correct, physicists have been able to dissect the atom into its component parts, and even into component parts of its component parts. In 1896, the French physicist Henri Becquerel first witnessed the weak force when he discovered radioactivity, stimulating other discoveries that led to a better understanding of the structure of the atom, nuclear decay, and the relationship between mass and energy. The British physicist Sir J. J. Thomson discovered the electron in 1897, evidence that the atom was not indivisible. Pierre and Marie Curie studied radioactive emissions and, in the process, discovered two new radioactive elements, polonium and radium. Most important, they determined that the emanations were a property of the atom rather than due to a chemical reaction or some other phenomenon. In 1911, the New Zealand–born physicist Sir Ernest Rutherford proposed a model for the structure of an atom that described the atom as mostly empty space but included a positively charged cluster of matter at the core. Today scientists know an atom consists of a nucleus containing both electrically uncharged neutrons and positively charged protons, particles that contribute practically all the mass to an atom, and that negatively charged electrons orbit the nucleus. The nucleus is packed into 1/10,000 of the total size of the atom; thus, most of the atom is empty space. Though Rutherford's model accounted for many observations, it did not explain the stability of an atom.

Enter the German physicist Max Planck, who today is considered the father of quantum physics. In a paper he published in 1900, "On the Distribution of Energy in a Normal Spectrum," Planck claimed that energy was exchanged between matter and electromagnetic waves in the form of tiny but discrete packages called quanta. His researches resulted in a constant that bears his name, *Planck's constant,* equal to $6.62606876 \times 10^{-34}$ joule-second (J·s), the ratio of energy in a quantum to the frequency of the wave. Because this value is so small, quantum effects are not observed in everyday phenomena, but they rule the subatomic world. In 1905, Einstein showed that light, and all electromagnetic waves, actually consisted of discrete packets, called photons, and related the energy of a photon to the frequency and to Planck's constant. The Danish physicist Niels Bohr applied quantum theory to the structure of the simplest element, hydrogen. He determined that electrons possessed specific quantities of energy and that they occupied certain energy levels within an atom. As an electron moved between energy levels, it either absorbed or emitted specific amounts of energy. Electrons could not exist between energy levels, just as a person cannot stand between steps on a staircase. Bohr's model explained the stability of atoms and why elements displayed unique spectra, the distribution of intensity of radiation emitted at different wavelengths. In 1923, a young Frenchman named Louis de Broglie submitted a revolutionary doctoral thesis demonstrating the dual nature of matter as both wavelike and particlelike.

All of these events shaped the development of quantum mechanics, though at the time no unifying principle pulled them all together. Two theoretical physicists credited with developing the mathematical framework that firmly established the field of quantum mechanics were the German Werner Heisenberg and the Austrian Erwin Schrödinger. Heisenberg put forth the uncertainty principle, a property of nature stating that one can never know the position and speed of a particle simultaneously. Schrödinger developed the wave equation that governs the behavior of the matter waves first described by de Broglie. The quantum world is characterized by probability; the chance of an event's happening can be calculated, but the actual occurrence of an event cannot be predicted. The quantum world also links events with the observer in a manner such that a property does not exist until it is observed; in other words, a particle can exist in numerous superimposed states at the same time, until it is observed to be in a particular position or have a particular speed.

INTEGRATION OF SPECIAL RELATIVITY AND QUANTUM THEORY

After the birth of quantum mechanics, physicists eagerly strove to integrate the theory of special relativity with quantum theory, the first step in the development of a theory of everything. The macroscopic and the subatomic worlds had been explained, but using completely different approaches. One step toward this yet unachieved end was the development of quantum field theories. The British physicist Paul Dirac began the process, and the results of his work suggested the existence of antiparticles, particles that have the same masses as matter particles but possess opposite electric charges. As Dirac predicted, antielectrons, or positrons, were discovered in 1932, and antiprotons were discovered in 1954. The U.S. physicists Julian Schwinger and Richard Feynman and the Japanese physicist Sin-Itiro Tomonaga built on Dirac's work to develop quantum electrodynamics (QED), the first quantum field theory. QED applied quantum mechanics to explain the interactions between radiation and charged particles and provided the best description of the structure of matter.

Electromagnetism explained how negatively charged electrons remained in association with the positively charged nucleus. In order to explain how the positively charged protons remained clustered together in the nucleus, physicists proposed the existence of a new fundamental force, which they called the strong nuclear force. This put the total number of fundamental forces at three: gravity, electromagnetism, and strong nuclear. The strong force was so named because it overcame the electromagnetic force, which in the absence of the strong force would have driven the protons away from one another. A fourth force, called the weak nuclear force, was proposed to explain the type of radioactive decay first witnessed by Becquerel and unexplainable by the other three forces. In beta decay, a neutron inside the nucleus turns into a proton, releasing an electron and an antineutrino in the process. These four forces are thought to explain every physical process that occurs in the entire universe.

Beginning in the 1930s, physicists started discovering other subatomic particles in addition to protons, neutrons, and electrons: muons, neutrinos, tau leptons, tau neutrinos, and more. With the advent of particle accelerators, physicists could smash together nuclei traveling at high speeds and examine the even tinier parts released. Quarks, the components that make up neutrons and protons, were discovered theoretically in the 1960s and named by the American physicist Murray Gell-Mann, who helped develop another quantum field theory called quantum chromodynamics (QCD), which described the effect of the strong force on quarks.

By 1973, physicists had developed the standard model, a theory of elementary particles and their interactions that successfully incorporates the special theory of relativity into quantum mechanics. This major advance in physics assigned fields to each species of elementary particle and incorporated the electromagnetic force, the weak force, and the strong force. In summary, the standard model says that all particles are either fermions, which have mass and compose matter, or bosons, which mediate interactions and carry forces between particles of matter. Fermions are grouped into three categories called generations, each consisting of two kinds of lepton (a negatively charged particle and an associated neutral neutrino) and two varieties of quark (the electrically charged fundamental building block of the hadrons, the heavier particles that include the proton and neutron). All the fermions are affected by the weak interaction. The electromagnetic interaction affects only the electrically charged ones. The quarks are the only fermions that undergo the strong interaction.

Leptons are extremely lightweight particles and include the electron and the associated electron neutrino, the muon and muon neutrino, and the tau and tau neutrino. Quarks are of six varieties, referred to as flavors (up and down, charm and strange, and top and bottom), where each flavor can exist in any of three states, called colors. Two up quarks and one down quark compose a proton, and two down quarks and one up quark compose a neutron. The strong force is carried by eight kinds of gluon, while the carriers of the weak force consist of three kinds of particle called intermediate bosons and designated W^+, W^-, and Z^0. The photon carries the electromagentic force.

At the same time that physicists were developing the standard model, in the late 1960s, they successfully combined the electromagnetic force and the weak force into an electroweak theory. Sheldon Glashow and Steven Weinberg, from the United States, and Abdus Salam, from Pakistan, contributed significantly to the development of the electroweak theory, receiving the 1979 Nobel Prize in physics for this work, and Gerardus 't Hooft and Martinus Veltman, both from the Netherlands, elucidated the quantum structure of the interactions.

As successful as it is in describing the interactions of particles of matter and three of the fundamental forces, the standard model, built on a quantum field theory framework, does not describe a quantum version of gravitational interactions. A problem with infinities keeps appearing, demanding ridiculous scenarios, such as an infinite force between two gravitons (the proposed particles that mediate gravitational interactions). Thus, general relativity, which is the theory of gravity, still stands alongside the standard model rather than being incorporated. Also, physicists have never observed one particle predicted by the standard model, the Higgs boson, believed to be responsible for giving mass to other particles. Many physicists consider the standard model as still too complicated. Certainly many of the model's features are arbitrary—though parameters such as the masses of the particles and the strengths of the interactions can be measured, the theory does not predict them mathematically. Thus, scientists continue to search for a theory of quantum gravity that will unify Einstein's theory of general relativity with quantum theory and link the cosmos with the subatomic world.

STRING THEORY

Though physicists have been unable to construct a working quantum theory of gravity, string theory, also called superstring theory, is being investigated as a possibility because it seems to predict the existence of the graviton, the particle thought to mediate the gravitational force. Einstein's equations for gravity emerged from the mathematics of string theory. According to string theory, instead of the innumerable particles described in the standard model, a single fundamental component of matter exists, the one-dimensional open or closed object called a string. The strings are extended objects rather than point particles. They are described by either tubes or sheets in space-time, depending on whether they are closed, as loops are, or open. The proposed strings would be extremely small, maybe Planck's length (10^{-35} m), and vibrate at different frequencies. Different vibration states would explain all the different elementary particles. Just as different vibrations on violin strings give rise to different musical pitches, one type of vibration would be characteristic of a quark, another would appear as an electron, and so on. The variations in tension of a string or in the energy of vibration would explain the different properties of different elementary particles. These strings could join with other strings or split into two, actions that would explain all of the forces. One surprising prediction of string theory was the existence of at least 10 dimensions, rather than the three familiar dimensions of space and the single dimension of time; the extra dimensions allow the strings to vibrate in enough different ways to explain the myriad observed elementary particles. The original string theory, which only described the bosons that carry the forces; developed into superstring theory, which also incorporated fermions that compose matter.

From the weakest to the strongest, the four forces are gravity, the weak nuclear force, the electromagnetic force, and the strong nuclear force. Even though gravity is the weakest, it is of infinite range and acts among all matter, whether electrically charged or not. Gravity is the force responsible for holding planets in orbit and keeping stars together to form galaxies. The weak nuclear force is a short-range force. This interaction causes neutrons to decay and is, thus, the driving force of some types of radioactive decay. The electromagnetic force is of infinite range. This is the force that holds negatively charged electrons in orbit around a positively charged nucleus, controls the formation of chemical bonds, and explains electromagnetic radiation. The strong nuclear force is of short range and is the force that holds quarks together to form neutrons, protons, and other nuclear particles and that holds nuclei together. One property unique to the strong force is that it becomes stronger as quarks move farther apart. (All the other forces grow weaker over increasing distance.) These four forces are thought to have all been subsumed in a single superforce very early in the development of the universe. Physicists hope to explain this single force eventually by the theory of everything. In order to understand how these came to be four distinct forces, one must consider how the universe itself formed.

Present understanding has the universe starting with what is called a big bang. Initially, the universe was a single point of infinitely dense, hot matter that exploded. Nothing more is known about this stage or what happened beforehand. After the explosion, the big bang, the universe continued to expand and to cool. At 10^{-43} second after the explosion, time and space came into existence and the four fundamental forces were combined into one superforce. At 10^{-35} second, the gravitational force split off from the other three, still combined to form what physicists call a grand unified force, and by 10^{-12} second, the other three had separated from each other. At 10^{-6} second, the temperature had cooled sufficiently to allow quarks to combine with one another to form neutrons, protons, and other nuclear particles. Within one second after the origin of the universe, most of the particles and antiparticles annihilated one another, leaving behind photons of light and the matter that composes the universe as it is today. Five minutes later, nuclei of light elements began to form, but it was still too hot for atoms to form. About 200,000 years later, the temperature allowed nuclei to hold on to electrons, forming atoms. Throughout, gravity was acting to pull masses together. Because most electrons had become incorporated into atoms, matter was now in an essentially electrically neutral condition and light was freed to travel great distances through space without being scattered by interactions with electrons, making the universe transparent. This cosmic background radiation—consisting of those same decoupled photons—can still be detected in the cosmos. When the universe was 1 billion years old, galaxies started to form, stars collapsed into supernovas, and heavier elements formed. At 9.3 billion years, new stars, planets, and the solar system formed. At present, about 13.7 billion years after the big bang, the universe has expanded to about 27 billion light-years across, and the four fundamental forces determine how the different kinds of matter interact with one another.

Going back in time, at extreme temperatures, these forces were indistinguishable. Technological limitations prevent present-day experimental examination of events beyond 10^{15} degrees Celsius, the temperature at which the electromagnetic and weak forces become indistinguishable and are unified. Theoretical evidence, however, suggests that at 10^{27} degrees Celsius, the strong force would be unified with the electroweak force to form a grand unified force. At temperatures approaching 10^{32} degrees Celsius, gravity and the grand unified force might have behaved similarly and merged with each other into one superforce, and all the quarks and leptons could be described as different faces of the same "superparticle" that would describe all matter.

According to string theory, at temperatures exceeding 10^{32} degrees Celsius, particles assume a one-dimensional stringlike nature. Equations set forth by string theory seem to describe all four fundamental forces. By the early 1990s, the excitement about superstring theory waned—five different string theories existed but did not seem to advance further. Then in 1995, a physicist from the Institute for Advanced Study at Princeton, Edward Witten, set string theory in a new framework, called M-theory. (*M* stands for numerous concepts including membranes, mystery, or magic.) He showed that the five existing theories were all different perspectives of one grander theory, renewing hope in string theory as a potential theory of everything.

That same year, 1995, the American physicists Andrew Strominger, David Morrison, and Brian Greene showed that the equations describing strings also described black holes, objects in space that have such a strong gravitational field that even light

cannot escape them. Black holes provide a unique system to examine the relationship between gravity and quantum mechanics because they are too compact to ignore quantum effects and are characterized by an immense gravitational force. One problem with which astronomers had been struggling was to explain where the entropy in a black hole originated. Using string theory to count the quantum states of a black hole helped solve the problem.

Superstring theory is still incomplete, and many critics do not believe it will successfully unify all the fundamental forces. One problem is that it makes no predictions, and no observational evidence, only mathematical evidence, currently supports it. Other theories that are not nearly as advanced as string theory but still attempt to reconcile quantum field theory with gravity include loop quantum gravity and twistor theory. Thus, the search for a theory of everything continues.

See also ACCELERATION; BIG BANG THEORY; CLASSICAL PHYSICS; EINSTEIN, ALBERT; ELECTROMAGNETIC WAVES; ELECTROMAGNETISM; ENERGY AND WORK; FARADAY, MICHAEL; FEYNMAN, RICHARD; GELL-MANN, MURRAY; GENERAL RELATIVITY; GRAVITY; HEAT AND THERMODYNAMICS; MASS; MATTER AND ANTIMATTER; MAXWELL, JAMES CLERK; MOTION; NEWTON, SIR ISAAC; NUCLEAR PHYSICS; OPTICS; PARTICLE PHYSICS; PLANCK, MAX; QUANTUM MECHANICS; RADIOACTIVITY; SPEED AND VELOCITY.

FURTHER READING

Falk, Dan. *Universe on a T-Shirt: The Quest for the Theory of Everything.* New York: Arcade, 2004.

Greene, Brian. *The Elegant Universe: Superstrings, Hidden Dimensions, and the Quest for the Ultimate Theory.* New York: W. W. Norton, 2003.

Schwarz, Patricia. The Official String Theory Web Site. Available online. URL: http://www.superstringtheory.com. Accessed July 25, 2008.

Weinberg, Steven. "A Model of Leptons." *Physical Review Letters* 19 (1967): 1264–1266.

Woit, Peter. *Not Even Wrong: The Failure of String Theory and the Search for Unity in Physical Law.* New York: Basic Books, 2006.

Thomson, Sir J. J. (1856–1940) British *Physicist* Sir J. J. Thomson entered physics at a critical point in the field's history. Earlier in the 19th century, physicists made many great discoveries in electricity, magnetism, and thermodynamics. Many physicists believed that the field, like an exhausted mine, would yield no more surprising results, but Thomson did not share this belief. His studies revolutionized physics by ushering in a new type of physics, nuclear physics, in 1897, when his discovery of electrons opened a door into a subatomic world that is still being studied today.

Joseph John Thomson was born on December 18, 1856, in Cheetham Hill, a suburb of Manchester, England. The son of a bookseller, Thomson entered Owens College, which is now Victoria University of Manchester, at the age of 14. While at Owens, he took several courses in experimental physics, and these classes whetted his appetite for the discipline. In 1876, he received a scholarship to Trinity College, at Cambridge University, where he obtained a B.A. in mathematics four years later. At Trinity, he had the opportunity to work in the Cavendish Laboratory, further developing the theory of electromagnetism set forth initially by James Clerk Maxwell, who proposed that a relationship exists between electricity and magnetism and that quantitative changes in one produced corresponding changes in the other. Thomson's insights into electromagnetism gained him prompt recognition by his peers, and they elected him fellow of the Royal Society of London in 1884. Soon afterward, he accepted an appointment as chair of physics at the Cavendish Laboratory.

Thomson's most significant research led him in 1897 to the conclusion that all matter, regardless of its source, is made up of identical, smaller particles inside each atom. Prior to his studies, a controversy existed within the scientific community about the nature of cathode rays, rays generated by applying voltage to a vacuum tube (cathode tube) with positive and negative electrodes at each end. The Germans believed that cathode rays were a side effect of residual ether in the cathode tube, whereas the British and the French believed the cathode rays were electrified particles generated by the voltage applied to the tube. Thomson set out to settle the dispute through a series of three experiments. In the first experiment, Thomson wanted to determine whether the negative charge associated with cathode rays could be separated from them by magnetism. Thomson constructed a cathode-ray tube that had cylinders with slits attached to electrometers, devices used to measure electrical output. After generating cathode rays, Thomson applied a magnetic field to the cathode-ray tube, causing the rays to bend away from the electrometers. If the negative charge had no direct relationship to the bending rays, Thomson hypothesized that the electrometer would still be able to measure a negative charge as it floated around in the tube. If the cathode rays and the negative charge were directly related to one another, the electrometer would not detect any charge. The electrometers did not register any charge, leading Thomson to conclude that the cathode rays and the negative charge were directly related. To determine whether the cathode rays were charged particles, Thomson designed

Sir Joseph John Thomson was a British physicist best known for his 1897 discovery of the electron, a negatively charged subatomic particle. *(Mary Evans/Photo Researchers, Inc.)*

a second experiment, in which he tried to deflect the cathode rays by using an electric field. In the past, others had tried to exploit this property of charged particles to decipher the nature of the cathode rays, but no one had been able to perfect a technique for creating a true vacuum within the cathode tubes before these studies. Thomson achieved this condition by using improved methodologies, and the electric field deflected the cathode rays, providing a convincing argument that the rays were particles. Thomson named these negatively charged particles *corpuscles*, although he later renamed them *electrons*, and he concluded that these particles were subatomic entities that originated from atoms in the electrodes. Additional studies confirmed that the cathode rays were particles carrying a negative electric charge, by using different colored gases and different metals as conductors. Thomson concluded from these studies that all matter and the atoms composing it contained electrons, and he gave further credence to this idea by spending the next three years developing alternative ways to isolate these particles. A third set of experiments followed as Thomson attempted to characterize these new particles. He designed an experimental procedure that would measure the charge-to-mass ratio of the electrons, again using cathode tubes to generate the particles and magnetic fields to deflect the rays. Thomson measured the amount of deflection that occurred under different conditions; the results permitted him to determine mathematically the amount of electric charge the electrons carried compared to the mass of the particles. Calculations of charge-to-mass ratios indicated that the particles were either very light or very charged. By the turn of the 20th century, the scientific community accepted Thomson's discoveries and insights. Thomson received the Nobel Prize in physics in 1906 for this body of work.

During the fruitful years of 1895 to 1914, Sir J. J. Thomson surrounded himself with gifted scientists from all over the world. He was an excellent teacher, who preferred to arouse enthusiasm for physics in his students rather than simply fill them with knowledge about the subject. Although trained as a mathematician, Thomson often encouraged the development of theories using intuitive model building and deductive reasoning rather than developing a lot of mathematical equations to explain experimental observations. Thomson took his role as an instructor very seriously, and he eagerly taught undergraduates, graduate students, and postgraduates, stressing the importance of teaching for a researcher as a means of perpetual review of the basics of the discipline. Thomson always encouraged students he advised to develop their own thoughts and theories about a new subject prior to reading other people's work; he believed that this approach to a new problem prevented any influence of others on assumptions one made in designing an experiment. Sir J. J. Thomson was an important figure in physics not only for his own body of work but also for the work that he inspired others to do. Seven Nobel Prize laureates worked under Thomson early in their career.

In 1904, Thomson gave a series of six lectures referred to as the Silliman Lectures at Yale University. Thomson unveiled one of the earliest models of subatomic structure, referred to as the "plum pudding" model. Taking into consideration the neutral state of most atoms, Thomson proposed that positively charged entities existed within an atom, which offset the negatively charged electrons. He envisioned that the electrons floated around in a soup, or cloud, of positively charged particles dispersed throughout the width and breadth of the atom. The electrons were not glued to any one position, so they could freely move and interact with each other, the positively charged particles that surrounded them, and particles from other atoms. So, much like raisins dotted throughout a pudding, electrons had no fixed position, making Thomson's model of atomic structure dynamic in nature. Sir Ernest Rutherford disproved

the plum pudding model in 1909 with his gold foil experiment, in which he discovered that most of the atom appeared to be empty space and a small, dense core (nucleus) contains an atom's full complement of positively charged particles. Later, Niels Bohr added to the Rutherford model of subatomic structure by proposing that electrons circle the nucleus in defined paths, or orbitals, much as the planets in our solar system travel around the Sun.

After the discovery of electrons and proposal of a model of subatomic structure, Thomson redirected his focus to characterizing positively charged rays called anode rays. He created an instrument that he called the parabola spectrograph to measure the effects of electrostatic and magnetic fields on these anode rays. The instrument and methodologies developed by Thomson led to the modern-day mass spectrometer, an instrument used by chemists to determine the mass-to-charge ratios of ions (atoms that carry a negative or positive charge). While trying to elucidate the character of these positive rays, Thomson discovered a means to separate different kinds of atoms and molecules from each other, and this process led to the discovery of isotopes, atoms with the same number of protons (positively charged subatomic particle) but different numbers of neutrons (subatomic particle with no charge) in the nucleus, prior to the outbreak of World War I.

Knighted in 1908, Sir J. J. Thomson continued to lead a productive existence both in and out of the laboratory. He married a fellow physics student, Rose Elizabeth Paget, in 1890, and they had a son and a daughter—George Paget Thomson and Joan Paget Thomson. His son later won his own Nobel Prize in physics in 1937 for studies that examined the wavelike properties of electrons. As an administrative head, Thomson led the highly successful Cavendish Laboratory until he stepped down in 1919 for Ernest Rutherford to assume the position. During Thomson's tenure, the Cavendish Laboratory received no government subsidies (grants) and no contributions from charitable organizations or industry, yet he oversaw the completion of two additions to the laboratories and a multitude of research projects by being creative with student fees and staff support. One staff member donated the necessary funds to obtain a piece of equipment critical to studying positively charged particles. Prior to leaving the laboratory's directorship, Thomson assumed the role of master of Trinity College, immensely enjoying the opportunity to mingle with young men interested in disciplines other than the sciences. Thomson was an avid fan of Cambridge cricket and rugby in his free time, although never an athlete himself. In his later years, he also developed an interest in botany and took long walks in the countryside, where he searched for rare botanical specimens to add to his gardens. He even wrote his autobiography, *Recollections and Reflections,* which was published in 1936. Sir J. J. Thomson died on August 30, 1940, in Cambridge, and his family buried him near Sir Isaac Newton at Westminster Abbey.

See also ATOMIC STRUCTURE; BOHR, NIELS; CLASSICAL PHYSICS; FISSION; NUCLEAR CHEMISTRY; NUCLEAR PHYSICS; PLANCK, MAX; QUANTUM MECHANICS; RADIOACTIVITY; RUTHERFORD, SIR ERNEST.

FURTHER READING

Dahl, Per F. *Flash of the Cathode Rays: A History of J. J. Thomson's Electron.* Washington, D.C.: Taylor & Francis, 1997.
The Nobel Foundation. "The Nobel Prize in Physics 1906." Available online. URL: http://nobelprize.org/physics/laureates/1906/index.html. Accessed July 25, 2008.

time   The concept of time is one of the most difficult concepts in physics, rivaled in the depth of its philosophical ramifications perhaps only by the concept of space. For the purpose of physics, however, one must ignore much of that difficulty. In order to make progress in physics, physicists cannot, in general, allow themselves to become mired in philosophical detail. So they make do with whatever general understanding is sufficient for carrying out the goals of physics.

One of the dominant characteristics of nature, one of which people possess powerful intuitive awareness, is becoming. Things do not stay the same—changes occur; situations evolve. That is what is meant by becoming. In physics, time can be described, even defined, as the dimension of becoming, where dimension is used in the sense of the possibility of assigning a measure. Time's measure is the answer to the question "When?" for an event. The answer to this question always takes the form of a single number, composed of, say, the date and "time" (in the sense of clock reading) of the event. Thus time involves an ordering of events into earlier and later, just as numbers are ordered into smaller and larger. Also, time involves a duration, an interval, between every pair of events, referred to as the elapsed time between the earlier event and the later one. Since a single number is required to answer the question "When?" time is said to be one-dimensional.

Avoiding the technicalities, though, time is generally viewed by physicists as the backdrop for becoming, for change and evolution. In physics, time appears as a parameter, or independent variable, usually denoted by t. The SI unit of time is the second (s). Other commonly used units are the minute (min), which equals 60 seconds; the hour (h), which contains 60 minutes

and thus 3,600 seconds; the day of 24 hours; and so on. One encounters also the millisecond (ms), which is a thousandth (10^{-3}) of a second; the microsecond (μs), one-millionth (10^{-6}) of a second; the nanosecond (ns), which equals a billionth (10^{-9}) of a second; and even tinier submultiples of the second.

Physical quantities are represented as functions of t, which give the value of their corresponding quantity for any time. As for becoming, however, as powerful as human intuition of it is, that aspect of nature is not comprehended by physics. Nothing in physics requires t to take any particular value or continually increase in value. This does not mean that physics will not eventually achieve a better understanding of the nature of time. But, for now, the nature of time remains largely beyond the grasp of physics.

Albert Einstein's special and general theories of relativity reveal that time and space are not independent of each other and that fundamentally they are best considered together as four-dimensional spacetime. These theories show, among other well-confirmed effects, that the rate of clocks depends on their state of motion and on their location in the gravitational field. Thus, time is not absolute in the sense of a universal clock that is applicable to all observers. Instead, the rate of one observer's clock or heartbeat, for instance, as observed by him or her, is different from that measured by another observer moving with respect to the first. In other words, the same clock runs at different rates for different observers.

Assume that a clock (or heart, etc.) is ticking off time intervals of duration T_0, as measured by an observer for whom the clock is at rest. Then, an observer moving at speed v relative to the first one, that is, an observer for whom the clock is moving at speed v, measures longer time intervals T for the same ticks (or heartbeats), where

$$T = \frac{T_0}{\sqrt{1 - v^2/c^2}}$$

The two time intervals are expressed in the same unit, speed v is in meters per second (m/s), and c denotes the speed of light in vacuum, whose value is 2.99792458×10^8 meters per second (m/s), or 3.00×10^8 m/s, to a reasonable approximation. In other words, a moving clock is measured as a slower clock. This effect is called time dilation. Similarly, a moving organism is observed to live longer than if it were at rest, and unstable elementary particles have longer lifetimes (or half-lives) when they decay in flight than when they are at rest.

As an example, let a clock tick off seconds, as observed by an observer with respect to whom the clock is at rest. Then $T_0 = 1.0$ s. Now, let a second observer fly by the first at 90 percent of the speed of light, $v/c = 0.90$. For the second observer, the clock is moving past her at 90 percent of the speed of light. Enter these data into the formula to see how much time the second observer observes for each of the clock's ticks:

$$T = \frac{1.0 \text{ s}}{\sqrt{1 - 0.90^2}} = 2.3 \text{ s}$$

So, in this case, the moving clock is observed to be running slow by a factor of 2.3.

Note that as v approaches c, T tends to infinity, and the clock appears to have almost stopped, the organism is in quasi-suspended animation, and the unstable particles almost never decay. In what is called the nonrelativistic limit, as v/c approaches 0, time dilation is negligible, reflecting the fact that in Newtonian mechanics time intervals are independent of the motion of observers. Time dilation is a mutual effect: each of two observers moving relative to each other measures the other's clock as running slower. This leads to what is known as the twin paradox, whereby a space-traveling astronaut ages more slowly than her twin who remains on Earth. This effect might play a role in future space travel.

Any device for measuring time is called a clock. A clock measures time by producing and counting (or allowing one to count) precisely equal time units. A pendulum swings with cycles of equal time durations. The most modern precision clocks, atomic clocks, count time units that are based on the frequency of the electromagnetic radiation emitted by atoms, when the latter undergo transitions between certain states of different energy.

But how can one be sure the time units generated by a clock are indeed equal? How does one know that a pendulum's swings are of equal duration or that the oscillations of an electromagnetic wave emitted by an atom have constant period? Clearly, the use of one clock to assure the equality of another clock's time units is circular reasoning.

Essentially, one is forced to make the assumption that a clock's time units are equal. This is not done for just any device, but only for those for which there exists good reason to think it true. In the example of the pendulum, at the end of each swing cycle, the bob returns to momentary rest at its initial position, whereupon it starts a new cycle. As far as one can tell, it always starts each cycle from precisely the same state. Why, then, should it not continue through each cycle in precisely the same manner and take precisely the same time interval to complete each cycle?

If one assumes that is indeed the case and the pendulum's time units are equal, a simple and rea-

sonable description of nature based on this equality ensues. Convinced that nature should be describable in simple terms, one then feels justified in making this assumption. For more sophisticated clocks, such as atomic clocks, the reasoning is similar, but correspondingly more sophisticated. One ends up defining the cycles of electromagnetic radiation as having equal duration. Thus, the time interval of one second is defined as the duration of 9,192,631,770 periods of microwave electromagnetic radiation corresponding to the transition between two hyperfine levels in the ground state of the atom of the isotope cesium 133.

See also ATOMIC STRUCTURE; CLOCK; EINSTEIN, ALBERT; ELECTROMAGNETIC WAVES; ENERGY AND WORK; GENERAL RELATIVITY; ISOTOPES; MEASUREMENT; SPECIAL RELATIVITY.

FURTHER READING
Davies, Paul. *About Time*. London: Penguin, 2006.
Scientific American Special Issue. *A Matter of Time*. 287, no. 3 (September 2002).

toxicology The 16th-century Swiss chemist Paracelsus gave us the most basic premise of the field of toxicology: "The dose makes the poison." Almost every chemical can be hazardous, depending upon the dose to which one is exposed. Toxicology is a branch of science that deals with the inherent hazards and associated risks of any chemical, either synthetic or naturally occurring. A toxicologist is a trained scientist who specializes in the understanding of the possible toxic effects at specific dosages. Once the hazard of a chemical has been identified, an assessment of the risk is performed. The appropriate quantitative assessment of the effects on human health and the environment are important to maintaining safe levels of the many materials we are exposed to every day via the air, water, and food supply.

In the early 1970s, the United States government enacted several pieces of legislation that regulated the production and disposal of many chemicals in an effort to protect human health and the environment. Establishment of the Clean Air and Clean Water Acts was the first step to maintaining a proper balance for sustainable industrial growth. This early legislation led to the establishment of the Environmental Protection Agency (EPA) and the Food and Drug Administration (FDA). Today both agencies, by law, require a variety of toxicological tests to be carried out on every new chemical, drug, and medical device being developed, and the agencies are also reassessing the safety of many chemicals that have been used for years. The types of required tests depend upon the class of chemical, the application for which it is being used, and the quantities that will be generated. Many questions need to be addressed: Will it enter into the food supply? Is it a drug that will be given to children? Or is it going to be used as an adhesive in the construction industry? The three scenarios would result in a different risk to human health and need to be addressed with different levels of testing.

The identification of a chemical's potential hazard is an important first step to understanding the associated risk of the material. The hazard is defined as the possible harm that could result from exposure to this material. To determine hazard, very high doses are given over a long period, and then toxicologists determine whether any harmful effects resulted. The risk is determined by considering the hazard, factoring in the means of exposure, and then applying safety margins so that there will be no harm to people's health. Driving a car has many potential hazards associated with it. The actual risk depends on how fast one drives and how safely. Because of this, laws exist that require car manufacturers to minimize the hazards, and additional laws mandate actions by drivers, so the overall risk is small.

Toxicologists who determine the potential hazard and risk of chemicals can specialize in a particular area. An environmental toxicologist studies chemicals for their potential harm to the environment and to the terrestrial and aquatic species that live in it. Developmental and reproductive toxicologists focus on the interaction of chemicals with the reproductive systems of species and evaluate any effects on the developing organism. Mammalian toxicologists deal with the interactions of chemicals specific to mammals, and immunotoxicologists deal with interactions of chemicals with the immune system. As with many scientific fields, the individual toxicologist typically specializes in a specific class of chemicals or focuses on a specific area of the body, such as the endocrine system or the liver. The rationale for the types of experiments that a toxicologist performs depends on whether the data are needed for regulatory reasons or for a greater scientific understanding of how a chemical is interacting, that is, regulatory toxicology or mechanistic toxicology. A regulatory toxicologist will conduct tests to meet the legal regulatory requirements for the EPA or FDA or other agencies. The experimental guidelines are predetermined, and once the testing is completed, the toxicologist submits a report to the agency recommending whether or not the chemical should be used in a particular formulation or product.

A mechanistic toxicologist takes a closer look at the biological mechanism by which certain classes of chemicals elicit their effects. Sometimes a chemical will have an effect on the liver, but it is not clear how the effect is occurring or whether this is a hazard that would be relevant to humans. A mechanistic toxicologist designs and carries out experiments specifically to address this question. Both types of toxicologists need to ask several questions in order to determine what hazard a chemical will pose. Once an understanding of the hazard is gained, a risk assessment is done. Several variables can factor into the risk assessment: potential timing and length of exposure, route of exposure, and dose level of the exposure.

Exposure to any material can be either acute or chronic. An acute exposure typically occurs through a single high dose, such as in an industrial accident where factory workers are exposed. A chronic exposure is seen with small repeated exposures, such as residual pesticides in the food supply or in the drinking water. Small doses of a harmful material given over years can result in a serious health risk, as will a short exposure to a high dose of a chemical. A toxicologist will evaluate both types of exposure. The same chemical can also result in two very different health effects, depending on the timing of the exposure.

In addition to the timing, the route of exposure is critical. The major routes of exposure, classified in order of producing the greatest effects, are the following: inhalation (via the lungs), oral (through ingestion), and dermal penetration (absorbed through the skin).

In terms of the dose of a given chemical, several different scenarios must be considered. For example, an industrial worker whose job is to produce an agrochemical used to grow apples will be working with very high levels of the chemical on a daily basis. Industrial hygiene practices are needed to protect the worker. The farmer who applies the chemical in a liquid form, spraying it on his or her trees once or twice a year, may receive a high acute inhalation dose if he or she does not wear appropriate personal protective equipment. An individual who eats the apple could potentially receive a small residual dose daily. Toxicologists need to predict all possible exposure scenarios and probable doses associated with each as part of the risk assessment. This is required by law in order to ensure that people are protected from inadvertent harm.

Once a chemical enters the body, several different variables come into play. Typically, the potency of any chemical is dependent on four factors: absorption, distribution, metabolism, and elimination (ADME). Absorption is the process by which a chemical passes through cells and enters the bloodstream. The route of exposure and the physiochemical properties of the chemical will determine the rate of absorption. For example, an inhalation exposure will typically result in a rapid absorption, as the blood vessels of the lungs are directly exchanging with the inhaled air. A dermal exposure will on average have a slower rate of absorption, as the layers of skin provide a relatively impenetrable barrier to the bloodstream.

Distribution is the manner in which the chemical is translocated throughout the body. Translocation typically occurs rapidly and depends on the blood flow and the exchange of the chemical from the blood into target cells or tissues. The nature of the compound will factor into the rate of distribution, as hydrophobic materials may passively diffuse through the cell membrane more easily than charged or hydrophilic materials.

Metabolism is primarily carried out by the liver, where a large family of cytochrome P450 enzymes is responsible for the oxidative metabolism of chemicals. The family of cytochrome P450 enzymes includes more than 20 subtypes. The overall objective of the metabolic enzymes is to transform the chemicals into a form that is more water soluble so the body can more easily eliminate the materials in the urine. Successive oxidations and hydrolysis reactions lead to metabolites that are more polar and more easily eliminated. The broad spectrum of enzymes can address the wide range of chemicals with which the body has contact. Classes of enzymes have relatively nonspecific criteria for binding, so that many materials can be effectively removed from the bloodstream. One unfortunate aspect of liver metabolism is that often the chemical can be more hazardous after the body's attempt to get rid of it. Metabolic activation can sometimes render a chemical more dangerous than the form in which it entered the body. The metabolism of many carcinogens (chemicals that can cause cancer) gives them the ability to cause the damage to deoxyribonucleic acid (DNA). The testing of chemicals must take into consideration both the parent chemical and the effect of any metabolites that would be formed.

The kidney is the primary route of elimination after the oxidative metabolism that occurs in the liver. Chemicals will be eliminated with the urine usually hours to days after an exposure, depending on their properties. Fecal elimination (occurring through biliary sources) and elimination through the lungs (from the rapid exchange of gases into the respired air from the bloodstream) are two minor sources of elimination. The rate at which materials are eliminated can vary. Hydrophobic materials can be distributed into fat stores or organs with high fat content, such as the brain, and are retained for long periods. As the compound is slowly excreted, the partitioning coefficients that are a physical characteristic of the chemical dictate the equilibrium of the material in the stored pools. Materials with poor water solubility and high vapor pressures may be exhaled rapidly and not bioaccumulate to any significant degree.

The way that toxicologists determine whether a material possesses associated hazards is through the tests that are conducted. Two examples of possible tests that could be carried out are the two-year bioassay and reproductive studies. In the two-year bioassay, an investigator administers the test material daily for the life span of the animal such as a rat, typically a two-year period. The objective of the test is to determine whether the chemical is a carcinogen, capable of causing cancer in the rat. At the completion of the study, every organ and tissue is evaluated for the incidence of tumors or disease. A group of rats that did not receive chemicals serves as the control group for comparison of the treatment groups. Careful statistical analysis of the data reveals whether or not the material causes an increase at any end point of the study. If a chemical results in an increase in liver tumors, for example, it may be classed as a carcinogen and will not be used in products to which people would be exposed. Reproductive studies are another type of test. Pregnant women are especially vulnerable to chemical exposure, as are babies. Experiments are designed to model exposure to the chemicals so the effect on the health of the offspring can be evaluated. Such experiments would include exposing animals through several reproductive cycles, encompassing two generations of animals. This allows for evaluation of the reproductive effects, as well as the developmental effects on the offspring.

If an effect is observed in a two-year bioassay or a two-generation reproductive study, a mechanical toxicologist will investigate the mechanism by which the chemical caused the effect. Further experiments will help the toxicologist unravel the mechanism of action and determine whether the information is relevant to humans.

The ultimate goal of toxicology is to understand how different chemicals interact with the human body and with the environment. The federal government is responsible for mandating and enforcing the appropriate testing of all chemicals used to make the many things people eat, drink, and use every day. The combined efforts of scientists and government increase the understanding and responsible regulation of chemical usage in order to protect both human health and the environment.

See also ATMOSPHERIC AND ENVIRONMENTAL CHEMISTRY; PHARMACEUTICAL DRUG DEVELOPMENT; SCIENTIFIC METHOD.

FURTHER READING

Klaassen, Curtis D. *Casarett and Doull's Toxicology: The Basic Science of Poisons,* 6th ed. New York: McGraw Hill, 2001.
Wexler, Philip. *Encyclopedia of Toxicology,* 2nd ed. San Diego: Academic Press, 2005.

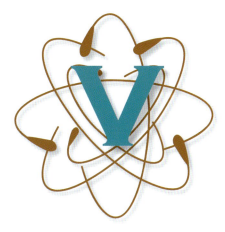

vectors and scalars A physical quantity that is characterized by both a magnitude, which is a non-negative number, and a direction is designated a vector. Velocity, acceleration, force, and electric field are vectors. As an example, the velocity of an aircraft can be specified by its magnitude, which is the aircraft's speed, say, 50.0 meters per second (m/s), and its direction, perhaps north by northwest and parallel to the surface of the Earth.

Vectors are represented graphically by arrows that point in the vectors' direction. The length of the arrow is proportional to the magnitude of the represented vector. So, it is common to speak of a vector as "pointing" and to refer to the "head," or "point," of a vector and to its "tail," or "base."

A vector is often denoted by a boldface symbol, such as **v**. That is the notation used in this book. Other notations, commonly used in handwriting, are $\vec{v}$ and $\underline{v}$. The magnitude of a vector is indicated by $|\mathbf{v}|$, $|\vec{v}|$, $|\underline{v}|$, or simply v. A vector's direction can be specified in any manner, but commonly by giving the three angles between its direction and the positive x-, y-, and z-axes of a Cartesian coordinate system, known as its direction angles and usually denoted (α, β, γ), respectively. Alternatively, direction cosines can be used to specify a vector's direction. They are the cosines of the respective direction angles (cos α, cos β, cos γ). A vector's components are the values of its perpendicular projections along the three axes and can be positive, negative, or zero. Thus, the x-, y-, and z-components of vector **v**, denoted $\mathbf{v}_{x, y, z}$ and (v_x, v_y, v_z), equal (v cos α, v cos β, v cos γ), where v represents the magnitude of **v**.

Taking again the example of velocity as vector **v**, let the x-axis point east, the y-axis north, and the z-axis directly upward. For the direction north by northwest and parallel to the Earth's surface, the direction angles are

$$(\alpha, \beta, \gamma) = (112.5°, 22.5°, 90°)$$

The corresponding direction cosines equal

$$(\cos \alpha, \cos \beta, \cos \gamma) = (\cos 112.5°,$$
$$\cos 22.5°, \cos 90°)$$
$$= (-0.383, 0.924, 0)$$

The magnitude of the velocity is v = 50.0 m/s, so the x-, y-, and z-components of the velocity are

$$(v_x, v_y, v_z) = (v \cos \alpha, v \cos \beta, v \cos \gamma)$$
$$= [(50.0 \text{ m/s}) \cos 112.5°,$$
$$(50.0 \text{ m/s}) \cos 22.5°,$$
$$(50.0 \text{ m/s}) \cos 90°)]$$
$$= (-19.1 \text{ m/s}, 46.2 \text{ m/s}, 0 \text{ m/s})$$

A physical quantity that is not characterized by direction is designated a scalar. Temperature, for instance, is a scalar. So are work, energy, and speed.

A number of operations are defined for vectors and scalars.

PRODUCT OF VECTOR AND SCALAR

One operation is multiplication of a vector **v** by a scalar s. The result of this operation, denoted $s\mathbf{v}$, is a vector whose magnitude equals $|s|$ times the magnitude of **v** (v):

$$|s\mathbf{v}| = |s|v$$

The direction of $s\mathbf{v}$ is the same direction as that of $\mathbf{v}$, if s is positive, and the opposite direction of $\mathbf{v}$, if s is negative. The components of $s\mathbf{v}$ are

$$(s\mathbf{v})_{x, y, z} = s(v_x, v_y, v_z) = (sv_x, sv_y, sv_z)$$

Using the velocity example, an aircraft flying at twice the velocity of the one given is traveling at 100 m/s, north by northwest and parallel to the Earth's surface. The components of its velocity are

$$2 \times (-19.1 \text{ m/s}, 46.2 \text{ m/s}, 0 \text{ m/s}) =$$
$$[2 \times (-19.1 \text{ m/s}), 2 \times (46.2 \text{ m/s}), 2 \times (0 \text{ m/s})]$$

$$= (-38.2 \text{ m/s}, 92.4 \text{ m/s}, 0 \text{ m/s}).$$

A particular case of the product of a vector and a scalar is the negative of a vector, which is the same as the vector multiplied by the number −1. The negative of vector $\mathbf{v}$, denoted $-\mathbf{v}$, is a vector with the same magnitude as $\mathbf{v}$ and pointing in the opposite direction. Expressed in terms of components, the components of $-\mathbf{v}$ are

$$(-\mathbf{v})_{x, y, z} = -(v_x, v_y, v_z) = (-v_x, -v_y, -v_z)$$

which are simply the negatives of the components of $\mathbf{v}$.

VECTOR ADDITION AND SUBTRACTION

Vectors of the same kind (such as two velocity vectors or two force vectors) can be added and subtracted. The sum of vectors $\mathbf{u}$ and $\mathbf{v}$, denoted $\mathbf{u} + \mathbf{v}$, is obtained graphically by placing the tail of $\mathbf{v}$ at the head of $\mathbf{u}$. The vector from the tail of $\mathbf{u}$ to the head of $\mathbf{v}$ is the desired sum. Since

$$\mathbf{u} + \mathbf{v} = \mathbf{v} + \mathbf{u}$$

the sum can be found just as well by placing the tail of $\mathbf{u}$ at the head of $\mathbf{v}$. Then, the vector from the tail of $\mathbf{v}$ to the head of $\mathbf{u}$ gives the same result as before.

The difference of vectors $\mathbf{u}$ and $\mathbf{v}$, denoted $\mathbf{u} - \mathbf{v}$, is found graphically by using the relation

$$\mathbf{u} - \mathbf{v} = \mathbf{u} + (-\mathbf{v})$$

to change the difference into a sum and then applying the procedure for adding vectors. The tail of vector -$\mathbf{v}$ is placed at the head of $\mathbf{u}$. The result is then the vector from the tail of $\mathbf{u}$ to the head of -$\mathbf{v}$.

In terms of components, the addition and subtraction of vectors are especially straightforward: one simply adds or subtracts the numbers that are the vectors' components. For addition,

$$(\mathbf{u} + \mathbf{v})_{x, y, z} = (u_x + v_x, u_y + v_y, u_z + v_z)$$

and, for subtraction,

$$(\mathbf{u} - \mathbf{v})_{x, y, z} = (u_x - v_x, u_y - v_y, u_z - v_z)$$

For example, say the components of $\mathbf{u}$ are $(3, 0, -2)$ and those of $\mathbf{v}$ are $(6, -1, -3)$. Then, the components of the vectors' sum and difference are

$$(\mathbf{u} + \mathbf{v})_{x, y, z} = (3 + 6, 0 + (-1), -2 + (-3)) = (9, -1, -5)$$
$$(\mathbf{u} - \mathbf{v})_{x, y, z} = (3 - 6, 0 - (-1), -2 - (-3)) = (-3, 1, 1)$$

SCALAR PRODUCT

Another operation is the scalar product of two vectors $\mathbf{u}$ and $\mathbf{v}$, denoted $\mathbf{u} \cdot \mathbf{v}$ and sometimes referred to as the dot product. The scalar product is a scalar, whose value equals the product of the vectors' magnitudes and the cosine of the smaller angle (less than or equal 180°) between their directions:

$$\mathbf{u} \cdot \mathbf{v} = uv \cos \phi$$

Here ϕ denotes the angle between the vectors' directions. This can be understood as the product of the magnitude of one of the vectors and the perpendicular projection of the other vector along the direction of the first:

$$\mathbf{u} \cdot \mathbf{v} = u(v \cos \phi) = v(u \cos \phi)$$

Note that for two vectors of fixed magnitudes, the value of their scalar product is greatest when they are parallel, with $\phi = 0°$ and $\cos \phi = 1$. Then $\mathbf{u} \cdot \mathbf{v} = uv$. When the vectors are antiparallel, with $\phi = 180°$ and $\cos \phi = -1$, their scalar product is maximally negative, $\mathbf{u} \cdot \mathbf{v} = -uv$. The scalar product of a pair of perpendicular vectors, with $\phi = 90°$ and $\cos \phi = 0$, equals 0. The scalar product possesses the property

$$\mathbf{u} \cdot \mathbf{v} = \mathbf{v} \cdot \mathbf{u}$$

Work forms an important example of a scalar product, since it involves the scalar product of force and displacement. When a constant force $\mathbf{F}$ acts along the straight-line displacement $\mathbf{d}$, the work W that the force performs equals $\mathbf{F} \cdot \mathbf{d}$. For example, let a constant force of magnitude $F = 11$ newtons (N) act on a body that is undergoing a straight-line displacement of magnitude $d = 1.5$ meters (m), where the angle between the direction of the force and that of the displacement is $\phi = 35°$. The work in joules (J) that is performed by the force on the body is then

$$W = \mathbf{F} \cdot \mathbf{d}$$
$$= Fd \cos \phi$$
$$= (11 \text{ N})(1.5 \text{ m}) \cos 35°$$
$$= 1.4 \text{ J}$$

The scalar product can conveniently be expressed also in terms of the vectors' components, as

$$\mathbf{u} \cdot \mathbf{v} = u_x v_x + u_y v_y + u_z v_z$$

For example, take for **u** and **v** the vectors that served in the example of vector addition and subtraction, (3, 0, -2) and (6, -1, -3), respectively. Then

$$\mathbf{u} \cdot \mathbf{v} = 3 \times 6 + 0 \times (-1) + (-2) \times (-3) = 24$$

The magnitude of a vector equals the square root of its scalar product with itself:

$$|\mathbf{v}| = \sqrt{\mathbf{v} \cdot \mathbf{v}} = \sqrt{v_x^2 + v_y^2 + v_z^2}$$

One can confirm this for the first velocity example. The sum of the squares of the velocity components is

$$v_x^2 + v_y^2 + v_z^2 = (-19.1 \text{ m/s})^2 + (46.2 \text{ m/s})^2 + (0 \text{ m/s})^2$$
$$= 2499.25 \text{ m}^2/\text{s}^2$$

and its square root gives

$$|\mathbf{v}| = \sqrt{v_x^2 + v_y^2 + v_z^2}$$
$$= \sqrt{(2499.25 \text{ m}^2/\text{s}^2)}$$
$$= 50.0 \text{ m/s}$$

(to the three-significant-digit precision of these calculations), which is indeed the magnitude of the velocity.

VECTOR PRODUCT

Yet a third operation is the vector product of two vectors, denoted **u** × **v** and sometimes called the cross-product. This is a vector whose magnitude equals the product of the two vectors' magnitudes and the sine of the smaller angle (less than or equal 180°) between their directions:

$$|\mathbf{u} \times \mathbf{v}| = uv \sin \phi$$

Note here that for two vectors of fixed magnitudes, the magnitude

of their vector product is greatest when they are perpendicular, with $\phi = 90°$ and $\sin \phi = 1$. Then, $|\mathbf{u} \times \mathbf{v}| = uv$. When two vectors are parallel or antiparallel, with $\phi = 0°$ or $\phi = 180°$ and $\sin \phi = 0$, their vector product vanishes.

The direction of **u** × **v** is perpendicular to the plane defined by the directions of **u** and **v**. The sense of this vector is such that if one imagines it pointing from one side of the **u**-**v** plane to the other, it is pointing toward the side from which the angle ϕ is seen as involving a counterclockwise rotation from the direction of **u** to the direction of **v** in the **u**-**v** plane. The vector product obeys the relation

$$\mathbf{u} \times \mathbf{v} = -(\mathbf{v} \times \mathbf{u})$$

Torque can serve as an example of use of the vector product. Let force **F** act on a particle or body, and denote by **r** the position vector of the point of application of the force with respect to reference point O. So, **r** is the spatial vector extending from reference point O to the point of application of the force. Then the torque of the force with respect to O, τ_O, is given by the vector product,

$$\tau_O = \mathbf{r} \times \mathbf{F}$$

For a numerical example, let the position and force vectors both lie in the x-y plane, with the force acting at the point 1.5 m on the positive x-axis. With the coordinate origin serving as reference point O, the position vector then has magnitude r of 1.5 m and points in the positive x direction. Assume a force of magnitude F of 3.0 N acting in the direction 45° from the positive x direction toward the positive y

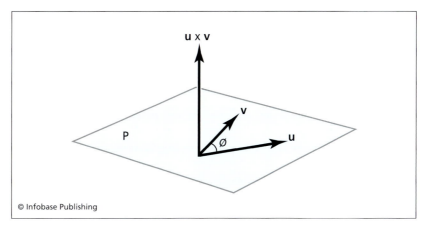

© Infobase Publishing

The vector product of vectors **u** and **v**. The plane P contains **u** and **v**. The smaller angle between **u** and **v** is denoted ϕ. The vector product of **u** and **v**, denoted **u** × **v**, is a vector perpendicular to P in the direction shown. The magnitude of **u** × **v** equals $uv \sin \phi$, where u and v are the magnitudes of **u** and **v**, respectively. For the vector product, **u** × **v** = -(**v** × **u**).

PHYSICS AND MATHEMATICS: A REMARKABLE PARTNERSHIP

by Joe Rosen, Ph.D.

It is mostly taken for granted that the laws and theories of physics are expressed in mathematical terms, that physics—and science in general—is couched in the language of mathematics. For an example, one might reach back to Johannes Kepler's laws of planetary motion. They are indeed formulated in terms of mathematics (including geometry), which include ellipses, areas, ratios, powers, and equalities. Then one might consider Sir Isaac Newton's laws of motion and law of gravitation, which form a theory—that is, an explanation—of Kepler's laws. They, too, are stated in mathematical language, as they speak of zero, straight lines, constant values, proportionality and inverse proportionality, magnitudes, directions, and more. And so on and on, through all fields of physics and through all times.

At its most fundamental level, what is going on might be embodied in this archetypal example. Physicists study some phenomenon, say, an object in free fall, and make a finite number of measurements. The number of measurements might be large but is still finite.

The measurements that are involved possess limited precision, which might be high but is still limited. After careful study, the investigators might notice that the numbers more or less match a very simple mathematical relation. In the case of free fall from initial rest, for example, the distance y that the falling object falls and the time t during which the object covers that distance seem more or less to obey the mathematical relation

$$y = \frac{1}{2} g t^2$$

where g is a constant that has the *same value* for all drops of all the bodies that were dropped.

Consider what has happened so far. The researchers performed *measurements* that gave *numerical* results. Instead, they might have recorded their thoughts about the falling objects, or they could have preferred to record their emotional reactions. They might have considered the comments of the bystanders or the state of the union during the objects' falls. But no: they aimed for numbers. *Why*? Then, having numbers to work with, they looked for *simple* mathematical relations among them. More complicated relations might well have better matched

their numbers, might have passed *through* all the points on their graphs, rather than only passing *near* them. Yet, they preferred to match to a *simple* relation, even if the match was less than perfect. *Why*?

But that is only the start. What the physicists next do requires terrific nerve. They take the simple mathematical relation they have found—which is based on an *approximate* matching to a *finite* number of measurements of *limited* precision—and propose it as a law that should *precisely* hold for *all* falls of *all* objects. *What justifies their belief that what they are doing makes sense*? Moreover, not only does it make sense, but it turns out to be correct! *How is it that nature is describable by mathematical relations at all, and furthermore, by simple ones*?

We are seeing here what Eugene Wigner called "the unreasonable effectiveness of mathematics in the natural sciences." *Why does mathematics work so well in understanding the world*? People have given thought to this issue, but no decisive and generally accepted answer has been forthcoming. Some say, "God is a mathematician." But that is no answer. It simply pushes the ques-

---

direction. Then, the angle ϕ between the two vectors is 45°. From the definition of the torque vector and of the vector product, it follows that the torque vector points in the positive z direction and has the magnitude

$$|\tau_O| = |\mathbf{r} \times \mathbf{F}|$$
$$= rF \sin \phi$$
$$= (1.5 \text{ m})(3.0 \text{ N}) \sin 45°$$
$$= 3.2 \text{ N·m}$$

The x-, y-, and z-components of $\mathbf{u} \times \mathbf{v}$ can be expressed in terms of the components of $\mathbf{u}$ and $\mathbf{v}$ as

$$(\mathbf{u} \times \mathbf{v})_{x, y, z} = (u_y v_z - u_z v_y, u_z v_x - u_x v_z, u_x v_y - u_y v_x)$$

For an example, take the same vectors $\mathbf{u}$ and $\mathbf{v}$ that served in earlier examples, $(3, 0, -2)$ and $(6, -1, -3)$, respectively. Then the components of $\mathbf{u} \times \mathbf{v}$ are

$$(\mathbf{u} \times \mathbf{v})_{x, y, z} = [0 \cdot (-3) - (-2) \cdot (-1), (-2) \cdot 6 - 3 \cdot (-3),$$
$$3 \cdot (-1) - 0 \cdot 6]$$
$$= (-2, -3, -3)$$

Similarly, the components of $\mathbf{v} \times \mathbf{u}$ are

$$(\mathbf{v} \times \mathbf{u})_{x, y, z} = (2, 3, 3)$$
$$= -(\mathbf{u} \times \mathbf{v})_{x, y, z}$$

which confirms the general relation

$$\mathbf{u} \times \mathbf{v} = -(\mathbf{v} \times \mathbf{u})$$

tion into another domain. The obvious response is "Why is God a mathematician?" Others suggest that scientists have *chosen* to treat nature quantitatively and mathematically, while alternative ways of approaching nature might be—some say, are—equally, if not more, successful. Perhaps. However, while the mathematical approach can evaluate its clear success by definite, objective criteria, other approaches do not seem to be able to offer anything close to definite, objective criteria for success.

The following could serve as a not unreasonable, very partial attempt to answer the question. Start with the assertion that scientists in general, and physicists in particular, use mathematics because that is the only tool they have, that is how their brains work. This statement is quite reasonable. Just ask a physicist whether he or she can suggest any other way of dealing with nature that might lead to meaningful understanding. More than likely, the question will evoke a very perplexed expression and no answer. Assume that, indeed, this is the way human brains work. Now use an argument based on evolution. The way human brains work cannot be very much out of tune with reality. Otherwise our ancestors could not have survived to pass on their genes. Furthermore, evolution refines the way human brains work, thereby improv-

ing humans' evolutionary advantage. Presumably, that means becoming more and more in tune with reality. This leads to the conclusion that the way humans do physics—mathematically—is the best way of dealing with nature, or at least very close to being the best way.

So it is plausible to think it is now understood that nature is best dealt with mathematically, that there is no point in even considering alternate modes of approach. If that is correct, it is a small step in the right direction. Yet, there remains the gaping chasm to cross: *Why is nature best dealt with mathematically*?

Here is another line of reasoning. It is not based on the way human brains work or on evolution, but rather on the character of science, on the fact that science is the human endeavor of attempting to achieve an *objective* understanding of nature. The demand of objectivity leads immediately to the necessity of a *rational* approach to nature, which means relying on the methods and tools of *logic*. Any other approach would involve *subjectivity*, which must be avoided in science to the extent possible. Objectivity also leads us to the quantification of—that is, assignment of numerical values to—physical effects. Any other way of considering, recording, reporting, or otherwise treating physical effects would involve subjectivity. Mathematics is the branch of

logic that deals with numerical quantities and with relations among them. The conclusion of this line of reasoning is that mathematics is used in science—and in physics, in particular—because of the objectivity of science.

If one accepts this argument, then, since science works so well, the question reduces to, *Why does the physical world possess objective aspects*? That leads to a very fundamental question of philosophy and this discussion will end here.

The main point is that while the enormous importance of mathematics in physics is indisputable, the reason for its importance is not at all obvious. Is it due, at least in part, to the properties of the human mind? Or could it result from the objective nature of science? And what is it telling us about nature itself and about reality? Clearly, there is still much to learn.

FURTHER READING

Feynman, Richard. *The Character of Physical Law*. Cambridge, Mass.: M.I.T. Press, 1965.

Oliver, David. *The Shaggy Steed of Physics: Mathematical Beauty in the Physical World*, 2nd ed. New York: Springer, 2004.

Wigner, Eugene P. *Symmetries and Reflections*. Cambridge, Mass.: M.I.T. Press, 1967.

FIELDS

Fields are physical quantities that depend on location. A vector field is a vector quantity that possesses values at points in space and whose magnitude and direction can vary from point to point. The electric and magnetic fields serve as examples of vector fields. A scalar field is a scalar quantity that depends on location, such as temperature.

The divergence of a vector field, $\mathbf{v}(x, y, z)$, is denoted div $\mathbf{v}$, or $\nabla \cdot \mathbf{v}$. It is a scalar field, given in terms of the components of $\mathbf{v}$ by

$$\text{div } \mathbf{v}(x,y,z) = \frac{\partial v_x(x,y,z)}{\partial x} + \frac{\partial v_y(x,y,z)}{\partial y} + \frac{\partial v_z(x,y,z)}{\partial z}$$

where the derivatives indicated by the ∂ sign ("curly d") are partial derivatives (i.e., derivatives with

respect to the indicated variable, x, y, or z, while treating the other two variables as constants). For an interpretation of the divergence, let the field, $\mathbf{v}(x, y, z)$, represent the velocity at every location in an incompressible fluid in flow. Then a nonzero value of div $\mathbf{v}$ at a location indicates that there is a source or sink of fluid at that location. If div $\mathbf{v} > 0$ at some point, fluid is entering the flow from outside the system (a source) there. For div $\mathbf{v} < 0$, fluid is disappearing from the flow (a sink) at that point.

On the other hand, every scalar field, $s(x, y, z)$, defines the vector field that is its gradient, denoted grad s, or ∇s. The x-, y-, and z-components of grad s are given by

$$(\text{grad } s)_{x,y,z} = \left(\frac{\partial s(x,y,z)}{\partial x}, \frac{\partial s(x,y,z)}{\partial y}, \frac{\partial s(x,y,z)}{\partial z} \right)$$

Take a temperature field, as an example, where $s(x, y, z)$ represents the temperature at every location. The direction of grad s at any location L indicates the direction at L in which the temperature has the greatest rate of increase. In other words, if the temperatures are compared at all locations on the surface of a small sphere centered on L, the temperature at the location in the direction of grad s from L will be the highest. The magnitude of grad s equals the rate of temperature increase in that direction.

From a vector field, $\mathbf{v}(x, y, z)$, it is also possible to obtain another vector field, called the curl of $\mathbf{v}$ and denoted curl $\mathbf{v}$, or $\nabla \times \mathbf{v}$, or sometimes rot $\mathbf{v}$. The x-, y-, and z-components of curl $\mathbf{v}$ are given by

$$(\text{curl } \mathbf{v})_{x,y,z} = \left(\frac{\partial v_z}{\partial y} - \frac{\partial v_y}{\partial z}, \frac{\partial v_x}{\partial z} - \frac{\partial v_z}{\partial x}, \frac{\partial v_y}{\partial x} - \frac{\partial v_x}{\partial y} \right)$$

The interpretation of the curl is more complicated than for the divergence and the gradient and beyond the scope of this text.

See also ACCELERATION; COORDINATE SYSTEM; ENERGY AND WORK; FLUID MECHANICS; FORCE; HEAT AND THERMODYNAMICS; ROTATIONAL MOTION; SPEED AND VELOCITY.

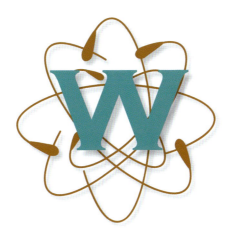

Watson, James (1928–) American *Biochemist*

James Dewey Watson, along with his colleagues Francis Crick, Maurice Wilkins, and Rosalind Franklin, changed the face of biochemistry when they elucidated the physical structure of deoxyribonucleic acid (DNA) in 1953. Less than a decade earlier Oswald Avery at the Rockefeller Institute in New York had shown that DNA alone could pass traits to another organism in an elegant experiment where harmless bacteria became virulent after being injected with DNA isolated from a virulent strain. Watson believed that the data suggested that DNA, not proteins, served as the physical structure within cells that passed traits from parents to progeny. Convinced that genes could not be understood until the biochemical nature and structure of nucleic acids were determined, Watson went to the Cavendish Laboratory at Cambridge University as a 23-year-old postdoctoral fellow in 1951. Upon his arrival, he met a 35-year-old doctoral candidate, Francis Crick, who also shared a desire to determine DNA's structure, and the two men began an intense intellectual collaboration working toward their goal. Over the course of the next two years, Watson and Crick put together bits and pieces of information about DNA and its physical properties that they gleaned from other people's work. Together, they knew that the basic building blocks of DNA were nucleotides composed of deoxyribose (a five-carbon sugar), a phosphate group, and one of four organic bases that contained both carbon and nitrogen atoms—adenine (A), cytosine (C), guanine (G), and thymine (T). Taking into consideration the results from base pair composition analyses, the hydrophilic and hydrophobic nature of the nucleotide components, and X-ray diffraction data that indicated the helical nature of DNA

as well as its dimensions, Watson and Crick used tin molecules from the Cambridge University machine shop and cardboard cutouts made by Watson himself to arrange and rearrange the components of DNA in an effort to figure out its structure. In early 1953, they correctly determined that the DNA molecule consisted of two polynucleotide strands running in opposite orientation to each other (antiparallel) that twisted to resemble a right-handed spiral staircase. Within the individual strands, a sugar-phosphate backbone faced the outside of the molecule with the nitrogenous bases facing inward. Hydrogen bonds between base pairs (A:T and G:C) linked the two individual strands together. Immediately, both men saw a feasible way for DNA to be replicated by unzipping the two complementary strands and using each one as a template to make a new DNA strand following base pair rules. James Watson, along with Francis Crick and Maurice Wilkins, received the 1962 Nobel Prize in physiology or medicine for discovering the structure of DNA and suggesting a feasible means for its replication. Watson went on to become an instrumental leader in the new field of molecular biology as a teacher and a researcher, capping off a stellar career by spearheading the Human Genome Project, whose aim was to sequence and map all the DNA in humans.

James Dewey Watson, born on April 6, 1928 in Chicago, was the only son of the businessman James Dewey Watson and Jean Mitchell Watson. Growing up, the young James developed a fascination for birds, which his father encouraged. A product of the American public school system, Watson attended Horace Mann Grammar School for eight years and South Shore High School for two years. He entered the University of Chicago at the age of 15 with

James D. Watson was an American biochemist best known for the determination of the three-dimensional structure of DNA with Francis Crick. *(National Library of Medicine)*

the intent of majoring in ornithology, or the study of birds. During his undergraduate years, Watson purposely avoided chemistry classes. In 1946, Watson changed his research focus from ornithology to genetics after reading Erwin Schrödinger's book *What Is Life?*, in which Schrödinger postulated the existence of a structural molecule that harbored all necessary genetic information. After graduating with a bachelor of science degree in zoology in 1947, Watson joined the laboratory of Salvador Luria, an Italian-born microbiologist, at Indiana University, where he began his doctoral work studying the effects of X-rays on the reproduction of bacteriophage (viruses that infect bacteria). Watson found himself on the cutting edge of genetics as a move from *Drosophila* systems to microbiological systems, led by Luria and others, including Max Delbruck (known collectively as the Phage Group), occurred. While Watson worked on his Ph.D., the Phage Group came to believe that they were close to identifying the physical nature of genes. Knowing that chromosomes consisted of two macromolecules—proteins and DNA (one type of nucleic acid)—they needed only to determine which of these two macromolecules served as the carrier of genetic information. First isolated in 1868, DNA appeared to be a boring molecule with a repetitive

structure too mundane to hold the key to heredity; in contrast, proteins assumed many different roles and many different structural configurations in the cell. After he received his doctorate in zoology in 1950, Watson decided to study nucleic acid chemistry to get at this physical entity he thought was responsible for heredity. The National Research Council awarded Watson a postdoctoral Merck Fellowship to examine the properties of nucleic acids and to study virus reproduction with the biochemist Herman Kalckar at the University of Copenhagen. In May 1951, Watson attended a meeting on the structures of biological molecules in Naples, Italy, where the biophysicist Maurice Wilkins (a New Zealander working in the physics department of King's College London) presented data from X-ray diffraction studies of DNA that suggested the molecule existed as a highly ordered structure. Watson immediately sensed that the structure of DNA could be solved, and he asked Luria to help him secure another postdoctoral position, with the chemist John Kendrew in the Cavendish Laboratory at Cambridge University, where researchers had recently determined the structure of the protein myoglobin.

Upon arriving at Cambridge, Watson met Francis Crick, a biophysicist who was performing X-ray diffraction studies of proteins. Crick also yearned to determine the structure of the molecule that stored genetic information, and they began a collaboration with the goal of determining DNA's structure. The basic building blocks, or nucleotides, of DNA consisted of a pentose sugar deoxyribose, a negatively charged phosphate group, and one of four nitrogenous bases. Two of the bases, cytosine and thymine (called pyrimidines), were single-ringed aromatic structures made up of nitrogen and carbon atoms; the other two bases, adenine and guanine (called purines), were double-ringed aromatic structures made up of nitrogen and carbon atoms. Preliminary X-ray crystallographic data suggested that DNA might exist as a helix with its strand or strands of DNA wrapped around themselves, like the recently discovered alpha helix motif in proteins, with 10 of these nitrogenous bases stacked on top of one another in each turn of the DNA. Measurements of the width of the DNA molecule from these studies seemed too wide for a single polynucleotide strand. With these facts in mind, Watson ordered a set of stick-and-ball models from the Cavendish machine shop, feeling that a careful examination of possible stereochemical configurations was key to their success, and the two men began tinkering with the physical structure, trying to incorporate all the known properties of DNA. An intuitive thinker with a background in physics, Crick worked out the mathematical equations necessary to solve DNA's structure using X-ray crystallography. Watson, who had a background in genetics, picked out the important results from biochemical studies of DNA. Their first model consisted of a triple-helical structure with three sugar-phosphate backbones running down the center of the molecule and the organic bases pointing outward, available for interactions with other molecules. They invited Wilkins and another X-ray crystallographer from King's College, Rosalind Franklin, to critique their model. A chemist by training, Franklin pointed out that hydrophilic moieties such as the negatively charged phosphate group tended to be on the outside of molecules, where they could interact with water molecules, and that hydrophobic molecules such as the nitrogenous bases would tend to be shielded from the aqueous environment. After this embarrassing failure to deduce the correct structure, the director of the Cavendish Laboratory, Sir Lawrence Bragg, told Watson and Crick to cease work on the project.

Complying with Bragg's request, Watson turned his attention to X-ray diffraction studies of the tobacco mosaic virus (TMV), honing his skills in interpreting crystallography data and subsequently determining the structure of TMV's coat protein in 1952. Francis Crick became well versed in hydrogen bonds, the type of bond now known to hold the two strands of DNA together in a double helix. Both men continued to keep up with new information about DNA structure. Fortuitously, the Austrian-American biochemist Erwin Chargaff visited England and shared his experimental data, demonstrating that the amount of adenine was always equal to the amount of thymine, and the amount of guanine was always equal to the amount of cytosine. Even in different species, the same pattern of base pair ratios held true. From Chargaff's ratios, Watson and Crick surmised that the nitrogenous bases existed in specific pairs in the DNA molecule. During this time, Jerry Donahue also showed Watson the correct tautomeric structure of the nitrogenous bases in vivo. After seeing a manuscript containing the attempt of his competitor Linus Pauling to solve the structure of DNA, Bragg gave Watson permission to resume work on elucidating the macromolecule's structure and encouraged collaboration with Wilkins and Franklin to get an edge in this highly competitive race. Franklin, who was preparing to transfer to another laboratory, refused to participate in a joint effort, although Wilkins agreed. A disgruntled Wilkins decided to show Watson results from Franklin's X-ray diffraction studies of B-form (physiological) DNA that clearly demonstrated the structure to be a double helix and that the calculated space group (a mathematical description of the inherent symmetry of a molecule that has been crystallized) from these studies indicated that two backbones ran in opposite directions to each other.

On his own, Watson figured out how the base pairs interacted with each other by playing with cardboard cutouts of the nitrogenous bases. He discovered when adenine made two hydrogen bonds with thymine and when guanine made three hydrogen bonds with cytosine, that both base pairs were identical in overall configuration and maintained a constant width to fill the space between the two sugar-phosphate backbones.

In February 1953, Watson and Crick solved the structure of DNA, a double helix that consisted of two strands of alternating sugar and phosphate groups connected by complementary base pairing (A with T and G with C) of the nitrogenous bases that lay perpendicular to the backbone like rungs of a ladder. The manner by which the two strands were linked through specific base pairing immediately implied to Watson and Crick a possible mechanism for DNA replication. They envisioned that the double helix would unzip itself, then use each of the strands as guides, or templates, to make new polynucleotide chains that would hydrogen bond to the template strands, resulting in two new DNA molecules. Within the span of a few months, after years of data accumulation and interpretation, Watson and Crick answered one of the most fundamental questions in genetics in two papers published in the April and May 1953 issues of the journal *Nature*, first outlining the structure of DNA and then proposing a means for its replication. They received a Nobel Prize in physiology or medicine in 1962 along with their colleague Maurice Wilkins as a reward for their efforts. Rosalind Franklin, who was responsible for key X-ray diffraction results, died of ovarian cancer in 1958, making her ineligible for the award.

After his success at Cambridge, Watson accepted a position as a research fellow in biology at California Institute of Technology, where he performed X-ray diffraction studies on the other nucleic acid, ribonucleic acid (RNA), until 1955. He joined the faculty at Harvard University, where he served as professor of biology until 1976. During his tenure at Harvard, Watson continued his studies of RNA, specifically messenger RNA (a temporary copy of genetic instructions used by the cell to make proteins), as he tried to elucidate the role of this nucleic acid in the mechanism of protein synthesis. In 1965, he wrote the first molecular biology textbook, *Molecular Biology of the Gene*, which is now in its fifth edition.

James Watson had a very busy year in 1968. He married Elizabeth Lewis, with whom he has two sons, Rufus Robert and Duncan James. In his professional life, he assumed the directorship of the Laboratory of Quantitative Biology at the prestigious Cold Spring Harbor Laboratory (CSHL) in Long Island, New York. Carrying on the CSHL mission of advancing genetic research, Watson expanded the laboratory's focus to include virology (study of viruses) and cancer research. With Watson at the helm, CSHL flourished in both its educational programs and research, becoming a world center in molecular biology. His book *The Double Helix* (1968), a personal account of the events, research, and personalities surrounding the determination of DNA's structure, also arrived in the bookstores, becoming a favorite among the general public because it exposed the humanness of scientific research.

In 1994, Watson assumed the role of president of CSHL, and, currently, he serves as chancellor to the facility. Thousands of scientists flock to CSHL each year to attend meetings to discuss the most recent research developments in biochemistry and molecular biology. From 1988 to 1992, James Watson served at the National Institutes of Health (NIH), first as associate director, then as director for the National Center for Human Genome Research, a division of NIH that oversaw the Human Genome Project. The goal of the project consisted of mapping and sequencing the 3 billion base pairs that constitute the human genome. (In 2003, scientists accomplished this goal.) Watson resigned from this position in 1992 over policy differences and alleged conflicts of interest (i.e., his investments in private biotechnology companies).

James Dewey Watson continues to be controversial and outspoken. A brilliant man, he has frank opinions on politics, religion, and the role of science in society that make people love him or hate him. Over the course of his career, he has received numerous prestigious awards, including the Presidential Medal of Freedom (1977) and the National Medal of Science (1997). He is also a member of the National Academy of Sciences and the Royal Society of London. After Watson and Crick elucidated the structure of DNA, an entire new world of research, the field of molecular biology, came into existence. In slightly more than 50 years since the landmark *Nature* paper, scientists have made huge advancements in understanding the processes of replication, transcription, and translation. They have isolated mutations responsible for many genetic diseases, identified people by their DNA fingerprints, and attempted to target mutated genes with gene therapy to repair them. Watson believed at the outset that if the molecular structure that carried genetic information could be identified, the possibilities in biochemistry and molecular biology would be limitless, and he was right.

See also BIOCHEMISTRY; CRICK, FRANCIS; DNA REPLICATION AND REPAIR; FRANKLIN, ROSALIND; NUCLEIC ACIDS; PAULING, LINUS; WILKINS, MAURICE.

FURTHER READING
Edelson, Edward. *Francis Crick and James Watson and the Building Blocks of Life (Oxford Portraits in Science)*. New York: Oxford University Press, 2000.
Watson, James D. *The Double Helix: A Personal Account of the Discovery of the Structure of DNA*. New York: Atheneum, 1968.
———. *Genes, Girls, and Gamow: After the Double Helix*. New York: Random House, 2002.
Watson, James D. and Francis H. C. Crick. "A Structure for Deoxyribose Nucleic Acid." *Nature* 171 (1953): 737–738.

waves A propagating disturbance is called a wave. A wave might propagate in a material medium or in a field. In the former case, called a mechanical wave, a disturbance is the condition of the matter's being locally out of equilibrium, such as the result of giving the lower end of a hanging rope a horizontal jerk. In regaining equilibrium at one location, the medium causes adjacent locations to depart from equilibrium. Then these, too, return to equilibrium, which causes their adjacent locations to depart from equilibrium, and so on. In this way, the disturbance travels through the medium by means of self-sustaining propagation. The elastic character of a medium (or, for fields, a property analogous to elasticity) causes the medium to tend to regain equilibrium when disturbed. Note that the medium itself does not propagate; it only undergoes local displacement about equilibrium at every point. Waves carry and transmit energy. The following are examples of mechanical waves: (1) Waves in a stretched rope or string: they might be brought about by hitting the rope or by plucking the string or rubbing it with a bow, as in playing a guitar or violin. The string or rope is disturbed by locally being displaced laterally from its equilibrium position. (2) Sound in air: this is a pressure wave. The sound source alternately compresses and decompresses the air, locally raising and lowering its pressure and density from their equilibrium values. (3) Surface waves on a liquid, such as ocean waves: the disturbance here is a local displacement of the surface from its equilibrium level.

An important case of waves in fields is electromagnetic radiation, which propagates as waves in the electromagnetic field. The propagating disturbance, in this case, is a deviation of the electric and magnetic fields from their equilibrium values. The electric and magnetic fields are coupled to each other in such a way that a change over time in one field generates the other. That brings about self-sustaining propagation of disturbances. Electromagnetic radiation includes visible light, ultraviolet and infrared radiation, X-rays, gamma rays, radio and television waves, and microwaves. Electromagnetic waves in vacuum all travel at the same speed, called the speed of light in vacuum and conventionally denoted c, which has the value 2.99792458×10^8 meters per second (m/s), rounded to 3.00×10^8 m/s.

TYPES OF WAVE

Waves are categorized according to the character of the propagating disturbance. In a scalar wave, the disturbance is in a scalar quantity (i.e., a quantity that does not possess a spatial direction), such as temperature, pressure, and density. An example is a sound wave in air, which can be described as a pressure wave. A longitudinal wave is one in which the disturbance is a local displacement along the direction of wave propagation, such as is produced in a rod by striking the end of the rod longitudinally. For another example, take a stretched-out Slinky and give the end a push or pull in the direction of the Slinky. A sound wave in air can also be described as a longitudinal wave, with the air undergoing longitudinal disturbance as a result of the pressure disturbance.

When the disturbance is a local displacement perpendicular to the direction of propagation, the wave is transverse. Such are waves in stretched strings. In the Slinky example, shake the end perpendicularly to the Slinky. Electromagnetic waves also are of this type. Besides those, there are additional types of wave. A torsion wave, for instance, propagates a twist displacement and is a version of a transverse wave. Waves need not be of any pure type. Waves through Earth, such as are produced by earthquakes, might consist of a mixture of types. Surface waves on a liquid involve a combination of longitudinal and transverse disturbances.

A transverse wave may possess the property of polarization. Inversely, a transverse wave is said to be unpolarized when the propagating disturbance occurs in all directions perpendicular to the direction of propagation, with no particular relation among the disturbances in the various directions. In a polarized wave, on the other hand, there are

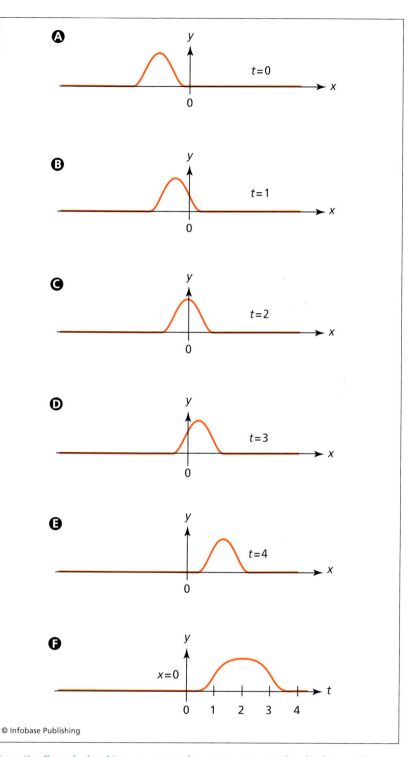

© Infobase Publishing

A vertically polarized transverse pulse wave propagating in the positive x direction. (a)–(e) Snapshots of the waveform for a sequence of times t. (f) The corresponding displacement at location $x = 0$ as a function of time.

constraints and relations among the disturbances in the various directions. In linear, or plane, polarization, for instance, the disturbance is confined solely to a single direction. An example is a horizontal stretched string in which a disturbance involving purely vertical local displacement (i.e., with no horizontal component) is propagated. There are also other kinds of polarization, such as circular and elliptical.

WAVE PROPAGATION

As a wave propagates, the medium might absorb energy from it, thus reducing its intensity. The wave is then said to be attenuated as it propagates.

When the transmission properties of a medium change rather abruptly along the path of propagation of a wave, such as when a wave is incident on an interface between two different media, the wave undergoes (generally partial) reflection. This is a breakup of the wave into two waves, one of which, the reflected wave, propagates back into the region of initial properties (into the initial medium), and the other, the transmitted wave, proceeds into the new-property region (through the interface into the second medium). If the direction of the transmitted wave is different from that of the original wave, the wave is said to be refracted. In the case of a well-defined interface, which is taken to be planar for this discussion, the directions of propagation of the reflected and refracted waves relate to that of the incident wave in the following manner. Consider a straight line perpendicular to the interface surface, called the normal, and another line that passes through the point where the normal penetrates the interface and is in the direction of propagation of the incident wave, the line of incidence. Denote the smaller angle between the two lines, the angle of incidence, by θ_i. The two lines define a plane, the plane of incidence. The direction of reflection is indicated by the direction of a line lying in the plane of incidence, passing through the penetration point, on the opposite side of the normal from the line of incidence, and making the same angle with the normal as does the line of incidence, θ_i. This angle is the angle of reflection.

The direction of refraction is similarly indicated by the direction of a line lying in the plane of incidence, passing through the penetration point, on the same side of the normal as the line of incidence, and making angle θ_r with the normal, the angle of refraction. According to Snell's law, named for its discoverer, the Dutch physicist and mathematician Willebrord Snell,

$$\frac{\sin \theta_i}{v_i} = \frac{\sin \theta_r}{v_r}$$

where v_i and v_r denote the propagation speeds of the waves in the medium of incidence (and reflection) and the medium of transmission (refraction), respectively. Note that if $v_r > v_i$ (i.e., the wave is incident on a medium of higher propagation speed), there is a range of incident angles θ_i—from a minimal angle, the critical angle, up to 90°—for which no refraction angle θ_r satisfies the relation. In that case, there is no transmission at all, a situation called total internal reflection. Such is the case, for example, when light waves in glass or water impinge on the interface with air. The critical angle, θ_c, is given by

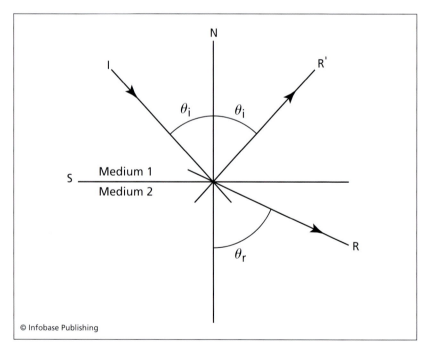

© Infobase Publishing

Reflection and refraction of a wave at the interface surface S between two media. The line of incidence I (in the direction of the incident wave), the normal N (perpendicular to the interface surface), the line of reflection R' (in the direction of the reflected wave), and the line of refraction R (in the direction of the refracted wave) all lie in the plane of incidence, which is the plane of the figure. The angles of incidence and refraction are denoted θ_i and θ_r, respectively. They are related through Snell's law. The angle of reflection equals the angle of incidence.

$$\sin \theta_c = \frac{v_i}{v_r}$$

In the case of light and other electromagnetic radiation, transparent materials are characterized by their index of refraction, or refractive index, n, which is defined as

$$n = \frac{c}{v}$$

where c denotes the speed of light in vacuum, with the approximate value 3.00×10^8 m/s, and v is the speed of light in the material. Then Snell's law takes the form

$$n_i \sin \theta_i = n_r \sin \theta_r$$

Here

$$n_i = \frac{c}{v_i} \text{ and } n_r = \frac{c}{v_r}$$

are the indices of refraction of the medium of incidence and the medium of transmission (refraction), respectively. The formula for the critical angle then takes the form

$$\sin \theta_c = \frac{n_r}{n_i}$$

PERIODIC WAVES

An important particular kind of wave is a periodic wave. It is produced when a medium (or the electromagnetic field) is disturbed periodically, that is, is made to oscillate in a repetitive manner, which generates a moving repetitive spatial pattern, or repetitive waveform, along the direction of propagation. At every point of the medium along the line of propagation, the medium undergoes local oscillation. At any instant, a snapshot of the whole medium shows a spatially repetitive waveform. As the waveform moves along, at all points the medium undergoes local, but well-correlated, oscillations. The oscillations at every point are characterized by their frequency, f, the number of cycles of oscillation per unit time in hertz (Hz); and their period, T, which is the time of a single cycle of oscillation in seconds (s) and equals $1/f$. The repetitive waveform, as revealed by a snapshot, has a characteristic repeat distance, the wavelength, λ, in meters (m). After a time interval of one period, every point of the medium completes a single full cycle of oscillation and returns to its initial state. During that time, the waveform moves such that it becomes indistinguishable from its initial appearance; that means it moves the repeat distance, a distance of a single wavelength. So, during time T, the wave propagates through distance λ. Its propagation speed, v, in meters per second (m/s), is then

$$v = \frac{\lambda}{T} = \frac{\lambda}{1/f} = f\lambda$$

This is a fundamental relation for all periodic waves. The speed of propagation is determined by the properties of the medium, while

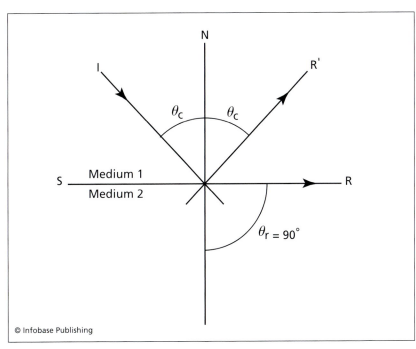

© Infobase Publishing

Critical-angle incidence of a wave at the interface surface S between two media. The line of incidence I (in the direction of the incident wave), the normal N (perpendicular to the interface surface), the line of reflection R′ (in the direction of the reflected wave), and the line of refraction R (in the direction of the refracted wave) all lie in the plane of incidence, which is the plane of the figure. The critical angle θ_c is that angle of incidence for which the angle of refraction θ_r equals 90°, which is the largest value it can have. At angles of incidence greater than the critical angle, there is no refracted wave and the incident wave is totally reflected, an effect called total internal reflection. The angle of reflection always equals the angle of incidence, both indicated by θ_c in the figure.

the frequency is the same as that of whatever is exciting the periodic wave. Together, they determine the wavelength through this relation.

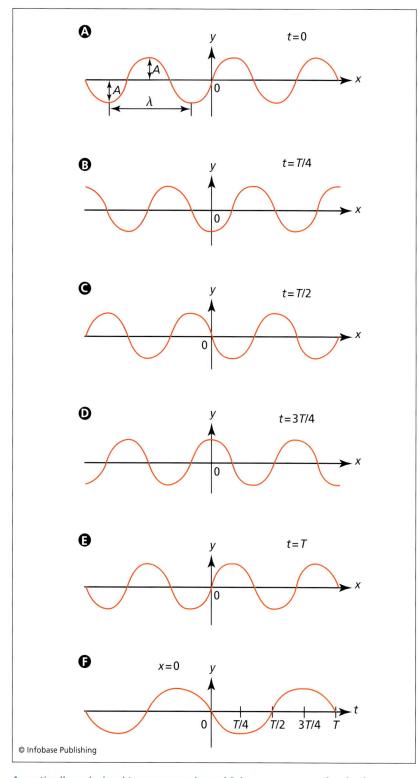

A vertically polarized transverse sinusoidal wave propagating in the positive *x* direction. (a) Wavelength λ and amplitude *A* are indicated. (a)–(e) Snapshots of the waveform for a sequence of times *t* at quarter-period intervals, where *T* denotes the period. (f) The corresponding displacement at location *x* = 0 as a function of time.

A particularly important kind of periodic wave is a sinusoidal, or harmonic, wave. At every point along the line of propagation, the medium's local oscillation has a sinusoidal dependence on time, that is, undergoes harmonic motion, while at any instant a snapshot of the whole medium shows a spatially sinusoidal waveform. For a wave traveling in the positive *x* direction, that can be expressed as

$$\text{Disturbance} = A \cos 2\pi \left(\frac{x}{\lambda} - ft \right)$$

where *A* denotes the wave's amplitude. The disturbance is a spatial and temporal sinusoidal oscillation between the limits of +*A* and -*A*. This can be seen by keeping *x* constant, resulting in sinusoidal time dependence of the disturbance at a fixed location, or by keeping *t* fixed, giving a sinusoidal waveform at a particular time. Another, equivalent form of this relation, one that shows the propagation speed, *v*, explicitly, is

$$\text{Disturbance} = A \cos 2\pi \frac{x - vt}{\lambda}$$

A surface in space that a wave has reached after propagating some time from its source is referred to as a wavefront. For example, in an isotropic medium (i.e., a medium in which the propagation speed is the same in all directions), the wavefronts of a wave emanating from a point source are concentric spherical surfaces. For a periodic wave, the oscillations produced by the wave on a wavefront are all in phase: that is, they are all at the same stage of their oscillation cycle.

INTERFERENCE

The combined effect of two or more waves passing through the same location at the same time is called interference. Under very special conditions, the waves might reinforce each other, a situation called constructive interference, or partially or completely cancel each other, which is destructive interference. In the general case, however, interference results in time-varying effects of no particular significance.

Consider interference between two sinusoidal waves. For maximal constructive interference to occur at some location, the waves must have the same frequency and reach the location in phase, meaning that the effects of the individual waves at the location are simultaneously at the same stage of their respective cycles (i.e., their phase difference is 0). Both waves disturb the medium upward at the same time, then downward at the same time, and so on, as an example. The individual effects then reinforce each other maximally, and the result is a local oscillation that is stronger than that produced by each wave individually. In a linear medium, in which the combined effect of several influences is the sum of their individual effects, the amplitude of oscillation equals the sum of the amplitudes of the two waves. This is referred to as the superposition principle.

For maximal destructive interference, the waves must possess equal frequencies, as before, but must reach the location in antiphase, that is, maximally out of phase, whereby the effects of the individual waves differ in time by half a cycle. In our example, when one wave disturbs the medium maximally upward, the other disturbs it maximally downward. Then the individual effects weaken each other. If they are of equal amplitude, they will totally cancel each other, and there will be no oscillation at that location. For unequal amplitudes in a linear medium, the amplitude of oscillation equals the difference of the two amplitudes.

Intermediate cases of interference occur when the wave frequencies are equal and the phase difference of the two waves reaching the location is between 0 and half a cycle.

DIFFRACTION

Inseparable from interference and actually caused by it, diffraction is the bending of waves around obstacles. It is a property of waves of all kinds. As an instant, portable demonstration of diffraction of light, move two adjacent fingers very close to each other, perhaps touching, so that there are tiny gaps between them, and look through such a gap at the sky or any other light source. If the gap is small enough, you will see a pattern of dark and light stripes or spots, depending on the shape of the gap. That is the diffraction pattern produced by the light bending around the edges of the gap. The bending of

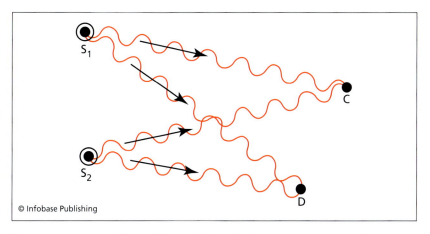

Two wave sources, S_1 and S_2, emit sinusoidal waves at the same frequency and phase. Waves arrive at C in phase, resulting in maximal constructive interference. At D waves arrive in antiphase (maximally out of phase), which produces maximal destructive interference.

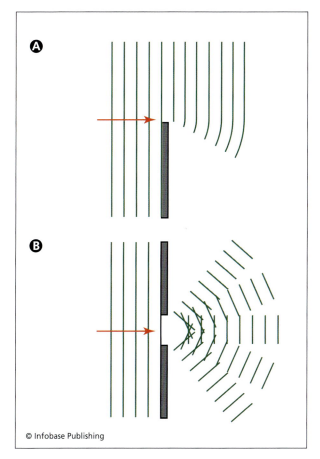

Diffraction is the bending of waves around obstacles. (a) Straight wavefronts moving in the indicated direction bend around a long obstacle. An example is ocean waves impinging on a jetty. (b) Straight wavefronts incident on a slit bend around the edges and fan out in certain directions.

waves around an obstacle is the result of constructive interference of nonobstructed waves, in directions that would be blocked by straight-line propagation.

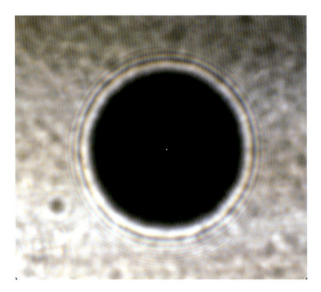

Because of diffraction and interference of light, the shadow of a circular opaque object possesses a system of concentric darker and lighter rings extending both outward and inward. At the very center of the shadow is a bright spot. *(sciencephotos/Alamy)*

Because of diffraction, the shadow of an opaque object is not sharp but has fringes at its edge, if it is examined in sufficient detail. In particular, the shadow of a circular object, for example, possesses a system of concentric darker and lighter rings extending both outward and inward. At the very center of the shadow is found a bright spot. Again because of diffraction, the image of an opening in an opaque obstacle, when light passes through it and strikes a screen, can have a very complex structure. In this way, diffraction limits the resolving power, the ability to form sharp images, of any optical system.

As an example, consider the case of light of wavelength λ passing through a narrow slit of width d and striking a screen at distance L from the slit. If the light is propagated in straight lines, the image on the screen would be a faithful image of the slit. However, as a result of diffraction, the image is spread out in the direction of the slit's width and has dark regions in it. At the center of the diffraction pattern is a strip of width $(2/L)(\lambda/d)$, which is flanked on both sides by strips of half that width separated by darkness.

Note in the example—and it is true in general—that the strength of the diffraction effect is proportional to the ratio of the wavelength of the waves to the size of the obstacle or of the opening (λ/d in the example). For a given geometry, diffraction becomes more pronounced for longer wavelengths,

while for the same wavelength, the effect is stronger for smaller obstacles and openings. As an easy demonstration of wavelength dependence, note that one can stand to the side outside an open doorway to a room and hear what is going on in the room while not being able to see inside. The sound waves emanating from the room, with wavelengths more or less in the centimeter-meter range, are diffracting around the edge of the doorway. The light waves from the room, on the other hand, whose wavelengths are about a million times shorter, around 5×10^{-7} meter, do not diffract effectively. You can use the finger demonstration to confirm the dependence of diffraction on opening size by varying the size of the gap and observing what happens to the diffraction pattern.

STANDING WAVES

A wave in a system can reflect from the system's boundaries, interfere with itself, and bring about a situation called a standing wave, in which the system oscillates while no wave propagation is apparent. The oscillation of a plucked stretched string (such as on a guitar), for instance, is in reality a standing wave, in which transverse waves are traveling along the string in both directions, reflecting from the ends, and interfering with each other. Those locations along a standing wave where the medium remains in equilibrium, that is, at rest, are termed nodes. Posi-

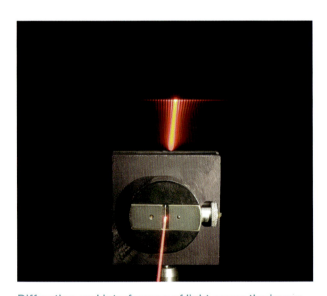

Diffraction and interference of light cause the image of a narrow slit opening in an opaque barrier to possess a complex structure. A light beam from a laser enters from the bottom center and impinges on the barrier, illuminating the adjustable slit. The slit's image is seen above the barrier. *(Edward Kinsman/Photo Researchers, Inc.)*

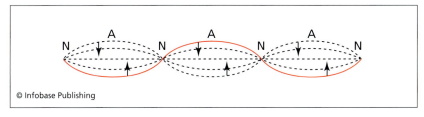

© Infobase Publishing

An example of nodes and antinodes of a standing wave on a taut string. Nodes, denoted N, are the locations at which the amplitude of oscillation is zero. Locations of maximal amplitude are the antinodes, labeled A.

tions of maximal oscillation amplitude, relative to nearby locations, are called antinodes. Examples are readily observed by shaking the end of a stretched rope transversely and finding those frequencies at which the shaking generates standing waves. The nodes and antinodes will be obvious. Their number is greater for higher standing-wave frequency.

See also ACOUSTICS; ELASTICITY; ENERGY AND WORK; EQUILIBRIUM; HARMONIC MOTION; OPTICS; PRESSURE; SPEED AND VELOCITY.

FURTHER READING

Elmore, William C., and Mark A. Heald. *Physics of Waves.* Mineola, N.Y.: Dover, 1985.

Pierce, John R. *Almost All about Waves.* Mineola, N.Y.: Dover, 2006.

Young, Hugh D., and Roger A. Freedman. *University Physics,* 12th ed. San Francisco: Addison Wesley, 2007.

Wilkins, Maurice (1916–2004) New Zealander *Biophysicist*   During the course of his prolific career as a physicist turned biophysicist, Maurice Wilkins made contributions to the fields of phosphorescence, radar technology, isotope separation, and X-ray diffraction. His most significant work centered on the elucidation of the structure of deoxyribonucleic acid (DNA) in 1953, a watershed event that revolutionized biology, biochemistry, and medicine. Using data from Wilkins and another research associate, Rosalind Franklin, in the King's College of London laboratory, James Watson and Francis Crick correctly surmised that DNA consisted of two polynucleotide chains wrapped around each other, similar to a gently twisted ladder. Alternating deoxyribose sugar residues and phosphate groups form a structural backbone for the DNA molecules serving as the rails, or sides of the ladder structure. Nitrogenous bases from one chain interact with specific nitrogenous bases from the other chain to form the rungs of the ladder. The specificity of the interactions between the nitrogenous bases provides a means for the faithful replication, or copying, of the genetic information within the DNA by using one of the two strands as a template. The importance of this discovery led to a Nobel Prize in physiology or medicine in 1962 for Watson, Crick, and Wilkins. Franklin received no formal recognition for her contribution to the joint efforts of the two laboratories in determining DNA structure because she had died of cancer in 1958, making her ineligible for the award.

Born on December 15, 1916, Maurice Hugh Frederick Wilkins spent the first six years of his life in the isolated community of Pongaroa, New Zealand. His father, Edgar Henry Wilkins, served as a physician in the School Medical Service. Wilkins attributed his later interest in scientific frontiers to the opportunities for exploration and discovery that his Irish parents gave him in New Zealand. When Maurice was six, his parents returned to England so his father could study preventative medicine. Maurice entered King Edward's School in Birmingham, where he showed himself to be adventurous and bright. After grammar school, Wilkins obtained a degree in physics from St. Johns College, Cambridge, in 1938, then he began a research associate position in the laboratory of Dr. John T. Randall, studying the luminescence of solids. For his doctoral studies, he examined the thermal stability of trapped electrons in phosphors (any substances that continue to glow after exposure to light or energized particles such as electrons), and he received his Ph.D. in physics in 1940. To contribute to the British war effort (World War II), he applied his energies to the improvement of cathode-ray tube screens for radar. (A cathode-ray tube is a vacuum tube with positive and negative electrodes at either end.) In 1943, while working for M. E. Oliphant, Wilkins moved to Berkeley, California, to join the Manhattan Project, an American wartime initiative exploring nuclear physics. At Berkeley, Wilkins used mass spectroscopy to separate different uranium isotopes for use in atomic bombs. Later in life, Wilkins asserted that he never had the sense of the catastrophic possibilities of nuclear weapons during this time; the Nazis were winning most facets of the war, so he and others were focused on gaining the technological upper hand at least. The ramifications of nuclear warfare became perfectly unacceptable to Wilkins in 1945, when the United States dropped Fat Man and Little Boy (two atomic bombs) on the Japanese cities of Hiroshima and Nagasaki, and he remained an ardent and vocal opponent of nuclear weapons for the rest of his life, even serving as president of the British Society for Social Responsibility in Science.

Maurice H. F. Wilkins was a biophysicist from New Zealand who contributed significantly to the elucidation of the structure of DNA in 1953. *(National Library of Medicine)*

In 1945, Wilkins returned to Great Britain, where he accepted a position as lecturer in physics at St. Andrew's University in Scotland. Still reeling from his role in the development of the atomic bomb, Wilkins decided to go into another branch of science where discoveries might possess a more positive impact. He read a book, *What Is Life?* (Macmillan, 1946), written by the theoretical physicist Erwin Schrödinger, in which Schrödinger speculated that genes (the carriers of genetic information inside the cell) were macromolecules consisting of three-dimensional arrangements of atoms. Schrödinger's book kindled Wilkins's interest in the biochemistry of genes. During this same period, John Randall asked Wilkins to join a newly formed research group, the Medical Research Council (MRC) Biophysics Research Unit, at King's College London. Thinking solid states physics dull, Randall envisioned tackling problems in biology with the methodologies of physics, creating a new field, biophysics, that interconnected two disciplines previously unrelated to one another to advance scientific discovery. At the age of 30, after seven years in physics, Maurice Wilkins accepted Randall's offer, moving to the field of biophysics in 1946. With no clear plan of attack,

Randall hoped that new insights about cell structure and function would be gleaned from applying physical techniques such as ultraviolet light microscopy and electron microscopy. In the years 1946–50, Wilkins had various projects that included studying the genetic effects of ultrasonics, developing a reflecting microscope that permitted the study of nucleic acids in cells, examining the position and grouping of purine and pyrimidine nitrogenous bases in DNA and tobacco mosaic virus (TMV), and determining the arrangement of virus particles in crystals of TMV using a visible light polarizing microscope. In the end, Wilkins began a study of isolated DNA and DNA in sperm cells using X-ray diffraction methods.

In preliminary studies, Wilkins produced thin strands of DNA containing highly ordered arrays of DNA ideal for good X-ray diffraction studies. At the outset of his studies, Wilkins did not know whether isolated DNA and physiological DNA in sperm heads would have the same configuration, but he considered it important to maintain an aqueous environment for both types of DNA. In the early summer of 1950, Wilkins and the graduate student Raymond Gosling developed an X-ray photograph indicating that DNA molecules created regular crystal-like structures. James Watson saw this photograph at a scientific meeting in Naples, Italy, in 1951, and it sparked his interest in DNA. Francis Crick, a friend of Wilkins's, impressed upon him the importance of DNA and his continuing studies. Wanting to get better pictures, Wilkins ordered a new X-ray tube and requested high-quality DNA from the laboratory of Rudolf Signer of Switzerland.

While Wilkins fine-tuned his methodologies during the summer of 1950, John Randall hired another research associate, Rosalind Franklin, for a three-year research fellowship. Randall originally envisioned that Franklin would perform X-ray crystallography studies on proteins such as collagen. Franklin's work in Paris delayed her arrival, giving Randall an opportunity to change his mind about her project. Before she arrived in Great Britain, Randall wrote to Franklin, suggesting that she take advantage of Wilkins's preliminary studies of DNA to try to elucidate the structure of DNA. Franklin arrived in the early part of 1951, while Wilkins was away on vacation. With the arrival of the new X-ray tube and Signer's DNA, Franklin started working on the DNA project, thinking it was hers alone. Franklin quickly discovered two forms of DNA: A-form (a dehydrated form) and B-form (the physiological form). When Wilkins returned, he wrongly assumed that Randall hired Franklin to be his assistant. Despite clearing up this misunderstanding, tension built between Wilkins, who was reserved and shy, and Franklin, who was aggressive and vocal. Because Franklin would not collaborate with Wilkins, they chose to perform parallel studies with little overlap. Wilkins spent most of his time studying A-form DNA, and Franklin spent most of her time studying B-form DNA.

In November 1951, Wilkins performed a series of X-ray diffraction studies that suggested DNA has a helical structure. On his diffraction photographs, he saw an X, which represents a helical structure to an experienced crystallographer. Wilkins shared his findings with Watson and Crick, who incorporated these findings into their first attempt to determine DNA structure. This first model had the sugar-phosphate backbone at the center of the molecule and the hydrophobic bases sticking out around the outside of the molecule. Franklin pointed out that hydrophilic (water-loving) molecules such as sugars and phosphates like to interact with the aqueous environment, and hydrophobic (water-hating) molecules hide from water, leading Watson and Crick to abandon model building for a while. Linus Pauling, the American chemist responsible for determining the structure of alpha helices in proteins, proposed a similar erroneous model of DNA. Pressure to determine the correct structure of DNA in the King's College laboratory became palpable during this period, yet neither Franklin nor Wilkins tried model building as a means to ascertain its structure. In the spring of 1952, Franklin asked to transfer her fellowship to another institution, and Randall granted her permission with the conditions that she not work on DNA after leaving the laboratory and that her work and data on DNA to date would become the property of the laboratory. She continued to work another year in the biophysics unit, further refining her crystallography studies of A- and B-form DNA.

In early 1953, Crick saw an MRC report submitted by Franklin that outlined her data, suggesting that phosphates were located on the external portion of the DNA molecule. A disgruntled Wilkins then shared with Watson one of Franklin's X-ray diffraction photographs of B-form DNA, which unquestionably demonstrated the double-helical nature of DNA. Taken together with results from Erwin Chargaff, who proved that nitrogenous bases in DNA have preferred partners (adenine with thymine and guanine with cytosine), and William Astbury, who measured the number of nucleotides, or building blocks, that exists in each turn, Crick and Watson developed the correct model of DNA—two strands of deoxyribose sugar-phosphate backbone running in opposite directions to one another (antiparallel) linked via hydrogen bonds between complementary base pairs in the core of the molecule, and the entire molecule twisting upon itself and similar to a spiral staircase. In their April 1953 *Nature* publication,

Watson and Crick acknowledged the importance of unpublished X-ray crystallography data performed by Franklin and Wilkins to the elucidation of DNA's structure. Both Wilkins and Franklin published X-ray data that supported the Watson and Crick model of DNA structure in the same issue of *Nature*. Later, Wilkins spent a great deal of energy repeating and, ultimately, extending much of Franklin's work, and he produced a great deal of evidence to support the helical model put forth by his friends. In 1962, Crick, Watson, and Wilkins received the Nobel Prize in physiology or medicine for determining the structure of DNA and suggesting a means for it to replicate itself. Rosalind Franklin was ineligible to receive the award because she had died of ovarian cancer, possibly as a result of prolonged X-ray exposure, in 1958. Crick and Watson might not have determined the structure of DNA if they had not been privy to the work of Wilkins and Franklin, but Wilkins took his relative anonymity compared to his friends in stride, knowing that very few significant scientific discoveries occurred at the hand of a single individual or group.

In 1959, Maurice Wilkins married Patricia Ann Chidgey, and they had four children: two daughters, Sarah and Emily, and two sons, George and William. Wilkins accepted a teaching position in the biophysics department of King's College, and he remained a member of its staff throughout his lifetime. He received the Albert Lasker Award, along with Crick and Watson, presented by the American Public Health Association in 1960, and he became a Companion of the Empire in 1962. In his last years, Wilkins wrote his autobiography, *The Third Man of the Double Helix*, and he continued to campaign against nuclear weapons. In his spare time, he enjoyed collecting sculptures and gardening. Maurice Wilkins died on October 5, 2004, only a few months after his friend Francis Crick passed away.

See also BIOCHEMISTRY; CRICK, FRANCIS; DNA REPLICATION AND REPAIR; FRANKLIN, ROSALIND; NUCLEIC ACIDS; PAULING, LINUS; WATSON, JAMES; X-RAY CRYSTALLOGRAPHY.

FURTHER READING

The Nobel Foundation. "The Nobel Prize in Physiology or Medicine 1962." Available online. URL: http://nobelprize.org/medicine/laureates/1962. Accessed July 25, 2008.

Wilkins, Maurice. *Maurice Wilkins: The Third Man of the Double Helix*. New York: Oxford University Press, 2005.

Wilkins, M. H. F., A. R. Stokes, and H. R. Wilson. "Molecular Structure of Deoxypentose Nucleic Acids." *Nature* 171 (1953): 738–740.

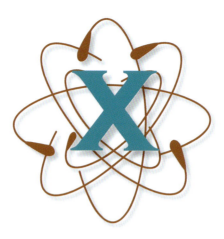

X-ray crystallography X-ray crystallography is a technique used to understand the material and molecular structure of a substance. Before X-ray crystallography, scientists studied crystals in terms of their macroscopic geometry. To decipher the arrangement of the atoms, crystallographers measured the angles of the crystal face relative to theoretical reference points (crystallographic axes) and established the symmetry of the crystal being studied. Crystallography now depends on the analysis of diffraction patterns (the way rays are scattered by the molecules they hit) that emerge from a closely packed, highly organized lattice, or crystal, targeted by a beam of short waves. Scientists most frequently use X-rays, a form of electromagnetic radiation, as the beam source, although neutrons and electrons (two kinds of subatomic particles possessing wavelike properties) are alternative sources.

X-ray crystallography is useful for studying a microscopic entity or molecule at the atomic level. Light microscopy permits the examination of gross features of microscopic objects by focusing light waves, or rays, of the visible light spectrum through a series of lenses to permit visualization of the object. At the atomic level, wavelengths of visible light are longer than not only the atoms but also the bonds linking the atoms to form a structure. X-rays have shorter wavelengths than the atoms being studied, making it possible to differentiate atomic positions relative to each other in a molecule. Because no lens exists that focuses X-rays, researchers require a cumulative effect of many atoms in similar configuration to detect the position and relationship of one atom to another, and atoms possess the same relative position and orientation throughout a crystal, giving rise to detectable signal.

Chemists frequently use X-ray crystallography to determine the structures of many types of molecules, including inorganic compounds, DNA, and proteins. Inorganic chemists study lattice structures of crystals, chemical formulas, bond lengths, and angles of complicated inorganic and organometallic compounds such as fullerenes and metalloporphyrins. Max Perutz and Sir John Kendrew elucidated the first protein structure, sperm whale myoglobin, using X-ray crystallography in 1958. Biochemists have determined another 30,000 protein structures of the 36,000 protein structures solved to date using this technique.

CRYSTALLIZATION

The first step in X-ray crystallography requires the creation of a crystal. A single molecule in solution cannot produce a detectable scattering pattern, but a crystal that contains infinite repeating arrays of a molecule generates a detectable diffraction pattern through the sum of multiple scattering events by atoms in the same relative position and orientation. Essentially, the collective effects of all the units that compose the crystal magnify the information, as the lens of a light microscope magnifies an image of a specimen. In general, smaller molecules crystallize more readily than larger molecules, and organic molecules, such as proteins, require gentler conditions (lower temperatures and less organic solvent) than inorganic molecules. To crystallize molecules, chemists frequently use microbatch and vapor diffusion techniques for inorganic and organic compounds. In both methods, crystallographers mix concentrated solutions of the protein with various solutions, typically consisting of a buffer to control the pH, a precipitating agent (polyethylene glycols, salts such as ammonium sulfate, or organic alcohols) to reduce its solubility, and other salts or additives, such as detergents or cofactors. Solutions containing the molecule of interest then concentrate slowly over days, weeks, or months.

In a specific type of vapor diffusion called hanging drop vapor diffusion, a chemist applies a small drop of solution containing the concentrated molecule, buffer, and precipitant to a glass coverslip. After suspending the inverted sample over a large reservoir containing a hygroscopic solution (a solution containing a substance that readily absorbs water from its surroundings, such as ethanol, methanol, glycerin, concentrated sulfuric acid, or concentrated sodium hydroxide), the investigator seals the chamber. As the water in the suspended drop evaporates to go to the solution in the reservoir, the molecule concentration slowly concentrates until small crystals (crystal nuclei) form, leading to large-scale crystal growth. If the concentration of the molecule of interest increases too quickly, molecules precipitate, or fall out of solution, forming a useless and amorphous mass that cannot be used. Typically, crystallographers screen hundreds or even thousands of conditions before a crystal of suitable quality forms. Crystallographers coax increasingly larger molecules to form crystals. For example, biochemists successfully crystallized the 8,500-kilodalton polio virus and solved its structure using X-ray crystallography. Yet, crystals do not always form. Impurities in the sample, fluctuations in the temperature, or multiple stable conformations of the molecules sometimes prevent crystal formation. A large number of biomolecules have yet to be crystallized for structural studies of high atomic resolution.

X-RAY DIFFRACTION

Two basic physical principles underlie the technique of X-ray crystallography. The first principle is that electrons scatter X-rays. The amplitude of the wave scattered by an atom is proportional to the number of electrons it contains; thus, an oxygen atom with eight electrons scatters X-rays eight times more strongly than a hydrogen atom, with one electron. Second, scattered waves recombine as a wave with the same frequency radiates from each atom hit by an X-ray. This phenomenon is referred to as Bragg's law because it was first described by the physicists Sir William Henry Bragg and his son, Sir William Lawrence Bragg, in 1912, after they performed experiments studying diffraction patterns of X-rays and neutrons. According to Bragg's law, each atom contributes to each scattered beam. If the scattered waves are in phase with one another (meaning the crests and troughs fall at the same points), the waves reinforce one another, intensifying the amount of

diffraction seen at a particular point (constructive interference). If the scattered waves are out of phase with one another (meaning the crests and troughs do not coincide), the waves cancel one another out, and no diffraction pattern is detectable at that point (destructive interference). The resulting wave interference patterns, both constructive and destructive, become the basis of diffraction analysis.

Any crystallographic study requires a source of illumination, a crystal, and a detector. X-rays are the most common form of illumination, but electrons and neutrons can also be used as illumination sources. The three types of radiation interact with the crystal specimen in different ways. X-rays disrupt the electron clouds surrounding atoms, or, in other words, they interact with the spatial distribution of the valence electrons. Because electrons are charged particles, both the atomic nuclei and the surrounding electrons cause the bombarding particles to scatter. Strong nuclear forces and magnetic fields within the atoms scatter neutrons that are used as illumination sources. When using X-rays as the source of illumination, a stationary anode source, a rotating anode source, or a synchrotron generates a narrow beam of X-rays of wavelength 1.54 nanometers (10^{-9} m). When the X-rays strike the crystal, part of the particles go straight through the crystal, and the interaction of the rays with the electron clouds scatters them in various directions. A piece of photographic film records the diffraction pattern as a blackening of the emulsion that is proportional to the number of scattered X-rays hitting that portion of the film. Crystallographers also use solid state electronic detectors, such as charge-coupled device detectors, to record diffraction patterns.

To perform an X-ray diffraction experiment, the crystallographer mounts the protein crystal in a capillary tube (part of a piece of equipment called the diffractometer) and positions the capillary tube in a precise orientation with respect to the X-ray source and the detector. The diffractometer moves the crystal in a defined path in front of the detector, or a piece of film, to collect a regular array of spots called reflections. Computers measure the intensity (darkness) of each spot, representing the degree of constructive and destructive interference at that location in the diffraction pattern. Crystallographers use these collected intensity measurements to construct an image of the molecule of interest. In light microscopy and electron microscopy, lenses focus the diffracted beams emanating from a sample to form an image directly, but no lens exists that focuses X-rays. To produce an image from a diffraction pattern, analyses require sophisticated mathematics and ongoing processes of modeling and refinement. For each spot, a complicated mathematical calculation called a Fourier transform determines a wave, or area, of electron density, whose amplitude is directly proportional to the square root of the measured intensity of the spot. When summed together, these mathematical relations help create an image. To determine whether the phases of these waves are constructive or destructive, crystallographers use diffraction patterns made by heavy atom references, such as uranium or mercury, placed at specific sites in the molecule being studied.

After the mathematical calculations, researchers generate electron density maps that give the intensity of electrons at different places within the crystal. A series of parallel sections stacked upon each other represents the three-dimensional electron density distribution. Contour lines, like the contour lines used by geological survey maps to show altitude changes, depict the electron density differences in each section. The limiting factor in these analyses is the resolution of the diffraction pattern. The purity of the sample, the quality of the crystal, and instrument limitations all affect resolution. With crystallography, higher resolution permits the distinction of two atoms from each other at very small distances between them. For the best structural studies, the level of resolution needs to allow groups of atoms to be delineated from each other at distances less than 4.0 nm, and individual atoms to be delineated from each other at distances less than 2.0 nm. From the electron density maps, crystallographers compare the experimental data to mathematically predicted electron density maps of diffraction patterns of a hypothesized structure of the molecule under study. These hypothetical models are created by series of educated guesses by the crystallographers that converge after multiple modeling and refinement steps to a point where the predicted patterns and observed patterns closely resemble one another. Using X-ray diffraction studies, biochemists have elucidated the structures of more than 30,000 proteins, providing insight into recognition and binding properties, function, and folding patterns of these molecules.

The entire process of data collection, the creation of electron density maps, fitting into an atomic model, and refinement steps, yields the best results using crystals because the mathematical methods for analysis apply specifically to wave patterns generated by orderly arrays. Crystallographers have gleaned a certain amount of information from diffraction studies using fibers rather than crystals. Fibers exhibit a degree of order that is sufficient to deduce the structure of simple molecules and that permits the elucidation of coarse features of more complicated molecules. Notably, James Watson and Francis Crick used X-ray diffraction studies of fibers, along with other experimental results, to determine the double-helical structure of deoxyribonucleic acid (DNA) in 1953.

See also BIOCHEMISTRY; CRICK, FRANCIS; FRANKLIN, ROSALIND; HODGKIN, DOROTHY CROWFOOT; INORGANIC CHEMISTRY; MATERIALS SCIENCE; NUCLEIC ACIDS; PAULING, LINUS; PERUTZ, MAX; PROTEINS; WATSON, JAMES; WILKINS, MAURICE.

FURTHER READING

Ladd, Mark F. C., and Rex A Palmer. *Structure Determination by X-Ray Crystallography*. New York: Springer, 2003.

APPENDIX I
CHRONOLOGY

ca. 5000 B.C.E. Egyptians develop the balance and a standard unit of weight

ca. 3000 B.C.E. Egyptians develop a standard unit of length

ca. 1450 B.C.E. Egyptians develop the water clock

ca. 550 B.C.E. Pythagoras, in Greece, studies acoustics, relating the pitch of a tone to the length of the wind instrument or of the string producing it

ca. 400 B.C.E. Democritus, in Greece, states that all matter is made up of "atoms" and empty space

ca. 370 B.C.E. Aristotle, in Greece, describes free fall but incorrectly claims that heavier bodies fall faster than lighter ones

ca. 270 B.C.E. Ctesibius of Alexandria, Egypt, invents an accurate water clock, in use until the Renaissance

ca. 260 B.C.E. Archimedes, in Greece, studies floating bodies and states his principle of buoyancy. He also states the law of the lever

ca. 60 B.C.E. Lucretius, in Greece, proposes the atomic nature of matter

ca. 62 C.E. Hero of Alexandria, Egypt, studies air pressure and vacuum

ca. 700 I Hsing, in China, develops a mechanical clock

ca. 1000 Abu Ali al-Hasan ibn al-Haytham (Alhazen), in Spain, studies optics and vision

ca. 1100 Abu L-Fath Abd al-Rahman al-Khazini, in Persia, proposes that gravity acts toward the center of Earth

ca. 1200 Jordanus de Nemore, in Germany, studies motion and explains the lever

1276 Roger Bacon, in England, proposes using lenses to correct vision

1284 Witelo, in Poland, describes reflection and refraction of light

1300s Italian craftsmen start building mechanical clocks

1305 Dietrich von Freiberg, in Germany, describes and explains rainbows

1546 Niccolò Tartaglia, in Italy, studies projectile motion and describes the trajectory of a bullet

1582 Galileo Galilei, in Italy, describes the motion of a pendulum, noticing that its period is constant and, for small amplitudes, independent of amplitude

1583–86 Flemish scientist Simon Stevin investigates hydrostatics and free fall

1590–91 Galileo investigates falling bodies and free fall

1600s Galileo develops the principle of inertia

1600 English scientist William Gilbert studies magnetism and its relation to electricity and describes Earth as a magnet

1609 Dutch lens maker Hans Lippershey invents the telescope. German astronomer Johannes Kepler presents his first and second laws of planetary motion

1622 Willebrord Snell presents his law of refraction of light

1636 French scholar Marin Mersenne presents his work on acoustics, including the calculation of the frequency of audible tones

1636 French mathematician René Descartes advances understanding of rainbows

1638 Galileo studies motion and friction

1640–44 Evangelista Torricelli researches hydrostatics and hydrodynamics

1657 Pierre Fermat proposes his principle of least time in optics

1660–62 Robert Boyle studies gases

1666–1704 Sir Isaac Newton actively studies a wide range of physics phenomena

1705 Edmund Halley predicts the return of the comet named for him

1714 Gottfreid Leibniz proposes the conservation of energy

1736–65 Leonhard Euler studies theoretical mechanics using differential equations

1738 Daniel Bernoulli investigates the theories of gases and of hydrodynamics

1743–44 Jean d'Alembert studies energy in Newtonian mechanics and proposes a theory of fluid dynamics

1751 American scientist Benjamin Franklin discovers that electricity can produce magnetism

1754 Joseph Black discovers "fixed air" (carbon dioxide)

1771 Joseph Priestley discovers that plants use carbon dioxide and create oxygen.

1772–88 French mathematician and physicist Joseph Lagrange investigates theoretical mechanics and proposes a new formulation of Newtonian mechanics

1776–84 French mathematician and physicist Pierre Laplace applies mathematical methods to theoretical physics, particularly to mechanics and electricity

1785 French physicist Charles Augustin de Coulomb proposes his law of electrostatics. French chemist Antoine Laurent Lavoisier develops the law of conservation of mass and names oxygen

1790 The French Academy of Sciences establishes the metric system of measurement

1797 British physicists Benjamin Thompson and Benjamin Rumford study the heat generated by work. French chemist Joseph Proust develops the law of definite proportions

1800 Italian physicist Alessandro Volta develops the electric battery

1801 English physicist Thomas Young demonstrates that light is a wave phenomenon

1802 French chemist Joseph Louis Gay-Lussac develops his gas law relating pressure and temperature

1802–05 British chemist John Dalton develops his atomic theory

1808 French scientist Étienne-Louis Malus discovers and investigates polarized light

1809 English scientist Sir George Cayley publishes his theoretical studies of aerodynamics, laying the foundation for flight

1811 Italian scientist Amedeo Avogadro describes gases in molecular terms

1814 German physicist Joseph von Fraunhofer discovers and investigates optical spectra. French physicist Augustin-Jean Fresnel explains light polarization in terms of light's wave nature

1817 German chemist Johann Dobereiner develops triads that demonstrate that the atomic mass of the second element is halfway between those of the first and third elements

1819 French chemists Pierre-Louis Dulong and Alexis Thérèse Petit show that the specific heat of an element is inversely proportional to its atomic weight

1820 Danish physicist Hans Christian Ørsted and French physicist André-Marie Ampère show that an electric current has a magnetic effect

1821 English physicist Michael Faraday, studying electromagnetism, introduces the concept of magnetic field

1822 German physicist Thomas Seebeck shows that an electric current can result from a temperature difference. French scientist Jean-Baptiste-Joseph Fourier publishes his work on heat, in which he makes use of calculus

1824 French engineer Nicolas-Léonard-Sadi Carnot publishes his analysis of heat engines, which leads to the laws of thermodynamics

1827 German physicist Georg Simon Ohm shows the proportionality of electric current and voltage, known as Ohm's law. English botanist Robert Brown discovers the motion of pollen grains suspended in a liquid, called Brownian motion

1831 English physicist Michael Faraday discovers electromagnetic induction, that magnetism can produce electricity. American physicist Joseph Henry invents the electric motor

1832 Michael Faraday studies and describes electrolysis

1840 English physicist James Prescott Joule develops the law of conservation of energy

1842 Austrian scientist Christian Doppler explains the dependence of the observed frequency on the motion of the source or observer, called the Doppler effect

1847 German scientist Hermann von Helmholtz expresses the conservation of energy in mathematical terms, the first law of thermodynamics

1848 English scientist Sir William Thomson (Lord Kelvin) describes absolute zero temperature

1850 German physicist Rudolf Clausius introduces the concept of entropy, which leads to the second law of thermodynamics. French physicist Jean-Bernard-Léon Foucault measures the speed of light in air, and later in water

1851 Foucault uses a huge pendulum, now known as a Foucault pendulum, to demonstrate the rotation of planet Earth

1854 German mathematician Georg Riemann describes the geometry of curved spaces, applied later by modern physicists to relativity and other problems

1856 Louis Pasteur describes fermentation by microorganisms

1859 Scottish physicist James Clerk Maxwell develops the kinetic theory of gases, based on a statistical treatment of the gas particles

1862 German physicist Gustav Robert Kirchhoff introduces the concept of a blackbody, whose later study

leads to the development of quantum mechanics

1863 English chemist John Newlands arranges the chemical elements in order of their increasing atomic masses, develops the law of octaves

1869 Russian physicist Dmitry Mendeleyev develops the periodic table based on atomic mass

1871 English physicist John William Strutt (Lord Rayleigh) mathematically relates the amount of scattering of light from particles, such as molecules, to the wavelength of the light, thus explaining the color of the sky

1873 Scottish physicist James Clerk Maxwell presents his set of equations, known as Maxwell's equations, which form a theoretical framework for electromagnetism and predict the existence of electromagnetic waves. Dutch physicist Johannes van der Waals modifies the ideal gas equation to take into account weak attractive intermolecular forces, called van der Waals forces

1874 Dutch chemist Jacobus van't Hoff works on the three-dimensional structure of carbon compounds

1877 English physicist John William Strutt (Lord Rayleigh) publishes his extensive work on acoustics, the science of sound waves

1879 Austrian physicist Josef Stefan shows experimentally that the rate of energy radiation from a body is proportional to the fourth power of the body's absolute temperature. English chemist William Crookes develops the cathode-ray tube

1883 Austrian physicist Ludwig Boltzmann explains Stefan's result as a property of blackbodies, known as the Stefan-Boltzmann law. This lays the foundation for the development of quantum mechanics

1887 German physicist Heinrich Rudolf Hertz observes the photoelectric effect, the emission of electrons from a metal that is irradiated with light, and discovers radio waves. American scientists Albert Abraham Michelson and Edward Williams Morley attempt to measure the motion of Earth in the ether, a proposed medium for the propagation of electromagnetic waves, with a negative result, which leads to the special theory of relativity in the 20th century

1893   German physicist Wilhelm Wien shows experimentally that the wavelength at which a blackbody radiates at maximal intensity is inversely proportional to the absolute temperature, known as Wien's displacement law

1895   German physicist Wilhelm Conrad Röntgen discovers X-rays

1896   French physicist Henri Becquerel discovers radioactivity in uranium

1897   British physicist Sir J. J. Thomson discovers electrons and develops his "plum pudding model" of the atom, where the electrons are imbedded in a positively charged sphere

1898   French chemist Marie Curie and French physicist Pierre Curie isolate radium, which is highly radioactive

1900   German physicist Max Planck proposes absorption and emission of radiation in discrete amounts, which introduces the quantum concept

1904   Sir Ernest Rutherford performs the gold foil experiment that demonstrates the existence of the atomic nucleus

1905   German-Swiss, later American, physicist Albert Einstein publishes his special theory of relativity. To explain the photoelectric effect, Einstein proposes that light consists of photons

1909   Belgian chemist L. H. Baekeland produces first plastic, known as Bakelite

1911   Dutch physicist Heike Kamerlingh Onnes discovers superconductivity. New Zealand–born British physicist Sir Ernest Rutherford discovers the structure of the atom. American physicist Robert A. Millikan uses his "oil drop method" to determine the charge of an electron

1911–12   Using X-ray methods, British physicists William Henry Bragg and William Lawrence Bragg and German physicist Max von Laue discover the atomic structure of crystals

1913   Danish physicist Niels Bohr proposes a "planetary model" of the hydrogen atom to explain the hydrogen spectrum, in which the electrons rotate around the nucleus in orbits as planets do the Sun. British physicist Henry Moseley rearranges the periodic table by atomic number

1915   Albert Einstein proposes his general theory of relativity

1916   American physicist Robert Millikan measures the value of the Planck constant, which characterizes all quantum phenomena

1919   English astronomer and physicist Arthur Eddington leads an expedition that measures the bending of starlight passing near the Sun during a total solar eclipse, which confirms the general theory of relativity

1922   Soviet mathematician and meteorologist Aleksandr Friedmann shows that the general theory of relativity predicts the universe is expanding

1923   French physicist Louis de Broglie proposes that matter possess wavelike properties. American physicist Arthur Holly Compton demonstrates, through the Compton effect, that electromagnetic radiation consists of photons

1925   German physicist Werner Heisenberg invents a matrix formulation of quantum mechanics

1926   Austrian physicist Erwin Schrödinger publishes his formulation of quantum mechanics in the form of the Schrödinger equation

1927   Werner Heisenberg proposes his uncertainty principle. American physicists Clinton Davisson and Lester Germer show that electrons possess wavelike properties. Leading to the big bang theory, Belgian astronomer Georges Lemaître states that the universe began its expansion from a tiny, hot state

1928   British physicist Paul Dirac derives the Dirac equation, which predicts the existence of antiparticles

1929   American astronomer Edwin Hubble demonstrates that all galaxies are receding from each other, indicating the expansion of the universe

1932 Indian-born American astrophysicist Subrahmanyan Chandrasekhar proposes that when a sufficiently massive star reaches the end of its life, it will collapse to a black hole. American physicists Ernest O. Lawrence and M. Stanley Livingston invent the cyclotron, a particle accelerator for investigating nuclei and elementary particles. American physicist Carl D. Anderson discovers the positron, the electron's antiparticle. British physicist James Chadwick discovers the neutron

1933 Swiss astronomer Fritz Zwicky studies the rotation of galaxies and shows that they must contain more mass than is visible, introducing the idea of dark matter

1934 French physicists Irène Joliot-Curie and Frédéric Joliot-Curie produce the first artificial radioactive isotopes

1935 Japanese physicist Hideki Yukawa proposes theory of the nuclear force, binding protons and neutrons into nuclei, which predicts the existence of mesons

1937 American physicists Carl D. Anderson and Seth Neddermeyer discover the muon in cosmic rays. English chemist Dorothy Crowfoot Hodkin finds the three-dimensional structure of cholesterol

1938 Soviet physicist Pyotr Kapitsa discovers that liquid helium exhibits superfluidity near 0 K. American physicist Hans Bethe explains the source of energy production in stars as nuclear fusion reactions

1938–39 Austrian physicists Lise Meitner and Otto Frisch explain that the German chemists Otto Hahn and Fritz Strassmann achieved nuclear fission by bombarding uranium with neutrons

1942 Using an electron microscope, based on the wavelike properties of electrons, Italian-American biologist Salvador Edward Luria images a 10^{-7}-m virus. A team lead by Italian-American physicist Enrico Fermi produces the first controlled nuclear fission chain reaction. The United States initiates the Manhattan Project to construct a nuclear fission (atomic) bomb

1943 As part of the Manhattan Project, the Los Alamos laboratory is built in New Mexico, directed by American physicist Robert Oppenheimer

1946 American chemist Willard Frank Libby invents the carbon 14 dating technique for determining when living organisms died. The first programmable digital computer, the ENIAC (Electronic Numerical Integrator and Comparator), starts operation

1947 A team directed by British astronomer Bernard Lovell completes construction of the first radio telescope. British physicist Cecil Frank Powell discovers the pion, predicted by Yukawa in 1935. American physicists John Bardeen, William Shockley, and Walter Brattain invent the transistor, a semiconductor device

1948 American physicists Richard Feynman and Julian Schwinger and Japanese physicist Sin-Itiro Tomonaga develop quantum electrodynamics (QED). The 200-inch reflecting telescope on Mt. Palomar in California commences operation

1949 German-American physicist Maria Goeppert Mayer and German physicist Hans Jensen model the nucleus as consisting of shells of protons and neutrons

1950 Swedish astrophysicist Hannes Alfvén reaches an understanding of the physics of plasmas (ionized gases), with relevance to space science and, later, to nuclear fusion. The United States Congress creates the National Science Foundation for the funding of basic research and science education

1951–52 American physicists Harold Ewen and Edward Mills Purcell observe the 21-cm radio signal from hydrogen atoms in space

1952–53 American physicist Charles H. Townes and, independently, Soviet physicists Alexander Mikhailovich Prokhorov and Nikolai Gennadiyevich Basov suggest that stimulated emission from molecules can be utilized to create intense, coherent microwave beams. Townes constructs the first maser (microwave amplification by stimulated emission of radiation), a forerunner of the laser (light amplification by stimulated emission of radiation)

1953 British scientists Francis Watson and James Crick determine the three-dimensional structure of deoxyribonucleic acid (DNA)

1954 The photovoltaic cell, a semiconductor device for generating an electric voltage from sunlight, is invented in the laboratories of the Bell Telephone Company

1956 American physicists Clyde Cowan, Frederick Reines, F. B. Harrison, H. W. Kruse, and A. D. McGuire discover the electron neutrino

1956–57 Chinese-American physicist Chien-Shiung Wu experimentally confirms the proposal by the Chinese-American physicists Tsung-Dao Lee and Chen Ning Yang that the weak force might not obey reflection symmetry

1957 American physicists John Bardeen, Leon Cooper, and Robert Schrieffer propose an explanation for superconductivity in terms of conduction by electron pairs. Bayer and General Electric develop polycarbonate plastics

1958 Japanese physicist Leo Esaki invents the tunnel diode, which exploits quantum tunneling

1959 Israeli physicist Yakir Aharonov and American physicist David Bohm predict that magnetic fields can affect particles in an observable, nonclassical way, when the particles do not pass through the field. Austrian molecular biologist Max Perutz determines the structure of hemoglobin

1960 The Aharonov-Bohm effect is observed. American physicist Theodore Maiman constructs the first laser, from a ruby crystal

1962 American physicists Leon M. Lederman, Melvin Schwartz, and Jack Steinberger discover the muon neutrino. British physicist Brian Josephson shows that electron pairs in superconductors can tunnel from one superconductor to another, an effect that underlies sensitive magnetic probes. The semiconductor laser is invented, allowing efficient production of light from electricity

1963 Dutch-American astronomer Maarten Schmidt discovers the first quasar (QUASi-stellAR radio source), a very distant object that appears similar to a star but radiates more than some galaxies

1964 American physicists Murray Gell-Mann and George Zweig independently propose the existence of quarks as components of protons, neutrons, and other hadrons

1965 American physicists Arno Penzias and Robert Wilson discover the cosmic microwave background

1967 American physicists Steven Weinberg and Sheldon Glashow and Pakistani physicist Abdus Salam independently propose unifying the electromagnetic and weak forces to a single, electroweak force

1967–68 British astronomers Jocelyn Bell and Anthony Hewish discover that certain stars, called pulsars, emit periodic radio pulses. American astrophysicist Thomas Gold explains pulsars as rotating neutron stars

1969 A group of American physicists, including Jerome I. Friedman, Henry Kendall, and Richard E. Taylor, discover experimental evidence for the existence of quarks inside protons

1970 Swiss company Hoffman-LaRoche files patent for liquid crystal displays

1970–73 Physicists develop the "standard model" of elementary particles, which includes the strong and electroweak forces

1972 American physicists Douglas Osheroff, Robert Richardson, and David Lee demonstrate superfluidity in helium 3 at very close to 0 K

1974 English physicist Stephen Hawking proposes that black holes can radiate particles and eventually evaporate. American physicist Kenneth Wilson develops a mathematical method, called renormalization group technique, for dealing with certain phase changes such as in crystals. American physicists Burton Richter and Samuel C. C. Ting and their groups independently discover the charm quark. General Motors introduces the catalytic converter

1974–77 American physicist Martin L. Perl and colleagues discover the tau particle

1975–77 Polish-American mathematician Benoit B. Mandelbrot introduces the concept of fractals, patterns in systems that are similar to themselves at all scales

1977 American physicist Leon Lederman and colleagues discover the bottom quark

1978 American astronomer Vera Rubin and others conclude from an analysis of the rotation of galaxies that the gravity from the visible stars is insufficient to prevent them from flying apart and that they must contain invisible matter, called dark matter

1979 French physicist Pierre-Gelles de Gennes publishes his work on the theories of polymers and liquid crystals

1980 German physicist Klaus von Klitzing discovers the quantum Hall effect, that the voltage that a current-carrying semiconductor at low temperatures in a magnetic field develops perpendicular to the magnetic field varies in discrete steps, an effect used to develop precise standards for electric properties. American physicist Alan Guth proposes adding inflation—a very brief period of extremely rapid expansion of the universe—to the big bang theory, in order to explain observations better

1981 German physicist Gerd Binnig and Swiss physicist Heinrich Rohrer invent the scanning tunneling microscope, which can image surfaces to the detail of individual atoms

1982 Chinese-American physicist Daniel Tsui and German physicist Horst Störmer discover the fractional quantum Hall effect, that electrons appear to carry a fraction of their normal charge, which is explained by the American physicist Robert B. Laughlin. French physicist Alain Aspect and collaborators report on their confirmation of quantum entanglement by performing the EPR experiment

1983 A team led by the Italian physicist Carlo Rubbia discover the W and Z bosons, the carriers of the weak force. American molecular biologist Kary Mullis develops the process of polymerase chain reaction (PCR)

1984 British geneticist Alec Jeffreys describes the technique of genetic fingerprinting

1985 English chemist Sir Harry Kroto and American chemists Richard Smalley and Bob Curl discover the structure of the buckminsterfullerene molecule

1986 Swiss physicist Karl Alexander Müller and German physicist Johannes Georg Bednorz discover high-temperature superconductors, materials that become superconducting at temperatures much higher than were previously known

1989 American astronomers Margaret Geller and John Huchra discover that the galaxies in the universe are located on thin sheets surrounding great voids that are empty of galaxies

1989–92 The U.S. National Aeronautics and Space Administration (NASA) launches the *Cosmic Background Explorer* (COBE) satellite, which maps the radiation from the sky in all directions, the cosmic microwave background, a remnant from the big bang. The cosmic microwave background is found to be very uniform and to correspond to the radiation of a blackbody at a temperature of 2.725 K. Tiny angular fluctuations in the radiation's generally uniform distribution are detected, indicating some nonuniformity in the universe at a very early age

1990 NASA launches the *Hubble* Telescope as a satellite above Earth's atmosphere to study the universe at high resolution

1993 The U.S. Air Force completes the Global Positioning System (GPS), allowing users on Earth to locate themselves and navigate

1995 American physicists Eric Cornell and Carl Wieman produce a Bose-Einstein condensate of 2,000 atoms at a temperature lower than 10^{-6} K, thus confirming a prediction of Bose-Einstein statistics. American geophysicists Xiaodong Song and Paul Richards demonstrate that Earth's solid inner core, with a diameter of 1,500 miles (2,400 km), rotates a little faster than the rest of the planet. The top quark is discovered by a group at Fermilab. Antihydrogen atoms, consisting of an antiproton and a positron, are created at the European Organization for Nuclear Research (CERN)

1997 Swiss physicist Nicolas Gisin and collaborators confirm quantum entanglement by performing the EPR experiment over a distance of almost seven miles (11 km). Matter is created from pure radiation in the production of electron-positron pairs by photons colliding with photons

1998 The Super Kamiokande experiment demonstrates that neutrinos undergo change of type ("neutrino oscillation") while in flight. The decay of the neutral kaon is shown to violate time-reversal symmetry. Observations of supernovae indicate that the expansion of the universe is accelerating

2000 A collaboration at Fermilab announces the detection of the tau neutrino

2003 Human Genome Project, a collaborative group of scientists from many nations, completes the sequence of the human genome

2006 U.S. president George W. Bush announces the Advanced Energy Initiative (AEI) to increase research on technology to reduce oil use for transportation, including hybrid vehicle batteries, ethanol, and hydrogen fuel cell vehicles and fueling stations. AEI also supports research into electricity production from clean coal, wind, and solar power

2008 The U.S. Senate introduces the Green Chemistry Research and Development Act of 2008 for the advancement of research into environmentally friendly chemicals by the National Science Foundation (NSF), National Institute of Standards and Technology (NIST), Environmental Protection Agency (EPA), and Department of Energy

APPENDIX II
GLOSSARY

abacus the earliest known calculating device, an abacus uses beads and strings to keep track of amounts

aberration any of various kinds of defects in lenses and mirrors that cause distortion of images

absolute of a measurement result for a physical quantity, observer independent; of temperature, expressed on a scale that is based on absolute zero temperature, the lowest conceivable temperature

absolute pressure the actual pressure, in contrast to gauge pressure

absolute temperature temperature expressed on the Kelvin scale in kelvins (K), in which absolute zero temperature has the value 0 and a degree has the same magnitude as on the Celsius scale, which is one one-hundredth of the difference between the boiling and freezing points of water

absolute zero the lowest conceivable temperature, 0 K, which can be approached but cannot be attained, and at which the random kinetic energy of the atoms, molecules, and so on, is at its minimum

absorption spectrometry an analytical chemistry technique that utilizes a spectrophotometer that passes a beam of light through a sample and detects the amount of light that is either absorbed by or transmitted through the sample

AC alternating current

acceleration the time rate of change of velocity, often denoted **a** (as a vector) or a (as its magnitude), whose SI unit is meter per second per second (m/s^2); in more restricted use, an increase of speed

accelerator a device for accelerating subatomic particles, ions, or nuclei to high kinetic energies

acid dissociation constant (K_a) the equilibrium constant for the dissociation of a weak acid

acid rain rain that has a lower pH than ordinary rainfall

acoustics the branch of physics that deals with sound and its production, propagation, and detection

action force together with a reaction force, one of a pair of forces between two bodies, each

force acting on a different body, that have equal magnitudes and opposite directions, according to Newton's third law of motion

adenosine triphosphate (ATP) commonly used cellular energy-carrying molecule composed of one unit of ribose, adenosine (a purine nucleotide), and three phosphate groups

adhesion the attraction between different materials

agrochemistry the branch of chemistry that studies the scientific optimization of agricultural products

air pollution the presence of unwanted particles and gases in the air that can adversely affect human health and the environment

alcohol fermentation an alternative form of fermentation that creates alcohol, usually ethanol, as an end point

aliphatic hydrocarbons carbon and hydrogen compounds that are straight-chained or branched compounds

alkali metals (group IA) the first group of representative elements, including lithium, sodium, potassium, rubidium, cesium, and francium

alkaline earth metals (group IIA) a group of representative elements, including beryllium, magnesium, calcium, strontium, barium, and radium

alkanes hydrocarbons that contain only single bonds

alkenes hydrocarbons that contain at least one double bond

alkynes hydrocarbons that contain at least one triple bond

alloy a fine, homogeneous mixture of substances, at least one of which is a metal, that has metallic properties; metals made up of more than one type of metallic element

alpha decay a mode of radioactive decay in which an alpha particle, or helium nucleus, is emitted; in alpha decay, the atomic number and mass number of the nucleus decrease by 2 and 4, respectively

alpha particle a helium nucleus

alpha ray a helium nucleus, especially as the product of radioactivity

alternating current (AC) the situation when the voltages and currents in an electric circuit have a sinusoidal dependence on time

AM amplitude modulation

amino acid the building block of proteins

ampere one of the SI base units, the SI unit of electric current, denoted A; the constant electric current that would produce a force of exactly 2×10^{-7} newton per meter length, if it flowed in each of two thin straight parallel conductors of infinite length, separated by one meter in vacuum

amplitude the maximal displacement from the center of oscillation

amplitude modulation (AM) the encoding of a transmitted signal in a carrier wave by varying the wave's amplitude

anabolism the use of cellular energy to build compounds

analysis of variance (ANOVA) a statistical test for mean equality between two or more groups

analytical chemistry the branch of chemistry that involves the study of the composition of substances

angular acceleration the time rate of change of the angular speed, often denoted α, whose SI unit is rad per second per second (rad/s^2)

angular momentum the rotational analog of linear momentum, a vector quantity expressed with respect to a point, often denoted L (as a vector) or L (as its magnitude), whose SI unit is kilogram-meter^2 per second (kg·m^2/s); the vector product of the position vector of a body with respect to some reference point and the body's momentum; also, the product of a body's moment of inertia and its angular velocity; the magnitude of angular momentum is the product of a body's moment of inertia and its angular speed

angular speed the magnitude of the angular velocity, often denoted ω, whose SI unit is radian per second (rad/s)

angular velocity the time rate of change of orientation (angle), often denoted ω (as a vector) or Ω (as its magnitude), whose SI unit is radian per second (rad/s)

anions negatively charged ions formed by nonmetals that take electrons from other atoms, causing them to have one or more additional electrons relative to the number of protons in the nucleus

anisotropic possessing different properties in different directions

anisotropy the state of possessing different properties in different directions

annihilation the mutual destruction of a particle and its antiparticle, resulting in pure energy in the form of two photons (gamma rays); the destruction of a photon and its conversion into a particle-antiparticle pair

anode the terminal, or electrode, of a pair of terminals or electrodes that is at the higher electric potential; the positive terminal or electrode of such a pair

anthracite coal the highest-quality and cleanest-burning variety of coal

antimatter ensembles of antiparticles

antinode any location along a standing wave at which the medium—or the electromagnetic field, for a standing electromagnetic wave—oscillates at maximal amplitude relative to nearby locations

antiparticle a particle that has the same mass, spin, and magnitude of electric charge as a matter particle, but opposite sign of electric charge

antipyretic fever-reducing substance

aqueous solution a homogeneous mixture with water as the solvent

Archimedes' principle states that a body immersed in a fluid, whether wholly or partially immersed, is buoyed up by a force whose magnitude equals that of the weight of the fluid that the body displaces

area the amount of surface, often denoted A, whose SI unit is square meter (m^2)

aromatic hydrocarbons carbon- and hydrogen-containing compounds with ring structures

Arrhenius acid a substance that adds a proton to water to produce H_3O^+

Arrhenius base a substance that produces OH^- in solution

aspirin acetylsalicylic acid, a member of a class of pain relievers known as nonsteroidal antiinflammatory drugs (NSAIDs)

asymmetry lack of symmetry

atmosphere a non-SI unit of pressure, denoted atm, equal to 1.013×10^5 pascals (Pa)

atmospheric and environmental chemistry the branch of chemistry that studies the impact of chemical substances on the environment

atom the smallest particle that retains the chemical properties of an element; consists of a compact nucleus, containing protons and neutrons, surrounded by electrons

atom bomb (atomic bomb) more aptly called fission bomb or nuclear-fission bomb, a bomb based on the energy released in an uncontrolled nuclear-fission reaction

atomic mass number also mass number, the number of nucleons (i.e., protons and neutrons) in a nucleus

atomic number the number of protons in a nucleus, giving the electric charge of the nucleus, and also the number of electrons in a neutral atom; the atomic number defines the chemical element

atomic radius   half the distance between the nuclei of two covalently bonded atoms of an element or two atoms in contact

atomic weight   the average mass of an atom in a representative sample of an element

Aufbau principle   states that electron orbitals are filled from the lowest- to the highest-energy states

average acceleration   the change of velocity divided by the elapsed time, often denoted $\mathbf{a}_{av}$ (as a vector) or a_{av} (as its magnitude), whose SI unit is meter per second per second (m/s^2)

average angular acceleration   the change of angular speed divided by the elapsed time, often denoted α_{av}, whose SI unit is radian per second per second (rad/s^2)

average angular speed   the angular displacement divided by the elapsed time, often denoted ω_{av}, whose SI unit is radian per second (rad/s)

average power   the work performed or energy transferred or converted divided by the elapsed time, often denoted P_{av}, whose SI unit is watt (W)

average speed   the total distance covered divided by the elapsed time, often denoted v_{av}, whose SI unit is meter per second (m/s); *not* the magnitude of the average velocity

average velocity   the net displacement divided by the elapsed time, often denoted $\mathbf{v}_{av}$, whose SI unit is meter per second (m/s)

Avogadro's law   the proportionality between the quantity (number of moles or molecules) of an ideal gas and its volume, at constant pressure and temperature

Avogadro's number   the number of particles in one mole of any substance, commonly denoted N_A, whose value is 6.0221415×10^{23} mol^{-1}

backbone model   structural representation of molecules that shows only the backbone structure of the molecule

ball-and-stick model   structural representation of molecules that depicts the bonding attachments between the atoms as well as the three-dimensional arrangements of the atoms in a molecule

bar   non-SI unit of pressure, denoted bar, equals 1.000×10^5 pascals (Pa), close to one atmosphere (atm)

barometer   any device for measuring atmospheric pressure

baryon   any of the set of relatively massive fermions of which the proton and the neutron are the lightest members

base dissociation constant (K_b)   measure of the equilibrium constant of a base

base quantities   in the International System of Units (SI), the seven physical quantities that are used to define the seven base units: length, mass, time, electric current, temperature, amount of substance, and luminous intensity

base units   in the International System of Units (SI), the seven units of the seven base quantities from which all other units are derived: meter, kilogram, second, ampere, kelvin, mole, and candela

battery   any device for storing chemical energy and allowing its release in the form of an electric current

BCS theory   named for its creators, John Bardeen, Leon Cooper, and Robert Schrieffer, a successful theory of superconductivity for superconductors of types I and II that explains the effect as conduction by paired electrons, called Cooper pairs, whose existence involves the participation of the material's crystal structure

becoming   nature's property of continual change and evolution

Bell's theorem   also Bell's inequalities, a statistical method of testing the validity of quantum theory in EPR experiments

Bernoulli's equation   for smooth, steady, irrotational flow of a nonviscous, incompressible fluid, an equation expressing the fact that the quantity $p + \frac{1}{2}\rho v^2 + \rho gy$ has the same value at all points along a flow tube, where p, ρ, v, y, and g represent the pressure, the density, the flow speed, the height, and the acceleration due to gravity, respectively

beta decay   a mode of radioactive decay in which a neutron in a nucleus converts to a proton, which remains in the nucleus, while an electron and an antineutrino are emitted; in beta decay, the atomic number of the nucleus increases by 1 and the mass number does not change

beta particle   an electron

beta ray   an electron, especially as the product of radioactivity

big bang   a proposed cosmic explosion that occurred some 14 billion years ago and brought the universe into existence

bilateral symmetry   symmetry under reflection through a midplane (for three-dimensional systems) or a midline (for two-dimensional systems)

binding energy   the energy equivalent of the mass defect of a nucleus, which is the amount of energy required to decompose a nucleus completely into its component nucleons (i.e., protons and neutrons)

biochemistry   the branch of chemistry that studies the chemistry of living things

bioenergetics   the study of the production and utilization of adenosine triphosphate (ATP)

biophysics   the field of science involved with the application of the understanding and methods of physics to biology

Biot-Savart law a relation that shows how all electric currents contribute to the magnetic field at every point in space

birefringence the phenomenon that the index of refraction of a transparent substance possesses different values in different directions

blackbody an object that absorbs all electromagnetic radiation that is incident upon it; also written *black body*

blackbody spectrum the distribution of intensity over wavelengths for the electromagnetic radiation from a blackbody

black hole an object whose gravity is so strong that nothing, not even light or other electromagnetic radiation, can escape from it

blue shift the Doppler shift of electromagnetic waves emitted by a source that is approaching the observer, in which the observed frequency is higher than the frequency of the source

Bohr model the first attempt to apply quantum ideas to the hydrogen atom, which overthrew certain classical notions

boiling the phase change from liquid to gas (vapor), also called vaporization

boiling point the temperature at which a liquid boils or a gas condenses, usually at atmospheric pressure

Boltzmann constant a fundamental physical constant in statistical mechanics, denoted k, whose value is $1.3806503 \times 10^{-23}$ J/K

bomb calorimetry operates on the same principle as constant pressure calorimetry but holds the test sample in a sealed container at a constant volume rather than a constant pressure

bond dissociation energy a measure of the amount of energy given off when a bond is broken

boost the transformation of velocity change

boost invariance the laws of physics are the same at all velocities

Bose-Einstein condensate a situation in which, at sufficiently low temperatures, very close to 0 K, a well-isolated collection of identical bosons become so mutually correlated that they lose their individual identities and form what amounts to a single entity

Bose-Einstein statistics the statistical rules governing any collection of identical bosons

boson any particle—whether an elementary particle or a composite particle, such as a nucleus or an atom—that has an integer value of spin, i.e., 0, 1, 2, . . .

Boyle's law the inversely proportional relation between the volume and pressure for a fixed quantity of ideal gas at constant temperature

breaking point the stress at which a material ruptures

Brønsted-Lowry acid a proton donor

Brønsted-Lowry base a proton acceptor

bronze an alloy made of tin and copper

Brownian motion the random movement of small particles suspended in a fluid

buffer a solution that resists a change in pH when additional amounts of acids or bases are added

bulk modulus in elasticity, the ratio of the bulk stress (the pressure) to the bulk strain (the relative change of volume), often denoted B, whose SI unit is pascal (Pa)

buoyancy the existence of an upward force, called a buoyant force, on a body, produced by the surrounding fluid in which the body is fully or partially immersed

buoyant force the upward force on a body produced by the surrounding fluid in which the body is fully or partially immersed

Calorie a dietary calorie, equal to 1,000 calories or a kilocalorie

calorie (cal) defined as the amount of energy required to raise the temperature of 1 gram of water by 1 degree Celsius

calorimeter the apparatus used for calorimetry, measures the amount of energy being given off by the chemical or physical change in the system to be measured as a change in temperature

calorimetry an experimental procedure to determine the amount of heat gained or lost during a chemical or physical change

candela one of the SI base units, the SI unit of luminous intensity, denoted cd; the luminous intensity in a given direction of a source that emits electromagnetic radiation of frequency 5.40×10^{14} hertz (Hz) at radiant intensity in that direction of 1/683 watt per steradian (W/sr)

capacitance qualitatively, the ability of objects to store electric charge or to maintain a charge separation; quantitatively, the ratio of stored charge to the potential of the object storing the charge or the ratio of separated charge to the voltage between the two conducting parts of the object maintaining the charge separation, commonly denoted C, whose SI unit is farad (F)

capacitor any device constructed for the purpose of possessing a definite value of capacitance for charge separation; the capacitance of a parallel-plate capacitor is given by $C = \kappa \varepsilon_0 A/d$ (all quantities in SI units), where A and d denote the plate area and separation, respectively; κ is the dielectric constant of the material separating the plates; and ε_0 is the permittivity of the vacuum

capillary a very narrow tube

capillary flow the spontaneous flow of liquids in very narrow tubes, called capillaries

carbohydrates a class of polymers whose main function inside the cell is as an energy source

carbon monoxide the colorless, odorless, tasteless gas (CO) that can be deadly, produced from the incomplete combustion of hydrocarbons

carcinogen a chemical substance that has been shown to cause cancer

Carnot engine an ideal, reversible heat engine that takes in heat at a higher temperature, discharges heat at a lower temperature, and performs work

Cartesian coordinate system a coordinate system consisting of mutually perpendicular straight axes bearing equal length scales

catabolism breaking down food sources to obtain cellular energy

catalyst a substance that increases the rate of a reaction without itself being consumed or altered in the reaction

cathode the terminal or electrode, of a pair of terminals or electrodes, that is at the lower electric potential; the negative terminal or electrode of such a pair

cathode ray a beam of electrons

cathode-ray tube (CRT) an image tube, such as for a television set or computer monitor, whose mode of operation is based on thermionic emission of electrons

cation positively charged ions produced when metals lose electrons

cellular membrane potential sum of electrical and chemical concentration gradients from one side of a cellular membrane to another

center of gravity the point at which the total weight of a body, the sum of the forces of gravity on all the components of the body, can be thought as acting

center of mass the point at which the mass of an extended body can be considered to be located for the purpose of Newton's second law of motion

centrifugal directed away from the center

centrifugation the method of separating components of a mixture based on their differential molecular weights or their densities by rapid spinning of the sample

centripetal directed toward the center

centripetal acceleration the center-directed acceleration of a body in circular motion, due to the continuously changing direction of the body's velocity, whose magnitude is often denoted a_c and whose SI unit is meter per second per second (m/s^2)

centripetal force the center-directed force acting on a body in circular motion and causing the body's centripetal acceleration, whose magnitude is often denoted F_c and whose SI unit is newton (N)

chain reaction usually in reference to nuclear reactions, a nuclear reaction that, once started, continues on its own until all the reacting nuclei are used up

charge the property of an elementary particle by which it affects and is affected by other elementary particles; often refers to electric charge in particular

Charles's law the proportionality between the temperature and volume for a fixed quantity of ideal gas at constant pressure

chemical engineering the application of chemical and physical properties of matter to the manufacture of a product

chemical equilibrium the state of a chemical reaction in which the concentrations of the reactants and the products of the reaction remain constant in time, so the reaction rates for the forward and the inverse reactions are equal

chemical potential energy potential energy in a chemical reaction stored in chemical bonds

chemistry the study of matter and the reactions it undergoes

chemist a scientist who studies matter and the reactions that it undergoes

chiral compounds that are optically active, meaning they can rotate plane-polarized light

chlorofluourcarbons (CFCs) synthetic compounds originally developed in the 1920s and 1930s as a safer alternative to refrigerants and coolants of the time

chromatography separation of a mixture based on attraction of the component parts to a column or chromatography paper

C invariance particle-antiparticle conjugation invariance

circuit a configuration of connected electric components; a closed circuit allows the flow of an electric current through it, while an open circuit does not

circular accelerator a particle accelerator that accelerates particles in a circular path, which is maintained by means of magnetic fields

circular motion the motion of a particle in a circular path

citric acid cycle a series of reactions in cellular respiration used to break down the carbon chain, producing carbon dioxide and releasing the energy of the bonds in the forms of adenosine triphosphate/guarosine triphospate (ATP/GTP) and electron carriers reduced nicotinamide adenine dinucleotide (NADH) and reduced flavin adenine dinucleotide ($FADH_2$); also referred to as the tricarboxylic acid (TCA) or Krebs cycle

classical domain the domain of nature for which classical physics can well approximate quantum physics, the domain characterized by lengths, durations, and masses that are not too small, that are in general larger than the atomic and molecular scales

classical physics among physicists, physics that does not take into account quantum effects; commonly, the physics of the 19th century, physics that does not involve relativity or quantum considerations

clock any device for measuring time by producing and counting (or allowing one to count) precisely equal time units

closed system a chemical reaction where system and surroundings are isolated

coefficient of expansion the ratio of the relative change of size of a sample of material to the change of temperature causing the size change, whose SI unit is inverse kelvin (K^{-1}), where the coefficient is of linear, area, or volumetric expansion for change of length, area, or volume, respectively

coherence the property of oscillations, such as of a periodic wave, that they possess a definite and regular phase relationship among themselves

cohesion the attraction of like molecules for each other; the forces that hold matter together

cold fusion the claimed creation of nuclear fusion at room temperature

collision an encounter in which two or more bodies approach each other, briefly interact with each other, and then possibly separate from each other

colloid heterogeneous mixture with intermediate particle size between approximately 1 and 100 nanometers

color in regard to elementary particles, a possible state of a quark, of which there are three: red, blue, and green

column chromatography mixture separation technique that utilizes a stationary phase placed in a column

combination reactions a class of chemical reactions that involve two reactants that join to produce a single product

combustion reaction a class of chemical reaction that involves the burning of a substance in the presence of oxygen

common ion effect the effect of one ion on the equilibrium of another related ion

completely inelastic collision a collision in which the colliding bodies stick together and do not separate after colliding

component the value of a vector's perpendicular projection along any one of the three axes, given by $v \cos \theta$, where v denotes the magnitude of the vector and θ is the angle between the vector and the axis under consideration; the x-, y-, and z-components of a vector equal ($v \cos \alpha$, $v \cos \beta$, $v \cos \gamma$), where (α, β, γ) are the vector's direction angles

Compton effect the increase in wavelength of electromagnetic radiation, specifically X-rays or gamma rays, when it is scattered by the more loosely bound electrons of matter

Compton scattering the Compton effect

Compton shift the value of the increase in wavelength of electromagnetic radiation, specifically

X-rays or gamma rays, when it is scattered by the more loosely bound electrons of matter

Compton wavelength of the electron the quantity $h/(m_e c)$, whose value is 2.4263×10^{-12} meter (m), where h, m_e, and c denote the Planck constant, the mass of the electron, and the speed of light in vacuum, respectively

computational chemistry a branch of chemistry that utilizes mathematical and computer modeling techniques to describe theoretically the structure of atoms, molecules, and the ways chemical reactions take place

concave describing a surface, curving inward, having the shape of the interior of a bowl

concentration the amount of solute per unit volume or mass of solution

concentration gradient the concentration difference between two locations divided by the distance between the locations

condensation the phase change from gas (vapor) to liquid

condensation reaction a chemical reaction in which two reactants are joined with the release of a water molecule

conductance the ability of an object to conduct electricity or heat; electric conductance is the inverse of the object's resistance and thus the ratio of the electric current flowing through the object to the voltage across it.

conduction the passage of charges (electric conduction) or heat (thermal conduction) through a material

conduction band the range of electron energies at which electrons can move freely throughout a solid and conduct electricity

conductivity a quantity that represents a material's ability to conduct electricity (electric conductivity) or heat (thermal conductivity), often denoted σ (Greek sigma, for electricity) or k (for heat), whose SI unit is inverse ohm-meter $[(\Omega \cdot m)^{-1}]$ (for electricity) or watt per meter per kelvin [W/(m·K), also written W/m·K] (for heat); a property of a metal

conductor a material that conducts, i.e., allows relatively easy passage of, electricity (electric conductor) or heat (thermal conductor); a solid whose conduction band is populated

conjugate acid the substance formed when a base gains a hydrogen ion

conjugate base the substance formed when an acid loses a hydrogen ion

conservation law a property of nature that some physical quantity in a suitably isolated system maintains a constant value over time

conservation of angular momentum a universal conservation law: in an isolated system the total angular momentum remains constant over time

conservation of charge a universal conservation law: in an isolated system the total electric charge remains constant over time

conservation of energy a universal conservation law: in an isolated system the total energy content remains constant over time

conservation of linear momentum a universal conservation law: in an isolated system the total linear momentum remains constant over time

conservation science new field that utilizes chemistry knowledge to rejuvenate works of art as well as authenticate works through analysis of the chemicals present

conservative force any force that performs zero total work on a body moving in a closed path; equivalently, any force such that the work that it performs on a body that moves between two points is independent of the path the body follows between those points

constant pressure calorimetry a technique used to measure the amount of heat given off during a chemical reaction performed in a vessel that is not sealed

constructive interference the interference of waves that results in an effect that is greater than the effect of each individual wave

contact force any force that is applied through direct contact of two bodies

continuity a characteristic of classical physics that processes occur in a continuous manner and the ranges of allowed values of physical quantities are continuous

continuity equation for the flow of an incompressible fluid, an equation expressing the fact that the volume of fluid flowing past any location along a flow tube during some time interval is the same as that passing any other location along the same flow tube during the same time interval; for the steady flow of a compressible fluid, an equation expressing the fact that the mass of fluid flowing past any location along a flow tube during some time interval is the same as that passing any other location along the same flow tube during the same time interval

control rods neutron-absorbing bars, often made of cadmium, that are used to stop a nuclear reaction

convection the transfer of heat through the motion of material

converging mirror a concave mirror, which causes parallel rays to converge

convex describing a surface, curving outward, having the shape of the exterior surface of a bowl

Cooper pair in the BCS theory of superconductivity, any of the electron pairs that conduct electricity in a type I or II superconductor in its superconducting state

coordinate any one of a set of numbers by which a point in space or in space-time is specified with respect to a coordinate system

coordinate system a grid in space or in space-time that serves as a reference for the specification of points—locations (in space) or events (in space-time)—by means of sets of numbers called coordinates

cosmic background radiation cosmic microwave background

cosmic microwave background electromagnetic radiation permeating the universe, whose source is thought to be the big bang

cosmic ray energetic particles, mostly protons, that impinge on Earth's atmosphere from space

coulomb the SI unit of electric charge, denoted C, equivalent to ampere-second (A·s)

Coulomb's law states that the forces between a pair of electric point charges q_1 and q_2 are attractive for opposite-sign charges and repulsive for charges of the same sign, where the magnitude of the force acting on each charge is given by $[1/(4\pi\kappa\varepsilon_0)]|q_1||q_2|/r^2$, with ε_0 the permittivity of the vacuum, κ the dielectric constant of the medium in which the charges are embedded (equals 1 for vacuum and effectively 1 for air), and r the distance between the charges

covalent bond a bond formed between two non-metals by the sharing of valence electrons

CP invariance the laws of physics are the same for elementary particles, except neutral kaons, as for their respective reflected antiparticles

CPT invariance the laws of physics are the same for all elementary particle processes as for their respective particle-antiparticle conjugated, reflected, and time-reversed image processes

cracking the process of cleaving the carbon-carbon bonds in high-molecular-weight hydrocarbons found in petroleum in order to produce more of the low-carbon-number molecules that are used as heating and fuel sources

critical field the strongest magnetic field in which a superconductor can be in a superconducting state

critical temperature the highest temperature at which a superconductor can be in a superconducting state

crystal a solid whose constituent particles—atoms, molecules, or ions—possess a regularly repeating spatial arrangement

crystallography the study of the structure of crystals in terms of the positions of their microscopic constituents—atoms, ions, and molecules—and of the defects of crystal structures

Curie principle the symmetry principle

Curie temperature the temperature above which a ferromagnetic material loses its ferromagnetism

curl of a vector field $\mathbf{v}(x, y, z)$, another vector field, denoted curl $\mathbf{v}$, $\nabla \times \mathbf{v}$, or rot $\mathbf{v}$, whose components are given by

$$(\text{curl } \mathbf{v})_{x,y,z} = \left(\frac{\partial v_z}{\partial y} - \frac{\partial v_y}{\partial z}, \frac{\partial v_x}{\partial z} - \frac{\partial v_z}{\partial x}, \frac{\partial v_y}{\partial x} - \frac{\partial v_x}{\partial y} \right)$$

current the flow of electric charge (electric current) or of thermal energy (heat current); the rate of passage of electric charge or of thermal energy, often denoted i or I (for electricity) of H (for heat), whose SI unit is ampere (A) (for electricity) or joule per second (J/s) (for heat)

cyclooxygenase an enzyme that is the molecular target for aspirin action that blocks the production of prostaglandins, thereby preventing inflammation and pain

cyclotron a type of circular particle accelerator

dark energy a hypothesized form of energy that pervades the whole universe and is responsible for the observed accelerating expansion of the universe

dark matter matter in the universe that does not radiate and reveal its presence in that way, such as cold dead stars

Davisson-Germer experiment an experiment first performed in the 20th century that demonstrated the wave character of electrons

Davy lamp a safety lamp used for miners developed by Sir Humphry Davy in 1816

DC direct current

deceleration a decrease of speed

decibel a non-SI unit of intensity level of sound

decomposition reactions a class of chemical reaction beginning with only one reactant and generally involving the addition of heat

definiteness a characteristic of classical physics that every physical quantity that is relevant to a physical system possesses a definite value at every time

degree of freedom a possible independent mode of motion of a system; in kinetic theory, an independent mode for an atom or molecule to possess a significant amount of energy

dehydration reaction a reaction forming a bond with the accompanying removal of water

density the mass of a unit volume of a substance, often denoted ρ (Greek rho), whose SI unit is kilogram per cubic meter (kg/m^3)

deposition the phase change from gas (vapor) to solid

destructive interference the interference of waves that results in an effect that is less than the effect of each individual wave

determinism a characteristic of classical physics that the state of a physical system at any instant uniquely determines its state at any time in the future and, accordingly, is uniquely determined by its state at any time in the past

diamagnetism the situation in which the induced magnetic field in a material partially cancels the applied magnetic field

dielectric constant the factor by which a medium reduces the magnitudes of the electric forces among electric charges embedded in it, often denoted κ (Greek kappa), which is unitless

diffraction the bending of waves around obstacles

diffusion the spreading-out process that results from the random motion of particles (atoms, molecules) of a substance; the scattering of light as it travels through a translucent medium or is reflected from a rough surface; the movement of substances from areas of high concentration to areas of low concentration across a semipermeable membrane

diffusion rate the net quantity of substance that is transferred by diffusion per unit time

diopter a non-SI unit of optical power, denoted D, equivalent to inverse meter (m^{-1})

dipole a separation of charge within a molecule due to the unequal sharing of electrons

Dirac equation combining quantum mechanics and the special theory of relativity, an equation to describe an electrically charged point particle with spin ½, such as the electron, that was found to predict the existence of antiparticles

direct current (DC) the situation when the electric currents flowing in a circuit maintain a constant direction, or, if currents are not flowing, the voltages involved maintain a constant sign

direction angles the three angles between a direction, such as of a vector, and the positive x-, y-, and z-axes, usually denoted (α, β, γ)

direction cosines the cosines of the direction angles of a direction, such as of a vector

director a term used when studying liquid crystals that represents the positional order that determines the likelihood of showing translational symmetry

discontinuity a characteristic of quantum physics that the evolution of a physical system, although proceeding continuously in time for some duration, might involve a discontinuous process; spontaneous transition or measurement

discreteness a characteristic of quantum physics that the range of allowed values of certain physical quantities can include, or possibly even consist wholly of, a set of discrete values, while according to classical physics that range is wholly continuous

dispersion the dependence of the speed of light in a transparent material, and thus also of the material's index of refraction, on the frequency of the light

displacement a change of position, whose SI unit is meter (m); in harmonic motion, the deviation from equilibrium; referring to an engine, the volume change of a single cylinder as its piston moves from smallest cylinder volume to largest, multiplied by the number of cylinders in the engine

dissipative force a force, such as friction, that converts nonthermal forms of energy into heat

divergence of a vector field **v**(x, y, z), a scalar field, denoted div **v** or ∇·**v**, and given in terms of the components of **v** by

$$\text{div } \mathbf{v}(x, y, z) = \frac{\partial v_x(x,y,z)}{\partial x} + \frac{\partial v_y(x,y,z)}{\partial y} + \frac{\partial v_z(x,y,z)}{\partial z}$$

diverging mirror a convex mirror, which causes parallel rays to diverge

DNA fingerprinting a technique used to distinguish between individuals of the same species by identifying distinctive patterns in their deoxyribonucleic acid (DNA) composition

Doppler effect the dependence of the observed frequency of a periodic wave on the motion of the observer or the source or both

Doppler shift the difference between the observed frequency of a periodic wave and the frequency of the source of the wave as a result of the motion of the observer and the source

double bond two bonds formed between the same two atoms, consisting of one sigma bond and one pi bond

double-replacement reactions a class of chemical reactions in which two ionic compounds exchange positive or negative ions

drag the retarding effect of forces of turbulence and viscosity acting on a body moving through a fluid

drag force the combination of the forces of turbulence and solid-fluid friction that act on a solid body moving through a fluid

ductile of a metal, capable of being pulled into wires

dynamics the study of the causes of motion and the causes of changes of motion

ECG electrocardiography, also denoted EKG

EEG electroencephalography

efficiency for a machine, the ratio of output work to input work, often expressed as a percentage

eigenfunction in quantum mechanics, the wave function is an eigenfunction of an operator, when the result of the operator's action on it is the same function multiplied by a number, that number being an eigenvalue of the operator

eigenstate the state of a quantum system specified by a wave function that is an eigenfunction of some operator

eigenvalue in quantum mechanics, the number that multiplies a function as the result of the action on the function of an operator of which the function is an eigenfunction

Einstein-Rosen bridge a possible, but yet undetected, feature of space-time, according to the general theory of relativity, that gives a "direct" connection between events, allowing time travel and "warp speed" space travel

elastic collision a collision in which kinetic energy is conserved: i.e., the total kinetic energy of the bodies emerging from the collision equals the initial total kinetic energy of the colliding bodies

elasticity the tendency of materials to resist deformation and, after being deformed, to regain their original size and shape after the cause of deformation is removed

elastic limit the maximal stress for which an elastic material will return to its original configuration when the stress is removed

elastic material a material that returns to its original configuration when the cause of deformation, the stress, is removed, for stresses up to the material's elastic limit

electric charge a property of certain elementary particles, whereby they exert nongravitational attractive or repulsive forces on each other that are independent of their relative velocities, i.e., electric forces; the amount of charge, which can be positive or negative, is commonly denoted q or Q and its SI unit is coulomb (C); the natural unit of electric charge is the magnitude of the charge of the electron, proton, etc.; is denoted e; and possesses the value $1.602176463 \times 10^{-19}$ C

electric current the net charge flowing past a point per unit time, commonly denoted I or i, whose SI unit is ampere (A)

electric dipole pair of equal-magnitude and opposite electric charges at some distance from each other or a charge configuration that is equivalent to that

electric dipole moment a quantity that measures the strength of an electric dipole, often denoted μ_e (Greek mu) (as a vector) or μ_e (as its magnitude), whose SI unit is coulomb-meter (C·m)

electric field a condition of space by which charged particles affect each other electrically; the mediator of the electric force; a vector field commonly denoted **E**, whose SI unit is newton per coulomb (N/C), equivalent to volt per meter (V/m); its value at any point is defined as the force on a unit (one-coulomb) electric point charge located at that point

electric field line a directed line in space whose direction at every point on it is the direction of the electric field at that point

electric flux through a surface, the integral over the surface of the component of the electric field perpendicular to the surface, often denoted Φ_e (Greek capital phi), whose SI unit is volt-meter (V·m); for a uniform electric field and a flat surface, the product of the surface area and the component of the field perpendicular to the surface

electric generator any device for converting mechanical energy to electric energy

electricity the phenomena that are based on the force between objects that is due only to their electric charges and is independent of their relative motion

electric polarization the alignment by an electric field of a material's natural atomic or molecular electric dipoles or the creation of aligned dipoles by an electric field in a material that does not normally possess such dipoles

electric potential also called potential, a scalar field whose value at any point is defined as the electric potential energy of a unit (one-coulomb) electric point charge located at that point, commonly denoted V, whose SI unit is volt (V)

electrocardiography (ECG) the measurement of the electrical activity of the heart via electrodes placed at various locations on the skin, also denoted EKG

electrochemical potential the sum of electrical and chemical concentration gradients from one side of a cellular membrane to another

electrochemistry the branch of chemistry that studies the energy transfers that take place during chemical reactions

electroencephalography (EEG) the measurement of the electrical activity of the brain via electrodes on the scalp

electrolysis the chemical decomposition of a substance by means of an electric current

electrolyte A substance that when dissolved in water conducts an electric current; most ionic compounds are electrolytes

electromagnet a device that acts as a magnet as a result of an electric current flowing through it

electromagnetic field a condition of space by which charged particles affect each other electromagnetically; the mediator of the electromagnetic force

electromagnetic induction the production of a potential difference or current by a changing magnetic flux

electromagnetic wave a propagating disturbance in the electromagnetic field

electromagnetism the force between objects that is due only to their electric charges; one of the four fundamental forces, it unifies electricity and magnetism

electromotive force emf

electron a negatively charged, spin-½ particle that is one of the constituents of atoms, found outside the nucleus, often denoted e^- or β^-

electron affinity the attraction of an atom for more electrons

electronegativity the attraction of an element for the electrons in a bond

electron orbitals the allowed energy states of electrons in the quantum mechanical model of an atom

electron transport chain a series of proteins in a membrane that transfer electrons

electron volt the amount of energy gained by an electron as it accelerates through a potential difference of one volt; equals $1.602176463 \times 10^{-19}$ J

electrostatics the study of the forces between charged particle that are at rest

electroweak force a force that unifies the electromagnetic and weak forces

electroweak theory a theory of the electroweak force

element a substance that cannot be separated into simpler substances by chemical means

elementary particle strictly, any of the particles that compose the ultimate constituents of matter and the particles that mediate the fundamental forces among them; more loosely, any subatomic particle

elute to remove a desired substance from a chromatography column

emf a potential difference created by a device such as a battery or produced by electromagnetic induction; the abbreviation of the old term *electromotive force*, though emf is not a force

emissions substances given off, especially in the context of pollution sources

emission spectrum a characteristic pattern of electromagnetic radiation given by a sample after it is energized in some way such as by heating or with electricity. When energized, the electrons in the sample become excited and move up to a higher energy level, then fall back down to their ground state and the electromagnetic radiation is emitted

endothermic reaction a chemical reaction that requires energy from the surroundings to be put into the system for a reaction to proceed

energy the capacity of a system to perform work, often denoted E, whose SI unit is joule (J); the ability to do work; in the case of most chemical reactions, the work done is in the form of changed heat

energy distribution function in statistical mechanics, a function of energy E that, when multiplied by the differential dE, gives the number of molecules per unit volume that have kinetic energies in the small range between E and $E + dE$

energy level in quantum mechanics, any of a discrete set of energy values that a system possesses; in electron configurations, the energy level indicates the distance of an electron from the nucleus

engine any device that converts energy into mechanical force, torque, or motion, in particular where the energy source is a fuel

engineering the application of the knowledge, understanding, and methods of science to practical purposes

enthalpy *(H)* the measure of heat lost or gained in a chemical reaction

entropy a measure of the degree of disorder of a system

eonology the study of wine or wine making

EPR abbreviation of Einstein-Podolsky-Rosen, refers to the 1935 article "Can Quantum-Mechanical Description of Physical Reality Be Considered Complete?" by Albert Einstein, Boris Podolsky, and Nathan Rosen

EPR experiment any experiment in which two particles are produced by the same source, fly apart from each other, and have their properties measured in a way that tests the validity of quantum theory

equilibrium a state of a physical system in which the system does not spontaneously evolve but rather maintains itself as it is (see chemical equilibrium)

equilibrium constant *(K$_{eq}$)* numerical value representing the ratio of reactants to products at the equilibrium point of a chemical reaction

equipartition of energy the principle whereby the contribution of each degree of freedom to the total average energy of a molecule is $\frac{1}{2}kT$, where k is the Boltzmann constant and T the absolute temperature

equipotential surface any surface in space such that at all points on it the electric potential has the same value; no work is required to move an electric charge along an equipotential surface

equivalence principle states that mass as it affects inertia is equal to mass as it is affected by gravitation; in another formulation, one cannot determine, by means of experiments carried out solely within a laboratory, whether the room is at rest and the masses in the lab are being affected by gravitation or the room is undergoing constant acceleration in a region where there is no net gravitational force

ether a physical medium that was proposed as the carrier of electromagnetic waves, whose existence has been disproved; also, a volatile, flammable organic liquid having the chemical formula $C_4H_{10}O$, or any organic compound that contains an oxygen atom covalently bonded to two hydrocarbon groups

eutrophication the death of a pond or other body of water due to an overabundance of nutrients that leads to an algal bloom, causing a lack of dissolved oxygen in the water

evaporation the change of state from a liquid into a vapor

event an occurrence at some location and at some instant, represented by a space-time point specified by its four coordinates (x, y, z, t)

event horizon a theoretical surface surrounding a black hole, such that a body or light ray originating within the surface will never pass through the surface and will descend inexorably toward the center of the black hole

excited state the state of a system that possesses more energy than its allowed minimum

exclusion principle also known as the Pauli exclusion principle, states that no more than a single fermion may exist simultaneously in the same quantum state. For example, no two electrons in an atom can have the same set of four quantum numbers; therefore, no more than two electrons (one of each spin direction, up and down) can exist in any atomic orbital

exothermic reaction chemical reactions that release energy from the system to the surroundings

expectation value in quantum mechanics, the average of the values obtained by repeated measurements of a physical quantity for a system in the same state

extended Mach principle states that the origin of the laws of nature lies in all the matter of the universe, so effectively the origin of the laws of nature lies in the distant stars

external-combustion engine an engine whose fuel is burned outside the body of the engine proper

Faraday's law for electromagnetic induction, states that the emf induced in a circuit equals the negative of the product of the number of loops of the circuit and the rate of change of the magnetic flux through the circuit

fermentation the anaerobic biochemical process that converts pyruvate to alcohol or an organic acid such as lactic acid

Fermi-Dirac statistics the statistical rules governing any collection of identical fermions

Fermi energy the highest particle energy in a system of identical fermions, when the system is close to absolute zero temperature

fermion any particle—whether an elementary particle or a composite particle, such as an atom or a nucleus—whose spin has a value that is half an odd integer, i.e., 1/2, 3/2, 5/2, . . .

ferromagnetism the strong response of certain materials to an applied magnetic field, by which the material becomes polarized as a magnet that strongly enhances the applied field and may even remain a magnet when the applied field is removed

Feynman diagram a pictorial representation of a process that takes part in the interaction of elementary particles

Fick's law the relation of the diffusion rate in steady state diffusion to the concentration difference $C_2 - C_1$, the distance L, the cross section area A, and the diffusion coefficient D:

$$\text{Diffusion rate} = DA\left(\frac{C_2 - C_1}{L}\right)$$

field a condition of space by which matter interacts with matter; any spatially extended physical quantity whose values generally depend on location, such as electric and magnetic fields

field emission the emission of electrons from a metal due to the effect of a strong external electric field pulling them off the surface of the material

field emission microscope a microscope for viewing the atomic structure of metals, based on the field emission effect

field force any force that is applied from a distance, including gravitational, electric, and magnetic forces

field line for a vector field, a directed line in space whose direction at every point on it is the direction of the field at that point

field theory a physical theory based on fields, such as Maxwell's theory of electromagnetism, which is based on electric and magnetic fields, combined as the electromagnetic field. Such a theory views the interaction of matter with matter as being mediated by a field or fields

first law of thermodynamics the increase in the internal energy of a system equals the sum of the work performed on the system and heat flowing into the system

Fischer projection structural representation of a three-dimensional molecule, especially a carbohydrate, in which the horizontal lines are assumed to project out of the page, while the vertical lines project into the page

fission the splitting of an atomic nucleus into two (usually, but possibly more) large parts, either naturally or induced by bombardment, most often with neutrons

fission bomb also called atom bomb or nuclear-fission bomb, a bomb based on the energy released in an uncontrolled nuclear-fission reaction

flavor a whimsical name for a variety of quark, of which there are six: up, down, charm, strange, top, and bottom

flow tube an imaginary tubelike surface in a flowing fluid, whose walls are parallel to streamlines

fluid a substance that flows, e.g., a liquid or gas

fluorescence the emission of light by a substance as the immediate effect of the substance's being irradiated by electromagnetic radiation of a frequency different from that of the emitted light or by particles such as electrons

flux of a vector field through a surface, the integral over the surface of the component of the field perpendicular to the surface; for a uniform vector field and a flat surface, the product of the surface area and the component of the field perpendicular to the surface

FM frequency modulation

focal length the distance from the surface of an optical component to its focal point, commonly denoted f, whose SI unit is meter (m), taken as positive or negative for a real or virtual focal point, respectively

focal point the point to which parallel rays are or seem to be made to converge by an optical system or component; a focal point can be real or virtual

focus the converging or apparent converging of rays to a point by an optical system or component

force that which causes a body's acceleration, or change of momentum, often denoted **F** or **f** (as a vector) or F or f (as its magnitude), whose SI unit is newton (N)

fossil fuels energy sources formed from the decomposition of organic compounds over a long period

Fourier transform infrared spectroscopy (FITR) an analytical chemistry technique that allows for the identification of unknown samples and the analysis of quality and purity of samples

fractal a pattern that is symmetric under change of scale

free-body diagram a diagram, useful for solving problems involving forces and torques, that shows only the body under consideration and all the forces acting on it

free fall motion affected only by the force of gravity

freezing the phase change from the liquid state to the solid state

freezing point the temperature at which a liquid freezes or a solid melts, usually specified at atmospheric pressure

frequency the number of cycles per unit time of a repetitive phenomenon, often denoted f or ν

(Greek nu), whose SI unit is hertz (Hz); for a wave, the number of wavelengths that pass a given point per unit time

frequency modulation (FM) the encoding of a transmitted signal in a carrier wave by varying the wave's frequency

friction mechanisms and forces that resist and retard existing motion and that oppose imminent motion, including adhesion, cohesion, surface irregularity, surface deformation, turbulence, and viscosity

fringe a band of light produced by the constructive interference of two or more light waves

functional group atom or group of atoms that contributes to the chemical and physical properties of a compound, such as the hydroxyl group of alcohols, carbonyl group of aldehydes and ketones, or carboxyl group of carboxylic acids

fundamental the lowest frequency in a harmonic sequence, so that all the frequencies of the sequence, called harmonics, are integer multiples of it

fundamental particle any of the particles that are the ultimate constituents of matter and the particles that mediate the fundamental forces among them

fusion in reference to nuclear reactions, the joining, or fusing, of lighter nuclei into heavier ones; in reference to phase change, the transformation from solid state to liquid state, melting

fusion bomb also called hydrogen bomb or nuclear-fusion bomb, a bomb in which a nuclear-fission bomb ignites an uncontrolled nuclear-fusion reaction, with concomitant release of energy

Galilei transformation any of a certain group of mathematical relations between an event's coordinates in any space-time coordinate system and the same event's coordinates in any coordinate system moving at constant velocity relative to the first, according to Newtonian mechanics

gamma decay a mode of radioactive decay in which a gamma ray (i.e., an energetic photon) is emitted; in gamma decay, neither the atomic number nor the mass number of the nucleus changes

gamma ray an energetic photon, especially as the product of radioactivity

gas the state of matter in which the atoms or molecules are neutral and the matter has neither a definite shape nor a definite volume, taking the shape and volume of its container; also called vapor

gas chromatography mixture separation technique in which the sample is vaporized before traveling through the column

gauge pressure the difference between the pressure in a container and the external (usually atmospheric) pressure

gauge symmetry symmetry under space-time-dependent changes that mix components of quantum fields

Gaussian surface the term for any closed surface used in Gauss's law

Gauss's law states that the electric flux through a closed surface equals the net charge enclosed within the surface divided by the permittivity of the medium or, in the absence of a medium, by the permittivity of the vacuum

Gay-Lussac's law the proportionality between the temperature and pressure of a fixed quantity of ideal gas at constant temperature

gel electrophoresis biochemical separation technique used for proteins ribonucleic acid (RNA), and deoxyribonucleic acid (DNA); samples are separated through a matrix by creating an electric current

generalized coordinate any member of the set of physical quantities needed to specify a state of a physical system

general relativity Albert Einstein's general theory of relativity

general theory of relativity proposed by Albert Einstein, a theory of gravitation that is expressed in terms of the curvature of space-time, whose basic postulate is that the laws of physics are the same in all reference frames, no matter what their relative motion

generation any one of three subdivisions of quarks and leptons, each containing two flavors of quarks and one from among the electron, muon, and tau together with its associated neutrino

geodesic motion force-free motion in curved space-time

geomagnetism the natural magnetism of planet Earth

geometric optics also called ray optics, the branch of optics that deals with situations in which light propagates as rays in straight lines and its wave nature can be ignored

global warming the predicted gradual warming of the Earth's average temperature due to an increase in the layers of greenhouse gases such as carbon dioxide in the atmosphere

gluon the spin-1 elementary particle, of which there are eight kinds, that mediates the strong force among quarks, denoted g

glycolysis the first stage in cellular respiration, which involves the breakdown of the six-carbon sugar glucose to two three-carbon pyruvate molecules with the concomitant production of adenosine triphosphate (ATP) and reduced nicotinamide adenine dinucleotide (NADH)

gold foil experiment Sir Ernest Rutherford's experiment, in which gold foil was bombarded with alpha particles and in which the nucleus was discovered

gradient of a scalar field $s(x, y, z)$, a vector field, denoted grad s or ∇s, whose components are given by

$$(\text{grad } s)_{x,y,z} = \left(\frac{\partial s(x,y,z)}{\partial x}, \frac{\partial s(x,y,z)}{\partial y}, \frac{\partial s(x,y,z)}{\partial z} \right)$$

grand unified force a proposed force that unifies the strong, electromagnetic, and weak forces

grand unified theory (GUT) a theory of a grand unified force

gravitation the force between objects that is due only to their masses; one of the four fundamental forces; gravity

gravitational field a condition of space by which particles affect each other gravitationally; the mediator of the gravitational force

gravitational lens the function of the intervening object in the effect, whereby light rays from a distant astronomical source are bent by an intervening massive astronomical object, such as a galaxy or a galaxy cluster, on their way to an observer on Earth, causing the source to appear distorted—as several sources or as a possibly incomplete ring around the intervening object, called an Einstein ring

gravitational mass mass as a measure of participation in the gravitational force; active gravitational mass is a measure of the capability of the body possessing it to serve as a source of the gravitational field, while passive gravitational mass measures the degree to which the gravitational field affects the body possessing it

gravitational wave a wave in the gravitational field, described also as a wave in the fabric of space-time

graviton a predicted spin-2 particle of gravitational radiation and the mediator of the gravitational force

gravity the force between objects that is due only to their masses; one of the four fundamental forces; gravitation

green chemistry environmentally friendly chemical processes

greenhouse effect the insulating effect of the atmosphere on the planet Earth when the Sun's energy is trapped by the atmosphere to maintain a warmer temperature

ground level ozone three covalently bonded oxygen atoms make up ozone; when found in the lower troposphere, contributes to smog

ground state a system's state of lowest energy

ground state electrons electrons in the lowest possible energy state, closest to the nucleus, and the most stable

group a vertical column in the periodic table; elements in groups have similar electron configurations and physical properties

GUT grand unified theory

Haber process industrial chemistry process for the formation of ammonia in the presence of a metal catalyst

hadron any elementary particle formed from quarks and antiquarks in various combinations

half-life the time required for half of a sample of radioactive material to decay

Hall effect the effect that materials carrying electric current in a magnetic field develop a voltage perpendicular to the current

halogens a group (Group VIIA) of representative elements including fluorine, chlorine, bromine, iodine, and astatine

harmonic any of an infinite sequence of frequencies that comprises all the integer multiples of the lowest frequency in the sequence, called a harmonic sequence, with the nth multiple being called the nth harmonic and the lowest frequency, the first harmonic, called the fundamental (frequency)

harmonic motion also called simple harmonic motion, oscillatory motion in which the displacement from the center of oscillation is described by a sinusoidal function of time

harmonic oscillator any system that can oscillate in harmonic motion, either literally or figuratively

harmonic sequence an infinite sequence of frequencies, called harmonics, that comprises all the integer multiples of the lowest frequency in the sequence

Hartree-Fock theory computational chemistry technique that utilizes the fundamental principle of molecular orbital theory, whereby each electron is located in its own orbital and is not dependent on interactions with other electrons in the atom

heat also called thermal energy, the energy of the random motion of the microscopic constituents of matter (its atoms, molecules, and ions), whose SI unit is joule (J)

heat capacity a measure of the amount of energy required to raise the temperature of a substance by a single unit, whose SI unit is joule per kelvin (J/K)

heat flow the passage of heat from one location to another

heat of fusion the amount of heat per unit mass required to melt material in its solid state at its melting point, which is the same quantity of heat per unit mass that must be removed from the material in the liquid state in order to freeze it at its melting point, often denoted L_f, whose SI unit is joule per kilogram (J/kg)

heat of transformation the quantity of heat per unit mass that is absorbed or released when a substance undergoes a phase transition, often denoted L, whose SI unit is joule per kilogram (J/kg)

heat of vaporization the amount of heat per unit mass needed to convert a substance at its boiling point from its liquid state to a gas, which is the same quantity of heat per unit mass that must be removed from the substance in its gaseous state in order to liquefy it at its boiling point, often denoted L_v, whose SI unit is joule per kilogram (J/kg)

Heisenberg uncertainty principle atomic structure principle stating that one cannot know precisely both the location and the momentum of an electron; states that the product of the uncertainties of certain pairs of physical quantities in quantum systems cannot be less than $h/(4\pi)$, where h is the Planck constant; such pairs include position-momentum and time-energy

Henderson-Hasselbach equation the concentration of acid-base pair that is present in the buffer solution determines buffering capacity according to the equation

$$pH = pK_a + \log([\text{base}]/[\text{acid}])$$

Hess's law states the amount of energy required to carry out a reaction that occurs in multiple steps is the sum of the required energy for each of the successive steps

heterogeneous mixture physical combination of substances that is not the same throughout

Higgs particle a predicted elementary particle that is assumed to be responsible for the masses of the other elementary particles

high-temperature superconductor any of a class of superconductors that are all copper-containing metallic oxides in ceramic form and whose critical temperatures range to above 130 kelvins (K)

hologram a two-dimensional record of three-dimensional information; most commonly, a two-dimensional representation of a three-dimensional scene, which can be viewed as the original scene and shows different aspects from different viewpoints

holography in general, the recording and storage of three-dimensional information in a two-dimensional medium and its reconstruction from the medium; in particular, the recording of three-dimensional scenes on photographic film in such a manner that the scenes can be viewed in their full three-dimensionality from the image on the film

homogeneity the state of possessing the same properties at all locations

homogeneous possessing the same properties at all locations

homogeneous mixture physical combination of substances that is the same throughout; also known as a solution

Hooke's law in elasticity, the proportionality of the strain to the stress, up to the proportionality limit

horsepower a non-SI unit of power, denoted hp, equals 746 watts (W)

hour a non-SI unit of time, denoted h, equals 60 minutes (min) and 3,600 seconds (s)

Hund's rule states that electrons being added to equivalent energy orbitals will add singly, one each into each orbital, before doubling up

hydrodynamics the field of physics that studies the flow of fluids

hydrogen bomb more aptly called fusion bomb or nuclear-fusion bomb, a bomb in which a nuclear-fission bomb ignites an uncontrolled nuclear-fusion reaction, with concomitant release of energy

hydrolysis a chemical reaction in which a reactant is broken down into two products with the addition of a water molecule

hydrophilic water-loving compounds; compounds that can be dissolved in water

hydrophobic water-fearing compounds; compounds that cannot be dissolved in water

hydrostatic pressure the pressure p at depth h beneath the surface of a liquid at rest, given by $p = p_0 + \rho gh$, where p_0 is the pressure on the surface of the liquid (often atmospheric pressure), ρ is the density of the liquid, and g is the acceleration due to gravity

hydrostatics the field of physics that studies the behavior of fluids at rest

hypersurface a three-dimensional subspace in (four-dimensional) space-time; a light cone is an example of a hypersurface

ideal gas the idealization of a gas whose constituents (atoms or molecules) possess zero volume and do not interact with each other or with the container walls except upon contact, whereupon the interaction is an instantaneous elastic collision

ideal gas law the relation among the pressure, volume, temperature, and quantity (number of moles or molecules) of an ideal gas

illuminance luminous flux density

illumination luminous flux density

image the representation of an object (i.e., a light source) by means of converging or apparently converging rays produced by an optical system or component; an image can be real or virtual

indeterminism a characteristic of quantum physics that, given an initial state of an isolated system, what evolves from that state at future times is only partially determined by the initial state

index of refraction of a transparent material, the ratio of the speed of light in vacuum to its speed in the material

indicator in chemistry, a compound that changes color in the presence of an acid or base, indicating the pH of a solution

industrial chemistry the application of chemical processes and understanding to all aspects of industrial processes

inelastic collision a collision in which kinetic energy is not conserved: i.e., the total kinetic energy of the bodies emerging from the collision is less than the initial total kinetic energy of the colliding bodies

inertia the tendency of a body to resist changes in its constant velocity motion or rest, measured by the body's mass

inertial mass mass as a measure of a body's inertia, of a body's resistance to changes in its velocity, i.e., its resistance to acceleration

inertial motion motion in the absence of forces

inertial reference frame a reference frame in which Newton's first law of motion is valid, in which force-free bodies remain at rest or move at constant velocity

inflation a proposed very brief period of extremely rapid expansion of the universe

infrared electromagnetic radiation with frequency somewhat lower than, and wavelength somewhat longer than, those of visible light and, concomitantly, whose photons have less energy than those of visible light

inhomogeneity the state of possessing different properties at different locations

inhomogeneous possessing different properties at different locations

inner transition metals elements found in the bottom two rows of the periodic table, they are divided into two periods, called the lanthanides and the actinides

inorganic chemistry a branch of chemistry that involves the study of chemical reactions in compounds lacking carbon

insulator a material that does not conduct electricity (electric insulator) or heat (thermal insulator); a solid in which the conduction band is not populated and the gap between the valence and conduction bands is greater than about 10 electron volts

integrated circuit a high-density electronic circuit containing many miniature circuit components—transistors, capacitors, resistors, etc.—that is manufactured in the surface layer of a thin plate of semiconductor material such as silicon

intelligent textiles synthetic textiles that are designed and engineered to fit a certain need or solve a particular problem

intensity the amount of energy flowing through a unit area per unit time, often denoted I, whose SI unit is watt per square meter (W/m^2)

intensity level a logarithmic function of sound intensity that correlates well with the perception of loudness, commonly expressed in decibels (dB)

interference the combined effect of two or more waves passing through the same location at the same time

intermediate vector boson any of the set of spin-1 elementary particles that mediate the weak force, comprising the particles denoted W^+, W^-, and Z^0

intermolecular forces attractive forces between atoms and molecules that hold molecules together but that are not true bonds

internal-combustion engine an engine whose fuel is burned within the body of the engine

International System of Units (SI) a unified, international system of units for physical quantities

interval the separation between a pair of events in space-time, analogous to the distance between a pair of points in ordinary space

invariance principle a statement that the laws of physics, or laws of nature, are invariant under a certain change

inverse beta decay also called electron capture, a mode of radioactive decay in which the nucleus captures an electron from outside the nucleus, which then combines with a proton to form a neutron, which remains in the nucleus, and a neutrino, which is emitted. In inverse beta decay the atomic number of the nucleus decreases by 1 and the mass number does not change

ion an electrically charged atom or molecule, one that has gained or lost one or more electrons relative to its neutral state

ionic bond an electrostatic interaction between a positively charged ion (cation) and a negatively charged ion (anion)

ionic radius half the distance across an ion of an element

ionization energy the measure of how easily an atom will lose an electron; the amount of energy required to remove an electron from an atom

irradiance radiant flux density

isotope a form of an element that has the same number of protons but a different number of neutrons

isotropic possessing the same properties in all directions

isotropy the state of possessing the same properties in all directions

joule the SI unit of energy and work, denoted J, equivalent to newton-meter (N·m)

Joule heat the heat produced in an object by the passage of an electric current through it

Joule's law states that the rate of heat production in an object caused by the passage of electric current through it equals the product of the voltage across the object and the value of the electric current

kaon a spin-0 particle contributing to the strong force that holds protons and neutrons together in atomic nuclei, consisting of a quark and an antiquark, denoted K^+, K^-, or K^0, depending on its electric charge

kelvin one of the SI base units, the SI unit of temperature, denoted K; 1/273.16 of the temperature interval between absolute zero and the triple point of water

kerosene hydrocarbon with 10–18 carbon atoms often used for heating

kilogram one of the SI base units, the mass of a platinum-iridium prototype kept at the International Bureau of Weights and Measures (BIPM) in Sèvres, France, denoted kg

kilowatt-hour a non-SI unit of energy, used mostly for electric energy, denoted kWh, equal to 3.6×10^6 joules (J)

kinematics the study of motion without regard to its cause

kinetic energy the energy that a body possesses that is due to its motion, often denoted K, KE, or E_k, whose SI unit is joule (J); for translational motion, one-half the product of the body's mass and the square of its speed; for rotational motion, one-half the product of the body's moment of inertia with respect to the rotation axis and the square of its angular speed

kinetic friction also called sliding friction, this is solid-solid friction, involving mostly adhesion and surface irregularity, that acts to retard existing motion

kinetic theory also called kinetic molecular theory, the explanation of properties of bulk matter in terms of the motion of its constituent particles (atoms, molecules, ions)

Kirchhoff's rules a set of rules by which one can find all the currents and voltages in any given circuit: (1) The junction rule states that the algebraic sum of all currents at a junction equals 0, or equivalently, the sum of currents entering a junction equals the sum of currents leaving it. (2) The loop rule states that the algebraic sum of potential rises and drops encountered as a loop is completely traversed equals 0, or equivalently, the sum of potential rises around a loop equals the sum of potential drops

Krebs cycle a series of reactions in cellular respiration used to break down the carbon chain, producing carbon dioxide and releasing the energy of the bonds in the forms of adenosine triphosphate/guanosine triphosphate (ATP/GTP) and electron carriers reduced nicotinamide adenine dinucleotide (NADH) and reduced flavin adenine dinucleotide ($FADH_2$); also known as the tricarboxylic acid cycle or the citric acid cycle

labile equilibrium also called neutral equilibrium, a state of equilibrium such that when a system is displaced slightly from it, the system remains in equilibrium

lactic acid fermentation anaerobic chemical process whereby the enzyme lactate dehydrogenase converts pyruvate to lactate and produces one molecule of oxidized nicotinamide adenine dinucleotide (NAD^+)

laminar flow also called streamline flow, a steady, smooth fluid flow, in which the path of any particle of the fluid is a smooth line, called a streamline

laser a device that produces light that is monochromatic, highly coherent, and relatively intense in a beam that is very parallel and propagates with little spreading

law a regularity of nature, by which predictions can be made

law of conservation of energy states that the total amount of energy in an isolated system must be accounted for; energy can neither be created nor destroyed

law of conservation of matter states that the total amount of matter in an isolated system is constant; matter can neither be created nor be destroyed

Lawson's criterion a relation, involving temperature, confinement time, and ion density, that determines whether a nuclear-fusion reaction will produce more energy than is required to create the reaction

leaching percolation through the soil

leaf electroscope instrument made up of a thin sheet of metal (gold or aluminum) attached to a metal rod; when an electric charge accumulates on the rod, the sheet of metal is repelled; in the presence of radiation, the area around the rod becomes ionized, neutralizing the effect of the electricity, and the metal sheet again associates with the metal rod

Le Chatelier's principle explanation of the equilibrium point of a chemical system; states that when a stress is applied to a system at equilibrium, the system will respond to that stress by shifting the equilibrium toward products or reactants to relieve that stress

length contraction the effect, according to special relativity, that the length of an object as measured by an observer with respect to whom the object is moving is less than the length of the object as measured by an observer with respect to whom the object is at rest

lens any optical device that modifies, by means of refraction and in a controlled manner, the direction of light (or other) rays passing through it

lens equation a mathematical relation for a lens that relates its focal length and the distances of an object and its image from the lens

lens maker's formula a mathematical formula that gives the focal length of a lens in terms of the radii of curvature of the lens's surfaces and the index of refraction of the material forming the lens

lepton any of the set of relatively lightweight, spin-½ particles comprising the electron, the muon, the tau, and their associated neutrinos

Lewis acid an electron pair acceptor

Lewis base an electron pair donor

Lewis electron dot structure representation of the valence electrons (those in the outermost energy level) of an atom

lifetime the average duration an unstable system exists before decaying

lift also called aerodynamic lift or hydrodynamic lift, the component of the force acting on a body due to the flow of a fluid past the body that is perpendicular to the direction of flow if the body were absent

light electromagnetic radiation that can be seen, sometimes including ultraviolet and infrared (which cannot be seen by humans)

light cone of an event in space-time, the locus of all points that possess a lightlike interval with that event, i.e., all points that can represent the event of emission of a photon that reaches the given event or the event of absorption of a photon that is emitted by the given event; equivalently, the three-dimensional hypersurface in space-time that is formed by all the photon world lines passing through the point of the given event

lightlike as describing a space-time interval or separation, an interval between two events, or space-time points, such that a photon emitted from the earlier event can reach the later event

lignite the lowest grade of coal; has the lowest carbon content and the highest sulfur content

linear accelerator referred to as *linac* for short, a particle accelerator that accelerates particles in a straight path

line of action the line in the direction of a force that passes through the force's point of application

lipids a class of hydrophobic ("water-fearing") covalent compounds that perform multiple functions within cells

liquefaction conversion of a gas to a liquid at high pressures

liquid the state of matter in which the matter has a definite volume but not a definite shape and assumes the shape of its container

liquid crystal a substance that combines the properties of a liquid (it flows) with those of a solid crystal (it contains molecules that are arranged in a regular manner)

locality a characteristic of classical physics that what happens at any location is independent of what happens at any other location, unless some influence propagates from one location to the other, and the propagation of influences occurs only at finite speed and in a local manner, such that the situation at any location directly affects only the situations at immediately adjacent locations

longitudinal wave a wave in which the propagating disturbance is in the direction of wave propagation; a sound wave can be described as a longitudinal wave

Lorentz transformation any of a certain group of mathematical relations between an event's coordinates in any space-time coordinate system and the *same* event's coordinates in any coordinate system moving at constant velocity relative to the first, according to special relativity

luminous flux the power transported through a surface by light, as corrected and evaluated for the light's ability to cause visual sensation in humans, whose SI unit is lumen (lm), equivalent to candela-steradian (cd·sr)

luminous flux density also called illuminance or illumination, the luminous flux through or into a surface per unit area of surface, whose SI unit is lumen per square meter (lm/m^2)

lyotropic liquid crystals liquid crystals that change phase on the basis of concentration as well as temperature

machine a device for making the performance of work easier by changing the magnitude or direction of a force or torque

Mach's principle states that the origin of inertia lies in all the matter of the universe, so effectively the origin of inertia lies in the distant stars

MAGLEV magnetic levitation

magnet any object that, when at rest, can produce and be affected by a magnetic field

magnetic dipole a pair of equal-magnitude and opposite magnetic poles at some distance from each other or any equivalent situation produced by an electric current loop

magnetic dipole moment a quantity that measures the strength of a magnet or of a current loop as magnet, often denoted $\boldsymbol{\mu}_m$ (as a vector) or μ_m (as its magnitude), whose SI unit is ampere-meter^2 (A·m^2)

magnetic domain   in ferromagnetic materials, a cluster of atoms or molecules with similarly aligned magnetic dipoles

magnetic field   a condition of space by which charged particles in relative motion affect each other magnetically; the mediator of the magnetic force; a vector field commonly denoted **B**, whose SI unit is tesla (T)

magnetic field line   a directed line in space whose direction at every point on it is the direction of the magnetic field at that point

magnetic flux   through a surface, the integral over the surface of the component of the magnetic field perpendicular to the surface, often denoted Φ_m (Greek capital phi), whose SI unit is weber (Wb); for a uniform magnetic field and a flat surface, the product of the surface area and the component of the field perpendicular to the surface

magnetic levitation (MAGLEV)   the use of magnetic forces to overcome gravity

magnetic resonance imaging (MRI)   a noninvasive imaging technique that applies the principles of nuclear magnetic resonance (NMR) to diagnosis of biological injuries, diseases, and conditions

magnetism   the phenomena that are based on the force between electrically charged objects that is due only to their charges and that is additional to the electric force between them; equivalently, the force between such objects that is due only to their charges and their relative velocity

magnification   of an optical system or component, the ratio of the size of an image to the size of its object

malleable   capable of being bent and shaped by hitting with a hammer or rolling out by rollers

mass   a quantity that measures the amount of matter, the inertia of a body, the body's participation in the gravitational force, and the body's inherent energy, often denoted m or M, whose SI unit is kilogram (kg)

mass defect   the difference between the sum of the masses of the individual nucleons (i.e., protons and neutrons) composing a nucleus and the mass of the nucleus as a whole

mass energy   the energy inherent to mass, according to the special theory of relativity, where $E = mc^2$, with E, m, and c denoting, respectively, the mass energy, the mass, and the speed of light in vacuum

mass flow rate   also mass flux, the mass of fluid flowing through a cross section per unit time, whose SI unit is kilogram per second (kg/s)

mass flux   mass flow rate

massless   not possessing rest mass (and assigned a nominal value 0 to its rest mass), although a massless particle possesses mass due to its energy; a massless particle cannot be brought to rest and moves at the speed of light; the photon and graviton are massless

mass number   also atomic mass number, the number of nucleons (i.e., protons and neutrons) in a nucleus

mass spectrometry   analytical chemistry technique in which ionized compounds are separated on the basis of the mass-to-charge ratio of the ions; also known as mass spectroscopy

mass spectroscopy   analytical chemistry technique in which ionized compounds are separated on the basis of the mass-to-charge ratio of the ions; also known as mass spectrometry

matter   broadly, anything that possesses rest mass; narrowly, as opposed to antimatter, ensembles of matter particles, which include the electron, proton, neutron, and others

Maxwell-Boltzmann statistics   the statistical rules governing any collection of distinguishable particles in thermal equilibrium

Maxwell's equations   a set of equations expressing the laws governing classical electromagnetism

mechanical advantage   for a machine, the ratio of the resistance force to the applied force, often expressed as a percentage

mechanical equilibrium   the state of a system in which the net external force and net external torque acting on the system both vanish

mechanics   the branch of physics that deals with the behavior of bodies and systems of bodies at rest and in relative motion under the effect of forces

Meissner effect   the effect whereby a material in the superconducting state does not suffer the presence of a magnetic field within itself and even expels a preexisting field when the material is cooled to its superconducting state while immersed in a magnetic field

melting   the phase change from solid state to liquid state, fusion

membrane potential   the sum of electrical and chemical concentration gradients from one side of a cellular membrane to another

meniscus   the surface of a liquid in a tube

mesogen   the fundamental part of the liquid crystal molecule that confers the liquid crystal properties on it

meson   any of the set of medium mass, integer-spin particles, which includes the pion and the kaon

mesoscale   the scale employed for nanotechnology, at a level of aggregates of atoms or billionths of meters in dimension, intermediate to the atomic and macroscopic scales

metabolism the sum of all biochemical reactions occurring within or by an organism that involve the utilization and production of energy

metal any of a class of materials that are generally lustrous (reflect light), ductile (can be drawn into sheets and wires), and malleable (can be shaped by pressure and hammering) and are good conductors of electricity and heat

metallic bond type of bond formed when a metal atom bonds with other like metal atoms

metallurgy the study of metals, their compounds, and their alloys; the science of studying metals and removing them from their ores

metals types of atoms with the following characteristics: low ionization energy, containing three or fewer valence electrons, luster, malleability, and ductility and are good conductors

meter one of the SI base units, the SI unit of length, denoted m; the length of the path traveled by light in vacuum during a time interval of 1/299,792,458 second

Michelson-Morley experiment an experiment carried out in the late 19th century that disproved the existence of the ether, a hypothetical material medium that was assumed to carry electromagnetic waves

millimeter of mercury a non-SI unit of pressure, denoted mm Hg, also called torr (torr), equal to 1.333×10^2 pascals (Pa)

minute a non-SI unit of time, denoted min, equal to 60 seconds (s)

mirror any optical device that modifies, by means of reflection and in a controlled manner, the direction of light (or other) rays impinging on it

mirror equation a mathematical relation for a spherical mirror that relates its focal length and the distances of an object and its image from the mirror

mobile phase the part of a chromatography column that is moving with the mixture to be separated

mobile pollution source pollution emitter that moves from place to place such as an automobile or a train

modulus of elasticity a quantity that represents the stiffness of a material, the extent to which the material resists deformation, whose SI unit is pascal (Pa)

molality the concentration unit that includes the number of moles of solute dissolved in a kilogram of solvent

molar heat capacity the amount of heat required to raise the temperature of one mole of a substance by one unit of temperature, often denoted C, whose SI unit is joule per kelvin per mole [J/(mol·K), also written J/mol·K]

molarity a unit of concentration that measures the amount of moles of solute per liter of solution; represented by the symbol M

molar mass the mass of a single mole of a substance, often denoted M, whose SI unit is kilogram per mole (kg/mol)

mole one of the SI base units, the SI unit of amount of substance, denoted mol; the amount of substance that contains as many elementary entities (such as atoms, molecules, ions, electrons, etc., which must be specified) as there are atoms in exactly 0.012 kilogram (kg) of the isotope carbon 12

molecular dynamics computational chemistry technique that simulates the conformations and interactions of molecules and integrates the equation of motion of each of the atoms

molecular orbital theory a sophisticated model of bonding that takes the three-dimensional shape of a molecule into account; the main principle of this theory involves the formation of molecular orbitals from atomic orbitals

moment of inertia the rotational analog of mass, expressed with respect to a rotation axis, often denoted I, whose SI unit is kilogram-(square meter) (kg·m^2); for a point particle, the product of the mass and the square of the distance of the particle from the axis; for a rigid body, the sum of the moments of inertia of all the particles composing the body with respect to the axis

momentum more precisely linear momentum, the product of an object's mass and velocity, often denoted p, whose SI unit is kilogram-meter per second (kg·m/s)

monochromatic light light that consists of only a single color, i.e., a single frequency

monosaccharides simplest form of carbohydrates

Mössbauer effect the achievement of sharp gamma-ray absorption and emission energies for nuclear transitions by anchoring the nuclei's atoms in a crystal—to eliminate recoil effectively—and reducing the temperature—to minimize the Doppler effect

motion change in spatial position or orientation occurring over time

MRI magnetic resonance imaging

multiverse a hypothetical supercosmos that contains the universe as well as other universes

muon a spin-½, negatively charged particle, similar to the electron but 207 times more massive, denoted μ or μ$^-$

natural polymers chemical substances made up of repeating monomers that exist naturally and include such compounds as proteins, nucleic acids, and polysaccharides

natural science science that deals with matter, energy, and their interrelations and transformations; usually refers to physics, chemistry, or biology

nature the material universe with which humans can, or can conceivably, interact

nematic phase type of liquid crystal that is closer to the liquid phase than the solid phase; it is still ordered relative to the director, yet it floats around as in a liquid and does not have any layering order

neural network a network, or circuit, of interconnected real neurons or of artificial neuronlike devices

neutrino any of three kinds of very light, spin-½, electrically neutral particles: electron neutrino, muon neutrino, and tau neutrino

neutron a neutral, spin-½ particle that is one of the constituents of atomic nuclei, consisting of three quarks, often denoted n; the mass of a neutron is approximately equal to that of a proton, but neutrons are uncharged

Newman projection molecular representation of complex molecules; the perspective is down the carbon-carbon bond with the front carbon represented by a point and the back carbon represented by a circle

newton the SI unit of force, denoted N, equivalent to kilogram-meter per second per second ($kg \cdot m/s^2$)

Newtonian mechanics the mechanics of Newton's laws of motion

Newton's first law of motion called the law of inertia, states that in the absence of forces acting on it, a body will remain at rest or in motion at constant velocity (i.e., at constant speed in a straight line)

Newton's law of gravitation states that two point bodies attract each other with forces whose magnitude is proportional to the product of the particles' masses and inversely proportional to the square of their separation

Newton's second law of motion states that the effect of a force on a body is to cause it to accelerate, where the direction of the acceleration is the same as the direction of the force and the magnitude of the acceleration is proportional to the magnitude of the force and inversely proportional to the body's mass

Newton's third law of motion called the law of action and reaction, states that when a body exerts a force on another body, the second body exerts on the first a force of equal magnitude and of opposite direction

nicotine an addictive chemical substance found in cigarettes

nitrogen cycle the natural use and regeneration of nitrogen-containing compounds from the atmosphere, the soil, and living organisms

nitrogen fixation the conversion of molecular nitrogen into nitrogen compounds useful to plants such as ammonium and nitrates, carried out by soil bacteria

nitrogen oxides (NO_x) refer to a class of covalent compounds formed between various numbers of nitrogen and oxygen atoms

nitroglycerine a highly explosive and unpredictable chemical with the formula $C_3H_5N_3O_9$

NMR nuclear magnetic resonance

noble gas a chemical element belonging to what is now the final group of representative elements (Group VIIIA, or IUPAC Group 18). This group contains helium, neon, argon, krypton, xenon, and radon. The noble gases have a filled outer energy level with two electrons in the s sublevel and six in the p sublevel. Since they have a full outer shell, these elements do not gain or lose electrons and are therefore considered unreactive

node any location along a standing wave at which the oscillating medium—or the electromagnetic field, for a standing electromagnetic wave—remains in equilibrium, i.e., at rest

nonconservative force any force that is not a conservative force, that performs nonzero total work on a body moving in a closed path

nonelectrolyte a substance that will not conduct electricity when dissolved in water

nonlocality a characteristic of quantum physics that the situations at different locations can be entangled, so that what happens at one affects what happens at the other in a manner that is not explainable in terms of an influence propagating from one to the other

normal boiling point the temperature at which a liquid boils or a gas condenses at the pressure of one atmosphere

normal freezing point the temperature at which a liquid freezes or a solid melts at the pressure of one atmosphere

normality unit of concentration of an acid in which the molarity of the acid is multiplied by the number of dissociable hydrogen atoms (equivalents) in the acid

normalization for a vector, conversion to a vector of unit magnitude in the same direction; for a wave function, imposition of the condition that the probability of the system's being anywhere is unity

north pole the pole of a magnet that is attracted to the magnetic pole of the Earth that is near its north geographic pole; the pole of a magnet from which magnetic field lines emerge

NSAIDs nonsteroidal antiinflammatory drugs

nuclear chemistry the branch of chemistry that involves the study of radioactivity

nuclear energy the energy found within the attraction between protons and neutrons, the subatomic particles in an atomic nucleus

nuclear engineering the application of nuclear physics to the design and construction of devices involving nuclear processes

nuclear fission the process in which a large atom (such as uranium) splits into two smaller atoms when bombarded with neutrons

nuclear-fission bomb also called fission bomb or atom bomb, a bomb based on the energy released in an uncontrolled nuclear-fission reaction

nuclear force the manifestation of the strong force as it acts between nucleons (protons and neutrons), possibly binding them to form atomic nuclei

nuclear fusion a method of producing energy by combining two smaller atomic nuclei into one larger nucleus

nuclear-fusion bomb also called hydrogen bomb or fusion bomb, a bomb in which a nuclear-fission bomb ignites an uncontrolled nuclear-fusion reaction, with concomitant release of energy

nuclear magnetic resonance (NMR) an analytical chemistry technique that determines the chemical composition and structure of substances on the basis of the motion of the atomic nuclei in a magnetic field

nuclear medicine the application of nuclear physics to medical purposes

nuclear physics the field of physics that studies atomic nuclei

nuclear reactor a device for maintaining and containing a controlled nuclear chain reaction

nucleic acids a class of polymers that are responsible for the storage of hereditary information within the cell and the transmittance of that information from the parent cell to any daughter cells it may produce

nucleon one of the components of an atomic nucleus: a proton or a neutron

nucleosome the basic unit of packaging of deoxyribonucleic acid (DNA) into chromatin

nucleus the central core of an atom, consisting of protons and neutrons, where most of the atom's mass resides; in biology, the membrane-bound organelle of eukaryotic cells that contains the genetic material

nuclide a type of nucleus characterized by both a particular atomic number and a particular mass number

nutrient cycling the constant use and regeneration of chemical elements through geological and life processes

object a source of light rays that enter an optical system or component; an object can be real or virtual

Ohm's law states that the voltage V across an electric circuit component is proportional to the electric current i flowing through the component, $V = iR$, where R denotes the component's resistance

oil reserves locations in the earth that contain large quantities of crude oil

open system a chemical reaction in which the system and surroundings can freely exchange material and energy

operator in quantum mechanics, a mathematical expression that represents a physical quantity and "operates" on a wave function

optical power of an optical system or component, the inverse of the focal length, often denoted D, whose SI unit is inverse meter (m^{-1}) but is commonly called diopter (D)

optics the study of the behavior of light, in particular, and of electromagnetic and other waves and radiation, in general

organic chemistry the branch of chemistry that involves the study of carbon-containing compounds

orientational order a term in liquid crystals that describes how likely it is that the mesogen will be aligned along the director

overtone any of the frequencies of a harmonic sequence that are higher than the fundamental (i.e., the lowest frequency of the sequence), where the first overtone is the second harmonic, the second overtone the third harmonic, etc.

oxidation the loss of one or more electrons by a substance during a reaction or an increase in the oxidation number of the substance

oxidation number the effective charge of an atom if it were to become an ion

oxidation-reduction reaction a chemical reaction in which one or more electrons are transferred between reactants; also called redox reactions

oxidizing agent the substance in an oxidation-reduction reaction that is reduced (gains electrons) and therefore is responsible for oxidizing the other component in the reaction

ozone a natural component of the atmosphere formed by covalent bonding of three oxygen atoms; ozone exists in two resonance forms with one double bond between one pair of oxygen atoms and a single bond between the others

pair production the process in which a sufficiently energetic photon (a gamma ray) converts in the vicinity of an atomic nucleus into a particle-antiparticle pair, most commonly into an electron and a positron

paper chromatography mixture separation technique that uses porous paper as the stationary phase

parabolic mirror more correctly called paraboloidal mirror, a mirror whose surface has the shape of a rotational paraboloid

paraboloidal mirror a mirror whose surface has the shape of a rotational paraboloid

parallel-axis theorem relates a body's moment of inertia with respect to any axis to its moment of inertia with respect to a parallel axis through its center of mass; states that if the moment of inertia of a body of mass m with respect to an axis through its center of mass is I_0, then its moment of inertia I, with respect to a parallel axis at distance d from the center-of-mass axis, is given by $I = I_0 + md^2$

parallel connection a method of connection of electrical components in which all of one of the two leads of every component are connected together, as all of the other leads of every component are connected together, assuring the same voltage across all the components

paramagnetism the situation in which the induced magnetic field in a material slightly enhances the applied magnetic field

parity a property of elementary particles related to their behavior under reflection

particle a discrete, localized entity, described by such characteristics as its position, mass, energy, and momentum; as distinguished from antiparticle, any of the components of ordinary matter such an electron, proton, or neutron, as well as others

particle-antiparticle conjugation invariance also called charge conjugation invariance; the laws of physics, except for the weak force, are the same for elementary particles as for their respective antiparticles

particle distinguishability a characteristic of classical physics that all particles are distinguishable from each other, at least in principle

particle indistinguishability a characteristic of quantum physics that identical particles are fundamentally indistinguishable

particle physics the field of physics that studies the properties of and forces acting among the elementary particles

particle pollution also known as *PM*, for particulate matter, a class of inhalable particles smaller than 10 microns (a micron equals one-millionth of a meter) in diameter, from various sources that are regulated by the Environmental Protection Agency (EPA)

parts per billion (ppb) unit of concentration that measures the number of parts of a solute present per billion parts of solvent

parts per million (ppm) unit of concentration that measures the number of parts of a solute present per million parts of solvent

pascal the SI unit of pressure, denoted Pa, equivalent to newton per square meter (N/m^2)

Pascal's principle states that an external pressure applied to a fluid that is confined in a closed container is transmitted at the same value throughout the entire volume of the fluid

Pauli exclusion principle states that no more than a single fermion may exist simultaneously in the same quantum state; e.g., two electrons within an atom cannot have exactly the same energy; named after the Austrian physicist Wolfgang Pauli

Peltier effect the production of a temperature difference between the junctions of a thermocouple when a current flows through it

percent mass/volume unit of concentration that measures the mass of the solute in grams per volume of solution in milliliters

percent volume/volume unit of concentration that measures the volume of solute per volume of solution

period a row of the periodic table, where physical properties vary across periods; the duration of a single cycle of a repetitive phenomenon, often denoted T, whose SI unit is second (s)

periodic wave a wave such that the medium in which the wave propagates—or, for an electromagnetic wave, the electromagnetic field—undergoes repetitive oscillation: at every point along the line of propagation, the medium oscillates locally, while at any instant, a snapshot of the whole medium shows a spatially repetitive waveform; a periodic wave is characterized by frequency and wavelength

permeability more fully called magnetic permeability, the constant μ that appears in the Biot-Savart law for the production of a magnetic field by an electric current or by a moving electric charge and in all other formulas for magnetic field production, whose SI unit is tesla-meter per ampere (T·m/A); the product of the relative permeability μ_r of the material and the permeability of the vacuum μ_0, whose value is $4\pi \times 10^{-7}$ T·m/A

permittivity a quantity that appears in Coulomb's law and in formulas related to the law, commonly denoted ε, equals $\kappa\varepsilon_0$, where κ is the dielectric constant of the medium and $\varepsilon_0 = 8.85418782 \times 10^{-12}$ C^2/(N·m^2) is the permittivity of the vacuum

perpetual motion the putative performance of any device that is claimed to produce free or almost-free energy while violating the first or second law of thermodynamics

pesticides chemicals applied to crops to eliminate harmful pests

pH the measure of the amount of hydrogen ions present in a solution, determined by the formula

$$pH = -\log [H^+]$$

phase in physics, a stage in the cycle of a repetitive phenomenon; in physics and chemistry, a state of matter; in physics and chemistry, a distinct, bounded component of a mixture

phase change phase transition, the transformation of matter from one state to another

phase transition the transformation of matter from one state to another

phlogiston a mysterious substance that many 18th-century scientists believed was released during burning

phonon the particle aspect of acoustic waves

phospholipid bilayer the membrane in cells, consisting of two layers of phospholipids

phospholipids two hydrophobic fatty acid chains attached to a hydrophilic ("water-loving") phosphate group

photoelectric effect the emission of electrons from metals as a result of illumination by light or other electromagnetic radiation

photon a spin-1 particle of electromagnetic waves and the mediator of the electromagnetic force, often denoted γ (Greek gamma)

physical chemistry the branch of chemistry studying the physical laws governing the motion and energy of the atoms and molecules involved in chemical reactions

physical optics also called wave optics, the branch of optics that takes into account the wave nature of light

physicist a person who has studied physics in depth and in breadth and has made physics his or her profession

physics the branch of natural science that deals with the most fundamental aspects of nature

piezoelectricity the production of voltage from pressure, and vice versa, typically in a crystal

pinhole camera design of one of the first photographic devices; the pinhole camera includes a light-sealed box with a hole the size of a pin or smaller

P invariance reflection invariance

pion a spin-0 particle carrying the strong force that holds protons and neutrons together in atomic nuclei, consisting of a quark and an antiquark, often denoted π^+, π^-, or π^0, depending on its electric charge

pitchblende the unrefined ore from which uranium is obtained

Planck constant a fundamental physical constant that characterizes quantum physics, denoted h, whose value is $6.62606876 \times 10^{-34}$ joule-second (J·s)

Planck length a length of about 10^{-35} meter (m) that is assumed to characterize the quantum nature of space

plane mirror a mirror whose shape is a flat surface

plane-polarized light visible light that has passed through a polarizing filter; only the light that vibrates in the same direction as the filter passes through the polarizing filter

plasma the state of matter that is similar to a gas, but with ionized atoms or molecules

plastic material an inelastic material, one that does not return to its original configuration when the cause of deformation, the stress, is removed

plastics polymers that can be molded and shaped into commercially useful products

pneumatic trough a device for the collection of gases made of a reaction vessel connected by a tube to a bottle inverted in a tub of liquid, either mercury or water

Poincaré transformation any of a group of transformations that form the most general mathematical relations between an event's coordinates in any space-time coordinate system and the *same* event's coordinates in any coordinate system moving at constant velocity relative to the first, according to special relativity

Poiseuille's law states that the volume flow rate of fluid flowing steadily through a pipe is given by $(\pi r^4/8\eta)(\Delta p/L)$, where r, L, η, and Δp represent the radius and length of the pipe, the viscosity of the fluid, and the difference in fluid pressures at the ends of the pipe, respectively

polar covalent bond a bond formed by the sharing of electrons between two atoms that do not have the same attraction for the electrons; a separation of charge is created between the two

polarization the property of a transverse wave that a constraint exists among the oscillations in various directions (which are all perpendicular to the direction of propagation); for example, in linear, or plane, polarization the disturbance is confined solely to a single direction

pole one of the two ends of a magnet from which magnetic field lines emerge (north pole) or into which magnetic field lines enter (south pole)

polyethylene polymer made by the addition of ethylene monomers; used in packaging

polyethylene terephthalate a polymer formed by condensation reactions; most commonly used in plastic soda and water bottles

polymer high-molecular-weight substance made up of a chain of smaller monomers in a repeating structure that produces a tough and durable product

polymerase chain reaction (PCR) a biochemical technique in which a small quantity of deoxyribonucleic acid (DNA) can be copied

polypropylene a polymer formed by addition reactions of propylene units and utilized in kitchenware, appliances, food packaging, and other containers

polysaccharides organic polymers made up of repeating monosaccharides; cellulose, starch, and glycogen are polysaccharides

polystyrene a thermoplastic polymer made from addition reactions of styrene monomers

polyurethane polymer formed by condensation reactions utilized in such applications as furniture upholstery foam, Spandex materials, and water-protective coatings for wood

polyvinyl chloride (PVC) thermoplastic polymer that results from addition reactions of vinyl monomers; one of the most commonly used plastics in the world

position vector of a point in space, the vector whose magnitude equals the point's distance from the origin of the coordinate system and whose direction is the direction of the point from the origin

positron the antiparticle of the electron, possessing the same mass, spin, and magnitude of electric charge as the electron, but with positive charge, often denoted e^+ or β^+

post-Hartree-Fock method a method of computational chemistry that is an improvement of the Hartree-Fock method. This computational method includes the repulsion caused by the electrons' interactions rather than averaging the repulsions

potential electric potential

potential difference between two points, the difference of the electric potential at one point and that at another, commonly denoted ΔV or V, whose SI unit is volt

potential energy energy of position or stored energy; any energy that a body possesses that is not due to the body's motion, often denoted U, PE, or E_p, whose SI unit is joule (J)

power the rate of performance of work or the rate of transfer or conversion of energy, often denoted P; its SI unit is watt (W)

precession the wobbling motion of a spinning body such that its axis of rotation sweeps out a cone

precipitate solid form of a compound formed in aqueous solution

precipitation the formation of an insoluble solid as a product of a chemical reaction

precipitation reaction type of chemical reaction in which one or more of the products formed is insoluble and falls out of solution as a precipitate

pressure the force acting perpendicularly on a surface per unit area of surface, often denoted p or P, whose SI unit is pascal (Pa)

prism an optical device made of a block of uniform transparent material with polished planar faces, used for changing the direction of light rays with or without inverting the transmitted image or for separating light into a spectrum

proportionality limit in elasticity, the greatest stress for which the strain is still proportional to the stress

prostaglandins a group of local hormones that were originally discovered in secretions from the prostate gland

protein a polymer of amino acids that is responsible for most of the cellular activity in a living organism

proton a positively charged, spin-½ particle that is one of the constituents of atomic nuclei, consisting of three quarks, often denoted p

pure tone a sound consisting of only a single frequency

pyroelectricity the production of a voltage across a crystal when the crystal is heated

qualitative analysis analytical chemistry analysis that concerns the identity and properties of components in a sample

quantitative analysis a technique common in analytical chemistry in which one studies the amount of each component in a sample substance

quantization the achievement of understanding of a physical system in terms of quantum physics, rather than of classical physics, and according to the rules of quantum mechanics

quantum an elementary unit of a physical quantity; an elementary, particle-like manifestation of a field

quantum chromodynamics (QCD) the quantum theory of the strong force

quantum cloning reproducing the quantum state of one system in another system by means of quantum entanglement

quantum cryptography quantum encryption

quantum dot a structure in a semiconductor crystal that confines electrons very closely in all three dimensions, with a size of the order of tens of nanometers

quantum electrodynamics (QED) the quantum theory of electromagnetism and the most accurate theory in existence in science

quantum encryption the secure transmission of secret information through the use of quantum entanglement

quantum entanglement two or more particles forming components of a single quantum state

quantum field theory the application of quantum mechanics to the understanding of fields, whose underlying idea is that all forces are described by fields that possess a particle aspect and all matter particles are described by fields, such that the principles of quantum physics and the rules of quantum mechanics are obeyed

quantum mechanics   the realization of the principles of quantum physics as they apply to actual physical systems

quantum number   a number characterizing the states in a sequence of discrete states of a quantum system

quantum physics   the most general and widely applicable physics that physicists have, physics characterized by the Planck constant $h = 6.62606876 \times 10^{-34}$ joule-second (J·s)

quantum theory   the theory of the quantum nature of the world

quark   a fractionally charged, spin-½ particle, appearing in six flavors and three color states, and three of which form every proton and neutron

radian   an SI supplementary unit, the SI unit of plane angle, denoted rad; $1/(2\pi)$ of a complete circle

radiant flux   the power transported by light or other electromagnetic radiation, whose SI unit is watt (W)

radiant flux density   also called irradiance or radiant power density, the radiant flux through or into a surface per unit area of surface, whose SI unit is watt per square meter (W/m^2)

radiant power density   radiant flux density

radiation   transmitted energy in the form of waves or particles; the transfer of heat by electromagnetic radiation

radical   a very reactive species that contains unpaired electrons; often undergoes spontaneous reactions and can be utilized in chain reactions; also called a free radical

radical reaction   type of chemical reaction involving atoms and molecules known as radicals or free radicals that contain lone unpaired electrons

radioactivity   the spontaneous transmutation, or decay, of an unstable atomic nucleus with a concomitant emission of radiation

radius of curvature   of a surface, the radius of the sphere of which the surface forms part

radius of gyration   of a rigid body with respect to an axis, the distance from the axis at which a point particle of the same mass as the body would have the same moment of inertia as does the body

rate law   the physical law that governs the speed at which reactants are converted into products in a chemical reaction

Rayleigh criterion   in wave optics, a criterion for minimal image resolution of two point light sources, which is that the center of the image of one source falls on the first diffraction ring of the second

ray optics   also called geometric optics, the branch of optics that deals with situations in which light propagates as rays in straight lines and its wave nature can be ignored

reaction engine   any engine that operates by ejecting matter, such as combustion products, in the rearward direction, causing a forward-directed reaction force on the engine and a consequent forward acceleration of the engine and the vehicle attached to it, according to Newton's second and third laws of motion

reaction force   together with an action force, one of a pair of forces between two bodies, each force acting on a different body, that have equal magnitudes and opposite directions, according to Newton's third law of motion

reaction rate   the speed at which a chemical reaction takes place

real focal point   the point at which parallel light rays are actually made to converge by an optical system or component

real image   an optical image formed by the actual convergence of light rays; a real image can be caught on a screen

real object   an actual source of light rays, from which the rays are diverging and entering an optical system or component

reciprocating engine   any engine that operates with a back-and-forth motion of an essential part

red shift   the Doppler shift of electromagnetic waves emitted by a source that is receding from the observer, in which the observed frequency is lower than the frequency of the source

reducing agent   the substance in an oxidation-reduction reaction that is oxidized (loses electrons) and therefore is responsible for reducing the other component in the reaction

reduction   the gain of one or more electrons by a substance during a reaction, or a decrease in the oxidation number of a substance

reference frame   a coordinate system that serves as a reference standard for position, rest, motion, velocity, speed, acceleration, inertia, and so on

reflection   the phenomenon in which some or all of a wave, impinging on the interface surface between two different media, propagates back into the medium from which it originates; a transformation that acts as reflection in a two-sided plane mirror

reflection invariance   the laws of physics, except the weak force, do not change under reflection

refraction   the change in the propagation direction of a wave upon passing through the interface surface between two media in which the wave possesses different speeds of propagation

refractometer device used to measure the amount of light that is refracted when it passes from air into a liquid

relative of a physical quantity, having an observer-dependent value, such as the velocity of a massive body; of a physical quantity, having a value that is expressed with reference to another physical quantity or entity, such as torque (with reference to a point in space) and relative permeability (with reference to the permeability of the vacuum)

relative permeability the factor μ_r that multiplies the permeability of the vacuum μ_0 in the Biot-Savart law for the production of a magnetic field by an electric current or by a moving electric charge and in all other formulas for magnetic field production, expressing the contribution to the net magnetic field of the magnetic field induced in the material

relative velocity the velocity of a moving object as observed in a reference frame that is itself moving

relativistic mechanics the mechanics of special relativity

representative elements elements within certain groups of the periodic table—Groups 1, 2, 13, 14, 15, 16, 17, and 18. The valence electrons of these elements are found in s and p orbitals

resistance the ratio of the voltage across an object to the electric current flowing through it, commonly denoted R or r, whose SI unit is ohm (Ω); depends on the size and shape of the object and on the material from which it is made, through the latter's resistivity

resistivity a property of any material that characterizes its intrinsic resistance to the flow of electric current, commonly denoted ρ (Greek rho), whose SI unit is ohm-meter ($\Omega\cdot m$); the resistance of a one-meter cube of a material, when a voltage is maintained between two opposite faces of the cube

resistor an electric device that is designed to possess a definite resistance and be able to function properly under a definite rate of heat production

resolving power the ability of an optical imaging system to exhibit fine detail that exists in the object being imaged

resonance the existence of one or more resonant frequencies for a system; any one of those frequencies

resonance structures representations of molecules that require more than one Lewis structure; the true structure lies between both Lewis structures

resonant frequency a frequency of a periodic external influence (such as an oscillating force, voltage, or magnetic field) that causes a system to respond strongly

rest mass the mass of an object as measured by an observer with respect to whom the object is at rest

restoring force the force that takes a system back toward equilibrium, when a system is displaced from a state of stable equilibrium

restriction enzymes enzymes that cleave deoxyribonucleic acid (DNA) at sequence-specific sites

Reynolds number the quantity $\rho v l/\eta$, used to characterize the flow of incompressible, viscous fluids as smooth (< about 2,000) or turbulent (> about 3,000), where ρ, η, and v represent the fluid's density, viscosity, and flow speed, respectively, and l is a length characteristic of the geometry of the system

RFLP (restriction fragment length polymorphism) the method of deoxyribonucleic acid (DNA) fingerprinting that Alec Jeffries developed in the early 1980s; involves comparing the number of minisatellites a particular individual has at a particular locus in the genome

ribbon diagrams molecular representations that highlight secondary structures such as alpha-helices and beta-sheets in large protein molecules

ribozymes catalytic ribonucleic acid (RNA) molecules that perform enzymatic roles in living organisms

rolling friction the friction, involving surface deformation, that acts to impede the rolling of a round body on a surface

root mean square in statistics, the square root of the mean (i.e., average) of the square of a quantity, denoted $\sqrt{(Q^2)_{av}}$ for any quantity Q, and often briefly denoted Q_{rms}

rotary engine any engine that involves solely rotational motion

rotation change of spatial orientation

rotation invariance the laws of physics are the same in all directions

scalar any physical quantity that is not characterized by direction, such as temperature and work

scalar product a scalar quantity obtained from any pair of vectors, whose value equals the product of the vectors' magnitudes and the cosine of the smaller angle (less than or equal to 180°) between the vectors' directions; also called dot product and denoted $\mathbf{u}\cdot\mathbf{v}$ for vectors $\mathbf{u}$ and $\mathbf{v}$

scalar wave a propagating disturbance in a scalar quantity, such as a temperature wave, a pressure wave, or a density wave

scanning electron microscopy (SEM) a visualization technique for evaluating the composition and structural characteristics of a substance at the surface

Schrödinger equation a differential equation for the wave function of a quantum system that determines the evolution and properties of the system

Schrödinger's cat an imaginary cat whose characteristic of being both alive and dead at the same time exposes a difficulty with quantum theory

science also natural science, the human endeavor of attempting to comprehend rationally—in terms of general principles and laws—the reproducible and predictable aspects of nature

second one of the SI base units, the SI unit of time, denoted s; the duration of 9,192,631,770 periods of the microwave electromagnetic radiation corresponding to the transition between the two hyperfine levels in the ground state of the atom of the isotope cesium 133

secondhand smoke cigarette smoke that leaves the end of the cigarette and exposes nonsmokers to the chemical components of cigarettes

second law of thermodynamics the entropy (measure of disorder) of an isolated system cannot decrease in time

Seebeck effect the production of a voltage between the junctions of a thermocouple when they are at different temperatures

selectively permeable membrane a biological membrane that allows only certain substances to pass through

semiconductor a solid in which the conduction band is not populated and the gap between the valence and conduction bands is around one electron volt, allowing electrons to be raised to the conduction band with relatively small amounts of energy

semiempirical methods computational chemistry techniques that utilize experimental data to correct their calculations of simplified versions of Hartree-Fock methods

series connection a method of connection of electrical components, with a lead of one component connected to a lead of the second, the other lead of the second component connected to a lead of the third, and so on, assuring that the same electric current flows through all components

shear the situation in which a pair of equal-magnitude, oppositely directed (i.e., antiparallel) forces act on material in such a way that their lines of action are displaced from each other

shear modulus in elasticity, also called rigidity modulus, the ratio of the shear stress (the magnitude of one of the pair of shear forces divided by the area parallel to it over which it acts) to the shear strain (the angle of deformation), often denoted S, whose SI unit is pascal (Pa)

SI the International System of Units

sidereal day the time for Earth to make a complete rotation about its axis with respect to the stars; about four minutes less than a solar day of 24 hours, the time for Earth to make a complete rotation about its axis with respect to the Sun

significant figures the digits in a number, representing the amount of a physical quantity, that carry information about the precision of that number; considered all of the measured digits and one estimated digit

simultaneity the occurrence of two or more events at the same time

single bond a type of bond in which a single pair of electrons are shared between two atoms. They form by the end-to-end overlap of s orbitals

single-replacement reactions a class of chemical reactions in which a lone metal atom replaces the positive ion of a compound

slow neutron a neutron that has been passed through paraffin and slowed down; slow neutrons have the potential for more collisions with nuclei, causing more radioactive emissions

smectic phase form of liquid crystal more like the solid phase with molecules ordered in layers and with some variability within the layers

Snell's law for the refraction of a wave at an interface, states that $\sin \theta_i / v_i = \sin \theta_r / v_r$, where θ_i and θ_r are the angles of incidence and refraction, respectively, and v_i and v_r denote the propagation speeds of the wave in the medium of incidence and the medium of transmission (refraction), respectively; alternatively, $n_i \sin \theta_i = n_r \sin \theta_r$, where n_i and n_r are the respective indices of refraction of the wave in the two media

solid the state of matter in which the matter has a definite shape and volume

solid phase peptide synthesis method of preparing proteins, polypeptides (long chains of amino acids), or smaller peptides by linking monomers by creating peptide bonds

solid state chemistry a branch of chemistry involved in the study of the physical and chemical properties of substances in the solid state

solubility the amount of solute that can dissolve in a unit volume or mass of solvent at a given temperature

solute the substance that is dissolved and is present in the solution in a lesser amount than the solvent

solvation the process of dissolving

solvent the substance that forms the major component of a solution, in which a solute is dissolved

sound a mechanical wave propagating through a material medium

south pole the pole of a magnet that is attracted to Earth's magnetic pole that is near its south geographic pole; the pole of a magnet into which magnetic field lines enter

space the three-dimensional expanse in which objects and events have location

space-filling model visual representation of the three-dimensional structure of a molecule in which each part of the model is the appropriate size relative to the size of the atom

spacelike as describing a space-time interval or separation, an interval between two events, or space-time points, such that a particle emitted from the earlier event and moving slower than the speed of light in vacuum cannot reach the later event; equivalently, an interval such that a photon (which moves at the speed of light in vacuum) that is emitted from the earlier event reaches the location of the later event after the later event actually occurs

space-time the four-dimensional merger of space and time; the arena for events, each of which possesses a location and a time of occurrence

spatial-displacement invariance the laws of physics are the same at all locations

spatial symmetry symmetry under spatial changes, such as displacements, rotations, and reflections

spatiotemporal symmetry symmetry under changes involving both space and time

special relativity Albert Einstein's special theory of relativity

special theory of relativity proposed by Albert Einstein, a supertheory expressed in terms of space-time, whose basic postulates are that the laws of physics are the same for all observers who move at constant velocity with respect to each other and that the speed of light in vacuum is one such law

specific gravity the ratio of the density of a substance to the density of water

specific heat capacity also called specific heat, the amount of heat required to raise the temperature of one unit of mass by one unit of temperature, often denoted c, whose SI unit is joule per kelvin per kilogram [J/(kg·K), also written J/kg·K]

spectrophotometer an instrument that separates light into its component frequencies (or wavelengths) and measures the intensity of each, also called a spectrometer or spectroscope

spectrophotometry an analytical chemical method that is based on the absorption and emission of electromagnetic radiation by a substance; also known as spectroscopy

spectroscopy an analytical chemical method that is based on the absorption and emission of electromagnetic radiation by a substance; also known as spectrophotometry; the study of the frequency components (colors) of light

spectrum an ordered array of the frequency components of a wave, especially of the colors of a light wave

speed the magnitude of the velocity, often denoted v, whose SI unit is meter per second (m/s)

speed distribution function in statistical mechanics, a function of speed v that, when multiplied by dv, gives the number of molecules per unit volume that have speeds in the small range between v and $v + dv$

speed of light usually refers to the speed of light in vacuum, which is the same for all electromagnetic waves and is a fundamental physical constant denoted c, whose value is 2.99792458×10^8 meters per second (m/s)

spherical mirror a mirror having the shape of part of the surface of a sphere; can be concave or convex

spin the intrinsic angular momentum of an elementary particle or atomic nucleus

stable equilibrium a state of equilibrium such that when the system is displaced slightly from it, the system tends to return to the equilibrium state

standard model the combination of the electroweak theory and quantum chromodynamics, a successful theory of the strong, weak, and electromagnetic forces and of the elementary particles

standing wave a situation in which a wave propagates within a system, reflects from the system's boundaries, and interferes with itself, so that the system oscillates while no wave propagation is apparent; a plucked guitar string exhibits a standing wave, for example

state of matter most commonly, one of the four states: solid, liquid, gas (vapor), and plasma

static friction solid-solid friction, involving mostly adhesion and surface irregularity, that is sufficient to prevent motion

stationary not changing in time in certain aspects

stationary pollution source a factory or other emitter of pollution that stays in one place

statistical mechanics the field of physics that studies the statistical properties of physical systems containing large numbers of constituents and relates them to the system's macroscopic properties.

Stefan-Boltzmann constant the constant appearing in the Stefan-Boltzmann law, denoted σ, whose value is 5.67051×10^{-8} W/(m^2·K^4)

Stefan-Boltzmann law the total power of electromagnetic radiation emitted by a blackbody at all wavelengths P is proportional to the fourth power of the absolute temperature T, $P = \sigma T^4$, where σ denotes the Stefan-Boltzmann constant

steradian an SI supplementary unit, the SI unit of solid angle, denoted sr; $1/(4\pi)$ of a complete sphere

stereochemical diagrams two-dimensional representations of the geometric arrangement of atoms in a molecule

steric number (SN) the total number of bonding and lone-pair electrons; contributes to the three-dimensional shape of the molecule

stimulated emission the emission of a photon by an atom in an excited state due to a passing electromagnetic wave whose photon energy matches that of the emitted photon

stoichiometry the mathematical representation of the amounts of substances in a chemical reaction

Stokes's law gives the magnitude of the viscous force that acts on a solid sphere moving at sufficiently low speed through a fluid as $6\pi r\eta v$, where r, η, and v represent the radius of the sphere, the viscosity of the fluid, and the sphere's speed, respectively

storage ring a device associated with a particle accelerator, in which previously accelerated particles are kept in circular orbits

strain the deformation of a material as a result of a stress

stratosphere the second layer of the Earth's atmosphere, above the troposphere

streamline the smooth-line path of any particle in a laminar fluid flow

streamline flow also called laminar flow, a steady, smooth fluid flow, in which the path of any particle of the fluid is a smooth line, called a streamline

stress the mechanical cause of a material's deformation, expressed as a force divided by an area, whose SI unit is pascal (Pa)

string theory also called superstring theory, a proposed theory of everything—a theory unifying all the fundamental forces—that involves exceedingly tiny objects, called strings, whose vibrational modes form the elementary particles

strip mining surface mining in which the surface layers of the earth are stripped away to reveal the coal below

strong force one of the four fundamental forces, a short-range force responsible for binding quarks together and for binding nucleons together to form nuclei

Student's t-test statistical test of mean equality

subatomic particle any of the particles that compose atoms or are similar to such particles and also particles that constitute such particles or mediate the forces among them

subbituminous coal form of coal with higher carbon content than lignite, but less than bituminous

sublimation the phase change from a solid directly to a gas or vapor without going through the liquid phase

sulfur dioxide a combustion product of sulfur-containing fuels such as coal and oil; contributes to the formation of acid rain

superconductivity the total lack of electric resistance that certain solid materials exhibit at sufficiently low temperatures, i.e., at temperatures below their respective critical temperatures

superconductor a material that possesses a superconducting state, in which it totally lacks electric resistance, at sufficiently low temperatures

superfluidity a low-temperature state of liquid matter that flows with no viscosity

supertheory a theory about theories, a theory that lays down conditions that other physics theories must fulfill; special relativity is an example of a supertheory

surface chemistry the atomic and molecular interactions at the interface between two surfaces

surface runoff the flow of water over land; surface runoff can carry with it substances from the land, including pollutants

surface tension the resistance of a liquid to increase of its surface area, quantified as force required to extend the surface, per unit length—perpendicular to the force—along which the force acts, often denoted γ, whose SI unit is newton per meter (N/m); skinlike attraction of water molecules to each other on the surface of a liquid due to hydrogen bonds

surroundings everything outside the experimental system

symmetric possessing symmetry

symmetry the possibility of making a change in a situation while some aspect of the situation remains unchanged, said to be symmetry under the change with respect to the unchanged aspect

symmetry principle also called Curie's symmetry principle or the Curie principle, states that the symmetry of a cause must appear in its effect; equivalently, any asymmetry in an effect must also characterize the cause of the effect; equivalently, an effect is at least as symmetric as its cause

synchrotron a type of circular particle accelerator

system the part of the reaction that one is studying

tangential acceleration the tangentially directed acceleration that is the time rate of change of the speed of a body in circular motion, whose magnitude is often denoted a_t and whose SI unit is meter per second per second (m/s^2)

tau a spin-½, negatively charged particle, similar to the electron but 3,480 times more massive, denoted τ

temperature a measure of the random kinetic energy of the microscopic constituents (atoms, molecules, ions) of a substance, often denoted T, whose SI unit is kelvin (K)

temporal symmetry symmetry under changes in time, time ordering, or time intervals

temporal-displacement invariance the laws of physics are the same at all times

tension stretching

terminal speed the final, constant speed of an object falling in a resistive medium

terminal velocity terminal speed

tesla the SI unit of magnetic field strength, denoted T, equivalent to weber per square meter (Wb/m^2)

textile chemistry the branch of chemistry that studies the design and manufacture of fibers and materials

theory an explanation for a law or laws

theory of everything (TOE) a theory that, if discovered, would unify all the four fundamental forces of nature

thermal contraction the contraction that most materials undergo as their temperature decreases

thermal efficiency the ratio of the work performed by an engine to the heat taken in, often expressed as a percentage

thermal expansion the expansion that most materials undergo as their temperature rises

thermionic emission the spontaneous emission of electrons from a hot metal, where the temperature endows some of the metal's electrons with sufficient thermal energy to leave the metal

thermochemistry the branch of chemistry involving the study of energy and energy transformations that occur during chemical reactions

thermocouple pair of wires of different metals that have both their ends respectively welded together; produces a voltage and electric current when the two junctions are at different temperatures (Seebeck effect) and creates a temperature difference when a current flows through it (Peltier effect)

thermodynamic equilibrium for systems to which thermodynamic considerations apply, a state of maximal entropy

thermodynamics the field of physical science that deals with the conversion of energy from one form to another while taking temperature into account and that derives relationships among macroscopic properties of matter

thermodynamics, first law of states that the increase in the internal energy of a system equals the sum of work performed on the system and heat flowing into the system

thermodynamics, second law of states that heat does not flow spontaneously from a cooler body to a warmer one; equivalently, the thermal efficiency of a cyclic heat engine is less than 100 percent; equivalently, the entropy of an isolated system cannot decrease over time

thermodynamics, third law of states that it is impossible to reach absolute zero temperature through any process in a finite number of steps; equivalently, a Carnot engine cannot have perfect efficiency; equivalently, at absolute zero temperature, all substances in a pure, perfect crystalline state possess zero entropy

thermodynamics, zeroth law of states that physical systems may be endowed with the property of temperature, which is described by a numerical quantity

thermotropic liquid crystal a type of liquid crystal that changes phase upon a change in temperature such as those used in thermometers with color readouts

thought experiment an imaginary experiment that, although it might not be technically feasible (at least not at the present time), does not violate any law of nature and can be used to illustrate a fundamental principle or problem

time the order of events by earlier and later; the dimension of, that is, the possibility of assigning a measure to, becoming; instant, as the time of occurrence of an event, usually denoted t, whose SI unit is second (s)

time dilation the effect, according to special relativity, that a clock is observed by an observer with respect to whom the clock is moving to run slower than it is observed to run by an observer with respect to whom the clock is at rest

timelike as describing a space-time interval or separation, an interval between two events, or space-time points, such that a particle emitted from the earlier event and moving slower than the speed of light in vacuum can reach the later event; equivalently, an interval such that a photon (which moves at the speed of light in vacuum) that is emitted from the earlier event reaches the location of the later event before the later event actually occurs

time-reversal invariance except for processes involving neutral kaons, the time-reversal image of any process that is allowed by nature is also a process allowed by nature

T invariance time-reversal invariance

titration a process that involves experimentally determining the concentration of a solution of acid with an unknown concentration by adding a known volume of a known concentration of base until the solution is neutral

TOE theory of everything

torque also called the moment of a force, a vector quantity that expresses the rotating, or twisting, capability of a force with respect to some point, often denoted τ (as a vector) or τ (as its magnitude), whose SI unit is newton-meter (N·m); the vector product of the position vector of the point of application of the force with respect to some reference point and the force vector; for rotation about an axis and a force perpendicular to the axis, the magnitude of a force's torque equals the product of the magnitude of the force and the perpendicular distance (which is also the shortest distance) between the force's line of action and the axis, the moment arm, or lever arm, of the force

torr a non-SI unit of pressure, denoted torr (torr), also called millimeter of mercury (mm Hg), equal to 1.333×10^2 pascals (Pa)

total internal reflection the effect whereby a light ray impinging on an interface surface with a medium of lower index of refraction than the medium in which it is propagating is completely reflected back into its original medium, with no transmission through the interface into the other medium

toxicologist a scientist who studies the impact on human health of chemicals and products

transcription the process of synthesizing a ribonucleic acid (RNA) message from a deoxyribonucleic acid (DNA) template

transformer an electric device for converting an alternating current (AC) voltage (the primary) to a different AC voltage (the secondary) or possibly to the same voltage

transistor a semiconductor device that serves in electronic circuits as a signal amplifier or as a switch

transition metals elements found in the center of the periodic table that possess the metallic properties of malleability, ductility, and conductivity

translation in physics, the change of spatial position; in biochemistry, the process of converting a messenger ribonucleic acid (RNA) code into an amino acid sequence to form a polypeptide

transmission electron microscopy (TEM) a visualization technique that allows the interior of materials (even cells) to be studied by sending an electron beam through thin sections (100 nm) of fixed samples

transverse wave a wave in which the propagating disturbance is perpendicular to the direction of wave propagation, such as an electromagnetic wave

tricarboxylic acid cycle a series of reactions in cellular respiration used to break down the carbon chain, producing carbon dioxide and releasing the energy of the bonds in the forms of adenosine triphosphate/guanosine triphosphate (ATP/GTP) and electron carriers reduced nicotinamide adenine dinucleotide (NADH) and reduced flavin adenine dinucleotide ($FADH_2$); also called the citric acid cycle or the Krebs cycle

triple bond chemical bond in which two atoms are connected by three covalent bonds; consists of one sigma bond and two pi bonds

triple point the unique set of conditions under which a substance's solid, liquid, and gas phases are in equilibrium and can coexist with no tendency for any phase to transform to another

troposphere the lowest layer of the Earth's atmosphere

tunneling a quantum effect in which particles can penetrate barriers that classical physics deems impassible

turbine a rotary engine that is powered by the flow of a fluid

turbulence fluid flow that is chaotic and rapidly changing, with shifting eddies and tightly curled lines of flow

turbulent flow chaotic and rapidly changing fluid flow, with shifting eddies and tightly curled lines of flow

twin paradox the apparent paradox of the lesser aging of a twin who takes a high-speed round trip in a spaceship than the aging of her brother who remains on Earth

type I superconductor any of certain superconductors that are pure metals, including aluminum, mercury, and niobium, for example, and that are characterized by critical temperatures below 10 kelvins (K) and relatively low critical fields, below about 0.2 tesla (T)

type II superconductor any of certain superconductors that comprise various alloys and metallic compounds, whose critical temperatures are more or less in the 14–20-kelvin (K) range and critical fields are higher than 15 teslas (T)

ultraviolet (UV) electromagnetic radiation with frequency somewhat higher, and wavelength somewhat shorter, than those of visible light and whose photons have more energy than those of visible light

uncertainty a characteristic of quantum physics that physical quantities do not possess any value until they are measured and then, in general, do not possess a sharp value; for a physical quantity of a quantum system, the spread of the values about the average of those values found by repeated measurements of the quantity, for whatever state of the system is being considered

uncertainty principle also called Heisenberg uncertainty principle, states that the product of the uncertainties of certain pairs of physical quantities in quantum systems cannot be less than $h/(4\pi)$, where h is the Planck constant; such pairs include position-momentum and time-energy

unified field theory a field theory that describes all known forces

unit a well-defined division of a physical quantity for use as a reference for specifying amounts of that quantity

unit operations a series of smaller individual steps in a chemical engineering process

universe   all that there is; the cosmos

unstable equilibrium   a state of equilibrium such that when the system is displaced slightly from it, the system tends to evolve away from the equilibrium state

vacuum   the absence of matter

valence band   in a solid, the range of lowest-energy states available to electrons, in which the electrons are immobile

valence bond theory   bonding theory of covalent compounds based on the overlap of electron orbitals, the allowed energy states of electrons in the quantum mechanical model

valence shell electron repulsion theory (VSEPR)   bonding theory based on three-dimensional geometrical approximations of covalent compounds determined by including electron pair repulsions

Van de Graaff accelerator   a type of linear particle accelerator that uses a single high voltage for acceleration

vapor   the state of a material in which the atoms or molecules are neutral and the matter has neither a definite shape nor a definite volume, taking the shape and volume of its container, also called gas; the gaseous state of a material that is generally a liquid at room temperature

vaporization   the phase change from liquid to gas (vapor) by the addition of heat, also called boiling

vat dyeing   the most common method of textile dyeing; involves adding the dye into a large vat of water and adding the fabric or the fiber

vector   any physical quantity that is characterized by both a magnitude, which is a nonnegative number, and a direction, such as velocity and force

vector product   also called cross product and denoted $\mathbf{u} \times \mathbf{v}$ for vectors $\mathbf{u}$ and $\mathbf{v}$, a vector obtained from any pair of vectors $\mathbf{u}$ and $\mathbf{v}$, whose magnitude equals the product of the vectors' magnitudes and the sine of the smaller angle (less than or equal 180°) between their directions, and whose direction is perpendicular to the plane defined by the vectors' directions, such that when the thumb of the right hand points in this vector's direction, the curved fingers indicate the sense of rotation through the above angle from $\mathbf{u}$ to $\mathbf{v}$

velocity   the time rate of change of position, often denoted $\mathbf{v}$ (as a vector) or v (as its magnitude), whose SI unit is meter per second (m/s)

Venturi effect   for smooth, steady, irrotational, horizontal flow of a nonviscous, incompressible fluid, the phenomenon that the pressure is lower where the flow speed is higher

virtual focal point   the point at which parallel light rays appear to be made to converge by an optical system or component, although no such converging is taking place

virtual image   an optical image formed by what appear to be converging light rays, although no such converging is taking place; a virtual image cannot be caught on a screen but can be seen and photographed

virtual object   what appears to be a source of diverging light rays entering an optical system or component, while in fact there exists no such source

viscosity   the resistance of fluids to flow

visible light   electromagnetic radiation with a wavelength between those of infrared radiation and ultraviolet radiation

vitreous state   a noncrystalline state of matter, such as the state of glass

voltage   potential difference

volume   the amount of space, often denoted V, whose SI unit is cubic meter (m^3)

volume flow rate   also volume flux, the volume of fluid flowing through a cross section per unit time, whose SI unit is cubic meter per second (m^3/s)

volume flux   volume flow rate

water cycle   the changes of state of the water molecule from a solid to a liquid to a vapor

watt   the SI unit of power, denoted W, equivalent to joule per second (J/s)

wave   a propagating disturbance, which is a continuous, nonlocalized effect, spread out over space

wavefront   a surface in space that a wave has reached after propagating some time from its source

wave function   in quantum mechanics, a mathematical function of generalized coordinates and time that is a solution of the Schrödinger equation for a particular system and contains all the information about the system that is available in principle, often denoted ψ or Ψ (lowercase or capital Greek psi)

wavelength   the length of one spatial cycle of a periodic wave, often denoted λ (Greek lambda), whose SI unit is meter (m); the distance of one complete wave cycle from crest to crest or trough to trough

wave optics   also called physical optics, the branch of optics that takes into account the wave nature of light

wave-particle distinction   a characteristic of classical physics that waves and particles are distinct from each other and bear no relation to each other

wave-particle duality   a characteristic of quantum physics that every wave has a particle aspect and every particle possesses a wave aspect

weak acids   acids that do not completely dissociate in water

weak force one of the four fundamental forces, a short-range force acting among elementary particles that is responsible for certain kinds of radioactivity

weber the SI unit of magnetic flux, denoted Wb

weight the force by which an object is attracted to Earth, in particular, or to any astronomical body, in general, whose SI unit is newton (N); the magnitude of such a force

Wien's displacement law for electromagnetic radiation from a blackbody, the wavelength of maximal intensity λ_{max} is inversely proportional to the blackbody's absolute temperature T, according to $\lambda_{max}T = 2.898 \times 10^{-3}$ m·K

work the result of a force acting along a displacement in the direction of the force, often denoted W, whose SI unit is joule (J)

work-energy theorem the work performed on an otherwise isolated system equals the increase of total energy content of the system

work function the minimal amount of energy required to cause an electron in a metal to leave the material, often denoted ϕ (Greek phi), whose SI unit is joule (J)

world line a continuous curve in space-time that consists of all the events that constitute the life of a massive particle or of a photon

wormhole Einstein-Rosen bridge

X-ray electromagnetic radiation with frequency (or wavelength) in the range between that of ultraviolet and that of gamma rays and, correspondingly, whose photons have energy greater than that of ultraviolet photons and less than the energy of gamma rays

X-ray crystallography an analytical chemistry technique used to determine the structure of chemical compounds by measuring the diffraction of X-rays through crystals of the substance

yield point the stress at which a material starts to flow, its strain increasing over time with no further increase of stress

Young's experiment an experiment first performed in the early 19th century that demonstrated the wave nature of light

Young's modulus in elasticity, the ratio of the tensile stress (the longitudinal force per unit cross section area) to the tensile strain (the relative change in length), often denoted Y, whose SI unit is pascal (Pa)

APPENDIX III
FURTHER RESOURCES

BOOKS

Adams, Steve. *Frontiers: Twentieth-Century Physics*. New York: Taylor & Francis, 2000. A survey of 20th-century physics written for high school students.

Ball, Philip. *The Elements: A Very Short Introduction*. New York: Oxford University Press, 2005. A simple introduction to each of the groups of chemical elements.

Breithaupt, Jim. *Teach Yourself Physics*. Chicago: NTC/Contemporary Publishing, 2002. A compact introduction to the key concepts, major discoveries, and current challenges in physics.

Bromley, D. Allan. *A Century of Physics*. New York: Springer, 2002. A tour from the last century of physics growth, impact, and directions. Numerous photos and illustrations.

Chapple, Michael. *Schaum's A to Z Physics*. New York: McGraw-Hill, 2003. Defines 650 key concepts with diagrams and graphs, intended for high school students and college freshmen.

Charap, John M. *Explaining the Universe: The New Age of Physics*. Princeton, N.J.: Princeton University Press, 2002. A description of the field of physics at the beginning of the 21st century.

Dennis, Johnnie T. *The Complete Idiot's Guide to Physics*. Indianapolis, Ind.: Alpha Books, 2003. A friendly review of high school level classical physics.

Dewick, Paul M. *Essentials of Organic Chemistry*. Hoboken, N.J.: John Wiley & Sons, 2006. An introductory textbook of organic chemistry.

The Diagram Group. *The Facts On File Physics Handbook*. Rev. ed. New York: Facts On File, 2006. Convenient resource containing a glossary of terms, short biographical profiles of celebrated physicists, a chronology of events and discoveries, and useful charts, tables, and diagrams.

Falk, Dan. *Universe on a T-Shirt: The Quest for the Theory of Everything*. New York: Arcade, 2002. A story outlining developments in the search for the theory that will unify all four natural forces.

Fisher, J., and J. R. P Arnold. *Chemistry for Biologists*. 2nd ed. New York: BIOS Scientific, 2004. An easy-to-read chemistry textbook that focuses on content relevant to the life sciences.

Fleisher, Paul. *Relativity and Quantum Mechanics: Principles of Modern Physics*. Minneapolis, Minn.: Lerner, 2002. An introduction to the concepts of relativity and quantum mechanics, written for middle school students.

Gonick, Larry, and Craig Criddle. *The Cartoon Guide to Chemistry*. New York: Harper Resource, 2005. A fun, simplified version of basic chemistry topics useful for high school chemistry students.

Grenthe, Ingmar, ed. *Nobel Lectures: Chemistry 1996–2000*. River Edge, N.J.: World Scientific, 2003. An excellent compilation of Chemistry Nobel Lectures from the years 1996–2000.

Griffith, W. Thomas. *The Physics of Everyday Phenomena*. 4th ed. Boston: WCB/McGraw-Hill, 2004. A conceptual text for nonscience college students.

Gundersen, P. Erik. *The Handy Physics Answer Book*. Detroit, Mich.: Visible Ink Press, 1999. Answers numerous questions about physics using a conceptual approach.

Hinchliffe, Alan. *Molecular Modeling for Beginners*. Hoboken, N.J.: John Wiley & Sons, 2003. A solid introduction to molecular structure and computational chemistry.

Holton, Gerald James, and Stephen G. Brush. *Physics, the Human Adventure: From Copernicus to Einstein and Beyond*. New Brunswick, N.J.: Rutgers University Press, 2001. Comprehensive introduction intended for nonscience college students. Difficult reading but covers a lot of material.

Interrante, Leonard V., Lawrence A. Casper, and Arthur B. Ellis. *Materials Chemistry: An Emerging Discipline*. Washington, D.C.: American Chemical Society, 1995. A college-level introduction to material science.

James, Ioan. *Remarkable Physicists: From Galileo to Yukawa*. New York: Cambridge University Press, 2004. Contains brief biographies of 50 physicists spanning a period of 250 years, focusing on the lives rather than the science.

Lavoisier, Antoine. *Elements of Chemistry.* New York: Dover, 1965. Reprint of the 1790 edition of the work by Antoine Laurent Lavoisier.

Leiter, Darryl J. *A to Z of Physicists.* New York: Facts On File, 2003. Profiles more than 150 physicists, discussing their research and contributions. Includes bibliography, cross-references, and chronology.

McGrath, Kimberley A., ed. *World of Physics.* Farmington Hills, Mich.: Thomson Gale, 2001. Contains 1,000 entries on concepts, theories, discoveries, pioneers, and issues related to physics.

Newton, David E. *Chemistry Today.* Phoenix, Ariz.: Oryx Press, 1999. An in-depth overview of the advances in the field of chemistry.

Rock, Peter A. *Chemical Thermodynamics.* Mill Valley, Calif.: University Science Books, 1983. A readable, concise introduction to chemical thermodynamics.

Sawyer, Donald T., Andrzej Sobkowiak, and Julian L. Roberts, Jr. *Electrochemistry for Chemists,* 2nd ed. New York: John Wiley & Sons, 1995. A through treatment of oxidation-reduction and electrochemistry.

Stevens, Malcolm P. *Polymer Chemistry: An Introduction.* 3rd ed. New York: Oxford University Press, 1999. An introductory text to polymers including history, polymerization processes, and industrial chemistry production of polymers.

Stillman, Maxson. *The Story of Alchemy and Early Chemistry.* New York: Dover, 1960. An interesting history of the development of chemistry as a discipline, appropriate for the high school student.

Trefil, James. *From Atoms to Quarks: An Introduction to the Strange World of Particle Physics.* Rev. ed. New York: Anchor Books, 1994. A primer on this complex subject written for general readers.

Whittaker, A. G., A. R. Mount, and M. R. Heal. *Physical Chemistry.* New York: Springer-Verlag, 2000. An introduction to all aspects of physical chemistry.

INTERNET RESOURCES

The ABCs of Nuclear Science. Nuclear Science Division, Lawrence Berkeley National Laboratory. Available online. URL: http://www.lbl.gov/abc/. Accessed July 22, 2008. Introduces the basics of nuclear science—nuclear structure, radioactivity, cosmic rays, antimatter, and more.

American Chemical Society. Available online. URL: http://portal.acs.org/portal/acs/corg/content. Accessed July 22, 2008. Home page of ACS. Includes useful resources under education and information about new areas in chemistry.

American Institute of Physics: Center for History of Physics. AIP, 2004. Available online. URL: http://www.aip.org/history/. Accessed July 22, 2008. Visit the "Exhibit Hall" to learn about events such as the discovery of the electron or read selected papers of great American physicists.

American Physical Society. A Century of Physics. Available online. URL: http://timeline.aps.org/. Accessed July 22, 2008. Wonderful, interactive time line describing major events in the development of modern physics.

———. Physics Central. Available online. URL: http://www.physicscentral.com/. Accessed July 22, 2008. Updated daily with information on physics in the news, current research, and people in physics.

Contributions of 20th Century Women to Physics. CWP and Regents of the University of California, 2001. Available online. URL: http://cwp.library. ucla.edu/. Accessed July 22, 2008. Highlights 83 women who have made original and important contributions to physics.

Fear of Physics. Available online. URL: http://www. fearofphysics.com/. Accessed July 22, 2008. Entertaining way to review physics concepts.

Jones, Andrew Zimmerman. "Physics." About, Inc., 2004. Available online. URL: http://physics.about. com. Accessed July 22, 2008. Contains regular feature articles and much additional information.

The Particle Adventure: The Fundamentals of Matter and Force. The Particle Data Group of the Lawrence Berkeley National Laboratory, 2002. Available online. URL: http://particleadventure. org/. Accessed July 22, 2008. Interactive tour of quarks, neutrinos, antimatter, extra dimensions, dark matter, accelerators, and particle detectors.

National Human Genome Research Institute home page. Available online. URL: http://www.genome. gov. Accessed July 22, 2008. Reliable and timely information on the progress and history of the Human Genome Project.

National Institutes of Health, Office of Science Education. Available online. URL: http://science. education.nih.gov/. Accessed July 22, 2008. A comprehensive site for science students and educators including lesson plans on many current science topics.

National Library of Medicine and the National Institutes of Health. National Center for Biotechnology Information home page. Available online. URL: http://www.ncbi.nlm.nih.gov. Accessed July 22, 2008. A resource for current research in biotechnology.

National Oceanic and Atmospheric Administration home page. Available online. URL: http://www.

noaa.gov. Accessed July 22, 2008. A useful site for all areas of environmental research.

Succeed in Physical Science. Ron Kurtus, Kurtus Technologies, and the School for Champions, 2003. Available online. URL: http://www.school-for-champions.com/science.htm. Accessed July 22, 2008. Lessons in a variety of physics and chemistry topics.

University of Colorado at Boulder. Physics 2000. Colorado Commission on Higher Education and the National Science Foundation. Available online. URL: http://www.colorado.edu/physics/2000/index.pl. Accessed July 22, 2008. A fun, interactive journey through modern physics.

Windows to the Universe team. Fundamental Physics. Boulder, Colo.: ©2000–04 University Corporation of Atmospheric Research (UCAR), ©1995–99, 2000. The Regents of the University of Michigan. Available online. URL: http://www.windows.ucar.edu/tour/link=/physical_science/physics/physics.html. Accessed July 22, 2008. Still under construction, this site will contain a broad overview of physics and already has many links to physics topics including mechanics, electricity and magnetism, thermal physics, and atomic and particle physics.

PERIODICALS

Accounts of Chemical Research
Published by the American Chemical Society
1155 16th Street N.W.
Washington, DC 20036
Telephone: (202) 872-4614
www.acs.org

American Scientist
Published by Sigma Xi, the Scientific Research
    Society
P.O. Box 13975
Research Triangle Park, NC 27709
Telephone: (919) 549-0097
www.americanscientist.org

Analytical Chemistry
Published by the American Chemical Society
1155 16th Street N.W.
Washington, DC 20036
Telephone: (202) 872-4614
www.acs.org

Biochemistry and Molecular Biology Education
Published by Wiley Subscriptions Services, a Wiley
    Company
111 River Street
Hoboken, NJ 07030
Telephone: (201) 748-8789
www.bambed.org

Chemical and Engineering News
Published by the American Chemical Society
1155 16th St., NW
Washington, D.C. 20036
Telephone: (202) 872-4600
www.cen-online.org

Chemical Engineering
Published by Access Intelligence Publishing
110 William Street
11th Floor
New York, NY 10038
Telephone: (212) 621-4674
www.che.com

Chemtracts
Published by Data Trace Publishing Company
110 West Road, Suite 227
Baltimore, MD 21204-2316
Telephone: (800) 342-0454
www.datatrace.com

Chemical week
Published by Access Intelligence Publications
110 Williams Street
New York, NY 10038
Telephone: (212) 621-4900
www.chemweek.com

Discover
Published by Buena Vista Magazines
114 Fifth Avenue
New York, NY 10011
Telephone: (212) 633-4400
www.discover.com

Environmental Science and Technology
Published by the American Chemical Society
1155 16th Street N.W.
Washington, D.C. 20036
Telephone: (202) 872-4582
www.acs.org/est

Inorganic Chemistry
Published by the American Chemical Society
1155 16th Street N.W.
Washington, D.C. 20036
Telephone: (202) 872-4614
www.acs.org

Issues in Science and Technology
Published by the University of Texas at Dallas
P.O. Box 830688
Mail Station J030
Richardson, TX, 75083-0688
Telephone: (800) 345-8112
www.issues.org

Journal of the American Chemical Society
Published by the American Chemical Society
1155 16th St., N.W.
Washington, D.C. 20036
Telephone: (202) 872-4614
www.acs.org

The Journal of Biochemistry
Published by Oxford University Press for the
    Japanese Biochemical Society
Ishikawa Building
Tokoyo, Japan
Telephone 81-3-3815-1913
www.jbsoc.or.jp

The Journal of Cell Biology
Published by Rockefeller University Press
1114 First Avenue
New York, NY 10065-8325
Telephone: (212) 327-8011
www.rockefeller.edu

Journal of Chemical Education
Published by the Division of Chemical Education,
    Inc., of the American Chemical Society
16 Hemlock Place
College of New Rochelle
New Rochelle, NY, 10805
Telephone: (856) 931-5825
www.jci.divched.org

Journal of Organic Chemistry
Published by the American Chemical Society
1155 16th Street
Washington, DC 20036
Telephone: (202) 872-4614
www.acs.org

Nature
Macmillan Building
4 Crinan Street
London N1 9XW
United Kingdom
Telephone: +44 (0)20 7833 4000
www.nature.com/nature

Physics Today
Published by the American Institute of Physics
Circulation and Fulfillment Division
Ste. 1NO1
2 Huntington Quadrangle

Melville, NY 11747
Telephone: (516) 576-2270
www.physicstoday.org

Science
Published by the American Association for the
    Advancement of Science
1200 New York Avenue, NW
Washington, DC 20005
Telephone: (202) 326-6417
www.sciencemag.org

Science News
Published by the Society for Science & the Public
1719 N Street NW
Washington, DC 20036
Telephone: (202) 785-2255
www.sciencenews.org

Scientific American
415 Madison Avenue
New York, NY 10017
Telephone: (212) 754-0550
www.sciam.com

SOCIETIES AND ORGANIZATIONS

American Association for the Advancement of Science (www.aaas.org) 1200 New York Avenue, NW, Washington, DC 20005. Telephone: (202) 326-6400

American Chemical Society (www.acs.org) 1155 16th Street NW, Washington, DC 20036. Telephone: (202) 872-4600

American Physical Society (www.aps.org) One Physics Ellipse, College Park, MD 20740-3844. Telephone: (301) 209-3200

American Society of Mechanical Engineers (www.asme.org) 3 Park Avenue, New York, NY 10016-5990. Telephone: (973) 882-1170

Institute of Electrical and Electronics Engineers (www.ieee.org) 3 Park Avenue, 17th Floor, New York, NY 10016-5997. Telephone: (212) 419-7900

The Minerals, Metals, & Materials Society (www.tms.org) 184 Thorn Hill Road, Warrendale, PA 15086-7514. Telephone: (724) 776-9000

Society of Physics Students (www.spsnational.org) American Institute of Physics, One Physics Ellipse, College Park, MD 20740-3843. Telephone: (301) 209-3007

APPENDIX IV
PERIODIC TABLE OF THE ELEMENTS

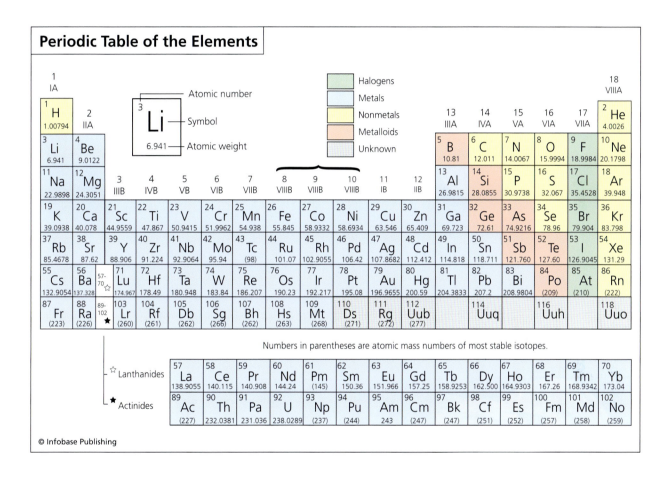

Periodic Table of the Elements

1 IA																	18 VIIIA
1 H 1.00794	**2** IIA											**13** IIIA	**14** IVA	**15** VA	**16** VIA	**17** VIIA	**2** He 4.0026
3 Li 6.941	**4** Be 9.0122											**5** B 10.81	**6** C 12.011	**7** N 14.0067	**8** O 15.9994	**9** F 18.9984	**10** Ne 20.1798
11 Na 22.9898	**12** Mg 24.3051	**3** IIIB	**4** IVB	**5** VB	**6** VIB	**7** VIIB	**8** VIIIB	**9** VIIIB	**10** VIIIB	**11** IB	**12** IIB	**13** Al 26.9815	**14** Si 28.0855	**15** P 30.9738	**16** S 32.067	**17** Cl 35.4528	**18** Ar 39.948
19 K 39.0938	**20** Ca 40.078	**21** Sc 44.9559	**22** Ti 47.867	**23** V 50.9415	**24** Cr 51.9962	**25** Mn 54.938	**26** Fe 55.845	**27** Co 58.9332	**28** Ni 58.6934	**29** Cu 63.546	**30** Zn 65.409	**31** Ga 69.723	**32** Ge 72.61	**33** As 74.9216	**34** Se 78.96	**35** Br 79.904	**36** Kr 83.798
37 Rb 85.4678	**38** Sr 87.62	**39** Y 88.906	**40** Zr 91.224	**41** Nb 92.9064	**42** Mo 95.94	**43** Tc (98)	**44** Ru 101.07	**45** Rh 102.9055	**46** Pd 106.42	**47** Ag 107.8682	**48** Cd 112.412	**49** In 114.818	**50** Sn 118.711	**51** Sb 121.760	**52** Te 127.60	**53** I 126.9045	**54** Xe 131.29
55 Cs 132.9054	**56** Ba 137.328	57-70 ☆ **71** Lu 174.967	**72** Hf 178.49	**73** Ta 180.948	**74** W 183.84	**75** Re 186.207	**76** Os 190.23	**77** Ir 192.217	**78** Pt 195.08	**79** Au 196.9655	**80** Hg 200.59	**81** Tl 204.3833	**82** Pb 207.2	**83** Bi 208.9804	**84** Po (209)	**85** At (210)	**86** Rn (222)
87 Fr (223)	**88** Ra (226)	89-102 ★ **103** Lr (260)	**104** Rf (261)	**105** Db (262)	**106** Sg (266)	**107** Bh (262)	**108** Hs (263)	**109** Mt (268)	**110** Ds (271)	**111** Rg (272)	**112** Uub (277)		**114** Uuq		**116** Uuh		**118** Uuo

Atomic number — Symbol — Atomic weight (3 Li 6.941)

Halogens
Metals
Nonmetals
Metalloids
Unknown

Numbers in parentheses are atomic mass numbers of most stable isotopes.

☆ Lanthanides

57 La 138.9055	**58** Ce 140.115	**59** Pr 140.908	**60** Nd 144.24	**61** Pm (145)	**62** Sm 150.36	**63** Eu 151.966	**64** Gd 157.25	**65** Tb 158.9253	**66** Dy 162.500	**67** Ho 164.9303	**68** Er 167.26	**69** Tm 168.9342	**70** Yb 173.04

★ Actinides

89 Ac (227)	**90** Th 232.0381	**91** Pa 231.036	**92** U 238.0289	**93** Np (237)	**94** Pu (244)	**95** Am 243	**96** Cm (247)	**97** Bk (247)	**98** Cf (251)	**99** Es (252)	**100** Fm (257)	**101** Md (258)	**102** No (259)

© Infobase Publishing

The Chemical Elements

(g) none
(c) nonmetallics

element	symbol	a.n.
carbon	C	6
hydrogen	H	1

(g) chalcogen
(c) nonmetallics

element	symbol	a.n.
oxygen	O	8
polonium	Po	84
selenium	Se	34
sulfur	S	16
tellurium	Te	52
ununhexium	Uuh	116

(g) alkali metal
(c) metallics

element	symbol	a.n.
cesium	Cs	55
francium	Fr	87
lithium	Li	3
potassium	K	19
rubidium	Rb	37
sodium	Na	11

(g) alkaline earth metal
(c) metallics

element	symbol	a.n.
barium	Ba	56
beryllium	Be	4
calcium	Ca	20
magnesium	Mg	12
radium	Ra	88
strontium	Sr	38

(g) none (c) metallics

element	symbol	a.n.	element	symbol	a.n.
aluminum	Al	13	scandium	Sc	21
bohrium	Bh	107	seaborgium	Sg	106
cadmium	Cd	48	silver	Ag***	47
chromium	Cr	24	tantalum	Ta	73
cobalt	Co	27	technetium	Tc	43
copper	Cu***	29	thallium	Tl	81
darmstadtium	Ds	110	titanium	Ti	22
dubnium	Db	105	tin	Sn	50
gallium	Ga	31	tungsten	W	74
gold	Au***	79	ununbium	Uub	112
hafnium	Hf	72	ununquadium	Uuq	114
hassium	Hs	108	vanadium	V	23
indium	In	49	yttrium	Y	39
iridium	Ir ****	77	zinc	Zn	30
iron	Fe	26	zirconium	Zr	40
lawrencium	Lr	103			
lead	Pb	82			
lutetium	Lu	71			
manganese	Mn	25			
meitnerium	Mt	109			
mercury	Hg	80			
molybdenum	Mo	42			
nickel	Ni	28			
niobium	Nb	41			
osmium	Os****	76			
palladium	Pd ****	46			
platinum	Pt ****	78			
rhenium	Re	75			
rhodium	Rh****	45			
roentgenium	Rg	111			
ruthenium	Ru****	44			
rutherfordium	Rf	104			

(g) pnictogen (c) metallics

element	symbol	a.n.
arsenic	As*	33
antimony	Sb*	51
bismuth	Bi	83
nitrogen	N	7
phosphorus	P**	15

(g) none (c) semimetallics

element	symbol	a.n.
boron	B	5
germanium	Ge	32
silicon	Si	14

(g) actinide (c) metallics

element	symbol	a.n.
actinium	Ac	89
americium	Am	95
berkelium	Bk	97
californium	Cf	98
curium	Cm	96
einsteinium	Es	99
fermium	Fm	100
mendelevium	Md	101
neptunium	Np	93
nobelium	No	102
plutonium	Pu	94
protactinium	Pa	91
thorium	Th	90
uranium	U	92

(g) halogens (c) nonmetallics

element	symbol	a.n.
astatine	At*	85
bromine	Br	35
chlorine	Cl	17
fluorine	F	9
iodine	I	53

(g) lanthanide (c) metallics

element	symbol	a.n.
cerium	Ce	58
dysprosium	Dy	66
erbium	Er	68
europium	Eu	63
gadolinium	Gd	64
holmium	Ho	67
lanthanum	La	57
neodymium	Nd	60
praseodymium	Pr	59
promethium	Pm	61
samarium	Sm	62
terbium	Tb	65
thulium	Tm	69
ytterbium	Yb	70

(g) noble gases (c) nonmetallics

element	symbol	a.n.
argon	Ar	18
helium	He	2
krypton	Kr	36
neon	Ne	10
radon	Rn	86
xenon	Xe	54
ununoctium	Uuo	118

a.n. = atomic number
(g) = group
(c) = classification

* = semimetallics
** = nonmetallics
*** = coinage metal
**** = precious metal

APPENDIX V
SI UNITS AND DERIVED QUANTITIES

DERIVED QUANTITY	UNIT	SYMBOL
frequency	hertz	Hz
force	newton	N
pressure	pascal	Pa
energy	joule	J
power	watt	W
electric charge	coulomb	C
electric potential	volt	V
electric resistance	ohm	Ω
electric conductance	siemens	S
electric capacitance	farad	F
magnetic flux	weber	Wb
magnetic flux density	tesla	T
inductance	henry	H
luminous flux	lumen	lm
illuminance	lux	lx

APPENDIX VI
MULTIPLIERS AND DIVIDERS FOR USE WITH SI UNITS

Multiplier	Prefix	Symbol	Divider	Prefix	Symbol
10^1	deca	da	10^{-1}	deci	d
10^2	hecto	h	10^{-2}	centi	c
10^3	kilo	k	10^{-3}	milli	m
10^6	mega	M	10^{-6}	micro	μ
10^9	giga	G	10^{-9}	nano	n
10^{12}	tera	T	10^{-12}	pico	p
10^{15}	peta	P	10^{-15}	femto	f
10^{18}	exa	E	10^{-18}	atto	a
10^{21}	zetta	Z	10^{-21}	zepto	z
10^{24}	yotta	Y	10^{-24}	yocto	y

APPENDIX VII
COMMON IONS

CATIONS

$^+1$	$^+2$	$^+3$
Ammonium (NH_4^+)	Barium (Ba^{2+})	Aluminum (Al^{3+})
Hydrogen (H^+)	Calcium (Ca^{2+})	Iron III (Fe^{3+})
Lithium (Li^+)	Magnesium (Mg^{2+})	Chromium III (Cr^{3+})
Sodium (Na^+)	Cadmium (Cd^{2+})	
Silver (Ag^+)	Zinc (Zn^{2+})	

ANIONS

-1	-2	-3
Acetate (CH_3COO^-)	Carbonate (CO_3^{2-})	Arsenate (AsO_4^{3-})
Bicarbonate (HCO_3^-)	Chromate (CrO_4^{2-})	Phosphate (PO_4^{3-})
Chlorate (ClO_3^-)	Dichromate ($Cr_2O_7^{2-}$)	
Cyanide (CN^-)	Hydrogen phosphate (HPO_4^{2-})	
Hydroxide (OH^-)	Oxide (O^{2-})	
Nitrate (NO_3^-)	Peroxide (O_2^{2-})	
Nitrite (NO_2^-)	Sulfate (SO_4^{2-})	
Perchlorate (ClO_4^-)	Sulfide (S^{2-})	
Permanganate (MnO_4^-)	Sulfite (SO_3^{2-})	
Thiocyanate (SCN^-)		

APPENDIX VIII
PHYSICAL CONSTANTS

Quantity	Symbol	Value
Bohr radius	$a_0 = h^2/(4\pi^2\, m_e e^2 k_e)$	$5.291\ 772\ 108 \times 10^{-11}$ m
Speed of light in vacuum	c	$2.997\ 924\ 58 \times 10^8$ m/s
Magnitude of electron charge	e	$1.602\ 176\ 53 \times 10^{-10}$ C
Electron volt	eV	$1.602\ 176\ 53 \times 10^{-10}$ J
Gravitational constant	G	$6.674\ 2 \times 10^{-11}$ N·m^2/kg^2
Planck constant	h	$6.626\ 069\ 3 \times 10^{-34}$ J·s
Boltzmann constant	k_B	$1.380\ 650\ 5 \times 10^{-23}$ J/K
Coulomb constant	$k_e = 1/(4\pi\varepsilon_0)$	$8.987\ 551\ 788 \times 10^9$ N·m^2/C^2
Electron mass	m_e	$9.109\ 382\ 6 \times 10^{-31}$ kg
		$5.485\ 799\ 094\ 5 \times 10^{-4}$ u
		$0.510\ 998\ 918$ MeV/c^2
Neutron mass	m_n	$1.674\ 927\ 28 \times 10^{-27}$ kg
		$1.008\ 664\ 915\ 61$ u
		$939.565\ 361$ MeV/c^2
Proton mass	m_p	$1.672\ 621\ 71 \times 10^{-27}$
		$1.007\ 276\ 466\ 88$ u
		$938.272\ 030$ MeV/c^2
Avogadro's number	N_A	$6.022\ 141\ 5 \times 10^{23}$ particles/mol
Gas constant	R	$8.314\ 472$ J/(mol·K)
		$0.082\ 057$ L·atm/(mol·K)
Atomic mass unit	u	$1.660\ 538\ 86 \times 10^{-27}$ kg
		$931.494\ 044$ MeV/c^2
Compton wavelength	$\lambda_C = h/m_e c$	$2.426\ 310\ 238 \times 10^{-12}$ m
Permeability of free space	μ_0	$4\pi \times 10^{-7}$ T·m
Permittivity of free space	$\varepsilon_0 = 1/(\mu_0 c^2)$	$8.854\ 187\ 817 \times 10^{-12}$ C^2/(N·m^2)

APPENDIX IX
ASTRONOMICAL DATA

Body	Mass (kg)	Mean Radius (m)	Orbital Period (Years)	Orbital Period (s)	Mean Orbital Radius (m)
Sun	1.991×10^{30}	6.96×10^8	—	—	—
Mercury	3.18×10^{23}	2.43×10^6	0.241	7.60×10^6	5.79×10^{10}
Venus	4.88×10^{24}	6.06×10^6	0.615	1.94×10^7	1.08×10^{11}
Earth	5.98×10^{24}	6.37×10^6	1.00	3.156×10^7	1.50×10^{11}
Mars	6.42×10^{23}	3.37×10^6	1.88	5.94×10^7	2.28×10^{11}
Jupiter	1.90×10^{27}	6.99×10^7	11.9	3.74×10^8	7.78×10^{11}
Saturn	5.68×10^{26}	5.85×10^7	29.5	9.35×10^8	1.43×10^{12}
Uranus	8.68×10^{25}	2.33×10^7	84.0	2.64×10^9	2.87×10^{12}
Neptune	1.03×10^{26}	2.21×10^7	165	5.22×10^9	4.50×10^{12}
Moon	7.36×10^{22}	1.74×10^6	27.3 days	2.36×10^6	3.84×10^8

Note: Pluto, which had long been considered a planet, has been recategorized. It now belongs to the class of astronomical bodies called plutoids, which are smaller than planets.

APPENDIX X
ABBREVIATIONS AND SYMBOLS FOR PHYSICAL UNITS

Symbol	Unit	Symbol	Unit	Symbol	Unit
A	ampere	g	gram	min	minute
u	atomic mass unit	H	henry	mol	mole
atm	atmosphere	h	hour	N	Newton
Btu	British thermal unit	hp	horsepower	Pa	pascal
C	coulomb	Hz	hertz	rad	radian
cd	candela	in	inch	rev	revolution
°C	degree Celsius	J	joule	s	second
cal	calorie	K	kelvin	sr	steradian
d	day	kg	kilogram	T	tesla
eV	electron volt	L	liter	V	volt
F	farad	lb	pound	W	watt
°F	degree Fahrenheit	ly	light-year	Wb	weber
ft	foot	m	meter	yr	year
G	gauss	mi	mile	Ω	ohm

APPENDIX XI
THE GREEK ALPHABET

Letter	Capital	Small	Letter	Capital	Small
Alpha	A	α	Nu	N	ν
Beta	B	β	Xi	Ξ	ξ
Gamma	Γ	γ	Omicron	O	o
Delta	Δ	δ	Pi	Π	π
Epsilon	E	ε	Rho	ρ	ρ
Zeta	Z	ζ	Sigma	Σ	σ, ς
Eta	H	η	Tau	T	τ
Theta	Θ	θ	Upsilon	Y	υ
Iota	I	ι	Phi	Φ	φ, ϕ
Kappa	K	κ	Chi	X	χ
Lambda	Λ	λ	Psi	Ψ	ψ
Mu	M	μ	Omega	Ω	ω

APPENDIX XII
SOME CONVERSIONS WITHIN THE METRIC SYSTEM

1 meter (m) = 100 centimeters (cm) = 1,000 millimeters (mm)
1 centimeter (cm) = 10 millimeters (mm) = 0.01 meters (m)
1 millimeter (mm) = 1000 micrometers (μm) = 1 micron (μ)

1 liter (L) = 1,000 milliliters (mL)
1 cubic centimeter (cc or cm^3) = 1 milliliter (mL)
1 milliliter (mL) = 1,000 microliters (μL)

1 kilogram (kg) = 1,000 grams (g)
1 gram (g) = 1,000 milligrams (mg)
1 milligram (mg) = 1,000 micrograms (μg)

APPENDIX XIII
COMMON CONVERSIONS FROM U.S. CUSTOMARY TO METRIC UNIT VALUES

Quantity	To Convert From	To	Multiply by (Rounded to Nearest 1,000th)
mass	pounds (lb)	kilograms (kg)	0.454
	ounces (oz)	gram (g)	28.350
length	miles (mi)	kilometers (km)	1.609
	yards (yd)	meters (m)	0.914
	feet (ft)	meter (m)	0.305
	inches (in)	centimeters (cm)	2.540
area	square feet (ft^2)	square meter (m^2)	0.093
	square miles (mi^2)	square kilometers (km^2)	2.590
volume	gallon (gal)	liters (L)	3.785
	quarts (qt)	liters (L)	0.946
	fluid ounces (fl oz)	milliliters (mL)	29.574

APPENDIX XIV
TEMPERATURE CONVERSIONS

In the Celsius scale, 0°C is the freezing point of water and 100°C is the boiling point. In the Fahrenheit scale, 32°F is the freezing point of water and 212°F is the boiling point. In the Kelvin scale, 0 K is absolute zero temperature and the boiling point of water is 273.15 K.

To convert temperature in degrees Celsius (T_C) to temperature in degrees Fahrenheit (T_F):

$$T_F = \frac{9}{5} T_C + 32$$

To convert temperature in degrees Fahrenheit (T_F) to temperature in degrees Celsius (T_C):

$$T_C = \frac{5}{9} (T_F - 32)$$

To convert temperature in degrees Celsius (T_C) to temperature in kelvins (T):

$$T = T_C + 273.15$$

To convert temperature in kelvins (T) to temperature in degrees Celsius (T_C):

$$T_C = T - 273.15$$

INDEX

Note: Page numbers in **boldface** indicate main entries; *italic* page numbers indicate photographs and illustrations.

A

abacus 290, *291*, 293
absolute, v. relative 623–624
absolute pressure 557
absolute temperature 311, 403
absolute zero 271, 311, 323
absorption, biological 667–668
absorption spectrophotometry 631
absorption spectrum 30–31, 631
AC. *See* alternating current
acceleration **1–3**, *2*, 246–247, 633
    centripetal *2*, 601–602
    due to gravity 2–3, *3*, 76
    inertial motion v. 281–282, 376
    Newton's laws of motion and 3, 168, 213, 246–247, 281, 432–434, 445
    as vector 669
accelerators 4, **4–5**, 276, 277, 499, *500*, 504, 630
accidents, laboratory 357–358
accuracy, of measurement 404–405
acetaminophen 34
acetylsalicylic acid. *See* aspirin
acid-dissociation constant 10
acid rain **5–8**, *7*, 18, 20, 35, 82, 333
acids **8–11**, 337
    conjugate 9–10, 62–63
    definitions of 8–9, 61–62, 337, 533–534
    gravimetry of 28
    nomenclature for 9, 340, *340*
    pH values of 8, 534, *535*
    strength of 9–10, 62–63
    titrimetry of 28–30, *30*, 536–537
acoustics **11–14**, 389–390
ACS. *See* American Chemical Society
actinides 421, 517–518
action 107
action and reaction, law of 169, 434, 445
action at a distance 399
action by continuous contact 399
action potential 413
activation energy 206, 210

activators 207
active magnetic screening 389
active medium, of laser 359
Adair, G. S. 521
Adams, John 561
addition reactions 490, 550
adenosine triphosphate (ATP)
    in amino acid metabolism 24
    as energy currency 44–45, 416, *416*, 417, 462, 464, 529
    in fatty acid metabolism 224
    synthesis of
        in cellular respiration 199, 417, *419*
        in citric acid cycle 104, 199, 413, 419, *419*
        in electron transport system 199–200, 413, 419, *419*
        in glycolysis 70, 199, 287–290, *288*, 413, 418–419, *419*
        in photosynthesis 413–414, 529, 531
adipose cells 224, *224*
adsorption 217
aerobic metabolism 46, 199, 287, 413
aerodynamics *240, 242, 242*
aerospace engineering 406
affinity chromatography 100–102, 615
A-form, of DNA 462, 464–465, *465*, 686–687
aging, free radicals and 575
Agricultural Research Service (ARS) 17
Agriculture, U.S. Department of (USDA) 16, 17
agrochemistry **14–17**, *91*
aircraft, and ozone layer 134
air pollution 6–8, **17–20**, 102–104
Air Pollution Control Act of 1955 19
Alaskan Arctic National Wildlife Refuge (ANWR) 21
alcaptonuria 27
alchemy 58, 338, 363, 364, 443
alcohol, as fuel 22
alcohol fermentation 95, 97–98, *98*, 420
algae bloom 469

aliphatic hydrocarbons 253, 491
alkali metals 142, 420–421, 516
alkaline batteries 186
alkaline earth metals 142, 421, 516–517
alkanes 491–492
alkenes 492
alkynes 492–493
allosteric effects 208, 522
alloys 422
ALMA. *See* Atacama Large Millimeter Array
alpha helix 513, 563
alpha particles/rays 38, *39*, 227–228, 407, 454–455, 577–578, 605
Alpher, Ralph 42, 125
alternating current (AC) 179, 304, 306
alternative energy sources 8, 19, **20–23**, *22*, 82, 302
Altman, Sidney 212, 593, *593*–595
AM. *See* amplitude modulation
American Arithmometer Company 291–292
American Centrifuge Project 79
American Chemical Society (ACS) 35, 90, 647
American Physical Society (APS) *544*, 544–545
American Society of Mechanical Engineers (ASME) 406, 451
American Society of Nondestructive Testing (ASNT) 451
American Society of Testing Methods (ASTM) 451
amino acid(s)
    essential 23–24, *24*, 562–563
    glucogenic 25, *26*
    ketogenic 25, *26*
    metabolic disorders of 25–27, 563
    metabolism of **23–27**, *25*, 104, 562–563
    nonessential 23–24, *24*, 25
    in protein composition 44, 562–563, 618
    structure of 23, 44, 562, *563*
ammonia 103, 216–218, 646–647
amorphous solids 622
AMPAC program 117